METHODS IN MOLECULAR BIOLOGY

For further volumes:
http://www.springer.com/series/7651

For over 35 years, biological scientists have come to rely on the research protocols and methodologies in the critically acclaimed *Methods in Molecular Biology* series. The series was the first to introduce the step-by-step protocols approach that has become the standard in all biomedical protocol publishing. Each protocol is provided in readily-reproducible step-by-step fashion, opening with an introductory overview, a list of the materials and reagents needed to complete the experiment, and followed by a detailed procedure that is supported with a helpful notes section offering tips and tricks of the trade as well as troubleshooting advice. These hallmark features were introduced by series editor Dr. John Walker and constitute the key ingredient in each and every volume of the *Methods in Molecular Biology* series. Tested and trusted, comprehensive and reliable, all protocols from the series are indexed in PubMed.

Editors
Andrés Aguilera
CABIMER
Universidad de Sevilla
Seville, Spain

Alexey Ruzov
Skryabin Institute of Bioengineering
Research Centre of Biotechnology RAS
Moscow, Russia

ISSN 1064-3745 ISSN 1940-6029 (electronic)
Methods in Molecular Biology
ISBN 978-1-0716-2476-0 ISBN 978-1-0716-2477-7 (eBook)
https://doi.org/10.1007/978-1-0716-2477-7

Cover Illustration Caption: An electron microscopy R-loop structure consisting of a thicker DNA-RNA hybrid and a thinner ssDNA strand, visualized by electron microscopy. Courtesy of Henriette Stoy and Massimo Lopes.

This Humana imprint is published by the registered company Springer Science+Business Media, LLC, part of Springer Nature.
The registered company address is: 1 New York Plaza, New York, NY 10004, U.S.A.

R-Loops

Methods and Protocols

Edited by

Andrés Aguilera

CABIMER, Universidad de Sevilla, Seville, Spain

Alexey Ruzov

Skryabin Institute of Bioengineering, Research Centre of Biotechnology RAS, Moscow, Russia

Humana Press

Preface

Double-stranded DNA–RNA hybrids and R-loops, the molecular structures comprised of such hybrids together with a displaced single DNA strand, are secondary nucleic acid conformations frequently formed in the genomes of all living organisms. Generally, both the terms (DNA–RNA hybrids and R-loops) are used interchangeably as, in the majority of biological contexts, DNA–RNA hybrids are likely to be present in the form of R-loops, even though this is not necessarily the case for certain physiological settings. R-loops or DNA-RNA hybrids can emerge as intermediates during a number of key biological processes such as retro-transposition, class switching recombination, telomere homeostasis, or certain steps of transcription of particular genes, therefore playing specific roles in cellular metabolism. However, in many cases, R-loops may also be unscheduled or abnormally persistent, representing an important source of genomic stress and instability. This adds an additional physiological relevance to these structures, provided that genome instability is a hallmark of cancer cells and frequently associated with a number of genetic diseases, including neurodegenerative conditions and cancer-prone syndromes. Altogether, the potential roles of R-loops in the regulation of cellular processes, and their strong influence on genome dynamics, have attracted a growing interest in these structures in different areas of molecular and cellular biology as well as biomedicine.

The existence of R-loops in vivo was first inferred via genetic methods that detected specific alterations of genome dynamics that could be reversed by expression of RNase H, a unique nuclease that degrades the RNA moieties of the hybrids. However, the development of molecular techniques for the detection of DNA–RNA hybrids helped expand greatly the field of R-loop biology. The availability of a monoclonal antibody that recognizes DNA–RNA hybrids, the mutagenesis methods allowing us to infer their analysis at the single-nucleotide level, as well as the development of different RNase H-derived tools, among other techniques, have significantly enriched our capacity to detect and purify DNA–RNA hybrids in different cellular contexts. This has expanded our knowledge of R-loop biology by the identification of specific determinants of R-loop homeostasis, including transcription, RNA processing, and DNA repair factors, as well as chromatin remodelers. Furthermore, this has led to the generation of genome-wide maps of the distribution of DNA–RNA hybrids in different cell lines. Together with the development of experimental techniques focused on the identification of proteins associated with DNA–RNA hybrids, these results have given R-loops and DNA–RNA hybrids a previously unforeseen physiological status implying their importance for multiple cellular processes.

In this book, we have compiled a series of laboratory protocols covering the most important aspects of R-loop biology. First, we focus on a range of different methods allowing for the detection of DNA–RNA hybrids, as well as their purification and visualization by electron microscopy. Most of the detection techniques described here rely on either immunofluorescence staining or immunoprecipitation of DNA–RNA hybrids using a specific monoclonal antibody. In addition, we present different methods based on the use of RNase H-derived tools to detect DNA–RNA hybrids in vitro and in vivo. Importantly, some of these protocols enable following the dynamics of R-loops in living cells. Given that a number of recent studies highlight RNA modifications as integral components of DNA–RNA hybrids potentially contributing to various aspects of their biology, this book contains several protocols that permit studying noncanonical RNA nucleotides in the R-loop context. Moreover, this volume encompasses a number of specific protocols devoted to the investigation of the topology of R-loops and their functional roles in the biology of mitochondria and telomeres. Finally, a large block of chapters is dedicated to different methods allowing genome-wide mapping of DNA–RNA hybrids in various organisms. It is important to note that a number of R-loop-related protocols already published in other volumes of the *Methods in Molecular Biology* series (vol. 1543, 1672, 1703, 1983, 2161) have not been included in this volume.

We expect that this compilation of protocols elaborated by leading experts in the field will be useful not only to the scientific community working on R-loop homeostasis but also to scientists studying such areas of molecular and cell biology as genome integrity, DNA replication and repair, chromatin remodeling, transcription, RNA processing, modification, and export, as well as to researchers elucidating the molecular mechanisms of cancer and genetic diseases.

Seville, Spain *Andrés Aguilera*
Moscow, Russia *Alexey Ruzov*

Contents

Contributors

ABDULKADIR ABAKIR • *Epigenetic Programme, Babraham Institute, Cambridge, UK*

YATHISH JAGADHEESH ACHAR • *IFOM (Fondazione Istituto FIRC di Oncologia Molecolare), Milan, Italy; Centre for DNA Fingerprinting and Diagnostics, Hyderabad, Telangana, India*

ANDRÉS AGUILERA • *Centro Andaluz de Biología Molecular y Medicina Regenerativa (CABIMER), Universidad de Sevilla-CSIC-Universidad Pablo de Olavide, Seville, Spain; Departamento de Genética, Facultad de Biología, Universidad de Sevilla, Seville, Spain*

ADELE ALAGIA • *Sir William Dunn School of Pathology, University of Oxford, Oxford, UK*

FAHAD ALENEZI • *General Department of Criminal Evidence, Ministry of Interior Affairs, Al-Dajeej, Kuwait*

KHELIFA ARAB • *Institute of Molecular Biology (IMB), Mainz, Germany*

UTTIYA BASU • *Department of Microbiology and Immunology, Vagelos College of Physicians and Surgeons, Columbia University, New York, NY, USA*

CHIARA BEGHÈ • *Sir William Dunn School of Pathology, University of Oxford, Oxford, UK*

JUAN C. CAÑAS • *Centro Andaluz de Biología Molecular y Medicina Regenerativa (CABIMER), Universidad de Sevilla-CSIC-Universidad Pablo de Olavide, Seville, Spain*

GIOVANNI CAPRANICO • *Department of Pharmacy and Biotechnology, Alma Mater Studiorum – University of Bologna, Bologna, Italy*

SUSANA M. CERRITELLI • *Division of Intramural Research, Eunice Kennedy Shriver National Institute of Child Health and Human Development, National Institutes of Health, Bethesda, MD, USA*

NA YOUNG CHEON • *Department of Biological Sciences, Ulsan National Institute of Science and Technology, Ulsan, Republic of Korea*

KARLENE A. CIMPRICH • *Department of Chemical and Systems Biology, Stanford University, Stanford, CA, USA*

MAGDALENA P. CROSSLEY • *Department of Chemical and Systems Biology, Stanford University, Stanford, CA, USA*

ROBERT J. CROUCH • *Division of Intramural Research, Eunice Kennedy Shriver National Institute of Child Health and Human Development, National Institutes of Health, Bethesda, MD, USA*

BJØRN DALHUS • *Department of Microbiology, Oslo University Hospital, Rikshospitalet, Oslo, Norway; Department of Medical Biochemistry, Institute for Clinical Medicine, University of Oslo, Oslo, Norway*

MARC DROLET • *Département de Microbiologie, Infectiologie et Immunologie, Université de Montréal, Montreal, QC, Canada*

ARIJIT DUTTA • *Department of Biochemistry and Structural Biology, University of Texas Health Science Center at San Antonio, San Antonio, TX, USA*

MARIANNA FERETZAKI • *Swiss Institute for Experimental Cancer Research (ISREC), School of Life Sciences, Ecole Polytechnique Fédérale de Lausanne (EPFL), Lausanne, Switzerland*

RITA VALADOR FERNANDES • *Swiss Institute for Experimental Cancer Research (ISREC), School of Life Sciences, Ecole Polytechnique Fédérale de Lausanne (EPFL), Lausanne, Switzerland*

MARCO FOIANI • *IFOM (Fondazione Istituto FIRC di Oncologia Molecolare), Milan, Italy; Università degli Studi di Milano, Milan, Italy*
XIANG-DONG FU • *Department of Cellular and Molecular Medicine, Institute of Genomic Medicine, University of California San Diego, La Jolla, CA, USA*
DANIEL GACKOWSKI • *Department of Clinical Biochemistry, Faculty of Pharmacy, Nicolaus Copernicus University in Toruń, Ludwik Rydygier Collegium Medicum in Bydgoszcz, Bydgoszcz, Poland*
MARÍA GARCÍA-RUBIO • *Centro Andaluz de Biología Molecular y Medicina Regenerativa (CABIMER), Universidad de Sevilla-CSIC-Universidad Pablo de Olavide, Seville, Spain*
GALINA GLOUSKER • *Swiss Institute for Experimental Cancer Research (ISREC), School of Life Sciences, Ecole Polytechnique Fédérale de Lausanne (EPFL), Lausanne, Switzerland*
BELÉN GÓMEZ-GONZÁLEZ • *Centro Andaluz de Biología Molecular y Medicina Regenerativa (CABIMER), Universidad de Sevilla-CSIC-Universidad Pablo de Olavide, Seville, Spain*
NATALIA GROMAK • *Sir William Dunn School of Pathology, University of Oxford, Oxford, UK*
CRISTINA GUILLÉN-MENDOZA • *Centro Andaluz de Biología Molecular y Medicina Regenerativa (CABIMER), Universidad de Sevilla-CSIC-Universidad Pablo de Olavide, Seville, Spain; Departamento de Genética, Facultad de Biología, Universidad de Sevilla, Seville, Spain*
MONIKA GULLEROVA • *Sir William Dunn School of Pathology, University of Oxford, Oxford, UK*
YAJING HAO • *Department of Cellular and Molecular Medicine, Institute of Genomic Medicine, University of California San Diego, La Jolla, CA, USA*
IAN J. HOLT • *Biodonostia Health Research Institute, San Sebastián, Spain; CIBERNED (Center for Networked Biomedical Research on Neurodegenerative Diseases, Ministry of Economy and Competitiveness, Institute Carlos III), Madrid, Spain; IKERBASQUE, Basque Foundation for Science, Bilbao, Spain; Universidad de País Vasco, Bilbao, Spain*
ANDRES KAECH • *Center for Microscopy and Image Analysis, University of Zurich, Zurich, Switzerland*
RUTH F. KETLEY • *Sir William Dunn School of Pathology, University of Oxford, Oxford, UK*
ARNE KLUNGLAND • *Department of Microbiology, Oslo University Hospital, Rikshospitalet, Oslo, Norway; Department of Biosciences, University of Oslo, Blindern, Oslo, Norway*
YOUNGHO KWON • *Department of Biochemistry and Structural Biology, University of Texas Health Science Center at San Antonio, San Antonio, TX, USA*
BRICE LAFFLEUR • *Department of Microbiology and Immunology, Vagelos College of Physicians and Surgeons, Columbia University, New York, NY, USA; INSERM U1236, University of Rennes, Etablissement Français du Sang, Rennes, France*
LI LAN • *Massachusetts General Hospital Cancer Center, Harvard Medical School, Charlestown, MA, USA; Department of Radiation Oncology, Massachusetts General Hospital, Harvard Medical School, Boston, MA, USA*
KEVIN S. LANG • *Department of Veterinary and Biomedical Sciences, University of Minnesota, Saint Paul, MN, USA*
JA YIL LEE • *Department of Biological Sciences, Ulsan National Institute of Science and Technology, Ulsan, Republic of Korea; Center for Genomic Integrity, Institute for Basic Science, Ulsan, Republic of Korea*
CONGHUI LI • *Department of Pathophysiology, School of Basic Medical Sciences, Wuhan University, Wuhan, China*

KUAN LI • *Center for Plant Biology, School of Life Sciences, Tsinghua University, Beijing, China, Tsinghua-Peking Center for Life Sciences, Beijing, China*

MIAOMIAO LI • *Department of Microbiology, Oslo University Hospital, Rikshospitalet, Oslo, Norway*

QIN LI • *Center for Plant Biology, School of Life Sciences, Tsinghua University, Beijing, China; Tsinghua-Peking Center for Life Sciences, Beijing, China*

SHUAI LI • *Center for Plant Biology, School of Life Sciences, Tsinghua University, Beijing, China; Tsinghua-Peking Center for Life Sciences, Beijing, China*

KAIWEI LIANG • *Department of Pathophysiology, School of Basic Medical Sciences, Wuhan University, Wuhan, China*

JUNGHYUN LIM • *Department of Microbiology and Immunology, Vagelos College of Physicians and Surgeons, Columbia University, New York, NY, USA; Department of Pharmacy, School of Pharmacy, Jeonbuk National University, Jeonju, South Korea*

JOACHIM LINGNER • *Swiss Institute for Experimental Cancer Research (ISREC), School of Life Sciences, Ecole Polytechnique Fédérale de Lausanne (EPFL), Lausanne, Switzerland*

MASSIMO LOPES • *Institute of Molecular Cancer Research, University of Zurich, Zurich, Switzerland*

JOEL LUETHI • *Institute of Molecular Cancer Research, University of Zurich, Zurich, Switzerland; Center for Microscopy and Image Analysis, University of Zurich, Zurich, Switzerland; Department of Molecular Life Sciences, University of Zurich, Zurich, Switzerland*

BRIAN LUKE • *Institute of Molecular Biology (IMB), Mainz, Germany; Institute of Developmental Biology and Neurobiology, Johannes-Gutenberg-University Mainz, Mainz, Germany*

JESSICA MARINELLO • *Department of Pharmacy and Biotechnology, Alma Mater Studiorum – University of Bologna, Bologna, Italy*

HOURA MERRIKH • *Vanderbilt University School of Medicine, Nashville, TN, USA*

CHRISTOF NIEHRS • *Institute of Molecular Biology (IMB), Mainz, Germany; Division of Molecular Embryology, DKFZ-ZMBH Alliance, Heidelberg, Germany*

JOHANNES RIEMANN • *Center for Microscopy and Image Analysis, University of Zurich, Zurich, Switzerland*

FABIENNE K. ROESSLER • *Institute of Molecular Cancer Research, University of Zurich, Zurich, Switzerland; Center for Microscopy and Image Analysis, University of Zurich, Zurich, Switzerland; Department of Chemical Engineering, Norwegian University of Science and Technology (NTNU), Trondheim, Norway*

ALEXEY RUZOV • *Institute of Bioengineering, Research Centre of Biotechnology RAS, Moscow, Russia*

KIRAN SAKHUJA • *Division of Intramural Research, Eunice Kennedy Shriver National Institute of Child Health and Human Development, National Institutes of Health, Bethesda, MD, USA*

KAVITHA SARMA • *Gene Expression and Regulation Program, The Wistar Institute, Philadelphia, PA, USA*

SÓNIA SILVA • *Centro Andaluz de Biología Molecular y Medicina Regenerativa (CABIMER), Universidad de Sevilla-CSIC-Universidad Pablo de Olavide, Seville, Spain; Departamento de Genética, Facultad de Biología, Universidad de Sevilla, Seville, Spain*

KONSTANTINA SKOURTI-STATHAKI • *MiNa Therapeutics, Translation & Innovation Hub, London, UK*

image analysis by providing access of multiple operators to the same set of raw images.

We have recently further adapted the classical electron microscopy approach to enable the direct visualization of R-loop structures on genomic DNA, allowing for unambiguous quantification of R-loop size, frequency and burden based on single molecules. Besides, simultaneous quantification of R-loops and replication intermediates allows to correlate R-loop levels with changes in replication fork architecture, advancing our understanding of how R-loops interfere with DNA replication. Importantly, this EM approach is optimized to maintain R-loop stability. By directly visualizing R-loops at single-molecule resolution and independently of DNA–RNA hybrid-specific antibodies, this approach complements available population assays, such as DNA–RNA immunoprecipitation (DRIP), immunofluorescence or dot blot approaches, to comprehensively study the accumulation of R-loops and their consequences on genome integrity.

This chapter will focus on the key experimental steps (Fig. 1) required to visualize and quantify R-loops on genomic DNA by EM, which include:

1. An EM-compatible genomic DNA extraction procedure from human cells that maintains R-loop stability.
2. An updated EM sample preparation procedure, including recent changes to the already published protocols [14–16].

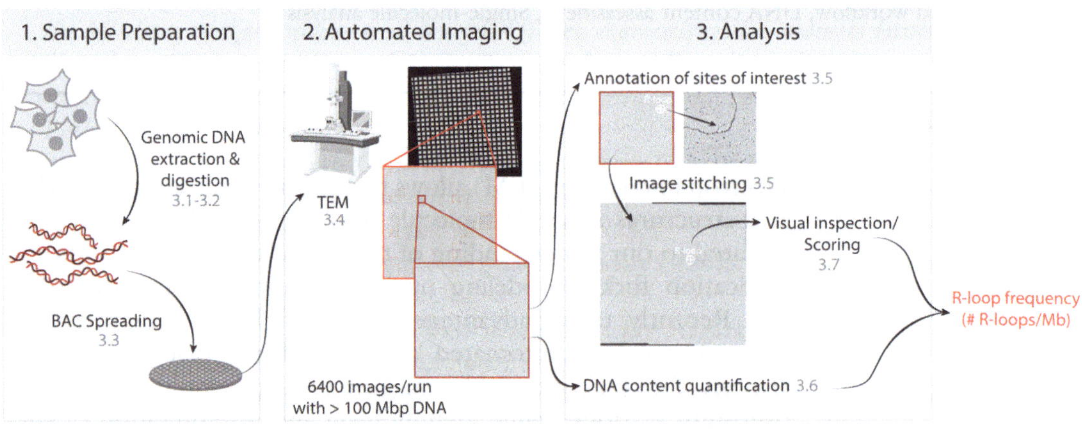

Fig. 1 TEM based quantification of R-loops on genomic DNA. This technique consists of three essential steps. (1) Sample preparation: genomic DNA is extracted, digested and subsequently spread on nickel grids. (2) Automated imaging: the grids are imaged in a two-step imaging setup to acquire large sets of images at high resolution. (3) Analysis: the acquired image sets are annotated for potential sites of interest, which are subsequently stitched and subjected to visual inspection to identify the R-loop structures. In parallel, the total DNA content is quantified to calculate R-loop frequency as # of R-loops/Mb

3. An automated image acquisition workflow, which simplifies the manual labor required for EM and increases the objectivity of this technique.
4. A publicly available image stitching algorithm to stitch large numbers of EM images efficiently and provide a larger context for each R-loop.
5. A workflow to quantify overall DNA content in the imaged area to normalize R-loop numbers to the overall DNA content and calculate R-loop frequencies in R-loops/MB.
6. A detailed guideline on how to reliably identify and quantify R-loops.

The described pipeline takes advantage of equipment and software that are available to us at the Center for Imaging of the University of Zurich. Where needed, the Notes discuss key factors to consider when adapting this procedure to different microscopes, associated software and alternative image stitching or analysis tools.

2 Materials

2.1 Genomic DNA Extraction

1. Ice-cold 1× PBS.
2. 1× TE buffer.
3. 20% SDS. Store at RT.
4. 20 mg/mL Proteinase K stock solution in ddH_2O. Store at −20 °C.
5. EM digestion buffer/QIAGEN buffer G2: 800 mM guanidine hydrochloride; 30 mM Tris–HCl pH 8.0; 30 mM EDTA pH 8.0; 5% Tween 20; 0.5% Triton X-100. Store at RT.
6. Chloroform–isoamyl alcohol 24:1. Store at RT under a ventilated hood.
7. Sodium acetate (NaOAc) as a 3 M stock, pH 5.2. Store at RT.
8. 100% and 70% ethanol.
9. 30 mL glass centrifugation tubes.
10. Eppendorf centrifuge 5810R, rotor A-4-81.
11. Sorvall Evolution RC, rotor HB-6 swinging (8000 rpm = 10,459 × *g*).
12. Eppendorf centrifuge 5424, rotor FA-45-24-11.

2.2 DNA Digestion

1. Restriction enzyme PvuII-HF. Store at −20 °C.
2. Restriction enzyme EcoRI-HF. Store at −20 °C.
3. 10× CutSmart buffer, diluted to 1× in ddH_2O.

4. 10 mg/mL RNase A, stock in ddH_2O. Store at −20 °C.
5. RNase III. Store at −20 °C.
6. RNase T1. Store at −20 °C.
7. RNase H, including 10× RNase H buffer. Store at −20 °C.
8. Silica Bead DNA Gel Extraction Kit.
9. 1× TE buffer.

2.3 Native DNA Spreading by the "BAC Method"

1. Carbon-coated nickel grids (Plano, Cat # G2400N), freshly prepared as described in [14, 16].
2. P2 pipette.
3. ddH_2O.
4. 10 mg/mL ethidium bromide (EtBr, freshly diluted 1:300 in ddH_2O for a 33.3 μg/mL working solution.
5. 100% Formamide (Sigma, Cat # 47680).
6. Benzyldimethylalkyl ammonium chloride (BAC) (Sigma-Aldrich, Cat # B6295) as a stock solution 0.2% w/vol in Formamide. The stock solution is freshly diluted 1:10 in 1× TE to create a BAC working solution.
7. 100% Ethanol.
8. Uranyl acetate (UAc, Fluka, Ca # 73943) as a stock solution of 5 mM in 5 mM HCl. Store at 4 °C. The stock solution is freshly diluted 1:10 in 100% EtOH to create a UAc working solution.
9. Round filter paper.
10. Tissue culture dishes, 150 × 20 mm.
11. Mica sheets (Plano, mica high grad quality V1, 25 × 76 mm, Cat # 56).
12. Graphite powder.
13. Dumont tweezer.

2.4 Automated Image Acquisition

1. Transmission electron microscope (TEM), where stage and camera control can be automated. The preferred setup used in this workflow consists of the Thermo Fisher Talos 120 transmission electron microscopy with a bottom mounted 4k × 4k pixel CMOS camera (BM Ceta) and the MAPS software package (*see* **Note 1** for alternatives).

2.5 Image Annotation and Image Stitching

1. Thermo Fisher Scientific MAPS Offline Viewer (*see* **Note 2**).
2. Image J.
3. A computer or virtual machine running Windows with at least 16 GB of RAM (*see* **Note 3**).
4. A Python environment ($\geq$ version 3.7) with the ForkStitcher application installed. The ForkStitcher application, its

installation instructions and its detailed usage instructions are available on https://github.com/jluethi/ForkStitcher (*see* **Note 4**).

2.6 DNA Content Quantification

1. Fiji.
2. Python environment (version 3.7).
3. MATLAB.
4. This workflow requires several different scripts and files, which are available on https://github.com/roessler-f/DNAQuantification, including detailed installation instructions for the necessary software and their dependencies (*see* **Note 5**).

2.7 Image Analysis and Interpretation

1. Image J or other software that enables contour length measurements using tiff files.
2. Standard statistic applications (e.g., Microsoft Office Excel or GraphPad).

3 Methods

3.1 Genomic DNA Extraction

1. Samples can be collected from asynchronously growing cells or at different time points in synchronization experiments (*see* **Note 6**). Every sample consists of 5–10 × 10^6 cells (corresponding to two 150 × 20 mm tissue culture dishes at 50–70% confluency for MCF7 cells). Collect the cells by scraping into cold 1× PBS and spin them down in a 50 mL falcon tube at 500 × *g* for 5 min.
2. Resuspend the cell pellet carefully in 1.6 mL TE, add 41.5 μL 20% SDS (final concentration 0.5%) and 50 μL 20 mg/mL proteinase K (final concentration 0.6 mg/mL) and incubate at 37 °C overnight (*see* **Note 7**).
3. Add 3 mL of EM digestion buffer containing 0.3 mg/mL proteinase K to the cell lysates and incubate at 37 °C for additional 1.5 h (*see* **Note 8**).
4. Add one volume of chloroform–isoamyl alcohol to the lysates. Invert vigorously but carefully for 30 times (the solution turns milky). Immediately transfer the sample into a glass centrifugation tube.
5. Centrifuge the samples at 4 °C for 20 min at 10,500 × *g* (phase separation).
6. Carefully transfer the upper phase into a new glass centrifugation tube using a cut 1 mL tip.

7. Add NaOAc pH 5.2 to a final concentration of 0.3 M, mix gently. Then add two volumes of 100% ethanol. Swirl the tube gently until the DNA precipitates (*see* **Note 9**).
8. Using a cut 1 mL tip transfer the DNA precipitate into a 1.5 mL Eppendorf tube, prefilled with 0.5 mL 70% ethanol.
9. Centrifuge at 20,000 × *g* for 3 min.
10. Wash DNA with 70% ethanol three times.
11. Air dry the DNA at 37 °C until the pellet begins to turn transparent. Make sure not to dry the DNA pellet too extensively (*see* **Note 10**).
12. Resuspend the pellet in 400–600 μL TE with slight rotation (*see* **Note 11**). Standard yield is 40–80 μg of genomic DNA for 5–10 × 10^6 cells.

3.2 DNA Digestion

1. Digest 10–15 μg of genomic DNA with 50 U PvuII-HF and 50U EcoRI-HF in 1× CutSmart buffer at 37 °C overnight (*see* **Note 12**).
2. To remove RNA moieties, split the digestion reaction into two equal reactions and digest with 0.25 mg/mL RNase A, 25 U/mL RNase T1, 25 U/mL RNase III ± 500 U/mL RNase H in RNase H buffer, supplemented with 0.5 mM NaCl, for 2.5 h at 37 °C (*see* **Note 13**).
3. The DNA is subsequently purified using the Silica Bead DNA Gel Extraction Kit (*see* **Note 14**): Add 3 volumes of binding buffer and 10–15 μL silica beads to the digestion reactions and incubate at 50 °C for 5 min (invert the tubes regularly to keep the beads in suspension). Centrifuge the samples on short for 5 s to pellet the beads. Carefully remove the supernatant. Wash the beads three times with ice-cold wash buffer without resuspending the beads to avoid sheering the DNA. Air-dry the pellet for 10 min. For elution, add 10–15 μL TE to the pellet and mix by flicking the tube, incubate at 50 °C for 5 min and subsequently centrifuge on short for 5 s. Collect the supernatant. Repeat the elution step and pool both eluates.
4. Digested and purified DNA can be stored at 4 °C (*see* **Note 15**).

3.3 Native DNA Spreading by the "BAC Method"

1. Visual illustration for the key steps of the DNA spreading as well as advice on troubleshooting are included in [14, 16].
2. For each DNA spreading distribute up to eight droplets (20 μL) of EtBr working solution on a piece of Parafilm and carefully place a carbon-coated grid on top of each drop, with the carbon-side facing down. Incubate the grids for 20–60 min. Prevent evaporation and light exposure by covering the grids with the lid of a 15 cm petri dish, covered in

aluminum foil. Just before spreading, drain the excess EtBr solution from each grid by contacting the filter paper, and place the dried grid (carbon-side down) on top of the filter paper.

3. Using a P2 pipette, mix at the very bottom of a 1.5 mL Eppendorf tube: 2 μL of Formamide and 0.4 μL BAC working solution.
4. Add 2 μL of genomic DNA (roughly 50–100 ng of DNA, predilute with 1× TE if too concentrated) in a separate drop to the side of the tube (*see* **Note 16**).
5. Pour approximately 20 mL (the minimal volume to cover the surface completely) of EM-grade water in the 15 cm petri dish; this is called the hypophase. Split a mica sheet (about 1 × 2 cm in size) and place it in the water as a ramp, with the freshly cleaved surface facing up.
6. Spin the tube containing the sample for a few seconds in a microfuge and aspirate it completely into a 10 μL tip. With a cotton swap sprinkle a generous boundary of graphite powder onto the water surface, as a half circle in close proximity to the mica ramp.
7. Pipette the entire sample volume out of the tip and let the droplet touch the ramp a few millimeters above the water surface. The droplet will immediately slide down the ramp and spread over the hypophase, pushing the graphite boundary backward which indicates the border of the monomolecular detergent film. The DNA content will be highest close to this border.
8. Using fine tweezers take a carbon-coated, EtBr-soaked grid from the filter paper (carbon-side down) and place it on the spreading surface close to the graphite border. Hold the tweezers as parallel to the water surface as possible and gently press the grid down for 1–2 s to ensure that the grid has sufficient and even contact with the water surface. Carefully and without disturbing the hypophase lift the grid up again (*see* **Note 17**) and incubate it for 15–20 s in 1 mL uranyl acetate working solution (best in a flat-bottom 20 mL tube). Wash the grid by dipping it 2–3 times into 100% EtOH and air-dry the grid (carbon-side up) on a fresh filter paper.
9. Wipe the tweezers with a napkin to remove EtOH remnants (*see* **Note 18**) and repeat **step 7** for the other grids, aiming to collect DNA from different regions of the BAC film.
10. The grids are subsequently subjected to flat angle-rotary shadowing with platinum as described in [14, 16].

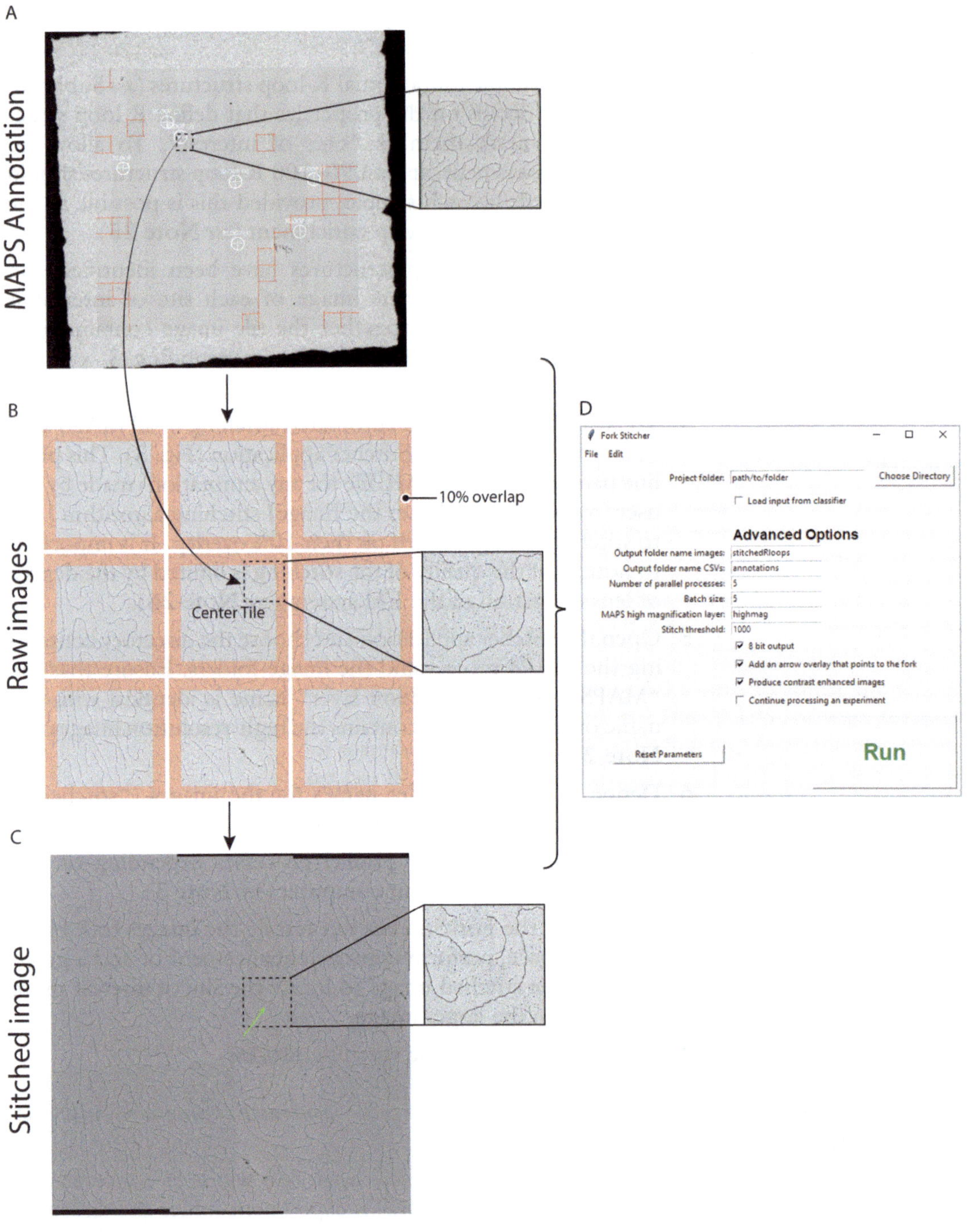

Fig. 3 Annotation and stitching workflow. (**a**) Using the MAPS offline viewer, R-loops are marked ("annotated") using the MAPS annotation feature. (**b**) To follow molecules that extend beyond the center tile, the surrounding tiles are identified. The 10% overlap to the neighboring image is marked in red. (**c**) Using this 10% overlap, the ForkStitcher workflow stitches the 3 × 3 images together and saves an optimized version containing an arrow (green) pointing to the original annotation. (**d**) The user interface of the ForkStitcher, where the user has to define the path to the project folder as well as adapt different optional parameters of the ForkStitcher. The stitching process is initiated by clicking "Run"

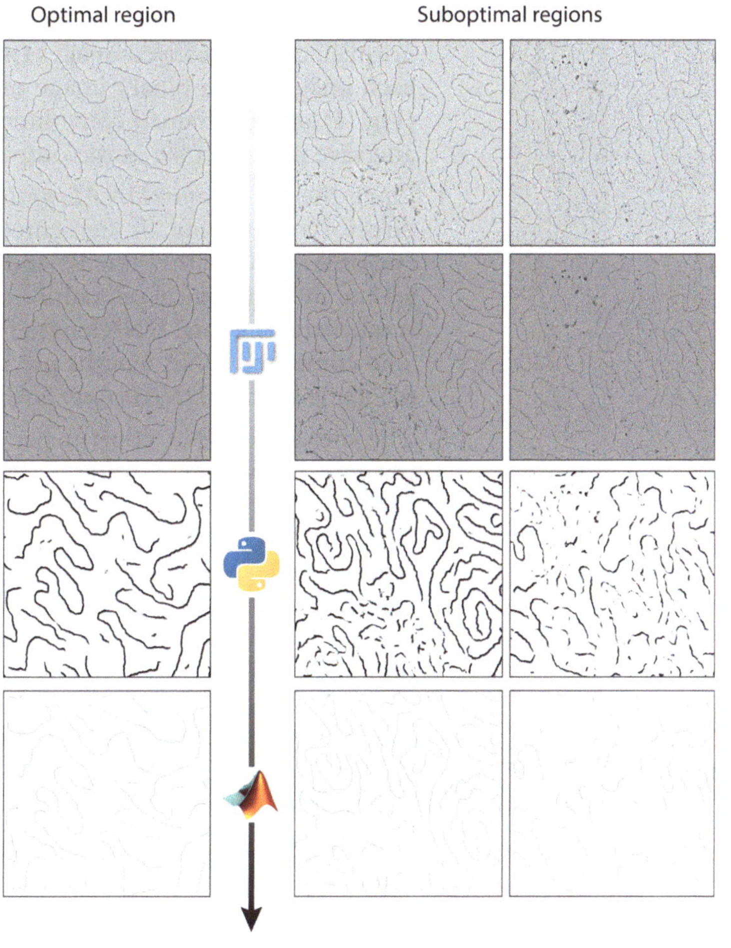

Tile set	DNA content [nm]
1	4498599.6821
2	4353529.5891
3	3342971.1767
...	

Fig. 4 DNA content quantification workflow. First, all acquired images are contrast-normalized and rescaled to a pixel size of 2 nm using Fiji. Afterward, the images are segmented into DNA (black) and non-DNA (white) parts using a trained machine-learning algorithm implemented in Python. The quality of the segmentation depends on the quality of the acquired regions (*see* comparison between optimal and suboptimal, dirty regions), it is therefore advisable to perform a quality check before image acquisition (*see* Subheading 3.4). After segmentation, the DNA parts are skeletonized and filtered for length using MATLAB. The skeletonized DNA parts are then used to determine the total DNA length of one tile set

directly comparing R-loop frequencies of individual samples within the same experiment. However, in order for these numbers to be as accurate as possible, it is essential to perform the R-loop annotation as thorough as possible, to ensure that most if not all the present R-loops are included in this analysis. In addition, only those tiles sets should be included in the calculation of total DNA content, which were actually screened for R-loop structures.

5. To calculate R-loop burden, calculate the total amount of base pairs involved in R-loop formation (i.e., sum of all R-loop lengths) and normalize to the overall DNA content (quantified in Subheading 3.6). R-loop burden is expressed as "bp of R-loops/total bp [%]" (*see* **Note 34**).
6. In addition to quantifying R-loop size, frequency and burden, R-loops can be examined for their location respectively to other known DNA structures, such as replication forks. Alternatively, an EM experiment, as it is described in this chapter, can allow to correlate R-loop frequency with replication fork architecture (detailed assessment of replication fork architecture is described in [14]) to build hypotheses on how R-loop accumulation genome wide can affect replication fork architecture.

4 Notes

1. The Talos 120 TEM/MAPS combination is a proprietary solution from Thermo Fisher Scientific; however, a comparable workflow can be achieved with alternative microscope–software combinations, provided they meet the following requirements. First, the microscope should be able to maintain stable imaging settings (including a stable illumination and a stable alignment or, alternatively, a robust automatic "on the fly" alignment feature) even during prolonged imaging periods (>12 h). Secondly, stage and camera control should be automated, allowing to set up a queue of imaging jobs. Thirdly, an integrated autofocus feature is favorable but not strictly necessary; if an autofocus is not available the focus needs to be set manually for every imaging job (i.e., tile set) during the initial setup. Finally, it is very useful if microscope settings for the low-resolution overview as well as the individual high-resolution tile sets can be saved as templates to increase productivity and reproducibility of this workflow.
2. The MAPS Offline Viewer is part of the proprietary MAPS software package from Thermo Fisher Scientific and hence only compatible with a Thermo Fisher Scientific TEM. Comparable offline viewers are available for different TEM setups; however, these solutions would not be compatible with the

ForkStitcher application (*see* Subheading 3.5), or the DNA content quantification workflow (*see* Subheading 3.6), presented in this protocol. In **Notes 4** and **5**, we provide advice on how these workflows could be adapted if necessary.

3. While the ForkStitcher runs on both Mac and Windows, the user interface is only compatible with Windows. The memory required to run the ForkStitcher depends on the chosen settings: The Stitcher can be run in parallel processes. Each process requires about 14 GB of memory. Thus, running five parallel processes (default) requires a computer with at least 70 GB of memory. On computers with lower amounts of memory (e.g., 16 GB), the "number of parallel processes" in the advanced options of the ForkStitcher should be decreased to 1.
4. The ForkStitcher implementation is specific for processing annotations made in the MAPS Viewer, it can however serve as an example to implement a comparable workflow for another system. The major steps that the ForkStitcher performs—and which a comparable workflow would need to include—are as follows.
   (a) Get information from the user about the location of the EM image data, the annotations and any parameters needed for stitching the images.
   (b) Parse the annotation data (e.g., for MAPS parse the MAPS XML file that describes the location of all annotations in the MAPS coordinates).
   (c) Calculate which image sites contain annotations and which surrounding sites would need to be stitched to them.
   (d) Stitch the central image containing an annotation with the surrounding images. For our workflow, we use a 3×3 stitching. The ForkStitcher can handle the edge cases if images at the edge of an acquisition contain an annotation and thus do not have neighbors on all sides. For the actual stitching part, the ImageJ stitching algorithms are very robust [17].
   (e) Postprocess the stitched image. Do a contrast enhancement, add an ImageJ overlay to point to the annotation in the large, stitched image and decide whether the image can be saved at a lower bit rate, for example, 8 bit.
5. The provided scripts execute all the steps described in Subheading 3.6 and shown in Fig. 4, but they can equally well serve as example code and can be adapted accordingly if this procedure is to be implemented on a different platform. The different steps of the workflow are depicted in Fig. 4.

6. This protocol has been developed using MCF7 cells undergoing sequential estrogen deprivation and subsequent stimulation [18]. This very specific manner of synchronization of estrogen-addicted cell lines results in a strong enrichment of R-loops, a prerequisite for efficient R-loop quantification by EM. Nevertheless, this protocol should be easily adaptable to other cell systems. However, the user should be aware that working with samples with expectedly low levels of R-loops could significantly increase the imaging and analysis time.
7. Alternative genomic DNA extraction protocols—for example [14]—could theoretically also work, have however not been properly validated for the stability of R-loop structures.
8. The incubation in EM digestion buffer is crucial to strip proteins from the DNA (since Phenol should be omitted during the extraction to prevent DNA nicking). However, extensively prolonging this incubation step was observed to result in fragmentation of the DNA, it is therefore advisable to stick to the indicated timing.
9. Salt should generally be avoided during DNA precipitation for EM preparation as it compromises the sample quality. In our experience, however, using NaOAc to help precipitating the DNA at this step does not compromise sample quality, since it is followed by several additional purification step throughout sample preparation.
10. Drying the DNA pellet too extensively can not only impede DNA resuspension afterward, but can also lead to more DNA crossings, hindering proper separation of DNA molecules during the spreading process later.
11. At this step, it can take hours to days for the DNA to be completely resuspended—to help resuspension, samples can be incubated at 37 °C or carefully (!!!) mixed by pipetting with a generously cut tip to avoid DNA shearing.
12. We observed that R-loop stability remains unaffected even during extensive DNA digestion [18] and decided to use a combination of restriction enzymes that results in fragment lengths between 3 and 15 kb.
13. Generally, free RNA complicates the EM analysis and should be removed by RNase A/III/T1 digestion. To reduce RNase A activity toward DNA–RNA hybrids, salt concentration should be adjusted to 500 mM using NaCl. We further advise to include the RNase H control for every experiment to ensure that the observed R-loop structures are reduced and hence RNase H–sensitive. Splitting the digestion reaction into two separate reactions for RNase digestion (+/− RNase H) ensures maximum comparability.

14. Comparison of R-loop stability during different DNA purification approaches has identified this kit to be superior to other purification approaches in terms of R-loop stability on digested DNA [18]. To avoid breaking DNA structures, the pellet should never be resuspended by vortexing or pipetting, but rather left intact or (as is the case during elution) carefully resuspended by flicking the tube.
15. Especially when attempting this technique for the first time, it is advisable to ensure that R-loops remain stable throughout the sample preparation for example by comparing R-loop levels on genomic vs digested, purified DNA on an S9.6 dot blot.
16. Formamide is a denaturing agent and essential to disentangle DNA molecules during spreading. However, to prevent extensive denaturation of the dsDNA into ssDNA, the exposure of the DNA to the formamide should be as short as possible.
17. If the hypophase is visibly disturbed, it is advisable to repeat **steps 3–7**, Subheading 3.3 for the remaining grids.
18. EtOH remanent will disturb the hypophase dramatically, wiping off the tweezers is hence essential to keep the hypophase stable throughout the spreading.
19. Automated electron microscopy takes time (about 16 h per sample) and generates large amounts of data (roughly 0.5–1 TB per sample). To avoid the acquisition of suboptimal samples, we recommend this quick and easy quality check prior to automated image acquisition. Besides, this quality check can be used to preselect an area of interest, which can significantly speed up the automated imaging setup later. These preselected areas of interest for automated imaging should have a flat carbon layer without any carbon folding, a homogeneous background without any dirt, good contrast and a decent DNA density that allows to clearly distinguish single molecules while at the same time providing sufficient sites of interest (3–5 sites/square). For the latter, we recommend to broadly check two or three squares for potential sites of interest (*see* Subheading 3.7 for details); this requirement can however be omitted for conditions of R-loop depletion (*see* **Note 25**).
20. A detailed, step-by-step guideline for setup and acquisition using the Talos 120 TEM in combination with the MAPS software package is freely available online (https://zmb.dozuki.com/). For questions concerning this setup, please contact info@zmb.uzh.ch.
21. The separation of low-resolution overview images and high-resolution tile sets into different MAPS layers is crucial to later apply the ForkStitcher application (Subheading 3.5). The default names for the layers are “Overview” and “highmag.”

22. It is advisable to orient the grid in the TEM holder in a manner that the grid bar orientation matches the camera chip (Fig. 2). The little mark on the edge of the grid can help to consistently orient the grid optimally.
23. Theoretically, setting the autofocus once per tile set (in the first tile) can be sufficient; however, in our experience, an additional autofocus step in the fourth tile, when illumination and stage movement have stabilized, improves the overall focus. Autofocusing in every tile—albeit possible—is not recommended, since it prolongs the imaging run significantly without significant improvement in image quality.
24. The offset necessary to achieve optimal underfocusing can vary. When attempting this imaging for the first time, it is advisable to perform a focus series to determine the optimal offset value. Once optimized this value should however be kept constant throughout the imaging experiment.
25. Since R-loop numbers can vary quite extensively between conditions of R-loop enrichment and R-loop depletion/resolution, we recommend using the amount of imaged DNA (as quantified in Subheading 3.6) for guidance: In conditions of R-loop enrichment by estrogen stimulation, around 300 Mb of genomic DNA had to be acquired to reach the required numbers of R-loops for statistical analysis. In control conditions, the required amount of genomic DNA went up to 600 Mb. In conditions of R-loop depletion/digestion, such as RNase H treatment, reaching 70–100 R-loops has proven challenging—for these conditions, we have therefore decided to image a comparable amount of genomic DNA as for the corresponding samples without RNase H digestion.
26. Depending on the automated microscope, the degree of lens distortion can change over time. In our Talos setup, a lens distortion correction calculated on day one is not very helpful to improve the image acquired on another day. As correcting lens distortion is not trivial, we decided not to implement it into our workflow. Thus, the stitching will not be pixel-perfect. A correction for the lens distortion could be added to this step. Alternatively, users can manually align individual images of interest where minor distortions would be an issue using an image editor and use the rest of the images with minor stitching artefacts.
27. The provided scripts run properly only when the images acquired for the different tile sets are saved in different subfolders ("…/experiment folder/tile set 1/", "…/experiment folder/tile set 2/", etc.).

28. It is important to rescale the acquired images to a pixel size of 2 nm as all the following steps in Subheading 3.6 are dependent on this assumption. The correct rescaling is achieved by defining the appropriate scaling factor in the Fiji macro script (variable "scalingFactor," line 9).
29. Ensure that the input folder, output folder and the path to the trained CNN files are defined correctly in the *PixelClassifier.py* file. More information about how and where to define them can be found in the GitHub repository.
30. It is advisable to check the quality of the segmented images before the first use of the whole DNA content quantification workflow. If the quality is not sufficient, you can try to improve it with the following two steps: (1) Change the parameters of the contrast-normalization step in the Fiji macro script (line 43) or (2) train the CNN on additional manually labelled images acquired with your own setup.
31. The factor 0.207 used to correct the DNA content for the 10% overlap applies to the image acquisition setup described in Subheading 3.4 (16 × 16 images, 10% overlap). If a different setup is used, the correction value c can be calculated with the following equation: $c = 2 \times (n_{\text{Tiles}} - 1) \times s/m$, with n_{Tiles} = number of images taken in one dimension (in our case: 16), s = overlap in pixels (in our case: 410 pixels) and m = pixel size of total tile set in one dimension (in our case: 59,386 pixels).
32. Since the flexibility of DNA can vary depending on the sample preparation, this conversion factor can vary as well. It is therefore advisable to determine this conversion factor at least once yourself by EM analysis of a plasmid of known size.
33. Sometimes single and double stranded structures are difficult to distinguish and depend on the local quality of the sample. It is therefore important to consider the quality of the surrounding DNA molecules, when evaluating a specific R-loop structure.
34. RNase H digestion has been observed to decrease R-loop frequency and/or R-loop size. In both cases, R-loop burden clearly decreases.

References

1. Sogo JM, Lopes M, Foiani M (2002) Fork reversal and ssDNA accumulation at stalled replication forks owing to checkpoint defects. Science 297:599–602
2. Lopes M, Foiani M, Sogo JM (2006) Multiple mechanisms control chromosome integrity after replication fork uncoupling and restart at irreparable UV lesions. Mol Cell 21:15–27
3. Quinet A, Tirman S, Jackson J, Šviković S, Lemaçon D, Carvajal-Maldonado D et al (2020) PRIMPOL-mediated adaptive

response suppresses replication fork reversal in BRCA-deficient cells. Mol Cell 77:461–474

4. Berti M, Cortez D, Lopes M (2020) The plasticity of DNA replication forks in response to clinically relevant genotoxic stress. Nat Rev Mol Cell Biol 21:663–651
5. Kolinjivadi AM, Sannino V, De Antoni A, Zadorozhny K, Kilkenny M, Técher H et al (2017) Smarcal1-mediated fork reversal triggers Mre11-dependent degradation of nascent DNA in the absence of Brca2 and stable Rad51 nucleofilaments. Mol Cell 67:867–881
6. Mukherjee C, Tripathi V, Manolika EM, Heijink AM, Ricci G, Merzouk S et al (2019) RIF1 promotes replication fork protection and efficient restart to maintain genome stability. Nat Commun 10:3287
7. Ray Chaudhuri A, Hashimoto Y, Herrador R, Neelsen KJ, Fachinetti D, Bermejo R et al (2012) Topoisomerase I poisoning results in PARP-mediated replication fork reversal. Nat Struct Mol Biol 19:417–423
8. Follonier C, Oehler J, Herrador R, Lopes M (2013) Friedreich's ataxia-associated GAA repeats induce replication-fork reversal and unusual molecular junctions. Nat Struct Mol Biol 20:486–494
9. Giannattasio M, Zwicky K, Follonier C, Foiani M, Lopes M, Branzei D (2014) Visualization of recombination-mediated damage bypass by template switching. Nat Struct Mol Biol 21:884–892
10. Zellweger R, Dalcher D, Mutreja K, Berti M, Schmid JA, Herrador R et al (2015) Rad51-mediated replication fork reversal is a global response to genotoxic treatments in human cells. J Cell Biol 208:563–579
11. Mijic S, Zellweger R, Chappidi N, Berti M, Jacobs K, Mutreja K et al (2017) Replication fork reversal triggers fork degradation in BRCA2-defective cells. Nat Commun 8:859
12. Lemacon D, Jackson J, Quinet A, Brickner JR, Li S, Yazinski S et al (2017) MRE11 and EXO1 nucleases degrade reversed forks and elicit MUS81-dependent fork rescue in BRCA2-deficient cells. Nat Commun 8:860
13. Mutreja K, Krietsch J, Hess J, Ursich S, Berti M, Roessler FK et al (2018) ATR-mediated global fork slowing and reversal assist fork traverse and prevent chromosomal breakage at DNA interstrand cross-links. Cell Rep 24:2629–2642
14. Zellweger R, Lopes M (2018) Dynamic architecture of eukaryotic DNA replication forks in vivo, visualized by electron microscopy. Methods Mol Biol 1672:261–294
15. Lopes M (2009) Electron microscopy methods for studying in vivo DNA replication intermediates. In: DNA replication. Springer, pp 605–631
16. Neelsen KJ, Chaudhuri AR, Follonier C, Herrador R, Lopes M (2014) Visualization and interpretation of eukaryotic DNA replication intermediates in vivo by electron microscopy. In: Functional analysis of DNA and chromatin. Springer, pp 177–208
17. Preibisch S, Saalfeld S, Tomancak P (2009) Globally optimal stitching of tiled 3D microscopic image acquisitions. Bioinformatics 25: 1463–1465
18. Stoy H, Zwicky K, Lang K, Krietsch J, Kuster D, Schmid J et al Direct visualization of transcription-replication conflicts reveals post-replicative DNA:RNA hybrids. under Revis. after peer Rev

Chapter 2

Detection of R-Loop Structures by Immunofluorescence Using the S9.6 Monoclonal Antibody

Konstantina Skourti-Stathaki

Abstract

This protocol describes a method of detection of R-loop structures by Immunofluorescence using the S9.6 antibody. R-loops are three-stranded nucleic acid structures that comprise the nascent RNA hybridized with the DNA template strand (RNA–DNA hybrid) leaving the nontemplate DNA strand single-stranded (ssDNA). R-loops are dynamic structures that have been linked to transcription-associated DNA damage and genomic instability in certain contexts but they also possess critical regulatory functions. They are direct products of transcription and they have been associated with transcriptional activation, repression and termination in cases of both protein-coding and noncoding genes. Visualizing and mapping R-loops has been a sought-after task over the last years. Next-generation sequencing of RNA–DNA hybrids, which are components of R-loops, using the S9.6 antibody, aims to detect R-loops genome-wide, whereas Immunofluorescence is performed to visualize R-loops in single cells. While mapping R-loops genome-wide is very important for identifying and studying their location-specific role, microscopy offers the advantage of spatial information and the ability to quantify them on a single cell level. In this chapter, I will describe the protocol I have used to image RNA–DNA hybrids in the nucleus of mammalian cells.

Key words R-loops, RNA–DNA hybrids, S9.6 antibody, Immunofluorescence, Single cells, Microscopy

1 Introduction

RNA is rapidly emerging as the smart molecule overshadowing its deoxy counterpart which is responsible for carrying the genetic information. However, when paired with DNA, an interesting connection emerges from the formation of RNA–DNA hybrids. DNA and RNA coexisted before the emergence of life on Earth [1] and RNA–DNA hybrids were first described in 1976 by R.W. Davies where he reported that RNA can stably hybridize with DNA in the presence of formamide [2]. During transcription, nascent RNA can hybridize naturally with the DNA template strand, leaving the nontemplate DNA strand single stranded. These structures are called R-loops, and their formation has been

Andrés Aguilera and Alexey Ruzov (eds.), *R-Loops: Methods and Protocols*, Methods in Molecular Biology, vol. 2528, https://doi.org/10.1007/978-1-0716-2477-7_2,

a subject of intense investigation over the last decade due to their involvement in various cellular processes.

For years R-loops were considered as transcriptional by-products and a threat to genome integrity [3]. Since then, we started realizing that R-loops can be a double-edged sword; deleterious structures when their formation is unscheduled, but also versatile molecules that can regulate a variety of cellular processes from gene expression and chromatin to targeting enzymes and DNA damage repair [4–7]. The tantalizing thought is that one structure can link all these seemingly unrelated processes within the same cell and can have different functions depending on the gene subset [8].

In science there is a long-standing saying that seeing is believing, and for R-loops 35 years have passed since they were first visualized by Meyer et al. using an electron microscope [9]. Since then, detecting R-loops accurately and specifically in single cells has indeed been an ever-sought task. Currently there are two methods for detection of R-loops by microscopy. The first relies on the use of the S9.6 antibody and is the one I will describe in this chapter (Fig. 1). The S9.6 antibody detects the RNA–DNA hybrid structure, and it is therefore structure specific rather than sequence specific [10]. The second method uses the catalytically inactive RNASEH1 expressing GFP-tagged plasmid. RNAse H1 is the enzyme that specifically degrades the RNA within the RNA–DNA hybrid, and therefore the catalytic mutant would bind the RNA–DNA hybrid but not cleave it [11]. Even though both methods can be employed to visualize R-loops in single cells, each has their own limitations but also conceptual differences. R-loops are transient, dynamic molecules with different turnover timelines depending on the gene context [8, 12], and while the catalytic mutant RNASEH1 would bind the RNA–DNA hybrids it is possible that it would only detect the stable RNA–DNA hybrids, which would be long-lived enough to be the substrate for RNAse H1 binding. Thus, capturing the dynamic nature of R-loops is potentially lost or reduced potentially losing the dynamic nature of R-loops and transcription. The S9.6 antibody is hybrid-specific and therefore a more direct tool for visualizing R-loops in single cells and capable of detecting stable and dynamic RNA DNA hybrids. One possible limitation to consider is that it has been suggested that it also has the ability to binds dsRNA. However, the affinity to dsRNA is three- to four fold lower [13] and S9.6 antibody shows a different staining pattern than that of J2 antibody which binds dsRNA [14] and Fig. 2. Nevertheless, it is important that the technical limitations of both approaches are considered in experimental design and appropriate control and dilution series of the antibodies used are carried out in each experimental cell line.

Another technical issue to be considered when visualizing nuclear structures and particularly transient RNA/DNA structures

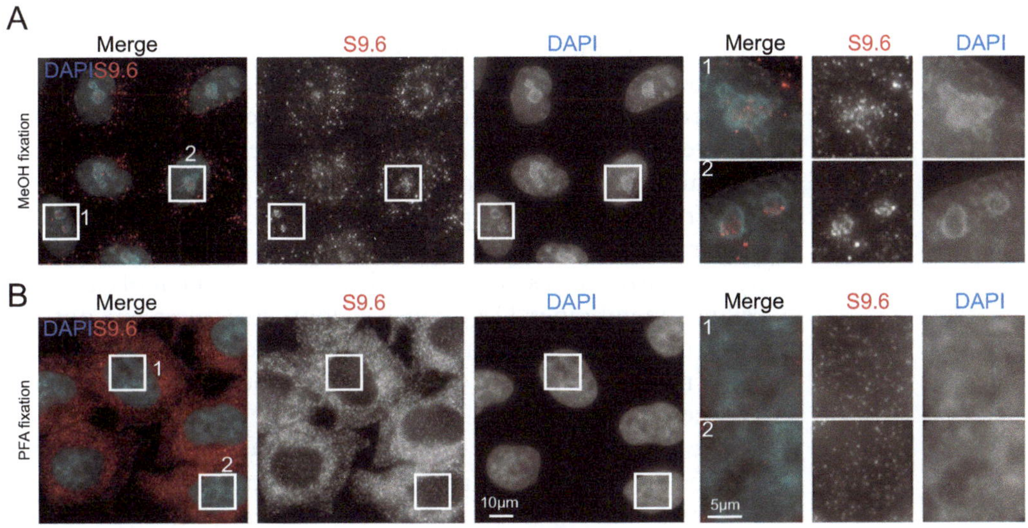

Fig. 1 Cellular localization of R-loops using the S9.6 antibody. Immunofluorescence imaging of R-loops with the S9.6 antibody using Methanol (MeOH; **a**) and paraformaldehyde (PFA; **b**) as fixing reagents. Fixation with methanol allowed for visualization of R-loops in HeLa cell nuclei. Enlarged boxes shown in right panels

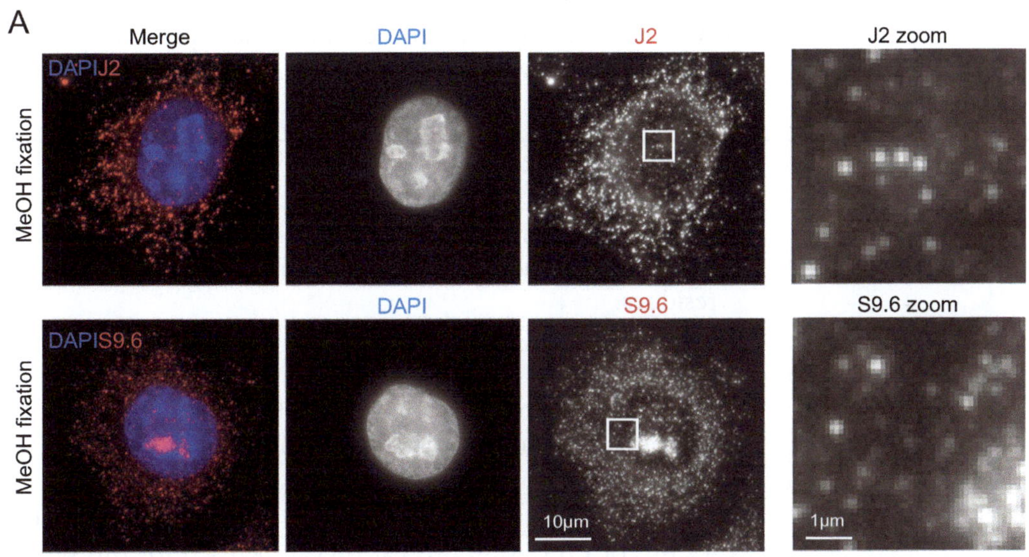

Fig. 2 Cellular localization of R-loops and dsRNA. Whole cell images showing immunofluorescence of dsRNA (J2-top panel) and R-loops (S9.6-bottom panel). Zoomed regions are also shown. Fixation with methanol revealed the different localization patterns of R-loops and dsRNA. Note the enriched staining of R-loops in the HeLa nucleoli

is the method of fixation. It is important that the fixation method is fast enough to fix the structure of interest but also effective in keeping the structure intact. The standard fixation method for Immunofluorescence is paraformaldehyde (PFA) or formaldehyde (FA), which keeps the cell membranes intact and creates covalent bonds between amine groups present of lipids and proteins. With PFA or FA, the cells would need to be permeabilized to allow antibodies to access intracellular structures and bind to their target epitopes. Another fixation method is the use of alcohol-based fixatives such as methanol which dehydrates and causes all cellular structures to collapse and precipitate the proteins. Methanol is preferred in nucleic acid protocols such as FISH, and chromosome preparations and has the added advantage that it does not require permeabilization as it utilizes the ability of alcohols to disrupt cellular membranes [15]. Since R-loop structures are not stable molecules such as proteins, but they can be highly variable due to their nature as nucleic acids, I did a side-by-side comparison between methanol (Fig. 1a) and PFA (Fig. 1b) fixation techniques before proceeding with the Immunofluorescence experiment. As shown, fixing with methanol leads to a better signal to noise ratio and retention of the specific structures corresponding to R-loops, while it reduces the unspecific binding that is observed with PFA and the S9.6 antibody (Fig. 1) [11, 14].

In this methods chapter, I will describe in detail the immunofluorescence approach using methanol as the fixation reagent and the S9.6 antibody as the method of detecting R-loops structures. This will be a step-by-step protocol that I have previously successfully used in mammalian cells [14].

2 Materials

1. PBS: 1× phosphate buffered saline—calcium and magnesium free. Use commercial PBS as homemade ones might have salt residues.
2. 0.1 M glycine for quenching; to be made fresh. Alternatively, 50 mM NH_4Cl can also be used.

   0.1 M glycine: Dissolve 0.37 g of glycine in 50 mL of PBS. A 50 mL Falcon tube can be used for this purpose.
3. 0.2% Triton-X 100 for permeabilization

   0.2% Triton-X: Resuspend 1 mL of 10% Triton-X 100 to 49 mL of PBS. A 50 mL Falcon can be used for this purpose. Vortex and invert tube multiple times to make sure, solution is mixed.
4. Mowiol mounting medium: It can be purchased or prepared in-house.

   For handmade preparation: Place 6.0 g of analytical grade glycerol in a 50 mL plastic tube and add 2.4 g of Mowiol 4-88 while stirring. After addition of 6 mL of dH_2O leave the solution for 2 h at room temperature. Add 12 mL of 0.2 M Tris-HCl pH 8.5, then incubate for 10 min at 50 °C while stirring. Clarify the Mowiol by centrifugation at 3000 × *g* for

15 min and then store at −20 °C as 0.5 mL aliquots in glass vials.

5. Tweezers: A pair of fine tipped number 7 Dumont Tweezers for handling coverslips.
6. Parafilm.
7. Whatman 3 mm paper.
8. Humid chamber: One 15 cm dish lined with Whatman 3 mm cut to fit inside the 15 cm dish, moist with water to create a humid chamber. Cut the Parafilm in a square (10 cm × 10 cm) that fits within the dish but leaves some of wet Whatman paper exposed. During the antibody incubations (primary and secondary) place the coverslips upside down (cells facing down) onto 85 μL droplets.
9. 6-well culture dishes for placing coverslips during fixing and washing steps.
10. Glass coverslips 18 mm grade 1.5H.
11. Glass SuperFrost slides.
12. DAPI stock concentration: 1 mg/mL.
13. S9.6 antibody: mouse antibody from hybridoma cells or commercially available (primary antibody). For this protocol it can be used antibody from the supernatant of mouse hybridoma cells at a dilution of 1:250.
14. Donkey anti-mouse IgG Alexa Fluor 647 (secondary antibody) stock concentration: 2 mg/mL.
15. Millipore water to wash the salt from PBS after the last wash of the secondary antibody.

3 Methods

3.1 Grow the Cells on Coverslips and Fixation

1. Grow cells in an appropriate density for your experiment and make sure cells are growing happily on 18 mm coverslips placed in 6-well dishes.
2. Fix cells with ice cold methanol (kept in −20 °C for at least 24 h) for 7 min in −20 °C (*see* **Note 1**). For this fixation step, after quickly adding the methanol place the 6-well dishes directly −20 °C freezer for 7 min.
3. After 7 min, rehydrate the cells by washing with PBS 3 times. For the washes, remove methanol and add 2 mL of PBS per well. Swirl in a double orbital shape /8 shape and leave at room temperature for 5 min. Repeat 3 times. Make sure wells do not dry at any time point and replace liquid in each well on at a time.

4. Coverslips with cells in PBS can be left at this stage in 4 °C fridge overnight (*see* **Note 2**).

3.2 Quenching, Permeabilization, and Primary Antibody

1. Quench the coverslips with cells with 2 mL of 0.1 M glycine (made fresh) for 10 min at room temperature.
2. Wash coverslips in PBS three times at room temperature. For each wash, add 2 mL of PBS per well, leave at room temperature for 5 min and repeat.
3. Permeabilize the cells with 0.2% Triton-X-100 for 7 min.
4. Wash 3 times as in **step 2**.
5. Create humidity chamber: take a 15 cm dish and use either side. Cut a Whatman 3 mm paper in similar dimensions of the 15 cm dish and spray it with water to wet it sufficiently and poor excess liquid. Cut Parafilm 10 cm × 10 cm and place it on top of humid Whatman paper. Make sure the surface is flat and stretched.
6. The primary antibody is S9.6 (mouse) used at 1:250 dilution in PBS (*see* **Notes 3** and **4**). Pipet 85 μL drops on top of the Parafilm in the humid chamber. Transfer the coverslips with fixed cells from the 6-well dish, making sure that the cells are facing down onto the antibody drops.
7. Cover the humidity chamber with a dark surface of place in a bench drawer and leave at room temperature for 1 h.
8. After the incubation, carefully pick the coverslips and return them to a fresh 6-well dish with PBS in them, make sure that the cells are now facing upward now.
9. Wash 3 times with 2 mL of PBS as previously.

3.3 Secondary Antibody, DAPI Staining and Mounting

1. Create another humidity chamber or remove the Parafilm with primary antibodies and place another clean Parafilm piece, ensure that the Whatman is still humid enough by applying more water.
2. Dilute secondary antibody (donkey anti-mouse Alexa Fluor 647) in PBS 1:1000.
3. Take 85 μL of the secondary antibody dilution and make small drops on the Parafilm in the humidity chamber as done with the primary antibody.
4. Transfer coverslips from the 6-well plates and place cells side facing down, onto the antibody drops. Incubate for 1 h.
5. Ensure the humidity chamber is not exposed to light and incubation happens in the dark.
6. At the end of the 1 h transfer coverslips to a 6-well plate and make sure to flip the coverslip so that the cells face up in the well.

7. Wash 2 times with PBS, with 5 min incubation. Make sure you keep the samples protected from light by covering the 6-well plates with aluminum foil for example.
8. To stain the DNA, incubate with DAPI staining solution (1: 1000 dilution in PBS) with PBS. DAPI can also be used during the secondary staining but then use 1:3000 to avoid background (*see* **Note 5**).
9. Wash 3 times with PBS, as before.

3.4 Mounting and Microscopy

1. Prepare a glass beaker with some Millipore water in it, this will be used to reduce salt deposition from the drying of the PBS.
2. Carefully take out each coverslip from the 6-well plates and dip briefly into the Millipore water.
3. Dry off excess liquid by touching the sides of the coverslip on a paper tissue.
4. Mount the coverslips by carefully placing them, cell side facing down onto a 10 μL drop of Mowiol mounting medium which is added to a microscope glass SuperFrost slide.
5. Gently remove any excess mounting medium with Whatman paper.
6. Leave the slides overnight at room temperature in dark to allow Mowiol medium to dry before taking images.
7. Use a 60×/NA 1.35 oil immersion objective on an upright microscope to take images (*see* **Notes 6** and 7).

4 Notes

1. PFA is strongly recommended to be avoided for fixation with the S9.6 antibody due to high background (Fig. 1).
2. As mentioned in the protocol, cells can be fixed with methanol, rehydrated and left overnight at 4 °C without affecting the success of the experiment.
3. For S9.6 antibody, with the appropriate quenching step and using the right dilution we found that blocking is not required, and that this led to reduction of the background. If high background is still present a blocking step can be included as well as dilution of the antibodies should be tested.
4. For multiple labeling and colocalization experiments, it is important to ensure that there is no cross-reactivity between the secondary antibodies used and hence we suggest using secondary antibodies that have been cross absorbed and species specific.

5. The DAPI staining can be used also together with the secondary antibody, in this case we recommend using at 1:3000 dilution. Alternatively for a quick stain, DAPI can also be included in the mounting medium. by include 1 μg/mL DAPI into mounting Mowiol medium.
6. As in vivo controls overexpression of RNase H1 can be performed in the cells by transfection 48 h prior to S9.6 Immunofluorescence. RNase H1 is the enzyme that cleaves RNA in RNA–DNA hybrids and the in-vivo over-expression should therefore abolish the signal. At the opposite end of the spectrum, knock-down of Senataxin (or another RNA/DNA helicase) or TOP1 should increase R-loops in vivo and therefore increase the S9.6 specific staining.
7. Staining with methanol leads to preservation of the nucleoli in human cells. A quick control of the efficiency and success of the method is to visualize the S9.6 at the nucleoli (which will be visible with the DAPI staining). R-loops are very prevalent in nucleoli where the rDNA repeats are and therefore if the method has worked successfully, S9.6 will have a beautiful, bright staining in the nucleoli (*see* Figs. 1a and 2-bottom panel).

References

1. Xu J, Chmela V, Green NJ, Russell DA, Janicki MJ, Góra RW, Szabla R, Bond AD, Sutherland JD (2020) Selective prebiotic formation of RNA pyrimidine and DNA purine nucleosides. Nature 582:60–66
2. Thomas M, White RL, Davis RW (1976) Hybridization of RNA to double-stranded-DNA: formation of R-loops. Proc Natl Acad Sci U S A 73:2294–2298
3. Aguilera A, Garcia-Muse T (2012) R-loops: from transcription byproducts to threats to genome instability. Mol Cell 46:115–124
4. Skourti-Stathaki K, Proudfoot NJ (2014) A double-edged sword: R loops as threats to genome integrity and powerful regulators of gene expression. Genes Dev 28:1384–1396
5. Santos-Pereira JM, Aguilera A (2015) R loops: new modulators of genome dynamics and function. Nat Rev Genet 16:583–597
6. Crossley MP, Bocek M, Cimprich KA (2019) R-loops as cellular regulators and genomic threats. Mol Cell 73:398–411
7. Niehrs C, Luke B (2020) Regulatory R-loops as facilitators of gene expression and genome stability. Nat Rev Mol Cell Biol 21:167–178
8. Skourti-Stathaki K, Triglia ET, Warburton M, Voigt P, Bird A, Pombo A (2019) R-loops enhance polycomb repression at a subset of developmental regulator genes. Mol Cell 73: 930–945
9. Meyer J, Stålhammar-Carlemalm M (1986) Visualization of RNA polymerase bound to R-loop molecules improves electron microscopic analysis of in vitro transcription. J Ultrastruct Mol Struct Res 96:189–193
10. Boguslawski SJ, Smith DE, Michalak MA, Mickelson KE, Yehle CO, Patterson WL, Carrico RJ (1986) Characterization of monoclonal antibody to DNA.RNA and its application to immunodetection of hybrids. J Immunol Methods 89:123–130
11. Crossley MP, Brickner JR, Song C, Zar SMT, Maw SS, Chédin F, Tsai MS, Cimprich KA (2021) Catalytically inactive, purified RNase H1: a specific and sensitive probe for RNA-DNA hybrid imaging. J Cell Biol 220: e202101092
12. Sanz LA, Hartono SR, Lim YW, Steyaert S, Rajpurkar A, Ginno PA, Xu X, Chédin F (2016) Prevalent, dynamic, and conserved R-loop structures associate with specific epigenomic signatures in mammals. Mol Cell 63: 167–178

13. Tan-Wong SM, Dhir S, Proudfoot NJ (2019) R-loops promote antisense transcription across the mammalian genome. Mol Cell 76:600–616
14. Skourti-Stathaki K, Kamieniarz-Gdula NJ, Proudfoot NJ (2014) R-loops induce repressive chromatin marks over mammalian gene terminators. Nature 516:436–439
15. Howe B, Umrigar A, Tsien F (2014) Chromosome preparation from cultured cells. J Vis Exp 83:e50203

Chapter 3

R-Loop Detection in Bacteria

Émilie Vlachos-Breton and Marc Drolet

Abstract

Early evidence for R-loop formation in vivo came from the study of *Escherichia coli topA* (topoisomerase I; topo I) null mutants. Assays with plasmids to detect RNase HI-sensitive hypernegative supercoiling or R-looped DNA were used in vitro and in vivo to demonstrate R-loop formation. In addition, these R-loop-dependent topological modifications of plasmid DNA were shown to correlate with severe growth and gene expression inhibition in *topA* null mutants that could be corrected by RNase HI overproduction. However, direct evidence for R-loop formation on chromosomal DNA from *E. coli* cells was only obtained recently by using the S9.6 antibody to detect RNA–DNA hybrids in dot-blot experiments. Here, we present a protocol for such experiments with a special emphasis on the procedure used for bacterial genomic DNA extraction and preparation including treatment with appropriate ribonucleases to eliminate RNA–RNA hybrids (that are also recognized by S9.6) as well as single-stranded RNA (ssRNA), in order to obtain a signal that is specific to stable RNA–DNA hybrids generated. Furthermore, we recommend that the results of such experiments be correlated with RNase HI-sensitive phenotypes.

Key words R-loop, S9.6, Bacteria, RNase H, Topoisomerases

1 Introduction

As transcription proceeds, the nascent RNA is normally displaced from the template strand in the transcription bubble. Soon after the discovery of negatively supercoiled DNA it was postulated that the free energy of such supercoiling should maintain the base pairing between the nascent RNA and the template DNA strand and, consequently, should interfere with the process of RNA displacement [1–3]. Indeed, a correlation was found between the level of negative supercoiling and the length of the RNA–DNA hybrid (R-loop, the template strand is paired with RNA, leaving the non-template strand unpaired) after transcription with *E. coli* RNA polymerase (RNAP). However, the formation of such RNA–DNA hybrids mostly occurred when protein denaturing agents were used to stop the transcription reactions, which suggested that RNAP possesses a "separator" function that counteracts the favorable free

Andrés Aguilera and Alexey Ruzov (eds.), *R-Loops: Methods and Protocols*, Methods in Molecular Biology, vol. 2528, https://doi.org/10.1007/978-1-0716-2477-7_3,

energy of negative supercoiling for RNA–DNA hybrid formation [2]. In fact, we now know that this "separator" function is provided by the RNA exit channel of RNAP [4].

However, extensive R-loop formation has been shown to occur during transcription of plasmid DNA in the presence of DNA gyrase, but in the absence of topo I, both in vitro and in vivo [5–8]. DNA gyrase, by introducing negative supercoiling promotes R-loop extension (this activity of DNA gyrase is particularly required for small supercoiling domains such as plasmids), whereas topo I, by relaxing transcription-induced negative supercoiling, inhibits the reannealing of the nascent RNA with the template strand behind the RNAP [8]. Topo I interacts with RNAP [9] and recent results explain how this interaction can promote the efficient relaxation of negative supercoiling behind RNAP [10]. Thus, the very high level of negative supercoiling immediately behind the moving RNAP, where the nascent RNA (coming out from the RNA exit channel) is still very close to the template strand, promotes DNA opening that facilitates the reannealing of the nascent RNA with the template strand.

R-loop formation in vitro and (in *topA* mutants) was demonstrated by showing RNase HI-sensitive hypernegative supercoiling of plasmid DNA or by gel retardation (and conformers formation) of R-looped plasmid DNA [5–8]. Such R-loop formation on plasmid DNA was corelated with growth inhibition [11], gene expression defects [12–14] and unregulated replication [15, 16] in *topA* null mutants when RNase HI was not overproduced. Unregulated replication from R-loops was first detected in *rnhA* (RNase HI) null mutants and allowed them to grow in the absence of the DnaA/oriC system for chromosomal replication initiation [17, 18]. The solid evidence for R-loop formation in *topA* null mutants offered us the opportunity to develop a dot-blot protocol with the S9.6 antibody to detect R-loop formation on bacterial chromosomal DNA, using *rnhA* and wild-type strains as positive and negative controls respectively [16].

To make sure that the protocol generates dot-blot signals revealing only stable R-loops formed, it is important to consider what has been learned from the early experiments about R-loop formation on negatively supercoiled DNA, as well as the substrate specificity of S9.6. First, RNase A, an endoribonuclease specific to ssRNA is added early in the extraction procedure. By degrading ssRNA, RNase A prevents its pairing with the template DNA strand upon denaturing RNAP during the extraction procedure (Fig. 1a, left). RNase A also degrades short, unstable, biologically irrelevant R-loops that may extend during genomic DNA extraction (Fig. 1a, right). Short RNA–DNA hybrids are more susceptible to helix breathing (and to its full destabilization) than longer ones, especially under low salt conditions, making their RNA susceptible to RNase A degradation. Second, another potential problem is related

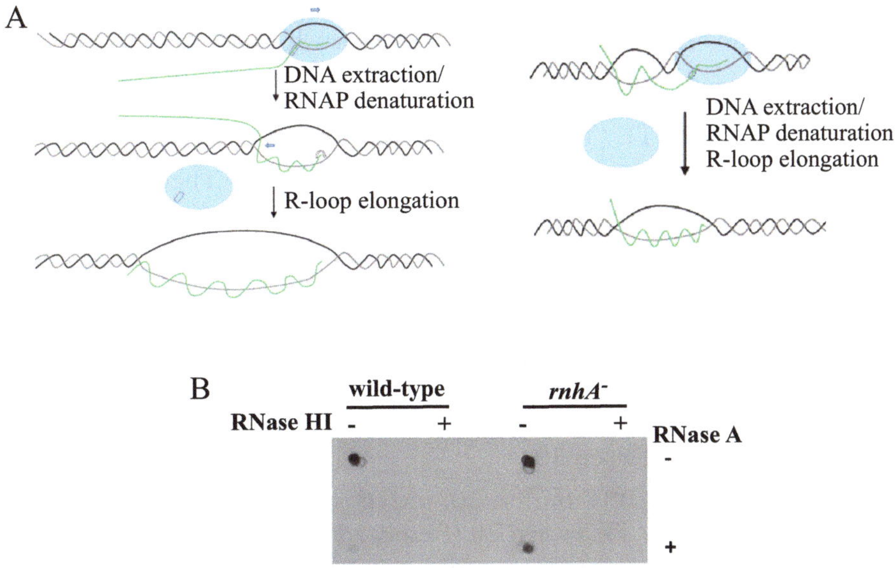

Fig. 1 Formation of R-loops during bacterial genomic DNA extraction. In (**a**) The denaturation of RNAP allows R-loop lengthening from the RNA–DNA hybrids in the transcription bubble. RNase A can prevent this reaction by degrading the nascent RNA (left). Short unstable R-loop may elongate during DNA extraction. RNase A can prevent this reaction by degrading the RNA of this short hybrid (right). (**b**) A dot-blot experiment with the S9.6 antibody showing that the addition of RNase A during genomic DNA extraction reveals as expected a much higher level of R-loops in *rnhA* mutants as compared to wild-type cells (compare + RNase A, -RNase HI samples from wild-type and *rnhA* strains). Note that RNase III was added to the samples to eliminate the signal related to the presence of RNA–RNA hybrids (*see* **Note 2**)

to the substrate specificity of the S9.6 antibody. Indeed, S9.6 binds RNA–RNA hybrids with only a five- to sixfold lower affinity than RNA–DNA hybrids [19]. Treatment of the extracted genomic DNA with RNase III (a dsRNA specific endoribonuclease) allows the degradation of stable RNA–RNA hybrids. Here, we present a protocol for dot-blotting with the S9.6 antibody that includes treatment of the bacterial genomic DNA with RNase A and RNase III, to allow the detection of stable R-loops formed in vivo.

2 Materials

2.1 Genomic DNA Extraction

1. LB medium supplemented with required antibiotics.
2. 250 mL Erlenmeyer flasks
3. Water bath shaker.
4. Spectrophotometer.
5. 50 mL disposable centrifuge tubes.

6. GenElute bacterial genomic DNA kit from Sigma Aldrich (Cat. NA2100) or similar kits (such as the QIAamp DNA mini kit from Qiagen).

2.2 Samples Preparation and Dot-Blotting

1. TE (Tris-EDTA, pH 8.0) buffer.
2. RNase III and its working buffer (Thermo Fisher Scientific, Ambion).
3. RNase HI (homemade purified or commercial sources, e.g., New England Biolabs).
4. Phenol–chloroform solution.
5. Ethanol 95%.
6. Amersham Hybond™ -N$^+$ membrane.
7. Tween 20.
8. TBST buffer: 20 mM Tris–HCl pH 7.4, 150 mM NaCl and 0.1% Tween 20 (freshly added).
9. Stratalinker 1800 UV Crosslinker or equivalent apparatus.
10. Powdered skim milk.
11. S9.6 antibody (home made preparation or commercial sources (Kerafast, Millipore Sigma or Thermo Fisher Scientific).
12. Orbital shaker.
13. Goat anti-mouse IgG (H + L) secondary antibody, HRP (Thermo Fisher Scientific).

2.3 Signal Revelation

1. SuperSignal™ West Pico PLUS Chemiluminescent Substrate (Thermo Fisher Scientific).
2. Amersham Imager 600 (Thermo Fisher Scientific) or equivalent apparatus.

3 Methods

3.1 Genomic DNA Extraction

1. Prepare overnight (ON) cultures of the *Escherichia coli* strains in LB medium. Add appropriate antibiotics and supplements.
2. Dilute the ON cultures to an OD_{600} of 0.01 in 30 mL LB medium (supplemented with required antibiotics and supplements) in a 250 mL Erlenmeyer flask. Incubate in a water bath at the appropriate temperature with vigorous shaking up to an OD_{600} of 0.4–0.5.
3. At this stage, 10 mL of the cell suspension are transferred to a 50 mL disposable centrifuge tube filled with ice and the cells are recovered by centrifugation at 4 °C.
4. Genomic DNA is extracted by using the GenElute bacterial genomic DNA kit from Sigma Aldrich (or the equivalent) exactly as described in the kit user Guide, except that the amount of RNase A added is reduced by 20-fold (*see* **Note 1**).

3.2 Samples Preparation and Dot-Blotting

1. Prepare two tubes (20 μL final volume in RNase III reaction buffer) containing between 100 ng and 1 μg of genomic DNA and 2 μL of RNase III (*see* **Note 2**). Add 2 μL of RNase HI to one of the tubes. Incubate the tubes for 1 h at 37 °C.
2. Perform phenol–chloroform extraction and ethanol precipitation. Resuspend in TE buffer.
3. Spot 10 μL of DNA solution (corresponding to 100 ng to 1 μg of genomic DNA) on the Hybond™ -N^{+} membrane.
4. UV-crosslink the membrane using the Stratalinker 1800 UV Crosslinker apparatus.
5. Block the membrane with 50 mL of a 5% milk solution in TBST buffer. Incubate for 1 h at room temperature on the orbital shaker.
6. Soak the membrane in 15 mL of 5% milk solution in TBST buffer containing 15–30 μg of the S9.6 antibody. Incubate overnight at 4 °C on the orbital shaker.
7. Rinse the membrane 3 times (5 min each) in 50 mL TBST buffer.
8. Soak the membrane in 15 mL of 5% milk solution in TBST buffer containing 40 μL of the secondary antibody (Goat anti-Mouse IgG (H + L) Secondary Antibody, HRP-conjugated). Incubate for 1 h at room temperature on the orbital shaker.
9. Rinse the membrane 4 times (5 min each) in 50 mL TBST buffer.
10. Rinse the membrane once for 2–3 min in 50 mL TBS buffer.

3.3 Signal Revelation

1. Prepare 12 mL of the "working solution" by mixing 1:1 the peroxide and luminol solutions (SuperSignal™ West Pico PLUS Chemiluminescent Substrate).
2. Soak the membrane in the "working solution". Incubate for 5 min at room temperature on the orbital shaker.
3. Remove the membrane and carefully eliminate the excess of "working solution".
4. The signal is revealed by using the Amersham Imager 600. Exposition for 1–10 s is normally sufficient to obtain a good signal. Alternatively, the signal can also be revealed by using an appropriate autoradiography film (e.g., HyBlot CL) with an exposition time varying between 5 to 30 s.

4 Notes

1. As stated in the introduction (*see* also Fig. 1a), RNase A is added during the first step of the genomic DNA extraction procedure, in order to degrade nascent RNA that may anneal

with the template strand following the denaturation of the RNAP. It is also added to degrade the RNA of short unstable biologically irrelevant RNA–DNA hybrids. However, using too much of RNase A may lead to the elimination of longer biologically significant RNA–DNA hybrids. The RNase A solution provided with the kit is at 10 mg/mL and it is stated in the kit to add 20 μL of this solution to the 180 μL cell suspension (final RNase A concentration is 1 mg/mL) for the 2 min treatment at room temperature. This is pretty high, as we found that RNase A at this concentration slightly reduced the dot-blot signal and significantly reduced the DRIP-qPCR and DRIP-seq signals (not shown) as compared to the addition of 20-fold less RNase A (50 μg/mL) (Fig. 1b).

2. As stated in the introduction, S9.6 also binds to RNA–RNA hybrids. In fact, we found that genomic DNA samples from *topA*, *rnhA*, and *topA topB* null cells not treated with RNase III, give a very strong dot-blot signal that is almost fully resistant to RNase HI. When the DNA samples are treated with RNase III, the strong dot-blot signal now obtained, is fully sensitive to RNase HI [16]. Thus, the very strong signal obtained in the absence of RNase III treatment likely reflects the accumulation of short truncated and highly structured rRNA intermediates in *topA* and *rnhA* mutants related to transcription blockage due to R-loops, as shown previously [3, 12]. With this RNase III treatment, spotting various amount of genomic DNA between 0.1 and 1 μg give good dot-blot signals that can be compared between the strains (for example, *see* Fig. 2).

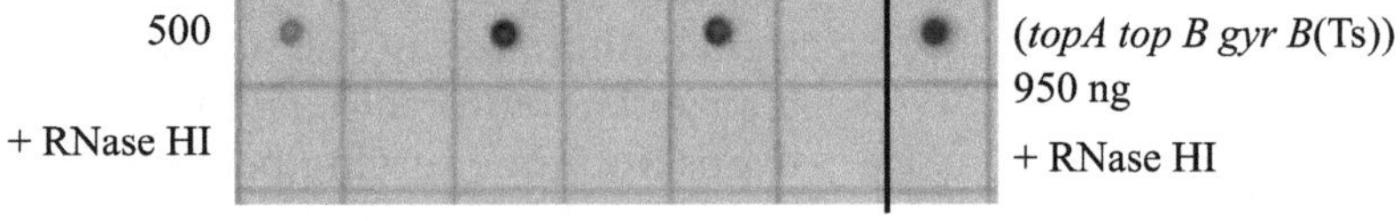

Fig. 2 Dot-blot experiment showing R-loop formation in *topA* and *topA topB* null mutants. RFM443 (wild-type), JB137 (*topA topB gyrB*(Ts)), DM800 (*topA gyrB225*), and JB305 (DMA800 *topB*) cells were grown in LB at 30 °C, genomic DNA extracted, and dot-blotting performed as described here. For JB305 two independent experiments were performed (lanes 2 and 3). +RNase HI means that 950 ng of genomic DNA were treated with RNase HI before spotting

Acknowledgments

We thank Fenfei Leng, Florida International University, for *E. coli* RNase HI and Michael Wilson, University of Toronto, for S9.6. This work was supported by a Discovery Grant from the Natural Sciences and Engineering Research Council of Canada (NSERC) to M.D. E.V.B. was supported by a Graduate Student Research Award from the NSERC.

References

1. Wang JC (1974) Interactions between twisted DNAs and enzymes: the effects of superhelical turns. J Mol Biol 87:797–816
2. Richardson JP (1975) Attachment of nascent RNA molecules to superhelical DNA. J Mol Biol 98:565–579
3. Drolet M, Broccoli S, Rallu F, Hraiky C, Fortin C, Massé E, Baaklini I (2003) The problem of hypernegative supercoiling and R-loop formation in transcription. Front Biosci 8: d210–d221. https://doi.org/10.2741/970
4. Sutherland C, Murakami KS (2018) An introduction to the structure and function of the catalytic core enzyme of *Escherichia coli* RNA polymerase. EcoSal Plus. 2018(1):10.1128/ecosalplus.ESP-0004-2018. https://doi.org/10.1128/ecosalplus.ESP-0004-2018
5. Drolet M, Bi X, Liu LF (1994) Hypernegative supercoiling of the DNA template during transcription elongation *in vitro*. J Biol Chem 269: 2068–2074
6. Phoenix P, Raymond MA, Massé E, Drolet M (1997) Roles of DNA topoisomerases in the regulation of R-loop formation *in vitro*. J Biol Chem 272:1473–1479
7. Massé E, Phoenix P, Drolet M (1997) DNA topoisomerases regulate R-loop formation during transcription of the *rrnB* operon in *Escherichia coli*. J Biol Chem 272:12816–12823
8. Masse E, Drolet M (1999) *Escherichia coli* DNA topoisomerase I inhibits R-loop formation by relaxing transcription-induced negative supercoiling. J Biol Chem 274:16659–16664
9. Cheng B, Zhu CX, Ji C, Ahumada A, Tse-Dinh YC (2003) Direct interaction between *Escherichia coli* RNA polymerase and the zinc ribbon domains of DNA topoisomerase I. J Biol Chem 278:30705–30710
10. Tan K, Zhou Q, Cheng B, Zhang Z, Joachimiak A, Tse-Dinh YC (2015) Structural basis for suppression of hypernegative DNA supercoiling by *E. coli* topoisomerase I. Nucleic Acids Res 43:11031–11046
11. Drolet M, Phoenix P, Menzel R, Massé E, Liu LF, Crouch RJ (1995) Overexpression of RNase H partially complements the growth defect of an *Escherichia coli* delta *topA* mutant: R-loop formation is a major problem in the absence of DNA topoisomerase I. Proc Natl Acad Sci U S A 92:3526–3530
12. Hraiky C, Raymond MA, Drolet M (2000) RNase H overproduction corrects a defect at the level of transcription elongation during rRNA synthesis in the absence of DNA topoisomerase I in *Escherichia coli*. J Biol Chem 275:11257–11263
13. Baaklini I, Hraiky C, Rallu F, Tse-Dinh YC, Drolet M (2004) RNase HI overproduction is required for efficient full-length RNA synthesis in the absence of topoisomerase I in *Escherichia coli*. Mol Microbiol 54:198–211
14. Baaklini I, Usongo V, Nolent F, Sanscartier P, Hraiky C, Drlica K, Drolet M (2008) Hypernegative supercoiling inhibits growth by causing RNA degradation. J Bacteriol 190: 7346–7356
15. Martel M, Balleydier A, Sauriol A, Drolet M (2015) Constitutive stable DNA replication in *Escherichia coli* cells lacking type 1A topoisomerase activity. DNA Repair 35:37–47
16. Brochu J, Vlachos-Breton É, Sutherland S, Martel M, Drolet M (2018) Topoisomerases I and III inhibit R-loop formation to prevent unregulated replication in the chromosomal Ter region of *Escherichia coli*. PLoS Genet 14(9):e1007668. https://doi.org/10.1371/journal.pgen.1007668
17. Kogoma T (1997) Stable DNA replication: interplay between DNA replication, homologous recombination, and transcription. Microbiol Mol Biol Rev 61:212–238
18. Drolet M, Brochu J (2019) R-loop-dependent replication and genomic instability in bacteria. DNA Repair 84:102693. https://doi.org/10.1016/j.dnarep.2019.102693
19. Phillips DD, Garboczi DN, Singh K, Hu Z, Leppla SH, Leysath CE (2013) The sub-nanomolar binding of DNA-RNA hybrids by the single-chain Fv fragment of antibody S9.6. J Mol Recognit 26:376–381

Chapter 4

Detection of R-Loops by In Vivo and In Vitro Cytosine Deamination in *Saccharomyces cerevisiae*

Juan C. Cañas, Andrés Aguilera, and Belén Gómez-González

Abstract

R-loops are transcriptional by-products formed by a hybrid of the nascent RNA molecule with its DNA template and the displaced nontemplate DNA strand. The single stranded nature of the displaced non-template strand makes it vulnerable to attack. This property is used in nature to cause directed mutagenesis and breaks by the action of the activation-induced cytosine deaminase (AID) enzyme and can thus be exploited to detect the presence of R-loops even when they form at low frequencies by overexpressing this enzyme in vivo or by in vitro treatment with the bisulfite anion, which further allows nucleotide resolution. This is of particular relevance given the fact that R-loops have the potential to hamper DNA replication and repair, threatening genome integrity. Here, we describe the protocols used in the yeast *Saccharomyces cerevisiae* to infer the presence of R-loops through increased AID-induced DNA damage, measured as increased recombination or Rad52 foci formation as well as to detect single R-loop molecules and determine their length at particular genomic sites via bisulfite treatment and amplification.

Key words DNA-RNA hybrid, R-loop, Bisulfite, Cytosine deamination, AID

1 Introduction

R-loops are secondary structures formed by the hybridization of the nascent RNA with its DNA template and are thus composed of a DNA–RNA hybrid plus a single stranded DNA (ssDNA) molecule displaced by the hybrid. DNA-RNA hybrids play several roles in cells, as they are involved in the replication of mitochondrial and bacterial DNA, and they are formed during eukaryotic DNA replication at every single Okazaki fragment (reviewed in [1]). They are also involved in the process of class switch recombination of immunoglobulin genes, where the Activation-Induced cytidine Deaminase (AID) enzyme is specifically targeted in B lymphocytes. However, their unbalanced formation is linked to genome instability, since they have the potential to hamper DNA replication and repair (reviewed in [1]). Thus, R-loops are considered active players in the dynamics of genome evolution.

Andrés Aguilera and Alexey Ruzov (eds.), *R-Loops: Methods and Protocols*, Methods in Molecular Biology, vol. 2528,
https://doi.org/10.1007/978-1-0716-2477-7_4,

A number of methodologies have been developed in order to directly or indirectly detect the presence of R-loops. For instance, the ability of a sequence to form DNA–RNA hybrids can be directly assayed in vitro by Electrophoretic Mobility Shift Assay (EMSA) [2]. To directly detect DNA-RNA hybrids in vivo, most tools rely on the capacity of proteins to recognize specifically such secondary structures. These include an inactive version of the enzyme that naturally degrades the RNA moiety within DNA-RNA hybrids (RNase H) [3] or its hybrid-binding domain fused to GFP (HBD-GFP) [4] and the S9.6 monoclonal antibody that reacts with DNA-RNA hybrids [5] (*see* ref. [6] for a detailed DNA–RNA hybrid ImmunoPrecipitation (DRIP) protocol), from which most of the available results derive. However, this antibody is also able to recognize double stranded RNA molecules, thus making the use of the RNase H, either by in vivo overexpression or by in vitro treatment, as well as the use RNase III enzyme, which specifically degrades dsRNA, indispensable to ensure the specificity of the detected signal [7, 8]. All these tools focus on the detection of the hybrid itself, but not the three stranded R-loop. In contrast, methods for indirect R-loop detection rely on the vulnerability of the displaced ssDNA. In this regard, R-loops can be inferred in vivo by the induction of DNA damage through the overexpression of the AID enzyme. This was initially shown in yeast R-loop accumulating mutants of the THO complex by increased recombination levels between direct repeats [9], but can also be detected by elevated Rad52 foci formation [10] (Fig. 1) or γ-H2AX foci formation in human cells [11]. AID overexpression has resulted in a successful tool to identify new factors involved in R-loop homeostasis such as the Sen1 helicase [12], histone mutants [10], and Mlp1 [13].

However, none of the mentioned direct or indirect methods allow determining the length covered by the R-loops since they detect these in a whole cell population via genetic tools, immunofluorescence or immunoprecipitation. In contrast, in vitro treatment with sodium bisulfite allows detecting the ssDNA within R-loops with nucleotide resolution. The bisulfite modification assay was originally developed and widely used for studying cytosine methylation patterns thanks to its ability to deaminate cytosines of ssDNA turning them into uracils if they are not methylated [14]. Its adaptation to detect R-loops was initially described in mammalian cells to detect physiological R-loops forming at the immunoglobulin class switch regions in the chromosomes of stimulated B cells [15] (*see* Chapter 5). Interestingly, this tool not only detects R loops versus DNA-RNA hybrids, but allows inferring the length of R-loops in single molecules by determining the length covered by consecutive bisulfite-induced cytosine deaminations in the displaced ssDNA, which would correspond to the length covered by the hybrid.

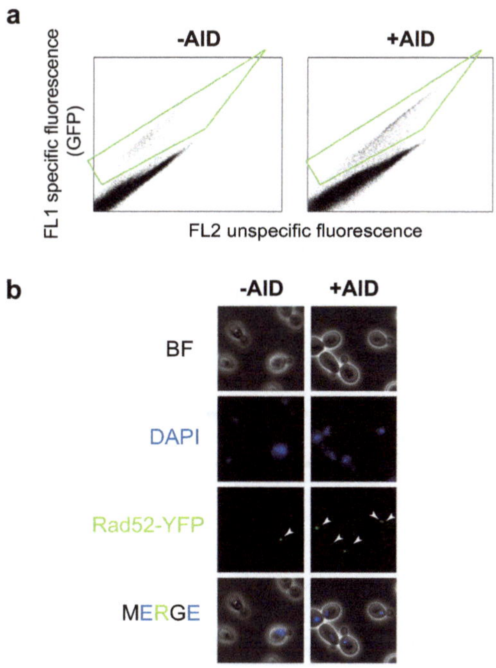

Fig. 1 Detection of R-loops by AID-induced hyperrecombination and Rad52 foci formation. (**a**) Example of the flow cytometry detection of GFP cells after AID overexpression (+AID) or not (−AID). The recombinant population emitting green fluorescence is enclosed in a green box. *y* axis, green fluorescence (FL1H); *x* axis, unspecific fluorescence (FL2H). (**b**) Example of cells with Rad52 foci visualized with the fluorescence microscope after AID overexpression (+AID) or not (−AID). White arrows point to the Rad52 foci. *BF* bright field

When the R-loop frequency is too low as to be amplified by standard PCR, converted primers that anticipate the deamination of cytosines can be designed to enrich in R-loop containing molecules (Fig. 2). These primers must contain guanine to adenine substitutions so that they can anneal to DNA sequences mutated by bisulfite. Given that converted primers are strand specific, this technique allows discerning R-loops from other DNA conformations such as double-stranded DNA (dsDNA) or ssDNA (Fig. 2). In dsDNA, both DNA strands are protected and bisulfite-treatment is innocuous. Bisulfite treatment of denatured DNA leads to cytosine deamination in both strands whereas only the nontemplate strand is sensitive to bisulfite-induced deamination in an R-loop conformation (Fig. 2). In addition to its initial development to determine hybrid length at the immunoglobulin genes of mouse B cells [15, 16] (*see* ref. 17 for a detailed protocol for measuring R-loop length at immunoglobuline genes in human cells), bisulfite deamination has been used to determine the strand specificity of RNA hybridization within R-loops in high-throughput sequencing experiments when combined either with S9.6 antibody prior to

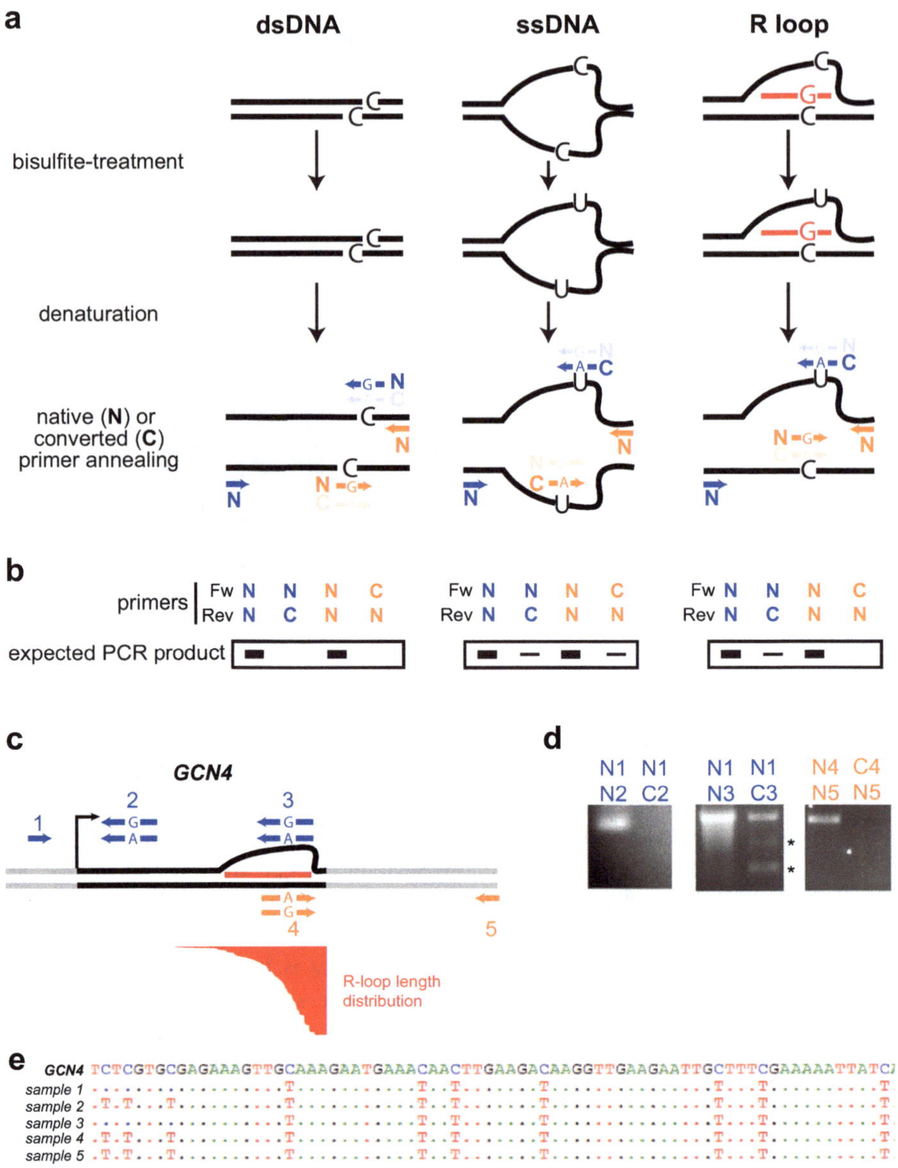

Fig. 2 Effect of bisulfite treatment and expected PCR products. (**a**) Vulnerability of different DNA conformations to bisulfite treatment. At DNA duplexes (dsDNA), both strands are protected by each other from the action of bisulfite. If the DNA is melted, both template and nontemplate strands will be single stranded (ssDNA) and thus vulnerable to bisulfite deamination that converts cytosine (C) to uracil (U). At R-loops, the template strand is protected by annealing with the complementary RNA (red line) and only the nontemplate strand will be attacked by bisulfite. Native (N) primers anneal to wild-type sequence of DNA and are represented with a plain arrow or an arrow containing a guanine (G). Converted (C) primers anneal to the bisulfite-induced mutated sequence and are represented with an arrow containing an adenine (A). Primers that do not anneal are represented by the lighter color. (**b**) Expected PCR products after gel electrophoresis with the indicated forward (Fw) or reverse (Rv) primers in DNA samples from cell cultures containing only dsDNA (left), dsDNA and melted DNA (middle), or dsDNA and R-loops (right). (**c**) Scheme of the *GCN4* gene and primers used to detect R-loops at different regions. A scale representation of the distribution of sequenced molecules with cytosine to thymine conversions after bisulfite treatment is shown at the bottom in red. (**d**) PCR products obtained by using the indicated native and converted primers at the *GCN4* gene region. Asterisks (*) indicate

DNA–RNA hybrid immunoprecipitation (bisDRIP-seq) [18] or with micrococcal nuclease fused to an inactive RNase H (BisMapR) [19].

Here, we describe different approaches to detect R-loops and to measure their length in specific genomic regions of the yeast *Saccharomyces cerevisiae* by using AID overexpression in vivo or by in vitro DNA treatment with sodium bisulfite coupled to PCR amplification and sequencing. These methods have been validated in wild-type and R-loop-accumulating yeast mutants [9, 10, 12, 13, 20].

2 Materials

2.1 Detection of R-Loops by AID Overexpression

1. The yeast strain of interest transformed with (1) a centromeric empty vector (negative control) or a plasmid containing the *AID* gene under the galactose promoter (such as pRS414GALAID and p413GALAID [9], carrying *TRP1* and *HIS3* genes as selection markers, respectively) and (2) either a centromeric plasmid with the recombination system of interest for the determination of recombination frequencies or the pWJ1344 plasmid [21], which contains the *RAD52* gene under its own promoter fused to the yellow fluorescent protein (YFP) marker, for the detection of Rad52 foci. *See* reference [22] from this series for a detailed description of available recombination systems and their use. They include the LlacZ system, which carries *leu2* truncated direct-repeats interrupted by the *lacZ* sequence [23] for the determination of recombination frequencies by counting colony-forming units on selective media (lacking leucine) or the GLG system [9], containing direct-repeats of the green fluorescent protein (GFP) coding gene interrupted by the bacterial *LacZ* sequence for the determination of recombination frequencies by flow cytometry (*see* **Note 1**).
2. Liquid and solid synthetic complete (SC) medium: 0.17% yeast nitrogen base (YNB), 0.5% ammonium sulfate, 2% glucose, 2% agar (only for solid media) and a mix of different requirements except those for plasmid selection (20 mg/L adenine sulfate, 20 mg/L uracil, 20 mg/L L-tryptophan, 20 mg/L L-histidine-HCl, 20 mg/L L-arginine-HCl, 20 mg/L L-methionine, 30 mg/L L-tyrosine, 120 mg/L L-leucine, 30 mg/L L-

Fig. 2 (continued) unspecific PCR amplification (*see* **Note 10**). Reproduced from reference [10] with permission from Elsevier (License Number 5153050988149). (**e**) Example of sequence alignments of sequenced molecules with cytosine to thymine conversions after bisulfite treatment and amplification of the *GCN4* gene by PCR in five different samples. The reference sequence is shown on top

isoleucine, 30 mg/L L-lysine-HCl, 50 mg/L L-phenylalanine, 100 mg/L L-glutamic acid, 100 mg/L L-aspartic acid, 150 mg/L L-valine, 200 mg/L L-threonine, 380 mg/L L-serine).

3. Liquid SGal medium: 0.17% yeast nitrogen base (YNB), 0.5% ammonium sulfate, 2% galactose, and a mix of different requirements as specified in SC medium.
4. 0.1 M KHPO pH 6.4 with 2.5% formaldehyde (added just before use).
5. 0.1 M KHPO pH 6.6.
6. 0.1 M KHPO pH 7.4.
7. 1 μg/mL DAPI.

2.2 Detection of R-Loops by Bisulfite Deamination

2.2.1 DNA Extraction by CTAB Procedure

1. Liquid YPAD medium: 1% yeast extract, 2% peptone and 2% glucose (dextrose) supplemented with adenine sulfate at a final concentration of 20 mg/L.
2. Spheroplasting buffer: 1 M sorbitol, 10 mM EDTA pH 8, supplemented in the moment of use with 0.1% β-mercaptoethanol and 20 mg/mL Zymolyase 20T (USB-Affimetrix, Santa Clara, CA, USA).
3. Solution I: 1.4 M NaCl, 100 mM Tris-HCl pH 7.6, 25 mM EDTA pH 8 and 2% CTAB (N,N,N-trimethylhexadecan-1-aminium bromide).
4. RNase A (10 mg/mL). Store at −20 °C.
5. Proteinase K (20 mg/mL).
6. Chloroform–isoamyl alcohol (24:1 vol:vol).
7. Solution II: 50 mM Tris-HCl pH 7.6, 10 mM EDTA pH 8 and 1% CTAB.
8. Solution III: 1.4 M NaCl, 1 mM EDTA pH 8, and 10 mM Tris-HCl pH 7.6.
9. Isopropanol and chilled 70% ethanol.

2.2.2 DNA Digestion

1. *Nde*I, *Not*I-HF, and *Xho*I enzymes and appropriate digestion buffers (*see* **Note 2**).
2. Chilled 100% and 70% ethanol.
3. 3 M sodium acetate.

2.2.3 Bisulfite Treatment

1. Hydroquinone (20 mM).
2. Sodium bisulfite (2.5 M and pH 5.2).
3. Wizard SV gel and PCR Cleanup System kit (Promega) (*see* **Note 3**).
4. NaOH 2 M.

5. Chilled 100% and 70% ethanol.
6. 3 M sodium acetate.

2.2.4 PCR Amplification

1. dNTPs (dATP, dGTP, dTTP, dCTP 10 mM each).
2. Native and converted specific primers (forward and reverse; 10 μM each) (*see* **Notes 4** and **5**).
3. FastStart High Fidelity PCR System (Roche).
4. Agarose gel: 0.8% (w/vol) agarose in TAE buffer (40 mM Tris, 20 mM acetic acid and 1 mM EDTA pH 8).
5. Ethidium bromide (10 mg/mL).

2.2.5 Cloning and Sequencing

1. pGEM-Teasy vector (Promega).
2. Wizard SV gel and PCR Cleanup System kit (Promega).
3. DH5α competent *E. coli* cells.
4. *Eco*RI enzyme and appropriate digestion buffer.
5. Agarose gel: 0.8% (w/vol) agarose in TAE buffer (40 mM Tris, 20 mM acetic acid, and 1 mM EDTA pH 8).
6. M13F and M13R primers.
7. LB plates supplemented with 0.1 mg/mL ampicillin, 0.02 mg/mL X-Gal (5-bromo-4-chloro-3-indolyl-β-D-galactopiranoside) and 0.1 mM IPTG (isopropyl-β-D-1-thiogalactopyranoside).
8. STETL: 8% sucrose, 5% Triton X-100, 50 mM Tris-HCl pH 8.0, and 50 mM EDTA pH 8.0 supplemented with 0.5 mg/mL lysozyme.
9. RNase A: 10 mg/mL.
10. Wooden sticks.
11. Isopropanol and cold 70% ethanol.
12. Free-nuclease water.

3 Methods

3.1 Detection of R-Loops by AID Overexpression

3.1.1 Determination of Recombination Frequencies by Plating in Selective Media

1. Streak three different transformants containing both the AID overexpression plasmid (or the empty vector) and the plasmid with the recombination system (such as pSCH204, carrying the LlacZ recombination system) onto SGal plasmid-selective plates (*see* **Note 1**).
2. Incubate the plates at 30 °C during 3–4 days until appropriate size of 2.5–3 mm diameter per colony is reached.
3. Dilute six different colonies of each transformant in a microtube containing 1 mL of sterile water.

4. Transfer 100 μL of this dilution into a microtube containing 900 μL of sterile distilled water to prepare a 1/10 dilution.
5. Repeat **step 4** in order to obtain 1/100, 1/1000, and 1/10,000 dilutions.
6. Plate 25 μL of each dilution in plates with selective media containing all requirements except for leucine to count for recombinant cells and in plasmid-selective plates to count for total cells.
7. Incubate plates at 30 °C for 3–4 days and count total and recombinant cells. Apply the dilution factor to estimate the number of total and recombinant cells in each colony.
8. Calculate the median recombination frequency obtained with the six colonies.
9. Estimate the recombination frequency as the average of at least three independent tests, each performed with six colonies from an independent transformant for each condition or strain to be tested.

3.1.2 Determination of Recombination Frequencies by Flow Cytometry

1. Streak six different transformants containing both the AID overexpression plasmid (or empty plasmid as a control) and the pGLG plasmid onto SC plasmid-selective plates.
2. Incubate the plates at 30 °C during 3–4 days until appropriate size of 2.5–3 mm diameter per colony is reached.
3. Transfer one colony of each transformant (approximately 10^6 cells) onto 5 mL of liquid SGal medium and let the cultures grow overnight at 30 °C.
4. Take 1 mL of the culture and centrifuge at 16,000 × *g* for 1 min.
5. Discard the supernatant and resuspend the pellet in 1 mL of distilled water.
6. Adjust the flow cytometer settings, optimizing its parameters. Create an FL1H and an FL2H (for GFP and unspecific fluorescence detection, respectively) dot plot and set detector gains and voltages.
7. Acquire the data with the flow cytometer and calculate the percentage of green fluorescence cells versus the total number of cells counted by determining the number of cells that fall above the diagonal in the FL1H-FL2H dot plot.
8. Determine the median recombination frequency from 6–12 transformants.

3.1.3 Detection of Rad52 Foci

1. Streak three different transformants containing both the AID overexpression plasmid (or empty plasmid as a control) and pWJ1344 onto selective SC plates.

2. Incubate the plates at 30 °C during 3–4 days until appropriate size of 2.5–3 mm diameter per colony is reached.
3. Inoculate a colony from each plate in 5 mL liquid SGal and incubate at 30 °C overnight to allow AID expression.
4. Transfer 1 mL of the culture to a microtube and centrifuge at 16,000 × *g* for 1 min.
5. Discard the supernatant and fix cells with 1 mL 0.1 M KHPO pH 6.4 with 2.5% formaldehyde (added just before use) during 10 min.
6. Centrifuge cells at 1000 × *g* for 3 min and discard the supernatant.
7. Resuspend the pellet in 1 mL of 0.1 M KHPO pH 6.6 and centrifuge cells at 1000 × *g* for 3 min.
8. Discard the supernatant and repeat **steps 6** and 7.
9. Resuspend the pellet in 1 mL of 0.1 M KHPO pH 7.4.
10. Centrifuge cells at 1000 × *g* for 3 min and discard the supernatant.
11. Permeabilize cells by resuspending the pellet in 80% ethanol during 10 min.
12. Centrifuge cells at 1000 × *g* for 3 min and discard the supernatant.
13. Resuspend the pellet in 50–100 μL of 1 μg/mL DAPI dissolved in distilled water to stain nuclei.
14. Count the cells that show visible foci from at least 200 cells in the fluorescence microscope and calculate the ratio of cells with Rad52 foci versus the total number of cells.

3.2 Detection of R-Loops by Bisulfite Deamination

3.2.1 DNA Extraction by the CTAB Procedure

1. Inoculate a colony from the strain of interest in liquid YPAD overnight on a shaker at 30 °C.
2. Dilute the culture to an OD_{600} of 0.3 in 50 mL of liquid YPAD and incubate during 3 h on a shaker at 30 °C to allow one round of replication (until an OD_{600} of 0.6–0.8 is reached).
3. Transfer the culture to a 50 mL tube, pellet by centrifugation at 1500 × *g* for 5 min at 4 °C and wash the pellet twice by adding 20 mL of cold water, centrifugation at 1500 × *g* for 5 min at 4 °C and discard supernatant.
4. Resuspend the pellet in 1 mL of cold water and transfer it into a new 2 mL microtube.
5. Spin cells at 16,000 × *g* for 1 min and discard the supernatant. Resuspend the pellet in 1 mL of spheroplasting buffer and incubate 1 h at 30 °C under soft shaking.
6. Centrifuge the 2 mL microtube and 5000 × *g* for 5 min at room temperature and remove the supernatant.

7. Wash the pellet by carefully adding 500 μL of cold water. Do not resuspend. Spin spheroplasts in a centrifuge and remove the water.
8. Resuspend cells in 400 μL of cold water by pipetting up and down with cut tips.
9. Once the pellet is resuspended, add immediately 500 μL of solution I along with 40 μL of RNase A (10 mg/mL). Mix by inverting the microtube and incubate at 50 °C for 30 min.
10. Add 40 μL of proteinase K (20 mg/mL) and incubate for 2 h at 30 °C under soft agitation (*see* **Note 6**).
11. Centrifuge samples at 5000 × *g* and room temperature for 5 min and then, transfer the supernatant to a new 2 mL microtube and keep also the pellet.
12. Incubate the pellet with 400 μL of solution III at 50 °C for 1 h. Then add 200 μL of chloroform–isoamyl alcohol (24:1) and centrifuge at 5000 × *g* for 5 min at room temperature. Save the supernatant in a new 2 mL microtube until **step 18**.
13. Mix the supernatant from **step 11** with 500 μL of chloroform: isoamyl alcohol (24:1) and then centrifuge at 5000 × *g* for 5 min.
14. Split the aqueous phase (500 μL approx.) into 2 new 2 mL microtubes using big cut tips.
15. Incubate at room temperature for 1 h with 2 vol of solution II.
16. Centrifuge at 16,000 × *g* and room temperature for 10 min and discard the supernatant.
17. Resuspend the resulting pellet in 250 μL of solution III and incubate at 37 °C for 2 h under soft agitation (*see* **Note 7**).
18. Join the sample from **step 17** with the supernatant from **step 12** and add 1 vol of isopropanol (1 mL approx.) to induce DNA precipitation. Mix carefully by inverting the microtube.
19. Centrifuge the microtube at 16,000 × *g* for 10 min at room temperature and discard the supernatant.
20. Wash the pellet with 500 μL of chilled 70% ethanol and remove it with a pipette after centrifugation at 16,000 × *g* for 5 min.
21. Dry pellet and add 100 μL of milli-Q water (*see* **Note 8**).

3.2.2 DNA Digestion

1. Digest all DNA with 50 U of *Nde*I, 50 U of *Not*I-HF, and 50 U of *Xho*I in the presence of the appropriate digestion buffer in a total vol of 250 μL (*see* **Note 2**).
2. Incubate the mix at 37 °C overnight.
3. Precipitate DNA with 3 vol of 100% ethanol (750 μL) and 1/10 vol of 3 M sodium acetate (25 μL).

4. Centrifuge the microtube at 16,000 × *g* for 10 min and discard the supernatant.
5. Add 500 μL of chilled 70% ethanol and centrifuge at 16,000 × *g* for 5 min.
6. Remove the supernatant and let the pellet dry.
7. Add 42 μL of milli-Q water (*see* **Note 9**).

3.2.3 Bisulfite Treatment

1. Mix the digested DNA from the previous step with 17.5 μL of 20 mM hydroquinone and 460.5 μL of 2.5 M sodium bisulfite pH 5.2.
2. Seal the mixture with mineral oil, cover the microtubes with foil and incubate it at 37 °C for 16 h in the dark.
3. Purify DNA using the Wizard SV gel and PCR Cleanup System kit (Promega) following manufacturer's instructions.
4. Elute DNA in 50 μL of milli-Q water preheated at 50 °C.
5. Desulfonate DNA by adding 86 μL of milli-Q water and 24 μL of 2 M NaOH and incubating at 37 °C for 15 min in the dark.
6. Precipitate DNA with 480 μL of 100% ethanol and 16 μL of 3 M sodium acetate and incubate at −20 °C for at least 20 min.
7. Centrifuge DNA at 16,000 × *g* for 10 min and discard the supernatant.
8. Wash with 500 μL of 70% ethanol and centrifuge again at 16,000 × *g* for 5 min.
9. Discard the supernatant and let the pellet dry.
10. Resuspend the pellet in 20–50 μL of milli-Q water.

3.2.4 PCR Amplification

1. Mix 20 ng of bisulfite-treated DNA with 1× FastStart Buffer, *GCN4* (native) forward and *GCN4* (native or converted) reverse primers 0.4 μM each, 0.2 mM dNTPs and 2.5 U of FastStart High Fidelity polymerase in a total vol of 100 μL (*see* **Notes 4** and **5**).
2. Run the following PCR program: initial denaturation at 95 °C for 5 min followed by 35 cycles of denaturation at 95 °C for 30 s, primer annealing at 45 °C for 30 s and DNA synthesis at 72 °C for 1 min and 45 s. A final extension step at 72 °C for 5 min is also recommended.
3. Run 2–3 μL of PCR amplification in an agarose gel 0.8% to check.

3.2.5 Cloning and Sequencing

1. Run the PCR products in a 0.8% agarose gel (*see* **Note 10**).
2. Cut the agarose band corresponding to the amplicon of the expected size and purify it with the Wizard SV gel and PCR

Cleanup System kit (Promega) following manufacturer's instructions.

3. Elute DNA in 50 μL of milli-Q water preheated at 50 °C.
4. Quantify DNA by NanoDrop or agarose gel and mix the necessary amount of DNA with pGEM-T easy vector (Promega) to get two molecules of DNA per one of pGEM-T easy vector in the presence of 1× T4 ligase buffer and 1 μL of T4 DNA ligase.
5. Incubate at 4 °C overnight and use this ligation mix to transform *E. coli* DH5 alpha competent cells.
6. Plate *E. coli* transformed cells in Lb Amp selective media supplemented with X-Gal and IPTG and incubate plates at 37 °C overnight.
7. Extract the plasmid of independent *E. coli* white transformants by picking a colony up into 2 mL of LB amp and grown it overnight at 37 °C.
8. Pellet bacterial culture and resuspend it in 110 μL of STETL.
9. Boil samples at 100 °C for 30 s and centrifuge at 16,000 × *g* for 15 min.
10. Remove the pellet with a wooden stick previously wet in 10 mg/mL RNase A.
11. Add 110 μL isopropanol, mix by tube inversion and centrifuge again at 16,000 × *g* for 15 min.
12. Remove supernatant and wash pellet by adding 500 μL of cold 70% ethanol and centrifuging at 16,000 × *g* for 5 min.
13. Let the pellet dry and resuspend it in 100 μL of free-nuclease water.
14. Digest 1 μL of plasmid DNA with 2 U of *Eco*RI enzyme in the presence of the appropriate digestion buffer in a total vol of 20 μL.
15. Run 10 μL of the digested plasmid in 0.8% agarose gel by electrophoresis in order to check that the fragment inserted in the plasmid presents the correct size.
16. Use the remaining nondigested plasmid for sequencing with M13F and M13R primers.
17. Align the resulting sequences with a wild-type sequence in order to detect cytosine to thymine conversions corresponding to the length of R-loop. The size will be considered starting from the primer used for sequencing until the last cytosine to thymine conversion.

4 Notes

1. The combined pLZGAID construct [10], carrying *TRP1* gene as a marker and containing both the human *AID* gene under the GAL promoter and the LlacZ recombination system, can be used instead of two plasmids.
2. The enzymes to be used should be adapted so that none of them cuts in the region to be tested. The enzymes presented here were chosen to detect DNA-RNA hybrids at the *GCN4* gene.
3. Next steps of bisulfite treatment require the employment of DNA purification kits. Depending on the kit, it might be required to digest the DNA to avoid its retention (and loss) in the kit column.
4. Primers used in this protocol were chosen based on previous data that demonstrate the presence of R-loops at the *GCN4* locus [10, 24, 25] (Fig. 2). Nonetheless, any other genomic regions can be studied by using a different pair of primers, in which case, it would be required to set up the PCR conditions.
5. The following primers can be used to amplify the region of the GCN4 gene: N1, TACCAATTGCTATCATGTACCCGT; N2, TTGGCAGTAGAAGTGGAAGCA; C2, TTAACAATAAAAA TAAAAACA; N3, AATGAAATCAGCGTTCGCCA; C3, AAT AAAATCAACATTCACCA; N4, TGGCGAACGCTGATTT CATT; C4, TAACAAACACTAATTTCATT; N5, TATC TAAACCTTAGCGTTTGCATTC).
6. Digestion with proteinase K during 2 h is sufficient for degrading most of proteins. However, it can be extended overnight in order to improve the efficiency of DNA extraction.
7. Resuspension can also be achieved at 30 °C overnight.
8. Pellet can be dried in a thermoblock at 37 °C for 10 min or at room temperature for longer. It is important not to resuspend the pellet by pipetting up and down, since it could damage the integrity of R-loops.
9. DNA in this step is too concentrated in 42 μL. Thus, it is convenient to incubate samples at 37 °C during 1 or 2 h in order to resuspend properly the DNA before continuing with bisulfite modification assay.
10. Given that converted primers do not contain guanines, but adenines instead, their melting temperature is low. This is likely to decrease PCR specificity making essential to perform PCR product purification from the agarose gel.

Acknowledgments

Research in A.A.'s lab is funded by the Spanish Ministry of Science and Innovation, the European Research Council and the Junta de Andalucía, plus the European Union (FEDER).

References

1. Garcia-Muse T, Aguilera A (2019) R loops: from physiological to pathological roles. Cell 179(3):604–618. https://doi.org/10.1016/j.cell.2019.08.055
2. Daniels GA, Lieber MR (1995) RNA:DNA complex formation upon transcription of immunoglobulin switch regions: implications for the mechanism and regulation of class switch recombination. Nucleic Acids Res 23(24):5006–5011. https://doi.org/10.1093/nar/23.24.5006
3. Chen L, Chen JY, Zhang X, Gu Y, Xiao R, Shao C, Tang P, Qian H, Luo D, Li H, Zhou Y, Zhang DE, Fu XD (2017) R-ChIP using inactive RNase H reveals dynamic coupling of R-loops with transcriptional pausing at gene promoters. Mol Cell 68(4):745–757.e745. https://doi.org/10.1016/j.molcel.2017.10.008
4. Bhatia V, Barroso SI, Garcia-Rubio ML, Tumini E, Herrera-Moyano E, Aguilera A (2014) BRCA2 prevents R-loop accumulation and associates with TREX-2 mRNA export factor PCID2. Nature 511(7509):362–365. https://doi.org/10.1038/nature13374
5. Boguslawski SJ, Smith DE, Michalak MA, Mickelson KE, Yehle CO, Patterson WL, Carrico RJ (1986) Characterization of monoclonal antibody to DNA.RNA and its application to immunodetection of hybrids. J Immunol Methods 89(1):123–130. https://doi.org/10.1016/0022-1759(86)90040-2
6. Garcia-Rubio M, Barroso SI, Aguilera A (2018) Detection of DNA-RNA hybrids in vivo. Methods Mol Biol 1672:347–361. https://doi.org/10.1007/978-1-4939-7306-4_24
7. Hartono SR, Malapert A, Legros P, Bernard P, Chedin F, Vanoosthuyse V (2018) The affinity of the S9.6 antibody for double-stranded RNAs impacts the accurate mapping of R-loops in fission yeast. J Mol Biol 430(3):272–284. https://doi.org/10.1016/j.jmb.2017.12.016
8. Silva S, Camino LP, Aguilera A (2018) Human mitochondrial degradosome prevents harmful mitochondrial R loops and mitochondrial genome instability. Proc Natl Acad Sci U S A 115(43):11024–11029. https://doi.org/10.1073/pnas.1807258115
9. Gomez-Gonzalez B, Aguilera A (2007) Activation-induced cytidine deaminase action is strongly stimulated by mutations of the THO complex. Proc Natl Acad Sci U S A 104(20):8409–8414. https://doi.org/10.1073/pnas.0702836104
10. Garcia-Pichardo D, Canas JC, Garcia-Rubio ML, Gomez-Gonzalez B, Rondon AG, Aguilera A (2017) Histone mutants separate R loop formation from genome instability induction. Mol Cell 66(5):597–609.e595. https://doi.org/10.1016/j.molcel.2017.05.014
11. Dominguez-Sanchez MS, Barroso S, Gomez-Gonzalez B, Luna R, Aguilera A (2011) Genome instability and transcription elongation impairment in human cells depleted of THO/TREX. PLoS Genet 7(12):e1002386. https://doi.org/10.1371/journal.pgen.1002386
12. Mischo HE, Gomez-Gonzalez B, Grzechnik P, Rondon AG, Wei W, Steinmetz L, Aguilera A, Proudfoot NJ (2011) Yeast Sen1 helicase protects the genome from transcription-associated instability. Mol Cell 41(1):21–32. https://doi.org/10.1016/j.molcel.2010.12.007
13. Garcia-Benitez F, Gaillard H, Aguilera A (2017) Physical proximity of chromatin to nuclear pores prevents harmful R loop accumulation contributing to maintain genome stability. Proc Natl Acad Sci U S A 114(41):10942–10947. https://doi.org/10.1073/pnas.1707845114
14. Frommer M, McDonald LE, Millar DS, Collis CM, Watt F, Grigg GW, Molloy PL, Paul CL (1992) A genomic sequencing protocol that yields a positive display of 5-methylcytosine residues in individual DNA strands. Proc Natl Acad Sci U S A 89(5):1827–1831. https://doi.org/10.1073/pnas.89.5.1827
15. Kao YP, Hsieh WC, Hung ST, Huang CW, Lieber MR, Huang FT (2013) Detection and characterization of R-loops at the murine immunoglobulin Salpha region. Mol Immunol 54(2):208–216. https://doi.org/10.1016/j.molimm.2012.11.009

16. Yu K, Chedin F, Hsieh CL, Wilson TE, Lieber MR (2003) R-loops at immunoglobulin class switch regions in the chromosomes of stimulated B cells. Nat Immunol 4(5):442–451. https://doi.org/10.1038/ni919
17. Yu K, Roy D, Huang FT, Lieber MR (2006) Detection and structural analysis of R-loops. Methods Enzymol 409:316–329. https://doi.org/10.1016/S0076-6879(05)09018-X
18. Dumelie JG, Jaffrey SR (2017) Defining the location of promoter-associated R-loops at near-nucleotide resolution using bisDRIP-seq. eLife 6. https://doi.org/10.7554/eLife.28306
19. Wulfridge P, Sarma K (2021) A nuclease- and bisulfite-based strategy captures strand-specific R-loops genome-wide. eLife 10. https://doi.org/10.7554/eLife.65146
20. Lafuente-Barquero J, Garcia-Rubio ML, Martin-Alonso MS, Gomez-Gonzalez B, Aguilera A (2020) Harmful DNA:RNA hybrids are formed in cis and in a Rad51-independent manner. eLife 9. https://doi.org/10.7554/eLife.56674
21. Lisby M, Rothstein R, Mortensen UH (2001) Rad52 forms DNA repair and recombination centers during S phase. Proc Natl Acad Sci U S A 98(15):8276–8282. https://doi.org/10.1073/pnas.121006298
22. Gomez-Gonzalez B, Ruiz JF, Aguilera A (2011) Genetic and molecular analysis of mitotic recombination in *Saccharomyces cerevisiae*. Methods Mol Biol 745:151–172. https://doi.org/10.1007/978-1-61779-129-1_10
23. Chavez S, Aguilera A (1997) The yeast HPR1 gene has a functional role in transcriptional elongation that uncovers a novel source of genome instability. Genes Dev 11(24):3459–3470. https://doi.org/10.1101/gad.11.24.3459
24. Castellano-Pozo M, Santos-Pereira JM, Rondon AG, Barroso S, Andujar E, Perez-Alegre M, Garcia-Muse T, Aguilera A (2013) R loops are linked to histone H3 S10 phosphorylation and chromatin condensation. Mol Cell 52(4):583–590. https://doi.org/10.1016/j.molcel.2013.10.006
25. Herrera-Moyano E, Mergui X, Garcia-Rubio ML, Barroso S, Aguilera A (2014) The yeast and human FACT chromatin-reorganizing complexes solve R-loop-mediated transcription-replication conflicts. Genes Dev 28(7):735–748. https://doi.org/10.1101/gad.234070.113

Chapter 5

Identification of RNA–DNA Hybrids Associated with R-Loops at the *IgH* Switch Sequence in Activated B Cells

Junghyun Lim, Brice Laffleur, Uttiya Basu, and Kefei Yu

Abstract

During transcription and replication, R-loops that contain RNA–DNA hybrids are generated across numerous genomic loci and contribute to many biological events. Using S9.6, a monoclonal antibody against RNA–DNA hybrids, accelerated the study of R-loop biology. An outpouring of recent studies has implicated various contributions of R-loop in physiological cellular functions. Earlier studies using non-denaturing sodium bisulfite probing also supported existence of DNA–RNA hybrids formation in mammalian cells. In activated B cells, RNA–DNA hybrids formation at *IgH* gene locus of B cells is crucial for class switch recombination that ensure the proper effector function of the antibody. Here, we describe the identification of R-loops associated with the *IgH* locus using RNA–DNA hybrid immunoprecipitation sequencing and nondenaturing sodium bisulfite probing. This will be helpful for future studies of R-loops status on whole genome as well as on *IgH* locus in B cells.

Key words Class switch recombination, B cells, DRIP-seq, Nondenaturing sodium bisulfite probing

1 Introduction

R-loop is a triple stranded nucleic acid structure composed of double stranded RNA–DNA hybrid and a displaced single-stranded DNA. RNA–DNA hybrids are known to be more stable than double strand DNA [1, 2] of the same sequences. R-loops are usually generated during transcription, and distributed widely across the genome. R-loop formation and stability depend on negative supercoiling [3–5], skewed GC content between the two DNA strands [6, 7], G-quadruplex formation potential of the displaced strand [8, 9], and other factors that are still poorly characterized. Since in vitro R-loop structure forming under the 70% formamide condition was first reported in 1976 [10], many studies have detected R-loops formation in vivo from different organisms. R-loops have been regarded as byproducts during transcription in the early days, however, later studies have uncovered

Andrés Aguilera and Alexey Ruzov (eds.), *R-Loops: Methods and Protocols*, Methods in Molecular Biology, vol. 2528,
https://doi.org/10.1007/978-1-0716-2477-7_5,

their diverse influence on physiological cellular functions. Programmed R-loop formation is involved in initiation of mitochondrial DNA replication [11, 12], DNA repair [13, 14], transcription initiation/elongation/termination [15–17], epigenetic regulation [18, 19], and immunoglobulin heavy chain class switch recombination (as in Fig. 1) [20–23]. Accumulation of R-loop structures can induce replication stress, chromatin alteration and genomic instability by unexpected double strand DNA breaks [18, 20, 24–27]. These imply that R-loops should be associated with various diseases including cancer. The level of R-loop formation at all specific regions on the genome are tightly regulated under physiological conditions. Recent studies have reported that alterations of RNA–DNA hybrids homeostasis are associated with human diseases, and identified many factors for balancing R-loop levels to maintain beneficial R-loops and/or to remove harmful R-loops (reviewed in [28, 29]). A large number of studies provided genome-wide mapping of R-loops using DNA–RNA hybrids precipitation (DRIP) with the S9.6 monoclonal antibody [30] and high throughput sequencing. In the past decade, methods for mapping of R-loops based on DRIP and NGS have been developed continuously for improved accuracy and resolution, and have helped a deeper understanding of R-loop functions in the nuclear processes. Here, we describe a basic method to extract genomic DNA and isolate fragments containing R-loops, as well as nondenaturing bisulfite probing following by DNA sequencing that provides near nucleotide resolution of the R-loop structure.

2 Materials

2.1 DRIP-Seq

1. Lipopolysaccharides from Escherichia coli O111:B4 (LPS): 20 mg/mL solution in water.
2. Recombinant murine Interleukin-4 (IL-4): 20 μg/mL solution in phosphate buffered saline.
3. Activated B cells (~70% confluent): cultured with 20 μg/mL of LPS and 20 ng/mL of IL-4.
4. Phosphate buffered saline (PBS): 150 mM sodium phosphate and 150 mM NaCl [pH 7.2].
5. 1× TE buffer: 10 mM Tris-Cl (pH 8.0), 1 mM EDTA.
6. 20% sodium dodecyl sulfate (SDS) solution in water.
7. Proteinase K.
8. Phenol–chloroform–isoamyl alcohol (25:24:1) (pH 8.0).
9. Chloroform.
10. 3 M NaOAc (pH 5.2).
11. Isopropanol.

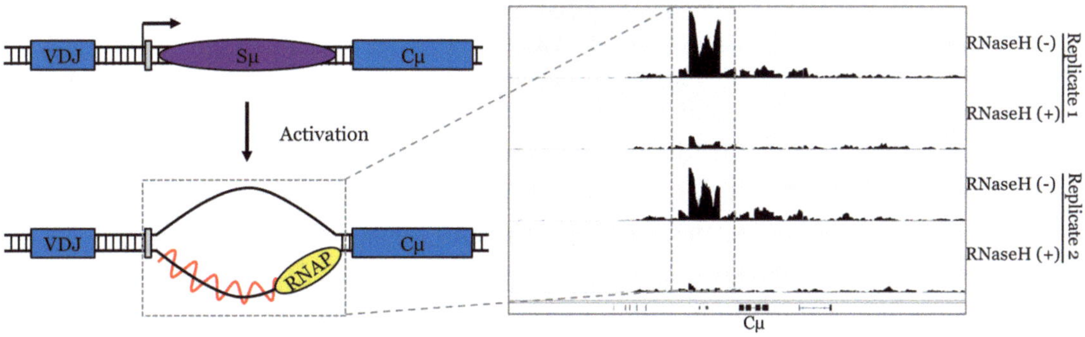

Fig. 1 RNA–DNA hybrids at Ig Sμ switch region. RNA–DNA hybrid peaks on IGV track are modified from DRIP-seq data [20]

12. 100% ethanol, 80% ethanol.
13. DNase/RNase-free water.
14. Restriction enzymes: BamHI, EcoRI, HindIII, NcoI, SpeI, XbaI.
15. RNase H.
16. RNase III.
17. 10× Antibody binding buffer: 100 mM $NaPO_4$ (pH 7.0), 1.4 M NaCl, 0.5% Triton X-100.
18. S9.6 antibody.
19. Protein A/G magnetic beads.
20. Glycogen.
21. Qubit dsDNA HS Assay Kit.
22. PCR primers for validation of IP quality.
    Sμ_P1_Fwd (GAGTACCGTTGTCTGGGTCACTCAC).
    Sμ_P1_Rev (CTGAAGTAGAGACAGCATCAGTACCTC).
    Sμ_P2_Fwd (ATCAGTAAGGAGGGACCCAGGCT).
    Sμ_P2_Rev (CTCGGATCTAAGCACTGTCCTTTGA).
23. PCR primers for spike-in control.
    Segmented Filamentous Bacteria_Fwd (GACGCTGAGG CATGAGAGCAT).
    Segmented Filamentous Bacteria_Rev (TGAATAACAAT CCGTGCCGTC).
    *Helicobacter Pylori*_Fwd (CTTAACCATAGAACTGCAT TTGA).
    *Helicobacter Pylori*_Rev (ATACTCATTGCGAAGGCGA CC).
    Universal Bacterial Primers_Fwd (ACTCCTACGGGAGG CAGCAGT).
    Universal Bacterial Primers_Rev (GCCAGCAGCCGCG GTAAT).

Lactobacillus on gut bacterial DNA _Fwd (AGCAGTA GGGAATCTTCCA).

Lactobacillus on gut bacterial DNA _Rev (CTCCATGTG-TAGCGGTG).

24. QIAquick Gel Extraction Kit.
25. Bioruptor Plus sonication device (Diagenode).
26. NEBNext Ultra II DNA Library Prep Kit (NEB).
27. NEB next adaptor for Illumina (NEB).
28. AMPure XP Beads (Beckman Coulter).
29. 2100 Bioanalyzer instrument (Agilent).

2.2 Nondenaturing Sodium Bisulfite Probing

1. Denature solution 0.2 N NaOH (add 20 μL of 1 N NaOH into 80 μL of water).
2. Neutralizing solution (200 mM Tris-HCl, pH 7).
3. Sodium hydroxide.
4. Hydroquinone.
5. Sodium bisulfite.
6. DNA purification resin (Promega), or DNA spin columns (Zymo Research).
7. Microcentrifuge.
8. Rotate mixer.
9. Fresh 2 M NaOH (1.25 mL): Dissolve 0.1 g of NaOH in 1.2 mL of MilliQ water (or weigh a pellet of NaOH and dissolve it in an appropriate amount of water).
10. Fresh 20 mM hydroquinone (1 mL). Dissolve 2.2 mg hydroquinone per mL of water.
11. Fresh 2.5 M Sodium Bisulfite, pH 5.2 (1 mL). Weigh 0.476 g of sodium bisulfite. Add 250 μL of 2 M NaOH. Add water to 1 mL. Vortex vigorously for ~1 min (*see* **Note 1**).

3 Methods

3.1 DNA–RNA Hybrid Immunoprecipitation and Sequencing

3.1.1 Nucleic Acids Extraction from Activated B Cells

1. Transfer 5×10^6 cells (from 70% confluent cultured cells) to 15 mL conical tube.
2. Pellet cells at 300 × *g* for 5 min at room temperature (RT) and remove supernatant.
3. Resuspend cells with 10 mL 1×PBS and pellet cells again. Repeat **step 2**.
4. Resuspend cells completely with 1.6 mL 1× TE and split cells into two 2 mL tubes containing 20 μL of 20% SDS and add 5 μL of proteinase K. Invert immediately each tubes several times and incubate them at 37 °C for overnight (*see* **Note 2**).

5. Add 800 μL (1 vol) of phenol–chloroform–isoamyl alcohol (PCI) to incubated cell lysate tubes and invert them carefully several times (*see* **Note 3**).
6. Centrifuge at 13,000 × *g* for 5 min and transfer nucleic acids on aqueous upper phase to new 2 mL tubes (*see* **Note 4**).
7. Repeat **steps 5** and **6**, and repeat **steps 5** and **6** with 800 μL (1 vol) chloroform instead of PCI.
8. Add 80 μL of 3 M NaOAc (pH 5.2) and 800 μL of isopropanol to each tube. Invert two tubes until forming white colored DNA clumps.
9. Take nucleic acids clumps by using cut tip from two tubes and combine them into a new 2 mL tube. Add 1.5 mL of fresh 70% EtOH and invert the tube gently several times. Incubate tube at RT for 5 min. Repeat this **step 3** times.
10. Transfer nucleic acids clump into a new tube and allow to air-dry. Add 100 μL of 1× TE and allow to dissolve at RT for 30 min (*see* **Note 5**).
11. Add 100 μL of 2× restriction enzyme cocktail and invert tube several times.
12. Incubate nucleic acids extract at 37 °C for 4 h and add 200 μL of 1× restriction enzyme cocktail.
13. Invert tube several times and incubate at 37 °C for overnight.
14. Add 1 vol PCI and invert tube several times and centrifuge at 13,000 × *g* for 5 min.
15. Transfer nucleic acids on aqueous phase into a new 1.5 mL tube and add 1 vol chloroform. Invert tube several times followed by centrifuge at 13,000 × *g* for 5 min.
16. Transfer nucleic acids on aqueous phase into a new 1.5 mL tube and add 0.1 vol of 3 M NaOAc (pH 5.2) and 2 vol of 100% ethanol. After at least 2 h incubation at −20 °C, centrifuge at 4 °C, 13,000 × *g* for 20 min.
17. Wash pellet with 70% ice-cold EtOH by centrifuge 13,000 × *g* at 4 °C for 5 min. Air-dry pellet.
18. Resuspend nucleic acids pellet with 200 μL of 1× TE and split into two new 1.5 mL tubes (*see* **Note 6**).
19. Add 5 μL RNase H into one aliquot tube and 1 μL of RNase III (to remove dsRNAs) into both aliquot tubes in 1× RNase H buffer condition. Incubate at 37 °C overnight (*see* **Note** 7).
20. Repeat **steps 14–17**.
21. Resuspend nucleic acids pellet with 100 μL of 1× TE and measure nucleic acids concentration of each RNase H treated or untreated nucleic acids.

3.1.2 DNA–RNA Hybrids Immunoprecipitation

1. Take 5 μg each of RNase H treated or untreated nucleic acids and dilute with 1× TE up to 440 μL vol (*see* **Note 8**).
2. Add 50 μL of 10× Ab binding buffer and 10 μg of S9.6 antibody or 10 μg of control normal IgG.
3. Rotate tubes at 4 °C overnight.
4. Add 30 μL of prewashed Protein A/G magnetic bead (*see* **Note 9**).
5. Rotate tubes at 4 °C for 2 h.
6. Place tubes on the magnetic rack and remove binding buffer after all magnetic beads bind to the tube's wall (*see* **Note 10**).
7. Remove tubes from magnetic rack and resuspend magnetic beads with 1 mL of 1× Ab binding buffer. Rotate tubes at RT for 5 min. Go to **step 6**.
8. Repeat **steps 6** and 7 four times. Remove tubes from the magnetic rack.
9. Add 200 μL of elution buffer with 5 μL of proteinase K and incubate at 50 °C for 1 h on a rotator.
10. Add 1 vol PCI and invert tube several times and centrifuge at 13,000 × *g* for 5 min.
11. Transfer RNA–DNA hybrids from aqueous phase into a new 1.5 mL tube and add 1 vol chloroform. Invert tube several times followed by centrifuge at 13,000 × *g* for 5 min.
12. Transfer RNA–DNA hybrids from aqueous phase into a new 1.5 mL tube and add 20 ng of glycogen, 0.1 vol NaOAc (pH 5.2) and 2.5 vol 100% ethanol. After 2 h incubation at –80 °C, centrifuge at 4 °C, 13,000 × *g* for 10 min (*see* **Note 11**).
13. Wash pellet with 80% ice-cold EtOH and air-dry.
14. Resuspend pellet with 50 μL 1× TE.
15. Store tubes at −20 °C before qPCR or library preparation for NGS (*see* **Note 12**).

3.1.3 Library Preparation

1. Quantify the total amount of immunoprecipitated DNA using the Qubit 1× dsDNA HS Assay Kit, and use 30 μL of each IP product for the library preparation.
2. Optional (**steps 2–4**): add internal controls ("spike-in") to normalize the number of reads after NGS. We recommend to use DNA from different species for spike-in (e.g., four different species of bacteria are used in this protocol).
3. Amplify specific targets by PCR for spike-in controls. Purify each PCR products using the QIAquick Gel Extraction Kit, quantify and make an equimolar mix of these PCR products using 1× TE for dilution.

4. Mix the IP products with the spike-in controls (*see* **Note 13**).
5. Sonicate the IP product (with or without the spike-in controls) using BioRuptor Plus sonication system (program set: three pulses at low intensity for 30 s).
6. Add 10 μL (if using spike-in) or 20 μL of 1× TE to complete the sheared DNA volume up to 50 μL. Use the NEB Next Ultra II DNA Library Prep Kit (NEB) for DNA end processing and adaptor ligation (follow exactly the kit protocol). Add 3 μL of NEB next ultra II end prep enzyme mix and 7 μL of NEB next ultra II end prep reaction buffer, incubate 30 min at 20 °C follow by 30 min at 65 °C on a thermocycler.
7. For adaptor ligation (NEB next adaptor for Illumina), first prepare a dilution of this adaptor in Tris/NaCl, pH 7.5–8.0 (typically 1:10 dilution when your input DNA is between 5 to 100 ng). Then add 2.5 μL of NEB next adaptor for Illumina (diluted if needed), 30 μL of NEB next ultra II ligation master mix, and 1 μL of NEB next ligation enhancer mix to the end processed DNA mixture (60 μL reaction product of **step 6**) and mix well. Incubate at 20 °C for 15 min in a thermocycler. Add 3 μL of USER® enzyme to the ligation mixture and incubate at 37 °C for 15 min.
8. Clean up the adaptor-ligated IP DNA fragments using AMPure XP Beads.
9. Mix the 15 μL of the adaptor-ligated DNA fragments, 25 μL of NEB next ultra II Q5 2× master mix, 5 μL of i7 primer and 5 μL of i5 primer and amplify DRIP DNA library (*see* **Note 14**).
10. Clean up the PCR reaction using AMPure XP Beads.
11. Quantify DNA concentration of each DRIP libraries using Qubit dsDNA HS Assay Kit and perform a Bioanalyzer to see the size distribution. Calculate the library concentrations in nM (regarding to the mass concentration and the average size) and prepare a working solution with all the combined libraries at 4 nM final into 5 μL using dilution buffer (10 mM Tris-HCl, pH 8.5 with 0.1% Tween 20).

3.1.4 Sequencing of DRIP Libraries

1. Prepare a 0.2 N NaOH denature solution (*see* **Note 15**).
2. Denature the DNA by combining 5 μL of the 4 nM library mix with 5 μL of 0.2 N NaOH, vortex briefly, spin down at 280 × *g* for 1 min, and incubate for 5 min at RT.
3. Neutralize denatured DNA by adding 5 μL of neutralizing solution, and vortex gently and spin down at 280 × *g* for 1 min.
4. Dilute the libraries to 20 pM by adding 985 μL of HT1 buffer (*see* **Note 16**).

5. Prepare the sequencing tube to load on NextSeq cartridge (follow kit instructions) and add PhIX positive control for sequencing quality test (*see* **Note 17**).
6. Thaw a NextSeq reagent cartridge, load the denatured DNA libraries with the PhIX, and run the sequencing.

3.2 Nondenaturing Sodium Bisulfite Probing

1. Digest 5–10 μg of purified genomic DNA with restriction enzyme(s) to generate fragments ≤10 kb without disrupting of the amplicon (the region to be PCR amplified for sequencing analysis).
2. Purify the digested DNA by phenol–chloroform extraction followed by ethanol precipitation. Use sodium acetate instead of glycogen during the ethanol precipitation step. Resuspend in 30 μL of 1× TE (pH 8.0).
3. Combine 457.5 μL of 2.5 M sodium bisulfite, 12.5 μL of 20 mM hydroquinone and 30 μL DNA sample in a microfuge tube (total volume = 500 μL).
4. Add mineral oil to seal the top to prevent evaporation.
5. Incubate the microfuge tube in the dark at 37 °C for 12–16 h.
6. Purify bisulfite converted DNA using 1 mL of Wizard resin (Promega) or a DNA spin column (Zymo Research) according to the manufacturer's instructions.
7. Adjust the volume of the eluted DNA to 96 μL with 1× TE, [pH 8.0].
8. Add 24 μL of 2 M NaOH.
9. Incubate at 37 °C for 15 min (desulfonation).
10. Purify desulfonated DNA by ethanol precipitation, or, Wizard resin (Promega) or DNA spin column (Zymo Research).
11. Recover DNA in 30 μL of 1× TE, pH 8.0 and use 1–2 μL as PCR template to amplify regions of interest (*see* **Notes 18–23**).
12. Store the rest of DNA in −20 °C.

4 Notes

1. A very small amount of sodium bisulfite crystals may remain undissolved; this will not affect the usage of the sodium bisulfite solution.
2. Do not use a pipette to avoid unexpected gDNA breaks. DNase and RNase free reagents and plastic wares were used for all step in DRIP-Seq experiment.
3. Do not vortex.
4. Use wide-orifice P1000 tips.

5. Add 100 μL 1× TE immediately after air drying. Do not overdry. Efficiency of DNA digestion can be checked before and after restriction enzyme treatment (Subheading 3.1.1, **steps 10** and **18**).
6. Do not overpipette while resuspending nucleic acids.
7. RNase H treatment is essential in this method because RNA–DNA hybrids that are sensitive to RNase H treatment should only be accounted as potential R-loops.
8. Use 5 μg nucleic acids for S9.6 antibody and control IgG each. To avoid DNA loss, we used the low binding 1.5 mL tubes from Subheading 3.1.2, **step 1** to Subheading 3.1.2, **step 6**.
9. Wash beads with 1× Ab binding buffer two times.
10. Use pipette and remove buffer carefully. Do not touch magnetic beads.
11. Incubate at least at –80 °C for 2 h or at –20 °C overnight. We used precipitated DNA for sequencing because reverse transcription-based strand specific DRIP-seq is challenging to perform in cells that have RNA processing/decay defects due to enrichment of dsRNAs or at GC-rich switch sequences.
12. IP quality has to be validated by qPCR using Sμ specific primers before the next steps. Fold enrichment compared to the Input should be clearly superior to the negative control (RNase H treatment and control IgG antibody).
13. We recommend a 1:10,000 ratio, with excess of IP product. For each condition add exactly the same amount of spike-in mix (e.g., 10 μL of diluted PCR products in 30 μL of DRIP IP product, use large volume to avoid pipetting error).
14. Please follow the kit instruction to set up best PCR cycles and temperatures. This step will tag your samples with specific barcodes from NEB Next Multiplex Oligos from Illumina (NEB), choose an appropriate combination of barcodes.
15. We follow instructions from Illumina NextSeq system to denature and to dilute the libraries.
16. HT1 buffer is contained in NEXTSeq reagent kit.
17. DRIP DNA libraries have high diversity, so only 1% of PhIX is enough.
18. Sodium bisulfite treatment severely damages DNA, possibly through the actions of free radicals [31]. Typically, up to 90% of the input DNA are damaged such that the recovered DNA are not suitable for subcloning, transforming *E.coli.* or being used directly for library preparation for Illumina or PacBio sequencing.
19. PCR of bisulfite-treated DNA is usually done with a set of primers flanking the region of interest. Taq DNA polymerase

or other thermal stable polymerases that are compatible with uracil containing template (e.g., Q5U from New England Biolabs) should be used for PCR. The PCR product is subsequently Topo-cloned (ThermoFisher) and sequenced (typically >30 clones) to identify patches of continuous C-to-U conversions (readout as C-to-T transition) in some of the clones. Alternatively, PCR amplicon can be analyzed by high-throughput long-read sequencing such as PacBio sequencing.

20. Native murine switch regions are long, GC-rich, repetitive, and difficult to PCR across the entire length. Therefore, the strategy described above is only applicable to engineered shorter switch regions or genomic loci that are amenable to PCR amplification.
21. A primer that is complementary to the anticipated bisulfite converted sequences can be used to selectively amplify alleles that adopt R-loop conformation [21]. Only R-loops covering the priming site will be amplified. Such primer is termed a "converted" primer. The usage of a converted primer only allows analysis of part of an R-loop (up to the priming site of the converted primer) and does not allow quantitation of the frequency of R-looped alleles unless sophisticated scheme is designed to control for the difference of PCR efficiency between converted and nonconverted primers.
22. Native bisulfite probing can be applied to S9.6-enriched nucleic acid to greatly facilitate R-loop analysis as most non–R-looped DNA fragments are excluded.
23. RNase H treatment prior to bisulfite conversion should be included as a negative control for all R-loop analyses.

Acknowledgments

U.B. is supported by grants from NIH/NIAID (1R01AI099195, 1RO1 AI143897-01A1 and R01AI134988), K.Y. is supported by funding from NIH (R01AI39039), and J.L. is funded by NRF-2021R1A2C1012477.

References

1. Roberts RW, Crothers DM (1992) Stability and properties of double and triple helices: dramatic effects of RNA or DNA backbone composition. Science 258:1463–1466. https://doi.org/10.1126/science.1279808
2. Huppert JL (2008) Thermodynamic prediction of RNA–DNA duplex-forming regions in the human genome. Mol BioSyst 4:686–691. https://doi.org/10.1039/B800354H
3. Massé E, Drolet M (1999) Escherichia coli DNA topoisomerase I inhibits R-loop formation by relaxing transcription-induced negative supercoiling*. J Biol Chem 274:16659–16664. https://doi.org/10.1074/jbc.274.23.16659

4. Tuduri S, Crabbé L, Conti C et al (2010) Erratum: Topoisomerase I suppresses genomic instability by preventing interference between replication and transcription. Nat Cell Biol 12: 1122–1122. https://doi.org/10.1038/ncb1110-1122
5. Manzo SG, Hartono SR, Sanz LA et al (2018) DNA topoisomerase I differentially modulates R-loops across the human genome. Genome Biol 19:100. https://doi.org/10.1186/s13059-018-1478-1
6. Ginno PA, Lott PL, Christensen HC et al (2012) R-loop formation is a distinctive characteristic of unmethylated human CpG island promoters. Mol Cell 45:814–825. https://doi.org/10.1016/j.molcel.2012.01.017
7. Ginno PA, Lim YW, Lott PL et al (2013) GC skew at the 5′ and 3′ ends of human genes links R-loop formation to epigenetic regulation and transcription termination. Genome Res 23: 1590–1600. https://doi.org/10.1101/gr.158436.113
8. Duquette ML, Handa P, Vincent JA et al (2004) Intracellular transcription of G-rich DNAs induces formation of G-loops, novel structures containing G4 DNA. Genes Dev 18(13):1618–1629. http://genesdev.cshlp.org. Accessed 31 Jul 2021
9. Magis AD, Manzo SG, Russo M et al (2019) DNA damage and genome instability by G-quadruplex ligands are mediated by R loops in human cancer cells. Proc Natl Acad Sci U S A 116:816–825. https://doi.org/10.1073/pnas.1810409116
10. Thomas M, White RL, Davis RW (1976) Hybridization of RNA to double-stranded-DNA: formation of R-loops. Proc Natl Acad Sci U S A 73:2294–2298. https://doi.org/10.1073/pnas.73.7.2294
11. Xu B, Clayton DA (1996) RNA-DNA hybrid formation at the human mitochondrial heavy-strand origin ceases at replication start sites: an implication for RNA-DNA hybrids serving as primers. EMBO J 15:3135–3143. https://doi.org/10.1002/j.1460-2075.1996.tb00676.x
12. Lee DY, Clayton DA (1998) Initiation of mitochondrial DNA replication by transcription and R-loop processing*. J Biol Chem 273:30614–30621. https://doi.org/10.1074/jbc.273.46.30614
13. Ohle C, Tesorero R, Schermann G et al (2016) Transient RNA-DNA hybrids are required for efficient double-strand break repair. Cell 167: 1001–1013.e7. https://doi.org/10.1016/j.cell.2016.10.001
14. Yasuhara T, Kato R, Hagiwara Y et al (2018) Human Rad52 promotes XPG-mediated R-loop processing to initiate transcription-associated homologous recombination repair. Cell 175:558–570.e11. https://doi.org/10.1016/j.cell.2018.08.056
15. Chen L, Chen J-Y, Zhang X et al (2017) R-ChIP using inactive RNase H reveals dynamic coupling of R-loops with transcriptional pausing at gene promoters. Mol Cell 68:745–757.e5. https://doi.org/10.1016/j.molcel.2017.10.008
16. Huertas P, Aguilera A (2003) Cotranscriptionally formed DNA:RNA hybrids mediate transcription elongation impairment and transcription-associated recombination. Mol Cell 12:711–721. https://doi.org/10.1016/j.molcel.2003.08.010
17. Skourti-Stathaki K, Proudfoot NJ, Gromak N (2011) Human senataxin resolves RNA/DNA hybrids formed at transcriptional pause sites to promote Xrn2-dependent termination. Mol Cell 42:794–805. https://doi.org/10.1016/j.molcel.2011.04.026
18. Powell WT, Coulson RL, Gonzales ML et al (2013) R-loop formation at Snord116 mediates topotecan inhibition of Ube3a-antisense and allele-specific chromatin decondensation. PNAS 110:13938–13943. https://doi.org/10.1073/pnas.1305426110
19. Castellano-Pozo M, Santos-Pereira JM, Rondón AG et al (2013) R loops are linked to histone H3 S10 phosphorylation and chromatin condensation. Mol Cell 52:583–590. https://doi.org/10.1016/j.molcel.2013.10.006
20. Laffleur B, Lim J, Zhang W et al (2021) Noncoding RNA processing by DIS3 regulates chromosomal architecture and somatic hypermutation in B cells. Nat Genet 53:230–242. https://doi.org/10.1038/s41588-020-00772-0
21. Yu K, Chedin F, Hsieh C-L et al (2003) R-loops at immunoglobulin class switch regions in the chromosomes of stimulated B cells. Nat Immunol 4:442–451. https://doi.org/10.1038/ni919
22. Wiedemann E-M, Peycheva M, Pavri R (2016) DNA replication origins in immunoglobulin switch regions regulate class switch recombination in an R-loop-dependent manner. Cell Rep 17:2927–2942. https://doi.org/10.1016/j.celrep.2016.11.041
23. Lim J, Giri PK, Kazadi D et al (2017) Nuclear proximity of Mtr4 to RNA exosome restricts DNA mutational asymmetry. Cell 169:523–

537.e15. https://doi.org/10.1016/j.cell.2017.03.043

24. Gan W, Guan Z, Liu J et al (2011) R-loop-mediated genomic instability is caused by impairment of replication fork progression. Genes Dev 25:2041–2056. https://doi.org/10.1101/gad.17010011
25. Herrera-Moyano E, Mergui X, García-Rubio ML et al (2014) The yeast and human FACT chromatin-reorganizing complexes solve R-loop-mediated transcription–replication conflicts. Genes Dev 28:735–748. https://doi.org/10.1101/gad.234070.113
26. García-Rubio ML, Pérez-Calero C, Barroso SI et al (2015) The fanconi anemia pathway protects genome integrity from R-loops. PLoS Genet 11:e1005674. https://doi.org/10.1371/journal.pgen.1005674
27. Costantino L, Koshland D (2018) Genome-wide map of R-loop-induced damage reveals how a subset of R-loops contributes to genomic instability. Mol Cell 71:487–497.e3. https://doi.org/10.1016/j.molcel.2018.06.037
28. Crossley MP, Bocek M, Cimprich KA (2019) R-loops as cellular regulators and genomic threats. Mol Cell 73:398–411. https://doi.org/10.1016/j.molcel.2019.01.024
29. García-Muse T, Aguilera A (2019) R loops: from physiological to pathological roles. Cell 179:604–618. https://doi.org/10.1016/j.cell.2019.08.055
30. Boguslawski SJ, Smith DE, Michalak MA et al (1986) Characterization of monoclonal antibody to DNA · RNA and its application to immunodetection of hybrids. J Immunol Methods 89:123–130. https://doi.org/10.1016/0022-1759(86)90040-2
31. Hayatsu H (1976) Bisulfite modification of nucleic acids and their constituents. In: Cohn WE (ed) Progress in nucleic acid research and molecular biology. Academic, pp 75–124

Chapter 6

Locus-Specific Analysis of Replication Dynamics and Detection of DNA–RNA Hybrids by Immuno Electron Microscopy

Henriette Stoy, Kevin S. Lang, Houra Merrikh, and Massimo Lopes

Abstract

DNA–RNA hybrids can interfere with DNA replication, but the underlying intermediates and molecular mechanisms have remained elusive. Here, we describe a single molecule approach that allows to monitor DNA–RNA hybrids locus-specifically in the context of ongoing replication. Using restriction digestion, gel electrophoresis and gel elution, this workflow allows to efficiently isolate replication intermediates and to study replication dynamics across a specific genomic locus. Here, we applied this procedure to isolate a bacterial genomic locus carrying an inducible transcription-replication conflict. Moreover, we combined electron microscopy with S9.6-Gold immuno-labeling to detect DNA–RNA hybrids on the isolated replication intermediates. With some limitations, this approach may be adapted to locus-specific replication analyses in different organisms.

Key words DNA replication, Locus-specific, Replication dynamics, Genomic locus isolation, Gel electrophoresis and elution, Electron microscopy, Immunolabeling, Direct detection of DNA–RNA hybrids

1 Introduction

Investigating progression and architecture of DNA replication forks is particularly important on genomic loci that present specific challenges to the replication process. Repetitive sequences have been successfully investigated on episomal systems, which can replicate autonomously, reach high copy numbers, and thereby enable molecular and structural biology investigations [1, 2]. However, single-molecule studies on replication interference by specific sequences should also consider the genomic context within the chromosome, which imposes additional technical challenges.

Replication dynamics at fragile sites and other interesting loci have been successfully studied coupling DNA combing and FISH, where the fragment of interest can be distinguished from all other

Andrés Aguilera and Alexey Ruzov (eds.), *R-Loops: Methods and Protocols*, Methods in Molecular Biology, vol. 2528, https://doi.org/10.1007/978-1-0716-2477-7_6,

genomic fragments via sequence-specific fluorescent probes [3–5] (Fig. 1a). Studying nuclear localization and chromatin composition of specific genomic loci has also been facilitated by insertion of specific sequences that can be tightly bound by proteins, such as LacI binding to *lacO* sites, Tus binding to *ter* sites, and more recently the CRISPR-Cas9 system [6–10] (Fig. 1b). However, the use of these systems for replication studies has been limited by the impact of protein binding itself on replication fork progression [11, 12]. Moreover, to date, isolation of locus-specific DNA intermediates via protein probes has not been successfully coupled to direct visualization of these structures.

Here, we describe an alternative strategy for the isolation of linear and replicating genomic fragments, which utilizes extensive restriction digestion, gel electrophoresis and gel elution of a fragment of interest (FOI; Fig. 1c). In essence, full genomic DNA is digested into relatively small fragments, using a cocktail of restriction enzymes that do not cut within the FOI, leaving it intact at a larger size than bulk genomic DNA. After gel electrophoresis, the FOI and its replication intermediates can then be eluted from a relevant region of the agarose gel. While describing key steps of this workflow, we discuss important limitations of this methodology when enriching for replicating fragments from specific loci, which stem from the unavoidable contamination of other genomic fragments in the portion of the gel used for FOI isolation.

In this chapter, we describe an application of this method in *Bacillus subtilis* to isolate a genomic fragment carrying an inducible transcription-replication conflict (TRC), where DNA–RNA hybrids are reportedly enriched and replication is known to slow down [13]. Visualization of the isolated intermediates by electron microscopy allows monitoring fork progression across the FOI by deriving replication maps from inspection and length measurements of individual replication intermediates [14]. We also describe how to use a customized gold-conjugated S9.6 antibody to detect DNA–RNA hybrids on the isolated locus-specific intermediates. While powerful and potentially revealing, this method is quite time-consuming and requires crucial internal controls. Moreover, it can only be applied under conditions of strong enrichment of DNA–RNA hybrids, as enabled in this case by the isolation of the TRC-containing fragment.

Appropriate modifications of the described approach and different immunolabeling reagents would in principle allow to investigate different events occurring during DNA replication and other DNA transactions, provided molecular probes for these events are available and validated.

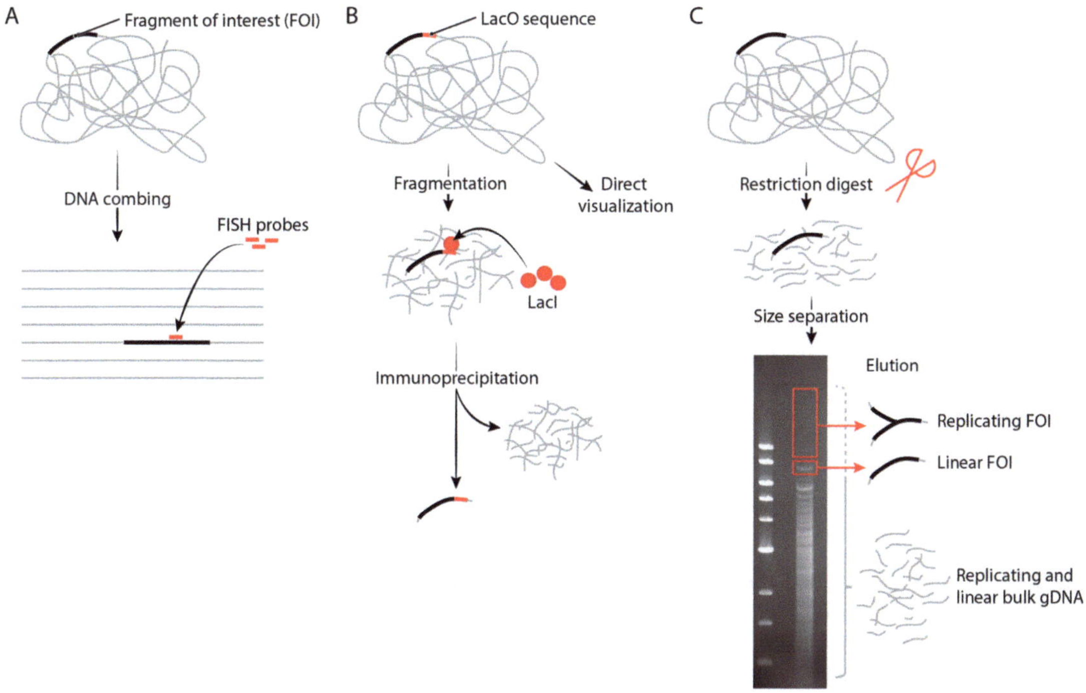

Fig. 1 Different approaches to study replication dynamics locus-specifically. (**a**) A fragment of interest (FOI, black) within the genome can be labeled using FISH probes within the context of the entire genome. (**b**) Labeling the FOI by insertion of a protein-detectable sequence (such as the *lacO* sequence) allows to either directly visualize the FOI or to specifically immunoprecipitate the sequence using the LacI protein. (**c**) The FOI-containing genomic DNA is subjected to restriction digest, which leaves the FOI as the largest fragment. During gel electrophoresis, the larger FOI (both linear and replicating) is separated from the contaminating, smaller genomic DNA, and can be recovered specifically by gel elution

2 Materials

2.1 Embedding and Lysis of Bacteria

1. Bacterial cultures: *Bacillus subtilis* strains HM1300 and HM2043, carrying an engineered head-on transcription-replication conflict (Fig. 2a) [13].
2. 2% Sodium azide: 20 mg/mL sodium azide in ddH_2O.
3. Ice-cold PBS.
4. 4,5′,8-Trimethylpsoralen (TMP) stock solution: dissolve 200 μg/mL TMP in Ethanol 100%. Stir extensively until the compound is dissolved completely. The solution can be stored at 4 °C for at least one year. Due to its DNA-modifying potential, TMP-containing solutions should be handled with gloves, lab coat, and protective glasses.
5. Low melting point agarose. Prepare 0.5% low melting point agarose in 100 mM EDTA in advance and store as 1.5 mL aliquots at 4 °C (*see* **Note 1**).

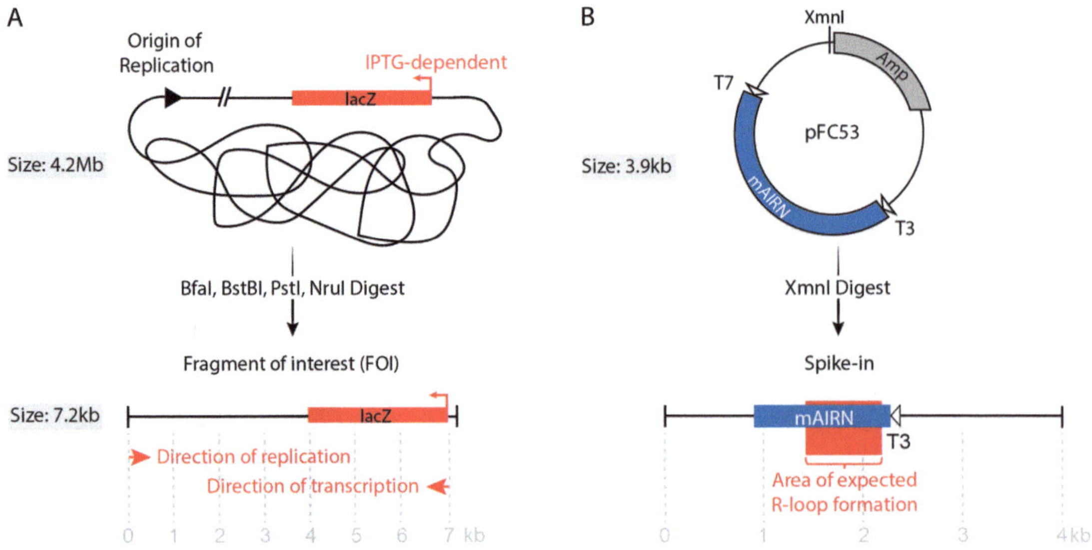

Fig. 2 DNA sequences that underlie this specific approach. (**a**) The fragment of interest (FOI) in this approach is an engineered transcriptional cassette, consisting of the LacZ gene under an IPTG-inducible promoter, inserted in a head-on orientation in respect to replication fork progression from the single origin of the *B. subtilis* genome. Digestion of the entire genome with BfaI, BstBI, PstI and NruI liberates the FOI as a 7.2 kb fragment. (**b**) To control for S9.6-Gold specificity and selectivity, in vitro transcribed pFC53 is used as a Spike-in during immuno electron microscopy. This plasmid carries an R-loop-prone region (mAIRN) under the T3 promoter. Linearization of pFC53 using XmnI positions the R-loop-prone sequence (red) asymmetrically on the fragment

6. Lysozyme, freshly prepared at 2 mg/mL in ddH_2O pH 7.0 (*see* **Note 2**).
7. Solution I (500 mM EDTA, 10 mM Tris-HCl, pH 7.5).
8. Solution II (5% sarcosyl, 0.4 M EDTA pH 7.5, 4 mg/mL proteinase K).
9. Wash solution (2 mM Tris-HCl, 1 mM EDTA, pH 8.0).
10. Preweighted 50 mL conical tubes.
11. Preweighted Eppendorf tubes.
12. Bio-Rad plug molds.
13. Biolink Crosslinker with monochromatic 365 nm lamps.
14. Heat block, heated to 95 °C.
15. Water bath, heated to 42 °C.

2.2 Extensive Genomic DNA Digestion

1. Restriction enzyme BfaI. Aliquot and store at −80 °C (*see* **Note 3**).
2. Restriction enzyme BstBI. Store at −20 °C (*see* **Note 3**).
3. Restriction enzyme PstI-HF. Store at −20 °C (*see* **Note 3**).
4. Restriction enzymes NruI-HF. Store at −20 °C (*see* **Note 3**).

5. 10× CutSmart buffer, diluted to 1× in ddH_2O.
6. Airtight container, tissues, inlay, Parafilm and tape to build a wet incubation chamber.
7. Small spatula.
8. Scalpel.

2.3 Isolation of a Single Genomic DNA Locus from the Bulk Genomic DNA by Electrophoresis

1. Low melting point NuSieve GTG Agarose. Store at RT.
2. 1× TBE.
3. Water bath, heated to 45 °C.
4. Electrophoresis apparatus (Biometry Horizon). Distance between electrodes 37 cm.
5. 1 kb ladder. Store at 4 °C.
6. 10 mg/mL ethidium bromide (EtBr). Store at 4 °C.
7. UV transilluminator.
8. Silica Bead DNA Gel Extraction Kit.
9. Eppendorf Concentrator 5301.

2.4 Preparation of R-Loop Carrying Spike-in DNA

1. Purified pFC53 plasmid [15] (Fig. 2b) (*see* **Note 4**).
2. T3 RNA polymerase, including 5× Transcription Buffer and 100 mM DTT. Store at −20 °C.
3. Ribonucleotide Solution Mix. Store at −20 °C.
4. Tween 20. Store at RT.
5. RNase A, stock 10 mg/mL in ddH_2O. Store at −20 °C.
6. RNase A working solution 0.1 mg/mL. Dilute RNase A stock 1:100 in ddH_2O.
7. Proteinase K stock solution 20 mg/mL in ddH_2O. Store at −20 °C.
8. 4,5′,8-trimethylpsoralen (TMP) stock solution: dissolve 200 μg/mL TMP in Ethanol 100%.
9. Biolink Crosslinker with monochromatic 365 nm lamps.
10. Glass plate, covered with Parafilm.
11. Freezing pack, flat 1.5–2 cm thick.
12. Chloroform–isoamyl alcohol 24:1. Store at RT under a ventilated hood.
13. Sodium acetate (NaOAc) as a 3 M stock, pH 5.2. Store at RT.
14. Ethanol: 100% and 70%.
15. 1× TE.
16. Restriction enzyme XmnI. Store at −20 °C.
17. 10× CutSmart buffer. Store at −20 °C.
18. Low melting point NuSieve GTG Agarose. Store at RT.

19. 1× TBE.
20. GelRed Nucleic Acid Stain. Store at RT.
21. Loading dye. Store at RT.
22. 1 kb ladder. Store at 4 °C.
23. UV transilluminator.
24. Silica Bead DNA Gel Extraction Kit.
25. NanoDrop.

2.5 RNase H Treatment

1. Conflict DNA (from Subheading 2.3).
2. Spike-in DNA (from Subheading 2.4).
3. RNase H. Including 10× RNase H buffer. Store at −20 °C.
4. Silica Bead DNA Gel Extraction Kit.
5. Eppendorf Concentrator 5301.

2.6 Immuno DNA Spreading

1. Conflict DNA (from Subheading 2.3) or the RNase H digested and purified DNA from Subheading 2.5.
2. Spike-in DNA (from Subheading 2.4) or the RNase H digested and purified DNA from Subheading 2.5.
3. S9.6-Gold conjugate 1 mg/mL, custom-made by Biosynthesis [14]. The S9.6 antibody (Kerafast Cat # ENH001) was conjugated to 6 nm gold beads and stored in a BSA-free buffer, compatible with EM visualization of nucleic acids.
4. S9.6-Gold conjugate working solution 0.1 mg/mL. Freshly dilute the S9.6-Gold conjugate stock 1:10 in 1× TE.
5. 1× TE.
6. Head block, set to 37 °C.
7. Glutaraldehyde, stock 25%.
8. Glutaraldehyde working solution 2.5%. Freshly dilute the Glutaraldehyde stock 1:10 in 1× TE.
9. Carbon-coated nickel grids (Plano, Cat # G2400N), freshly prepared as described in [16].
10. P2 pipette.
11. ddH_2O.
12. Ethidium bromide (EtBr), 10 mg/mL stock, diluted 1:300 in ddH_2O for a 33.3 μg/mL working solution.
13. Formamide 100% (Sigma, Cat # 47680).
14. Benzyldimethylalkyl Ammonium Chloride (BAC) (Sigma-Aldrich, Cat # B6295) as a stock solution 0.2% w/v in Formamide. The stock solution is freshly diluted 1:10 in 1× TE to create a BAC working solution.
15. Ethanol 100%.

16. Uranyl acetate (UAc, Fluka, Ca # 73943) as a stock solution of 5 mM in HCl 5 mM. Store at 4 °C. The stock solution is freshly diluted 1:10 in 100% EtOH to create a UAc working solution.
17. Round filter paper.
18. Tissue culture dishes, 150 × 20 mm.
19. Mica sheets (Plano, mica high grad quality V1, 25 × 76 mm, Cat # 56).
20. Graphite powder.
21. Dumont tweezers.

2.7 Image Acquisition

1. Transmission Electron Microscope, connected to a computer-driven CCD camera. Minimal resolution required is 1k × 1k if the camera is mounted on the bottom port; when mounted to the side 35 mm port a resolution of 2k × 3.5k is preferred.
2. Software for Camera Control and image acquisition (e.g., Digital micrograph).

2.8 Analysis and Interpretation

1. Image J or other software that enables contour length measurements on *.tiff files.
2. Standard statistic applications (e.g., Microsoft Office Excel or GraphPad).

3 Methods

3.1 Embedding and Lysis of Bacteria

1. Grow *B. subtilis* cultures to an OD of 0.3–0.4. Once the OD is reached add 2% sodium azide to a final concentration of 0.02% (1:100 dilution) to kill the cells.
2. Melt prepared 0.5% LMP agarose aliquots (in a heat block at 95 °C). Equilibrate agarose temperature to 42 °C in a water bath. Calculate in advance how much agarose is needed for all samples to melt and equilibrate sufficient agarose.
3. Load the *B. subtilis* culture into a preweighted 50 mL conical tube.
4. Centrifuge at 5500 × *g* for 3 min to collect cells in a compact pellet. Decant most of the supernatant, leaving behind less than 1 mL.
5. Resuspend the cell pellet in the remaining supernatant and transfer into a preweighted Eppendorf tube.
6. Centrifuge at 5000 × *g* for 30 s. Remove the supernatant completely with a pipette. If necessary, centrifuge again to remove as much liquid as possible.

7. Determine the weight of the cell pellet (*see* **Note 5**). 10 mL of *B. subtilis* culture at an OD of 0.3 should weight approximately 20 mg.
8. Wash the cell pellet three times with ice-cold PBS.
9. To crosslink, resuspend the cell pellet in PBS, add TMP to a final concentration of 10 μg/mL and irradiate for 2 min with UV 365 nm monochromatic light.
10. Wash the cell pellet three additional times with ice-cold PBS.
11. Place 12 μL 2 mg/mL lysozyme on the wall of each Eppendorf tube away from the cell pellet and lay the tube on its side preventing premature contact between the lysozyme and the cell pellet.
12. Add 180 μL of the melted and equilibrated LMP 0.5% agarose to the cell pellet (*see* **Note 6**).
13. Use a P200 pipette to resuspend the cell pellet in the agarose and to mix in the lysozyme. Pipette the mixture up and down for 8–10 times until the content is completely homogeneous. Avoid introducing bubbles.
14. Immediately transfer the mixture into BioRad plug molds.
15. Incubate the plug molds at 4 °C for about 30–40 min to let the agarose solidify. Cover the molds with a plastic box to prevent the agarose plugs from drying out on the top (due to possible air circulation).
16. Once solidified, push the plugs out of the molds into fresh scintillation vials, containing 1 mL of solution I. Incubate o/n at 37 °C to digest the cell wall.
17. The next day, add 400 μL of solution II. Incubate at 50 °C for 5 h (*see* **Note** 7). Invert the tubes twice during this incubation. The plugs should turn clear during this incubation.
18. To wash out the proteinase K, replace the buffer with 5 mL 1× wash solution. Incubate for 4 h to o/n at 4 °C with agitation. Repeat this wash step at least eight times before proceeding with DNA digestion (Subheading 3.2).
19. Plugs can be stored in wash solution at 4 °C for at least a year.

3.2 Extensive Genomic DNA Digestion

1. Transfer the plugs from the storage solution to a Parafilm coated surface and carefully remove any remaining storage buffer with a tissue (*see* **Note 8**).
2. Using a clean scalpel, cut a 5 mm thick slice of the plug and incubate it in excess 1× CutSmart buffer with slight shaking for at least 5 h, with the first 4 h at RT and the last 1 h at 37 °C (*see* **Note 9**).

3. Assemble the wet chamber: fix a piece of Parafilm with tape on an inlay, fill the airtight container with wet tissues and place the inlay on top. Prewarm the wet chamber for 1 h at 37 °C.
4. Transfer the plugs onto the Parafilm within the wet chamber; ensure that plugs from different samples are sufficiently separated, as plugs can move slightly during the digestion. Carefully remove any remaining CutSmart buffer with a tissue.
5. Mix 69 μL of 1× CutSmart with 9 μL of each of the four restriction enzymes (*see* **Note 10**). Pipette the digestion mixture onto the plug.
6. Carefully close the wet chamber and incubate it at 37 °C o/n.
7. The next morning, carefully remove the digestion mixture with a tissue and replace it with a new digestion mixture of the same composition (*see* **Note 11**). Incubate at 37 °C for an additional 3 h.
8. Remove the digestion mixture with a tissue and proceed with **step 2** in Subheading 3.3.

3.3 Isolation of Single DNA Loci from the Bulk Genomic DNA by Electrophoresis

1. Prepare 300 mL of 0.5% LMP agarose in 1× TBE: dissolve the agarose by heating in a microwave and subsequently incubate in a 45 °C water bath for at least an hour (*see* **Note 12**).
2. Transfer the plug slices from Subheading 3.2, **step 8** onto the comb and leave to dry for 30 min (*see* **Note 13**). Leave an empty well between the marker and the first plug slices, as well as in between plug slices.
3. Place the gel tray at 4 °C and ensure that it is perfectly horizontal. Place the comb with the dried gel slices in the gel tray and carefully pour the LMP agarose into the gel tray (*see* **Note 14**).
4. Let the gel solidify for 1 h. Then place it in the gel tank and fill the tank with 2 L of cold 1× TBE.
5. Load 25 μL of 1 kb ladder in the first lane and 25 μL of 6× loading dye to the last well to track migration even hours later.
6. Run the gel at 4 °C at 50 V for 21–24 h to allow sufficient separation between the conflict locus and the rest of the genomic DNA.
7. Stain the gel with 0.3 μg/mL EtBr for 1 h and visualize on a UV transilluminator (*see* **Note 15**).
8. Identify the regions of interest (Fig. 3a). While the linear FOI is usually very well visible on an EtBr-stained gel, the replicating FOI fraction should contain barely any visible signal. If the complete digest was successful, the genomic DNA (gDNA) should accumulate most strongly below the linear FOI; however, it is nearly impossible to remove the gDNA contamination completely from the upper parts of the gel—traces of

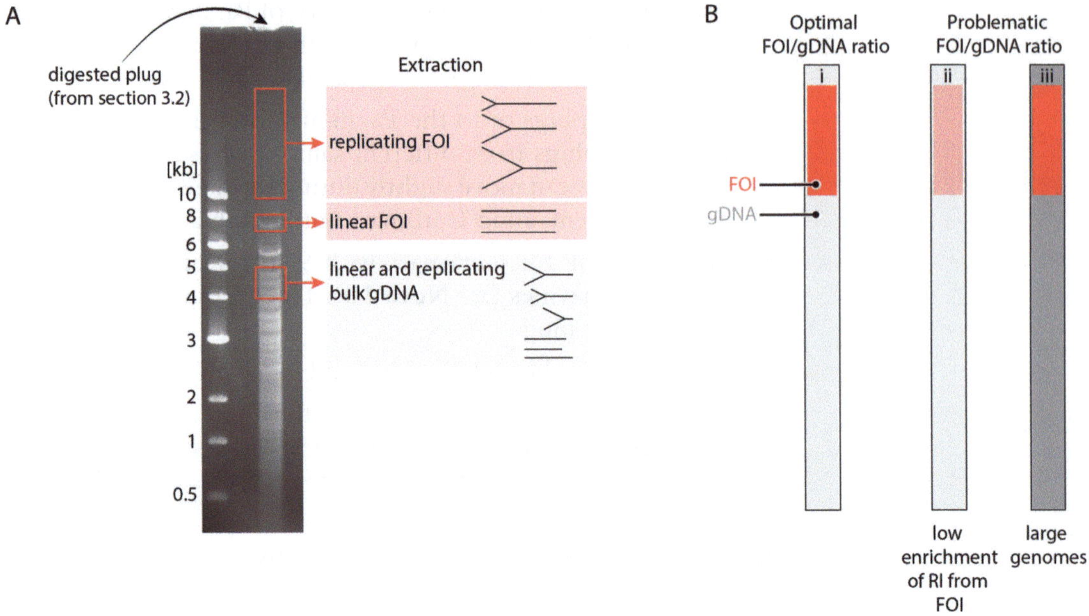

Fig. 3 Gel electrophoresis-based enrichment of the fragment of interest for electron microscopy analysis. (**a**) The digested plug (from Subheading 3.2) is poured into an LMP gel to size separated the containing gDNA fragments by gel electrophoresis. The linear FOI is visible right below the 8 kb marker, while the replicating FOI accumulates as an "invisible smear" above the linear FOI. As a control, a fraction of bulk gDNA (containing both linear and replicating fragments) is coeluted; this bulk gDNA is eluted from an area within the gel, which is sufficiently separated from the linear FOI. (**b**) Critical for the success of this approach is the sufficient enrichment of FOI (red) over gDNA contamination (grey) to reliable distinguish the FOI from the gDNA contamination. Unfortunately, even upon near-perfect digestion traces of contaminating gDNA will always coelute with the FOI, this should however not significantly affect the identification of FOI (optimal FOI–gDNA contamination ratio). In problematic cases, however, the ratio of FOI and gDNA contamination can further be decreased, preventing unambiguous identification of the FOI (problematic FOI/gDNA contamination ratio). Left: an insufficient stalling or slowing of replication forks while progressing through the FOI will lead to only few replicating FOI and a generally low enrichment of replicating FOI over background gDNA contamination. Right: The digestion of large genomes would increase the background of contaminating gDNA to an extend that even upon strong enrichment of replicating FOI, the ratio of FOI/gDNA contamination is too low to allow reliable identification of FOI from contaminating gDNA

gDNA contamination will always be coeluted with the linear and replicating FOI. It is therefore crucial for the success of this technique that the ratio between the FOI and this contaminating gDNA is optimal (Fig. 3b). In addition to the replicating and linear FOI fractions, we always excise a fraction of the bulk gDNA from a distinct region of the gel as a control (Fig. 3a). This control fraction will contain linear and replicating molecules from other regions of the genome.

9. Cut the regions of interest and transfer the gel pieces into separate tubes.

10. Extract the DNA from the gel using the Silica Bead DNA Gel Extraction Kit, according to manufacturer's instructions with minor adjustments (*see* **steps 11–14**).
11. Weight the gel and dissolve it in 0.5 v/w TBE conversion buffer and 2 v/w binding buffer at 50 °C for roughly 5 min.
12. Add 5 μL of silica beads and incubate at 50 °C for an additional 5 min (invert the tubes regularly to keep the beads in suspension) (*see* **Note 16**).
13. Centrifuge the samples in a tabletop centrifuge to maximum 7500 x *g* for 5 s to pellet the beads. Carefully remove the supernatant. Wash the beads three times with ice-cold wash buffer without resuspending the beads, to avoid sheering the DNA.
14. Airdry the pellet for 10 min to allow any residual ethanol from the wash buffer to evaporate (this would interfere with the spreading in Subheading 3.6). For elution, add 5 μL TE to the pellet and mix by flicking the tube, incubate at 50 °C for 5 min and subsequently centrifuge to maximum 7,500 x *g* for 5 s. Collect the supernatant. Repeat the elution step and pool both eluates (*see* **Note 17**).
15. When necessary, concentrate the samples using a vacuum concentrator (*see* **Note 18**).

3.4 Preparation of R-Loop Carrying Spike-in DNA

1. Set up the in vitro transcription reaction with 4 μg of pFC53 and 20 U/μg T3 RNA polymerase in 1× transcription buffer, supplemented with 20 mM DTT, 0.05% Tween 20, and 50 μM rNTPs, at 37 °C for 45 min.
2. Inactivate the in vitro transcription reaction at 65 °C for 10 min.
3. Add 5 μL RNase A working solution (final concentration 15 μg/mL) and incubate at 37 °C for 30 min.
4. Add 2 μL Proteinase K (final concentration 1.25 mg/mL) and incubate at 37 °C for 30 min.
5. Insert five monochromatic 365 nm lamps in the crosslinker.
6. Place a Parafilm-covered glass plate onto of a freezing pack into the crosslinker.
7. To cross-link the plasmid DNA, add TMP to a final concentration of 10 μg/mL and incubate in the dark for 5 min to allow TMP to intercalate into the DNA.
8. Transfer the in vitro transcription reaction (now mixed with the TMP) as a drop onto the Parafilm (*see* **Note 19**) and irradiate the sample for 3 min.
9. Transfer the reactions into fresh Eppendorf tubes.

10. To purify the plasmid DNA, add one volume of chloroform–isoamyl alcohol to the lysates. Invert vigorously but carefully for 30 times (the solution turns milky).
11. Centrifuge in a tabletop centrifuge at RT for 10 min at 16,000 × *g* (Phase separation).
12. Carefully transfer the upper phase into a new Eppendorf tube.
13. Mix the upper phase with NaOAc pH 5.2 to a final concentration of 0.3 M. Then add two volumes 100% ethanol, invert and incubate at −20 °C for several hours up to o/n to facilitate precipitation.
14. Centrifuge at 4 °C for 30 min at 16,000 × *g*.
15. Immediately and carefully remove the supernatant (*see* **Note 20**).
16. Add 70% EtOH and centrifuge at 4 °C for 10 min at 16,000 × *g*. Immediately and carefully remove the supernatant.
17. Repeat **step 16**.
18. Airdry the pellet at 37 °C for 5–10 min.
19. Resuspend the plasmid DNA in a small volume of TE (usually around 25 μL) (*see* **Notes 21** and **22**).
20. Digest 1–2 μg of transcribed and purified plasmid DNA with 40 U/μg XmnI in 1× CutSmart buffer for 3 h at 37 °C (*see* **Note 23**).
21. Prepare a 0.9% LMP agarose/TBE gel, containing 1:10,000 GelRed (*see* **Note 24**). Let solidify for 1 h.
22. Mix the digestion reaction with loading dye, load on the solidified 0.9% LMP agarose gel and run at 50–80 V until sufficient separation is achieved.
23. Excise the band corresponding to the linear plasmid DNA; for pFC53/XmnI this band is expected around 4 kb.
24. Extract the linear plasmid DNA from the gel using the Silica bead Gel Extraction Kit as described in Subheading 3.3, **steps 10–14**. Determine the concentration of the extracted material using a NanoDrop.

3.5 RNase H Treatment (See Note 25)

1. Mix 70–100 ng of Spike-In DNA (from Subheading 3.4) (*see* **Note 26**), 5 μL of extracted conflict DNA (from Subheading 3.3) and 10 U RNase H in 1× RNase H buffer and incubate at 37 °C o/n.
2. Purify the DNA using the Silica Bead DNA Gel Extraction Kit, according to manufacturer's instructions with minor adjustments.

3. Add 3 volumes of binding buffer and 5 µL silica beads to the digestion reactions and incubate at 50 °C for 5 min (invert the tubes regularly to keep the beads in suspension).
4. Centrifuge the samples in a tabletop centrifuge on short for 5 s to pellet the beads. Carefully remove the supernatant. Wash the beads three times with ice-cold wash buffer without resuspending the beads, to avoid sheering the DNA.
5. Airdry the pellet for 10 min. For elution, add 5 µL TE to the pellet and mix by flicking the tube, incubate at 50 °C for 5 min and subsequently centrifuge on short for 5 s. Collect the supernatant. Repeat the elution step and pool both eluates (usually the final volume here is around 9 µL; if this is not the case, fill up to 9 µL with TE).
6. Proceed with DNA spreading (Subheading 3.6) as soon as possible. Extended storage is not recommended at this point.

3.6 Immuno DNA Spreading

1. Mix 70–100 ng of Spike-In DNA (from Subheading 3.4) with 5 µL of extracted conflict DNA (from Subheading 3.3). The final volume should not exceed 9 µL (*see* **Note 27**). For RNase H digested controls, use the already mixed and purified samples from Subheading 3.5, **step 6**.
2. Add 1 µL of S9.6-Gold working solution to the Spike-In/ Conflict DNA and mix carefully by flicking the tube. Incubate at 37 °C for 1 h (*see* **Note 28**).
3. Prepare the Glutaraldehyde working solution as fresh as possible and add 0.8 µL to the DNA/S9.6-Gold reaction to crosslink the S9.6-Gold conjugate to the DNA. Mix carefully by flicking the tube and incubate at 37 °C for exactly 15 min (*see* **Note 29**).
4. Distribute up to eight droplets (20 µL) of EtBr working solution on a piece of Parafilm and carefully place a carbon-coated grid on top of each drop, with the carbon-side facing down. Incubate the grids for 20–60 min. Prevent evaporation and light exposure by covering the grids with the lid of a 15 cm Petri dish, covered in aluminum foil. Just before spreading, drain the excess EtBr solution from each grid by contacting the filter paper, and place the dried grid (carbon-side down) on top of the filter paper (*see* **Note 30**).
5. Using a P2 pipette, mix at the very bottom of a 1.5 mL Eppendorf tube: 1 µL of Formamide and 0.2 µL BAC working solution.
6. Add 5 µL of the S9.6-Gold labeled and crosslinked DNA in a separate drop to the side of the tube. Handle with care since the drop can easily slide down.

7. Pour approximately 20 mL (the minimal volume to cover the surface completely) of EM-grade water in a 15 cm Petri dish: this is called the hypophase. Split a mica sheet (about 1 × 2 cm in size) and place it in the water as a ramp, with the freshly cleaved surface facing up.
8. Spin the tube containing the sample for a few seconds in a microfuge and aspirate it completely into a 10 μL tip. With a cotton swap sprinkle a generous boundary of graphite powder onto the water surface, as a half circle in close proximity to the mica ramp.
9. Pipette the entire sample volume out of the tip and let the droplet touch the ramp a few millimeters above the water surface. The droplet will immediately slide down the ramp and spread over the hypophase, pushing the graphite boundary backward which indicates the border of the monomolecular detergent film. The DNA content will be highest close to this border.
10. Using fine tweezers take a carbon-coated, EtBr-soaked grid from the filter paper (carbon-side down) and place it on the spreading surface close to the graphite border. Hold the tweezers as parallel to the water surface as possible and gently press the grid down for 1–2 s to ensure that the grid has sufficient and even contact with the water surface. Carefully and without disturbing the hypophase lift the grid up again and incubate it for 15–20 s in 1 mL uranyl acetate working solution (best in a flat-bottom 20 mL tube). Wash the grid by dipping it two to three times into 100% EtOH and airdry the grid (carbon-side up) on a fresh filter paper.
11. Wipe the tweezers with a napkin to remove EtOH remnants and repeat **step** 7 for the other grids, aiming to collect DNA from different regions of the BAC film.
12. The grids are subsequently subjected to flat angle-rotary shadowing with platinum as described in [16].

3.7 Image Acquisition

1. The grids can be analyzed using any transmission electron microscope. The DNA molecules should be clearly distinguishable from the homogeneously granular background. While most DNA molecules will be exclusively double stranded, with a thickness of about 10 nm, some molecules can contain sections of ssDNA, which are noticeably thinner (5–7 nm). If the sample quality is not satisfactory—for example, residual protein bound to the DNA, heterogeneous shadowing, dirty background—we recommend to repeat the experiment, starting either from the DNA spreading or—if necessary—from earlier steps.

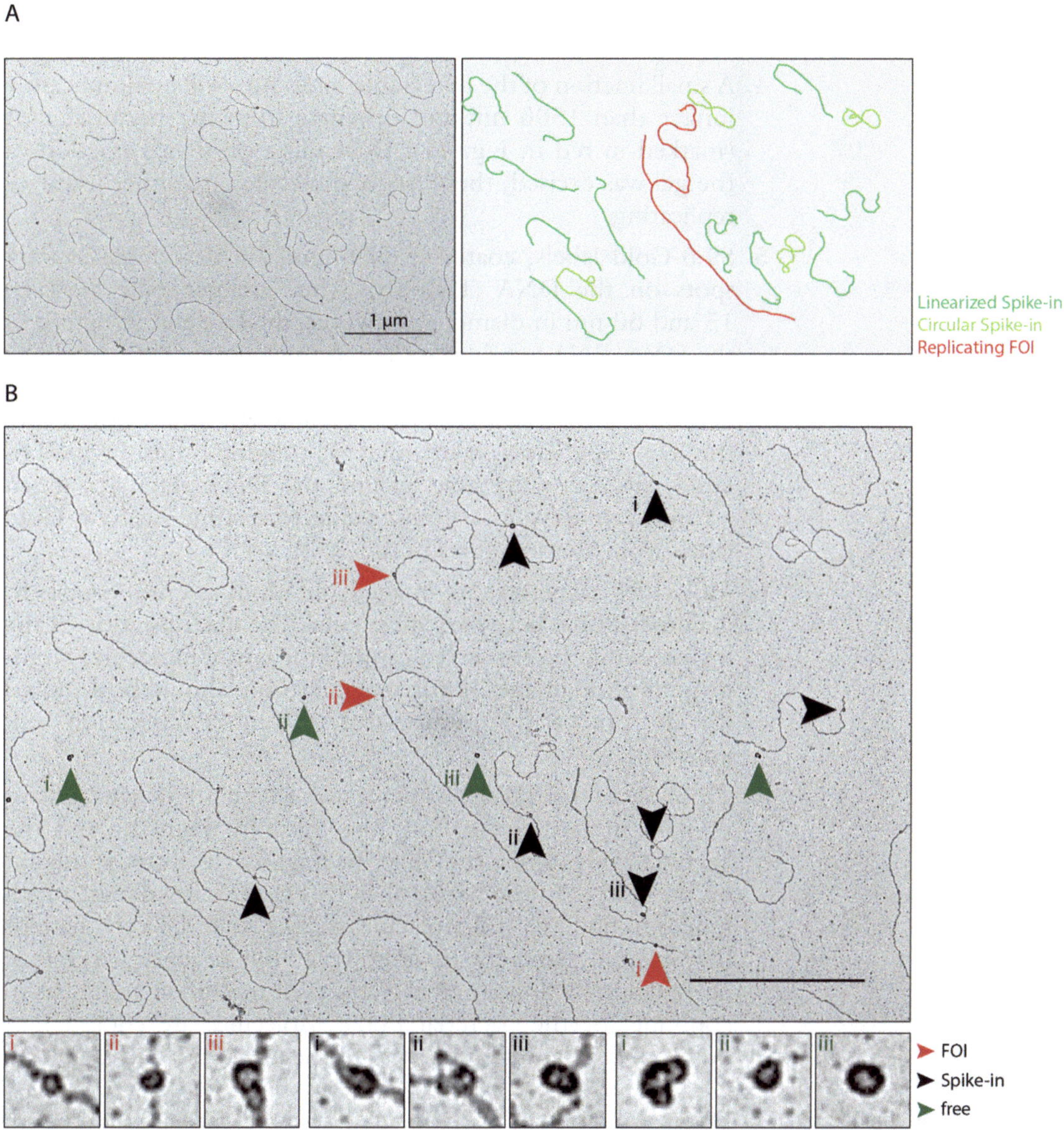

Fig. 4 Electron microscopy of immunolabeled *B. subtilis* samples. (**a**) Left: Overview of an electron microscopy sample of *B. subtilis* replicating FOI mixed with Spike-in DNA and immunolabeled with S9.6-Gold. Right: even without exact contour measurements, Spike-in DNA (green) can be distinguished from *B. subtilis* FOI (red). It is important to note, that the Spike-in DNA is far more abundant than the bacterial FOI. (**b**) Identification of S9.6-Gold particles bound to the bacterial FOI (red), the Spike-in DNA (black) or free of DNA (green). Bottom: high magnification images of the S9.6-Gold labels marked in the overview image above

2. To find molecules of interest, screen the grids at a magnification of 21–48 k (Fig. 4).
3. The majority of the visible molecules will originate from the XmnI-digested Spike-in plasmid and can be identified by the homogeneous length of 4000 bp (corresponding to roughly

1500 nm, marked in green in Fig. 4a). Due to the linearization, these plasmid molecules should be mostly linear.

4. A small fraction of the detectable molecules will be significantly longer than 1500 nm and originate from the bacterial FOI (marked in red in Fig. 4a). Depending on which fraction of the gel was excised, these larger molecules are either linear or replicating.
5. S9.6-Gold labels, coated in platinum, are detectable as dark spots on the DNA (Fig. 4b). They are generally between 15 and 60 nm in diameter; however, the size and structure of the DNA–RNA hybrid can significantly influence the label size (*see* **Note 31**). The actual gold particle can sometimes be detected as an intensely black, completely round dot within the S9.6-Gold label; its diameter is roughly 6 nm. Free S9.6-Gold particles detectable around the DNA (Fig. 4b, green arrows) can serve as a useful guide to distinguish S9.6-Gold labels from abnormalities in the DNA.
6. Molecules of interest are imaged and saved as dm3 formats to maintain all the necessary meta data. The magnification of the image should be chosen to capture the entire molecule. High-magnification images (180k magnification) should be acquired for any potential immuno label as well as replication fork junctions.
7. To allow statistical assessment, we advise to image at least 70–100 molecules for both the Spike-in fragment and the bacterial FOI. Since the Spike-in fragment is more abundant on the grid than the bacterial FOI (Fig. 4a), it will take much longer to find 70–100 molecules for the bacterial FOI than for the Spike-in fragment (it is however not necessary to image more than 100 Spike-in fragments even though imaging has to continue for the bacterial FOI). Molecules are chosen randomly and from different areas of the grid.

3.8 Analysis and Interpretation

1. Open the dm3 files in Image J to perform contour length measurements on each molecule and score the presence and exact location of S9.6-Gold labels.
2. For each Spike-in fragment, determine the length, as well as the position of the S9.6-Gold label as shown in Fig. 5a. The Spike-in fragments should have a size of 1500 nm ± 10% (Fig. 5b) (*see* **Note 32**). Under optimal labeling conditions, 30–50% of Spike-in fragment should carry a detectable S9.6-Gold label. These labels can be quite big (Fig. 4b), frequently preventing proper visualization of the underlying R-loop structure recognized by the antibody. Due to the asymmetric digestion pattern of the Spike-in fragment (Fig. 2b), the R-loop-forming region of the plasmid (the mAIRN gene) is positioned roughly

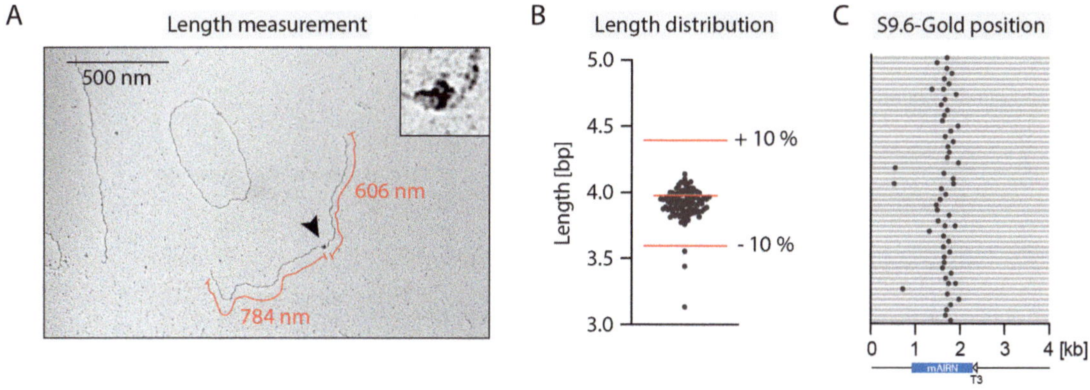

Fig. 5 Analysis of the Spike-in DNA to determine S9.6-Gold binding specificity. (**a**) Representative EM image of a Spike-in fragment, including the contour measurements (red) from the fragment ends to the S9.6-Gold label. (**b**) Length distribution of all analyzed Spike-in fragments. Even though the fragment lengths can vary depending on the spreading, the general spread should stay within a 10% deviation. (**c**) S9.6-Gold positions within the Spike-in displayed across multiple fragments. A majority of the S9.6-Gold labels is detected specifically within the R-loop-prone region of the Spike-in plasmid (compare Fig. 2b)

350–800 nm away from the closest fragment end, which is also where specific S9.6-Gold binding is expected. Whether an S9.6-Gold label is located within the mAIRN gene is determined by measuring the shortest distance to a fragment end. In conditions of optimal S9.6-Gold binding, >85% of the labels should be within this AIRN region (Fig. 5c). No detectable R-loop should remain without an S9.6-Gold label (*see* **Note 33**). Upon RNase H digestion, all S9.6-Gold labels should be lost, and the Spike-in fragment should appear homogeneously double stranded.

3. Depending on which gel fraction was extracted, the bacterial conflict fragment is either linear or replicating. In the latter case, these DNA molecules show the classical structure consisting of yet-to-be replicated parental strands, two already replicated daughter strands and potentially a fourth regressed arm (Fig. 6a) (*see* **Note 34**). To distinguish parental from daughter strands, measure the length of all strands; daughter strands should be equally long (± 10%). The total fragment length of these replication intermediates equals the sum of parental and daughter strand (Fig. 6a). To identify the population of bacterial FOI, plot all imaged replication intermediates by total fragment length (Fig. 6b); a distinct population of bacterial FOI around the expected size range is clearly detectable. However, this is only technically possible upon conditions of enrichment for RIs from the FOI, while it may prove difficult or impossible if forks proceed unperturbed on the FOI, or if the size of the genome is too large (Figs. 3b and 6b). Replication intermediates which fit the expected size range can then be

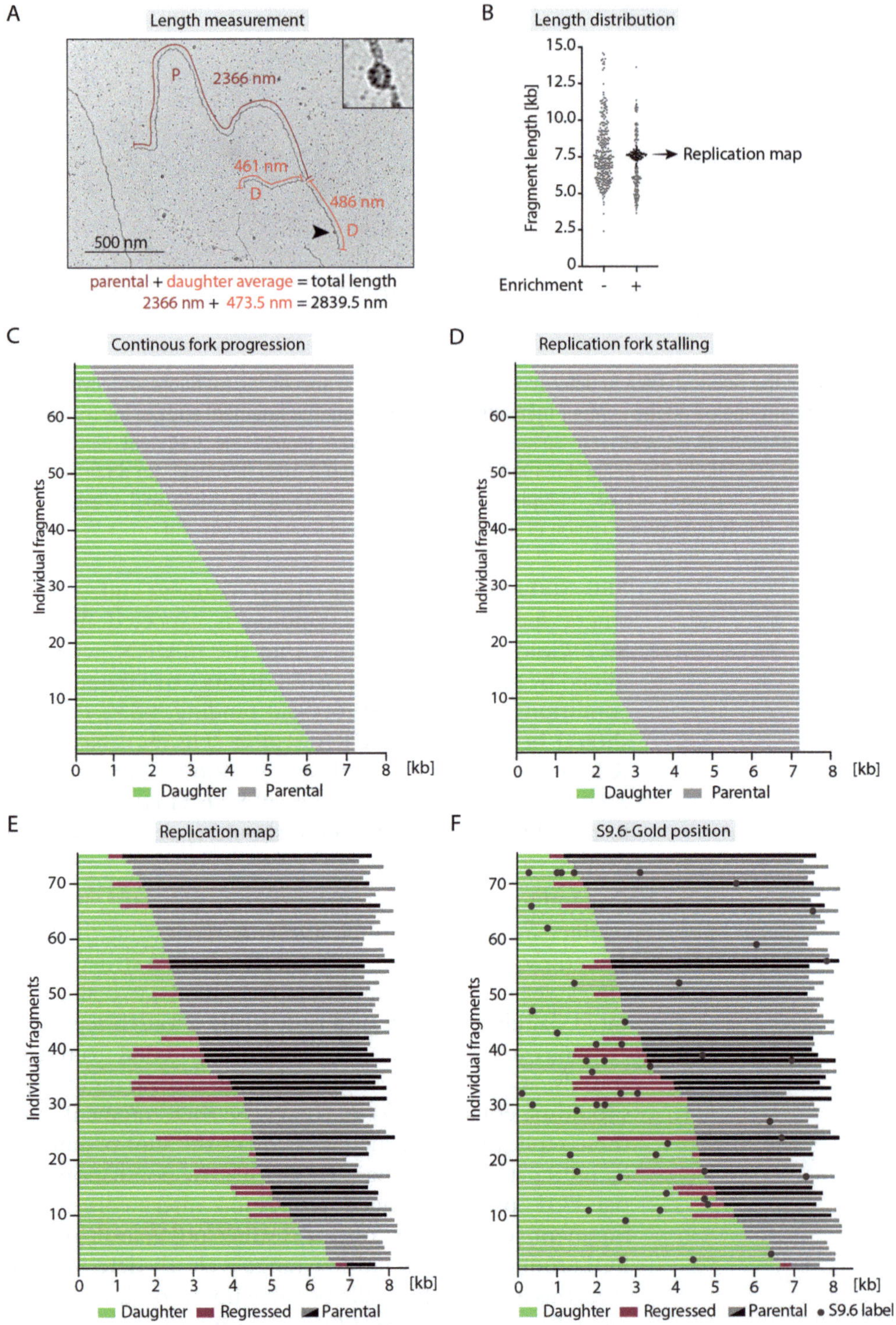

Fig. 6 Analysis of the replicating FOI from *B. subtilis*. (**a**) Representative EM image of a replication intermediate originating from the bacterial FOI, including the contour measurements (red) to distinguish parental and daughter strands and to calculate the total fragment length. (**b**) Length distribution of all imaged replication intermediates upon control condition (− enrichment), where replication forks are not sufficiently stalled or slowed throughout the FOI, and upon condition of fork interference (+ enrichment), where replication forks are

sorted into replication maps (*see* **Note 35**) to extract information on replication fork progression and stalling from single replication intermediates (Fig. 6c); a prevalent stalling event would result in the accumulation of replication intermediates with a defined daughter strand length (Fig. 6d). Reversed forks can optionally be indicated within these maps and sorted depending on the location of the junction prior to reversal, by subtracting the regressed arm from the parental strand; marking the length of the regressed arm in a separate color, allows to determine not only the site but also the extent of fork backtracking associated with reversal (Fig. 6e).

4. To determine where S9.6-Gold binds within these replication intermediates, note the strand (daughter, parental or regressed) to which the S9.6-Gold conjugate bound as well as the distance to the fragment end. These labels can then be marked within the replication maps to provide a visual representation of S9.6-Gold binding within replicating DNA (Fig. 6f).

4 Notes

1. Never reuse already melted and resolidified aliquots. Melted agarose should always be cooled down to 42 °C before adding it to the cells.
2. Even though lysozyme has an optimal activity at pH 6.2, it is active over a wide range of pH (6.0–9.0). We recommend using it at pH 7.0.
3. The combination of restriction enzymes should be chosen wisely to digest the entire genome into fragments smaller than the FOI. In silico predictions (*see* for example https://www.molbiotools.com/restrictionanalyzer.php) can assist the design of this digestion cocktail, but the choice of the right computational tool largely depends on the size of the genome to be digested. In case a reasonable combination of enzymes is

Fig. 6 (continued) stalled. A distinct accumulation of bacterial FOI is only detectable in conditions of enrichment. Replication intermediates enriched for the FOI (indicated in black) can then be sorted into replication maps. (**c**) Exemplary replication map of continuous, unperturbed fork progression through an FOI. Green: daughter strand; grey: parental strand (**d**) Exemplary replication map of replication fork stalling within an FOI. Green: daughter strand; grey: parental strand (**e**) Replicating FOI in (**b**) are sorted into a replication map. Reversed forks are sorted according to the original site of interference; the degree of reversal (= length of the regressed arm) is indicated in magenta. (**f**) S9.6-Gold labels can be mapped within these replication maps to correlate DNA–RNA hybrid formation and replication fork progression and architecture within the FOI

not available to enable the following steps of isolation, consider engineering the FOI to remove cutting sites.

4. The pFC53 plasmid carries the R-loop-prone promoter region of the mouse AIRN gene under the control of the T3 promotor. It can in principle be replaced by any plasmid that carries an R-loop-prone region under a promoter compatible with in vitro transcription (i.e., T3, T7, or SP6 promoter). The size of this plasmid should, however, be sufficiently different from the size of the genomic locus of interest to enable clear distinction between the genomic locus and the Spike-in plasmid, based on length measurements on electron micrographs (*see* Subheading 3.8).
5. If cell pellets between samples differ strongly in weight, resuspend in 100 μL LB/0.02% sodium azide. Calculate how much of the volume to use for each sample to get equal cell pellet weights, centrifuge those volumes again to gather the pellet. Once pellets are equal in weight, proceed with **step 8**.
6. When processing multiple samples, perform **steps 12–14** one sample at a time to avoid samples solidifying prematurely.
7. This incubation can be extended to o/n if necessary.
8. The agarose plug tends to stick to the tissue tightly, so make sure not to touch the plug with the tissue or use a spatula to secure the plug.
9. This step is important to replace the storage buffer in the plug entirely with the CutSmart buffer. Prewarming the plug prior to digestion will further improve digestion efficiency.
10. When digesting multiple plugs simultaneously, prepare a separate digestion mix for each plug to ensure that every plug gets the same volume of restriction enzymes.
11. In our experience, this additional digestion step is crucial to achieve a near-complete digestion of the genomic DNA.
12. Equilibrating the temperature to 45 °C will ensure that the plug slice will stay intact when pouring the gel.
13. Drying the plug on the comb will prevent it from floating off when pouring the gel. In addition, it can help to "glue" the plug onto the comb using 10 μL of the prepared LMP agarose.
14. It is advisable to pour the LMP agarose at a constant speed from the bottom edge of the gel to avoid that the plug slices float off the comb.
15. Work quickly to minimize the UV exposure.
16. This step should be done in Eppendorf tubes—if the volume of the gel slice exceeds 400 mg, dissolve it in a falcon tube and split into multiple Eppendorf tubes once the gel slice is completely dissolved.

17. For gel slices that exceeded 400 mg, the eluates from the different Eppendorf tubes are combined at this step.
18. The overall yield from the replicating fraction is generally very low and not detectable by NanoDrop. However, the replicating fraction is distributed in a very large gel piece (*see* Fig. 3a), which is split up in multiple extraction reaction (*see* Subheading 3.3, **step 12**)—the resulting eluate volume is usually between 40 and 60 μL. In this case, we recommend concentrating the sample to roughly 10 μL and use half of this for RNase H digestion (Subheading 3.5) and the other half directly for immunolabeling (Subheading 3.6). For the linear, nonreplicating fraction this step is usually not required since the volume is already roughly 10 μL.
19. By irradiating the samples "as drops"—rather than keeping them in Eppendorf tubes—we can ensure full and homogeneous sample irradiation. In our experience, irradiation of samples in Eppendorf tubes is less efficient than this drop-method.
20. The pellet can be instable and barely visible.
21. To help resuspension we recommend placing the samples for 30 min at 37 °C; gently shaking can help as well.
22. To confirm successful R-loop formation, it is advisable to run a gel shift assay or an S9.6 dot blot, comparing nontranscribed and transcribed samples [14].
23. Linearization of the plasmid DNA is necessary to allow for mapping the position of R-loop binding by S9.6-Gold, which will be crucial to determine S9.6-Gold specificity (*see* Subheading 3.8). In the case of the pFC53 plasmid, XmnI digestion positions the R-loop site asymmetrically on the plasmid (Fig. 2b). When using a different plasmid, it is strictly necessary to choose a unique restriction enzyme that achieves this asymmetric positioning as well. When analyzing the bulk *B. subtilis* fraction, which is roughly 4–5 kb in size (Fig. 3a), the digestion of pFC53 has to be changed, to enable proper distinction of the Spike-in from genomic DNA; we recommend to digest pFC53 with DraI and BamHI to generate a 2.4 kb pFC53 fragment, carrying the R-loop asymmetrically.
24. The agarose percentage of the LMP gel should be adjusted depending on the plasmid size that is used to ensure optimal separation of the linear from the circular plasmid.
25. While RNase H digestion converts R-loop structures into duplex DNA, it converts DNA–RNA hybrids into ssDNA gaps. Excessive ssDNA accumulation can, however, interfere with DNA binding to the carbon-coated grids. It is hence important to keep in mind that RNase H digestion of material

enriched in DNA–RNA hybrids can prevent efficient EM visualization of these replication intermediates [14].

26. In our experience, 70 ng of DNA is the minimum amount required for efficient S9.6-Gold labeling and DNA spreading. Attempting this procedure with less DNA will strongly affect the DNA density on the EM grid and the overall S9.6-Gold binding.
27. If the volume of the Spike-In and conflict DNA exceeds 9 μL, carefully concentrate using the Eppendorf concentrator.
28. When processing multiple samples, it is recommended to do labeling and spreading steps one sample at a time. When more experienced, samples can be staggered in 15 min intervals.
29. The timing of **steps 2** and **3** should be tightly coordinated with the preparation for DNA spreading in **steps 4–11** to ensure that neither the cross-linking reaction nor the grid incubation with EtBr (**step 5**) is prolonged beyond the indicated times.
30. Visual illustration for the key steps of the DNA spreading as well as further advice on troubleshooting is included in [16, 17].
31. While smaller hybrids are usually detected as rather small S9.6-Gold labels, longer hybrid structures can be bound by multiple S9.6-Gold particles or, alternatively, bound multiple times by the same S9.6-Gold particle (one gold particle can be conjugated to multiple S9.6-antibodies), generating larger aggregates. In addition, the presence of a displaced strand (i.e. R-loop), which then collapses around the S9.6-Gold particle, can further enlarge these platinum-coated structures.
32. During DNA spreading, DNA stretching can vary slightly, resulting in up to 10% deviation from the calculated size.
33. If R-loops remain unlabeled, increase the antibody concentration. If the S9.6-Gold conjugate binds outside the expected area, either decrease the antibody concentration or—if the DNA content on the grids is low—increase the Spike-in amount used for labeling.
34. Detailed instructions on how to identify these four-way junctions (i.e. reversed forks) are provided in [16].
35. We recommend determining the average of the two daughter strands and then sort the replication intermediates from smallest to largest daughter strand average.

References

1. Follonier C, Oehler J, Herrador R, Lopes M (2013) Friedreich's ataxia-associated GAA repeats induce replication-fork reversal and unusual molecular junctions. Nat Struct Mol Biol 20(4):486–494

2. Follonier C, Lopes M (2014) Combined bidimensional electrophoresis and electron microscopy to study specific plasmid DNA replication intermediates in human cells. Methods Mol Biol 1094:209–219
3. Drosopoulos WC, Vierra DA, Kenworthy CA, Coleman RA, Schildkraut CL (2020) Dynamic assembly and disassembly of the human DNA polymerase δ holoenzyme on the genome in vivo. Cell Rep 30(5):1329–1341
4. Madireddy A, Kosiyatrakul ST, Boisvert RA, Herrera-Moyano E, García-Rubio ML, Gerhardt J et al (2016) FANCD2 facilitates replication through common fragile sites. Mol Cell 64(2):388–404
5. Norio P, Schildkraut CL (2001) Visualization of DNA replication on individual Epstein-Barr virus episomes. Science 294(5550): 2361–2364
6. Ishimoto R, Tsuzuki Y, Matsumura T, Kurashige S, Enokitani K, Narimatsu K et al (2021) SLX4–XPF mediates DNA damage responses to replication stress induced by DNA–protein interactions. J Cell Biol 220(1): e202003148
7. Schmidtmann E, Anton T, Rombaut P, Herzog F, Leonhardt H (2016) Determination of local chromatin composition by CasID. Nucleus 7(5):476–484
8. Agelopoulos M, McKay DJ, Mann RS (2012) Developmental regulation of chromatin conformation by Hox proteins in Drosophila. Cell Rep 1(4):350–359
9. Pereira B, Chin SF, Rueda OM, Vollan HK, Provenzano E, Bardwell HA et al (2016) The somatic mutation profiles of 2,433 breast cancers refines their genomic and transcriptomic landscapes. Nat Commun 7:11479
10. Viollier PH, Thanbichler M, McGrath PT, West L, Meewan M, McAdams HH et al (2004) Rapid and sequential movement of individual chromosomal loci to specific subcellular locations during bacterial DNA replication. Proc Natl Acad Sci 101(25):9257–9262
11. Willis NA, Chandramouly G, Huang B, Kwok A, Follonier C, Deng C et al (2014) BRCA1 controls homologous recombination at Tus/Ter-stalled mammalian replication forks. Nature 510(7506):556–559
12. Doi G, Okada S, Yasukawa T, Sugiyama Y, Bala S, Miyazaki S et al (2021) Catalytically inactive Cas9 impairs DNA replication fork progression to induce focal genomic instability. Nucleic Acids Res 49(2):954–968
13. Lang KS, Hall AN, Merrikh CN, Ragheb M, Tabakh H, Pollock AJ et al (2017) Replication-transcription conflicts generate R-loops that orchestrate bacterial stress survival and pathogenesis. Cell 170(4):787–799 e18
14. Stoy H, Zwicky K, Lang K, Krietsch J, Kuster D, Schmid J et al Direct visualization of transcription-replication conflicts reveals post-replicative DNA–RNA hybrids. Under revision after peer review
15. Ginno PA, Lott PL, Christensen HC, Korf I, Chédin F (2012) R-loop formation is a distinctive characteristic of unmethylated human CpG island promoters. Mol Cell 45(6):814–825
16. Zellweger R, Lopes M (2018) Dynamic architecture of eukaryotic DNA replication forks in vivo, visualized by electron microscopy. Methods Mol Biol 1672:261–294
17. Neelsen KJ, Chaudhuri AR, Follonier C, Herrador R, Lopes M (2014) Visualization and interpretation of eukaryotic DNA replication intermediates in vivo by electron microscopy. In: Functional analysis of DNA and chromatin. Springer, pp 177–208

Chapter 7

RNase H1, the Gold Standard for R-Loop Detection

Susana M. Cerritelli, Kiran Sakhuja, and Robert J. Crouch

Abstract

RNase H1 has become an essential tool to uncover the physiological and pathological roles of R-loops, three-stranded structures consisting of and RNA–DNA hybrid opposite to a single DNA strand (ssDNA). RNase H1 degrades the RNA portion of the R-loops returning the two DNA strands to double-stranded form (dsDNA). Overexpression of RNase H1 in different systems has helped to address the questions of where R-loops are located, their abundance, and mechanisms of formation, stability, and degradation. In this chapter we review multiple studies that used RNase H1 as an instrument to investigate R-loops multiple functions and their relevance in health and diseases.

Key words RNase H1, R-loops, RNA–DNA hybrids, Transcription–replication conflicts, DNA damage, Replicative stress

1 Introduction

R-loops form when the nascent RNA emerging from the RNA polymerase (RNAP) during transcription hybridizes to the template DNA strand, leaving the nontemplate strand as ssDNA. In the last 10 years, R-loops have gained prominence as important regulators of cellular processes, participating in functions as varied as initiation of DNA replication in bacterial ColE1-type plasmids [1, 2], immunoglobulin class switch recombination (CSR) [3], completion of mitochondrial DNA (mtDNA) replication [4], facilitating transcription termination [5] and initiation of the repair of DNA double-strand breaks (DSB) [6, 7]. However, unscheduled R-loops can be a source of genome instability, which are normally prevented by engaging the RNA as soon as it surfaces from the RNAP with ribosomes in bacteria and with proteins related with splicing, RNA transport and RNA modification in eukaryotes. Defects or delays in these processes may facilitate R-loop formation resulting in DNA damage [8, 9]. The discussion about conflicts between DNA replication and transcription began in 1995 [10], but at that time there was little knowledge about R-loops.

Andrés Aguilera and Alexey Ruzov (eds.), *R-Loops: Methods and Protocols*, Methods in Molecular Biology, vol. 2528, https://doi.org/10.1007/978-1-0716-2477-7_7,

Although a clear association between R-loops and genomic stress has been difficult to establish, it is likely that R-loops can act as replication fork barriers, creating transcription–replication collisions (TRCs) [11–13] and inducing replication stress (RS) by arresting DNA synthesis [14, 15], which could generate aberrant processing and the formation of DSBs [16, 17]. DNA breaks could also occur in the ssDNA component of the R-loops, exposing them to reactive oxygen species and ssDNA nucleases and susceptible to DNA damage [18]. Likewise, R-loops might cause transcription stress by blocking the progress of the RNAP, in a similar way as they impede replication. Transcription-coupled nucleotide excision repair (TC-NER), which normally removes bulky lesions in DNA that prevent transcription might cleave at the R-loops producing DNA breaks and damage [17]. The best support for a role of R-loops in inducing DNA damage comes from studies that utilize RNase H1, an enzyme that processes R-loops by cleaving at the RNA strand of the RNA–DNA hybrid [19]. In vivo overexpression of RNase H1 in different systems from *E. coli* to humans, partially suppresses a variety of genome instability phenotypes [20–22], which helped establishing a relation between R-loops and DNA damage.

There is a second RNase H in cells, RNase H2, which also cleaves at the RNA moiety of the R-loop, in addition to initiating the removal of ribonucleotides in DNA [19, 23]. RNase H2 is a heterotrimeric enzyme, making it less amenable to manipulation than the monomeric RNase H1, and consequently most of the studies using RNase H as a reporter for R-loops utilize RNase H1. However, the presence of RNase H2 in cells and its processing activity could affect the conclusions reached in some systems.

It is obvious that in diploid cells the number of R-loops at any one locus is limited to the DNA copy number. In nondividing cells there are only two copies of any locus. In contrast, cycling cells after DNA replication have twice as many DNA loci as cells in G1. In theory, R-loops can increase from none to no more than two or four. Most studies we describe later are performed on a mixture of cells at various stages of the cell cycle. An increase of loci-specific R-loops can be due to an increment in one cell or to more cells in the population with R-loops. Typically, when a phenotype is suspected of being related to R-loops, overexpression of RNase H1 is used to establish the association, although in most cases only partial correction of the defect under investigation is obtained. The lack of complete phenotypical reversion might indicate that the R-loops are not normally resolved by RNase H1 but by the myriad of other means of resolving R-loops. Two examples in which R-loops are not normal substrate for RNase H1 or RNase H2 are the ones formed during immunoglobulin isotype switching via CSR and those formed at G-rich pause sites for 3′ mRNA polyA addition. Lieber's group demonstrated R-loop formation during CSR by treating purified genomic DNA isolated from cells undergoing isotype switching with *E. coli* RNase HI and by the susceptibility of cytosine residues to bisulfite modification in the displaced strand [3]. Subsequent studies have shown that other enzymes, not RNase H1, are responsible for CSR [24]. Interestingly, when we overexpressed RNase H1 in mouse B cells, removal of the RNA of the hybrid permitted greater access of the transcribed strand to modification by the AID enzyme without affecting the frequency of CSR [25]. R-loops are critical for Senataxin-mediated transcription termination [5]. Coupling between Senataxin R-loop removal and the exonuclease Xrn2 allows for proper 3′-end formation. Overexpression of RNase H1 can remove the hybrid prematurely leading to poor termination and readthrough [5], as described below.

In this chapter, we summarize different studies that have used expression of RNase H1 from different organisms to uncover R-loop formation, accumulation and degradation and their role in physiological and pathological functions. This is a very active field; thus, we apologize if we omitted any studies that cover RNase H1 overexpression to explore R-loop functions.

2 Using RNase H1 as a Reporter for the Detection of R-Loops

One of the best tools for using RNase H1 as a reporter for the detection of R-loops was developed in our lab [26] to determine the subcellular localization of the RNase H1 enzyme in mammals. We fused the complete cDNA sequence of the mouse *Rnaseh1* gene to the green fluorescent protein (GFP) gene, cloned in the plasmid pEGFP-N2, and transfected African green monkey kidney-derived Cos-1 cells or HeLa cells with this construct. By confocal microscopy, we were able to determine that RNase H1 localizes to both the nucleus and the mitochondria (Fig. 1a). By fusing different regions of the *Rnaseh1* gene to the GFP sequence we concluded that the first 26 amino acids of the RNase H1 protein contain the mitochondrial targeting signal (MTS) (Fig. 1b) [26]. These studies suggested the presence of R-loops or stable RNA–DNA hybrids in mitochondria, removal of which are required for embryonic development, as mice depleted of the *Rnaseh1* gene die in utero [26]. Subsequent work demonstrated an essential role for RNase H1 in completing mtDNA replication [4].

The transcript from the *RNASEH1* gene contains the information for the mitochondrial and nuclear forms of RNase H1. As shown in the HEK393 cells [27], translation of the mRNA occurs by a complex leaky-scanning mechanism producing a much greater amount of the nuclear protein. We found that overproduction of the mitochondrial enzyme is detrimental for cell viability, indicating the importance of using only the protein starting at the M27 codon (nuclear form) for studies of R-loops.

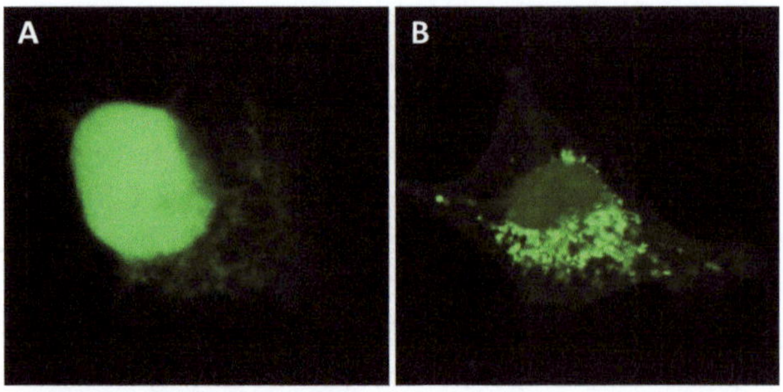

Mitochondrial and nuclear forms of RNase H1-GFP **RNase H1 MTS-GFP**

Fig. 1 Representative images from confocal microscopy of RNase H1-GFP fusion proteins. (**a**) Plasmid containing the full-length human *Rnaseh1* cDNA sequence fused to the GFP gene expresses the mitochondrial and the nuclear isoform of RNase H1, with the nuclear form being in excess. (**b**) Plasmid with the mitochondrial targeting sequence (MTS) of the *Rnaseh1* gene fused to the GFP expresses a protein that localize to the mitochondria. GFP fusion plasmids were transfected into Human Osteosarcoma cells using Nucleofection and plated and incubated for 24 h in a 6 well plate with a coverslip. Confocal Microscopy images were taken from fixed cells on the coverslips

We did not imagine the extensive use that the RNase H1-GFP fusion construct was going to enjoy and the many remarkable findings it was going to uncover, but soon it became clear that it had the potential for many other purposes, the most important being the gold standard for detecting R-loops. GFP-tagged proteins have many advantages. They can be used for confocal microscopy detection without the need of additional antibody treatments and immunodetection. Flow cell cytometry can purify and quantify GFP positive cells and quantitative image-based cytometry (QIBC) allows for quantitative measurement of GFP intensity for each individual cell. Different iterations of the original RNase H1-GFP transiently expressed construct have made these tools more useful, such as conditional expression of RNase H1, stable expression, and deletions and mutations of the RNase H1 sequence. Other constructs that fused RNase H1 to different tags have been successfully used in R-loop detection, immunoprecipitation studies, and RNA–DNA sequencing (Fig. 2). Below, we explore some of them, considering the processes they helped investigate.

3 Defects of Transcription and mRNA Processing Factors Lead to R-Loop Accumulation and DNA Damage

The first system to show that overexpression of RNase HI prevents the deleterious effect of R-loops was established in *E. coli* [20]. A multicopy plasmid that increases the levels of RNase HI about

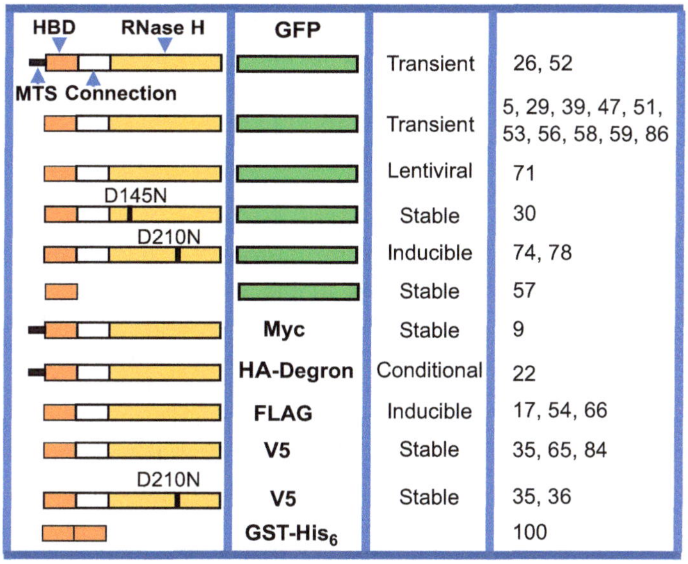

Fig. 2 Schematic representation of the RNase H1 constructs used as reporters for the detection of R-loops. The different regions of RNase H1 are color-coded, representing the mitochondrial targeting signal (MTS) as a black line, the hybrid-binding domain (HBD) as an orange rectangle, the connection domain as a white rectangle and the RNase H domain as a yellow rectangle. In some cases, inactive RNase H with catalytic site mutations D145N or D210N were utilized. The different forms of the enzyme were fused to unique tags and expressed in mammalian cells transiently from a plasmid or from a lentiviral vector or stably, which in turns could be inducible or conditionally expressed. In the right panel are presented the references to the manuscripts that utilized each construct. Not all the constructs are represented here for lack of space and because not all the fusion proteins were described in sufficient detail in some manuscripts

tenfold, partially complemented the growth defects of the *topA* mutants, indicating that the slow growth observed in the absence of TOPO I was due to deleterious R-loops, and that a central role of TOPO I is preventing transcription-associated R-loops. To confirm this, a plasmid was used that expressed the *rnhA* gene under the control of the IPTG inducible *tac* promoter. In this way, the growth rate of the *topA* mutant strain could be regulated by adjusting the IPTG concentration. This pioneer work paved the way to multiple studies in different systems relating R-loops to DNA damage.

Aguilera's group has made remarkable contributions to the understanding of transcription-associated recombination, initially by using a plasmid that overexpressed yeast RNase H1 under the control of the GAL promoter [21], which was originally developed in the Crouch lab. Mutants of the *S. cerevisiae* THO/TREX complex, which binds the mRNA as it emerges from the RNAPII and thus prevents its association with the transcribed DNA strand, showed hyperrecombination phenotypes that can be partially reversed by growing cells containing the *GAL-RNH1* gene in galactose media. They developed a model that connects recombination, transcription, and mRNA metabolism through R-loops accumulation [21].

Subsequent studies, also using overexpression of RNase H1, expanded on the findings that mRNA-processing factors play important roles in preventing DNA damage by suppressing R-loop formation. Li and Manley [9] described SR proteins, involved in pre-mRNA splicing, as required to prevent R-loop-mediated genomic instability. Defects of SR proteins induced hypermutations, DSB, and DNA rearrangements. Stable overexpression of Myc-tagged human RNase H1 suppressed DNA fragmentation, hypermutation and cell-cycle arrest in cells depleted of ASF/SF2, a member of the SR family of pre-mRNA splicing factors [9], leading to the conclusion that R-loops were the culprits creating these defects. Paulsen et al. [22] used a similar approach to demonstrate that a prevalent role of mRNA-processing factors is preventing DNA damage caused by R-loops. They made a construct for stable expression of an RNase H1-HA-tagged enzyme in HeLa cells that also had a FKBP L106P degron to obtain conditional protein removal via the addition of the small molecule Shld1 [28]. Paulsen et al. [22] found that cells depleted of mRNA processing factors have a decrease of the phosphorylation of H2AX, an early mark of DNA damage, when RNase H1 was overexpressed. The Cimprich lab expanded on these findings by using a FLAG-tagged RNase H1 stably expressed and under the control of the TetO promoter in HeLa cells [17]. They showed that depletion of several mRNA processing factors or inhibition of Topoisomerase I (Top I) induced R-loop accumulation and DNA damage which can be reversed by RNase H1 overexpression. The formation of DSB in cells depleted of the helicases Aquarius and Senataxin, the splicing factor ASF/SF2 or the inhibition of Top I could be prevented by expressing tetracycline-inducible RNase H1 or by knocking down XPG, a member of the TC-NER factors [17]. They suggested a novel role for the TC-NER in removing R-loops that cause transcription stress, in addition to their canonical function of eliminating bulky lesions in DNA that block transcription. According to this model, TC-NER would mediate R-loop-induced genome instability. Previously, it was shown that inhibition of transcription by camptothecin (CPT) favors R-loop formation and DSB in post-mitotic primary neurons and noncycling HeLa cells [29], suggesting that the DNA damage was independent from DNA replication. Expression of an RNase H1-GFP tagged construct decreased the induction of phosphorylated H2AX, a marker of DSB.

Tresini et al. [30] showed that when DNA lesions block transcription, RNAPII pausing induces displacement of assembled cotranscriptional spliceosomes and the formation of R-loops, resulting in alternative splicing. R-loops signal the activation of the ATM kinase to initiate the DNA Damage Response (DDR). When transcribed genes are damaged, the ATM response does not seem to require DSB formation, as it is the case of ATM canonical role of stimulating the DNA repair of DSB [31]. Catalytically inactive RNase H1(D145N)-GFP fusion protein stably overexpressed in human dermal fibroblast localized to UV-irradiated actively transcribed genes, suggesting R-loop formation after DNA damage. ATM activation and spliceosome mobilization followed, but overexpression of wild-type RNase H1 prevented the ATM response and subsequent spliceosome reorganization [30]. Noncanonical ATM activation after DNA damage in transcribed genes seems to affect alternative splicing and the sequence of mature transcripts, suggesting the intriguing possibility that in response to DNA damage there is alteration of mRNA maturation to favor the formation of proteins involved in DDR.

Mutations in multiple splicing factors are associated with myelodysplastic syndromes (MDSs), which give a predisposition to develop acute myeloid leukemia (AML), other cancers of hematopoietic nature and of other origins [32–34]. Chen et al. [35] tested the role of R-loops on the defects associated with mutated splicing factors using a novel method called RChIP [36] (*see* Subheading 9). They found that mutations in *SRSF2* and *U2F1* splicing factors accumulate R-loops which preceded the DNA damage observed when cells entered S-phase [35]. By stably expressing RNase H1 with a V5 tag, HEK293T cells with mutations in *SRSF2* and *U2F1* showed a partial rescue of the cell growth defects and a decrease in γ-H2AX foci [35]. Using an MDS-L cellular model, they replicate the effect of RNase H1 overexpression in HEK293T cells containing mutant splicing factors. A knockin mouse heterozygous for the splicing factor *Srsf2*(P95H) mutation, which recapitulate the MDS phenotype [37, 38], have impaired proliferation of early hematopoietic progenitor cells that can be partially reversed by infecting the cells with a retrovirus expressing V5-tagged *RNASEH1*. These studies suggest a role for R-loops in MDS induced by splicing factor mutations [35], and together with other works described in this section indicate a direct relation between splicing defects and R-loop formation.

4 R-Loops as Regulators of Transcription

Transcription termination in some genes occurs when RNAPII pauses at a G-rich region preceded by a polyA addition site. A construct that overexpresses GFP-tagged mouse RNase H1 in HeLa cells, provided evidence that R-loops are formed at G-rich transcription termination regions facilitated by RNAPII pausing.

Senataxin, a helicase that unwinds RNA–DNA hybrids, enables RNAPII release and allows trimming of the 3′-end of the transcript for polyA addition [5]. Senataxin depletion induced R-loop accumulation at the G-rich 3′-end of genes preventing efficient termination. The level of increase overexpression of RNase H1 was inversely correlated with termination efficiency and directly correlated with transcription readthrough, suggesting that R-loops are required for the transcription termination of these genes, and that Senataxin is involved in their resolution. At G-rich RNAPII pausing termination regions, but not at genes that use different termination methods, R-loops induce antisense transcription, which generates dsRNA triggering the RNA interference (RNAi) response [39]. Overexpression of GFP-tagged RNase H1 reduced R-loops, antisense RNA and DICER, a member of the RNAi machinery, occupancy [39]. RNAi facilitates gene silencing across the genome by recruiting the G9a histone lysine methyltransferase and promoting an H3K9me2 repressive mark and heterochromatin formation [40]. Overexpression of RNase H1-GFP demonstrated that both R-loops and the H3K9me2 mark are important for transcription termination. They concluded that a group of human genes, characterized by possessing G-rich 3′-end regions, share a termination mechanism initiated by R-loop formation around the RNAPII pause region, followed by heterochromatin establishment through the synthesis of antisense RNA, formation of dsRNA and RNAi processing which leads to repressive H3K9me2 mark and strong termination [39].

In mouse embryonic stem cells, Polycomb group (PcG) complexes repress transcription of developmentally regulated genes to maintain expression patterns established during cell commitment [41, 42]. PcG complexes are essential to silence CpG-rich promoter genes using a mechanism that requires transcription but only at low levels [43–46]. Skourti-Stathaki et al. [47] found that in a subset of PcG target sequences, R-loops formed and were important for transcription repression. Transiently overexpressed RNase H1-GFP construct reduced the R-loop signal over the PcG repressed genes leading to their activation. They showed that the repression at R-loop-positive PcG target signals is mediated by the recruitment of PcG complexes, uncovering a relation between R-loops and PcG regulation. However, overexpression of RNase H1 had the opposite effect at active genes, reducing their expression. This agrees with reports demonstrating an activator role of R-loops at CpG-rich active gene promoters [48–50], suggesting that R-loops might have distinct functions in different genomic contexts.

R-loops acting as activators of transcription were reported to induce antisense transcription at promoter regions to form long noncoding RNAs (lncRNAs) [51]. By mapping R-loops in a strand specific manner, they observed a concentration of R-loops at the transcription start sequence (TSS) of many genes, which correlated with nascent transcription in both directions around the promoter region as measured by Chromatin-bound RNA sequencing (RNA-seq). Transient overexpression of GFP-tagged RNase H1 selectively reduced antisense transcripts but has little effect in sense, protein-coding transcripts. They concluded that R-loops induce the synthesis of antisense lncRNA at many promoter regions by recruiting the transcription preinitiation complexes. Overexpression of RNase H1 reduced the accumulation of preinitiation complexes only at antisense transcription initiation. These data suggest a new role for R-loops as antisense promoters.

How an R-loop can activate transcription at certain loci and inhibit at others is a question that needs further investigation. Features such as the stability and length of R-loops, their interactions with downstream complexes and their accessibility to RNase H and other R-loop processing factors might be contributing factors that determine R-loop functioning as activators or repressors of transcription.

5 R-Loop Induced DNA Damage Due to Transcription–Replication Collisions

Multiple reports have established a correlation between R-loop accumulation and genome instability associated with cancer development resulting from upregulated transcription and proliferation. Kotsantis et al. [52] explored the effect of increased transcription on oncogene-induced RS, and found that oncogenes such as $HRAS^{V12}$ promote RNA synthesis and R-loop accumulation, leading to the slowing of the replication fork progression and subsequent RS and DNA damage, all the hallmarks of TRCs. Human RNase H1-GFP transiently overexpressed in mammalian cells harboring $HRAS^{V12}$ decreased R-loop accumulation and improved replication fork progression and the reduction of DSB, suggesting that the oncogene-induced RS is caused by R-loops [52]. Similarly, oncoprotein EWS-FLI1 expressed in Ewing sarcoma cells drives increased transcription and R-loop accumulation associated with RS, which can be reversed by transient transfection with RNase H1-GFP construct [53]. Estrogen hormonal (E2) stimulation induces increased transcription and the formation of R-loops at E2 dependent genes [54]. An ER-positive breast epithelial MCF7 cell line, upon E2 stimulation showed an increased accumulation of DSBs, which could be partially reversed by stable expression of a doxycycline-inducible Flag-tagged RNase H1, also demonstrating a correlation between R-loop formation and DNA damage [54]. Use of the Flag-tagged RNase H1 revealed that the orientation of the R-loops with respect to the replication fork controls the extent of the DNA damage induced by TRCs. R-loops are more damaging when replication fork and transcription collide head-on than when replication and transcription are codirectional [11]. Depletion of Top I in human cells induced slow replication fork progression due to TRCs [55]. These cells presented a slight increase in R-loop accumulation but only those associated with head-on TRCs showed DNA damage signals [56]. Transiently expressed RNase H1-GFP fusion protein reversed the slow replication fork progression of Top I depleted cells, suggesting an effect of RNase H1 on specific harmful R-loops that form during head-on TRCs and stalled forks [56].

BRCA2 can be found at the intersection of DNA damage created by R-loops blocking both transcription and replication. BRCA2-depleted cells accumulate R-loops in both replicating and nonreplicating cells [57]. BRCA2 is a member of the Fanconi Anemia (FA) pathway, involved in removing replication blocks to prevent genome instability and because it can bind to the branched structure formed by a transcription R-loop, it could also prevent genome instability resulting from R-loops. A construct consisting of the 52 amino acids of the RNA–DNA hybrid-binding domain (HBD) of RNase H1 fused to an enhanced GFP (EGFP) helped to demonstrate an interaction between BRCA2 and R-loops. HBD-EGFP binds but cannot process R-loops, rather it might make them more stable, providing a source of RS. They found an increase in DSB in cells overexpressing HBD-EGFP and depleted of BRCA2 [57]. Chromosomal DNA breaks are, in general, a breast cancer susceptibility factor and hallmark of BRCA2-defective cells. The association of DNA breaks and R-loops has lead others to suggest a possible connection between R-loops and cancer susceptibility. The authors expanded their search to include other FA pathway members and found that the increased DSB observed in cells depleted of FANCD2 or FANCA could be reduced by overexpressing human RNase H1-GFP fusion protein [58]. They concluded that R-loops are inducing DNA breaks in FA defective cells and that a role of the FA pathway might be protecting against R-loop-mediated TRC. Since the main role of FA pathway is removing interstrand crosslinks (ICLs) in DNA, an intriguing possibility is that the FA pathway might also eliminate ICLs in RNA–DNA hybrids. Schwab et al. [59] also found that accumulation of RNA–DNA hybrids is associated with FA deficiencies, and went further to determine that FANCM, a member of the FA pathway with dsDNA translocase and branch migration activities could unwind RNA–DNA hybrids structures returning the DNA to its dsDNA form thereby preventing R-loop induced RS. As before, overexpression of RNase H1-GFP prevented accumulation of R-loops in FA defective cells and reversed the slow progression of the replication fork and DNA breaks observed in these cells [59].

RNAPII pauses after making a short transcript at the promoter region, and requires a second phosphorylation event, regulated by BRD4, to engage in productive elongation [60]. Inhibition of BRD4 prevents transcription [61–64], and triggers RS by inducing the formation of R-loops at promoter regions with perturbed transcription elongation [65]. Loss or inhibition of BRD4 caused the usual phenotypes associated with TRCs, such as slowing of replication fork progression, DSB formation and phosphorylation of H2AX. These effects were connected to R-loop accumulation because, as in other situations, overexpression of wild-type RNase H1, in this case with a V5 tag, partially suppressed these phenotypes [65]. Similarly, HeLa cells treated with molecules that inhibit or degrade BRD4, showed increased DSB, DNA damage and apoptosis that could be partially eliminated by doxycycline-induced FLAG-tagged RNase H1 expression [66].

6 Increase TERRA R-Loops in Telomeric Disfunction

TERRA, lncRNAs important for telomere length regulation [67, 68], are predicted to form RNA–DNA hybrids with the telomere DNA template strand. ICF (Immunodeficiency Centromeric Instability and Facial anomalies) syndrome type I is a rare autosomal recessive disorder characterized by telomere shortening and increased levels of TERRA RNA [69, 70]. ICF cells showed DSB at chromosome ends, characteristic of TRC due to R-loops, resulting in telomere loss. Expression of RNase H1-GFP at eightfold higher than wild-type levels by using a lentiviral vector in ICF cells decreased TERRA associated RNA–DNA hybrids concomitantly with a significant reduction in DSB at the telomere regions [71]. These studies established a connection between an increase in TERRA RNA and accumulation of telomeric R-loops leading to DNA damage and telomeric dysfunction.

7 RPA as Sensor of R-Loops and Initiator of DNA Damage Response

Replication protein A (RPA) is a complex that binds ssDNA originating during RS or DNA damage and signals the initiation of DDR [72]. RPA is found associated with transcribed regions [73], suggesting the presence of R-loops and that RPA might bind to the ssDNA displaced by the RNA–DNA hybrid of the R-loop. To prove this, Nguyen et al. [74] expressed a GFP-tagged nuclear mutant RNase H1 in cells. A catalytically inactive form of RNase H1 (D210N) [75] fused to GFP colocalizes with RPA and S9.6 foci, created by the S9.6 antibody specific for RNA–DNA hybrids [76], suggesting that RNase H1 and RPA might bind simultaneously to R-loops [74]. In the course of determining the region of RNase H1 responsible for its interaction with RPA, they found that the HBD [77] is critical for RPA binding and identified specific residues within the HBD required for the interaction. RPA binding stimulates the RNase H1 activity toward RNA–DNA hybrids, but the binding and degradation of R-loops was reduced when mutants in the HBD that lack interaction with RPA were present. This indicates that RNase H1 needs to interact with RPA for optimal R-loop removal [74]. RPA is a general sensor of DNA damage, activating the DDR, and this work demonstrates that R-loop sensing and regulation of RNase H1 are part of the DDR initiated by RPA. Another lab [78] observed an increase in R-loop accumulation in cells that have exhausted the RPA pool due to RS. They used stable doxycycline-inducible wild-type and inactive RNase H1 (D210N) fused to GFP to detect RNA–DNA hybrid by automated high-content microscopy and observed an increase of R-loops in cells depleted of their RPA pool leading to RS. This work also suggests that RPA stimulates R-loop processing by RNase H1.

RPA association with R-loops is also important for a mitosis-specific noncanonical ATR pathway, which is independent of DNA replication and damage but requires R-loops at centromeric regions and Aurora A, the key mitotic regulator [79]. Accurate chromosomal segregation is mediated by RNAPII active transcription at centromeres during mitosis [80, 81], which could facilitate R-loop formation. The role of R-loops in this pathway might be providing its ssDNA portion as a nucleation platform for RPA filament formation and subsequent ATR activation. Overexpression of RNase H1 diminished ATR activation, while the overexpression of the catalytically inactive D210N dominant negative RNase H1 mutant enhanced ATR and RPA phosphorylation. Together, these results indicate that stable R-loops are needed for induction of the mitosis-specific RPA-ATR cascade, resulting in Aurora A activation and accurate chromosomal segregation.

8 R-Loops and Chromatin Remodeling

Polybromo 1 (*PBRM1*) encodes a subunit of the Polybromo associated BAF complex (PBAF), one of the Switch/sucrose nonfermentable (SWI/SNF) chromatin remodeling complexes [82]. Defects in PBRM1 are found frequently in cancer cells [83], and are associated with RS and R-loop accumulation. Cells depleted of PBRM1 are hypersensitive to inhibitors of PARP and ATR, which compromise the DDR [84]. V5-tagged RNase H1 stably overexpressed in *PBRM1* knockout cells partially restored viability in the presence of PARP inhibitors, suggesting that increased R-loops might be the cause of the DNA damage and genome instability observed in PBRM1-deficient cells [84].

SWI/SNF chromatin remodeler complexes share a BRG1 subunit that provides the main ATPase activity required for chromatin remodeling. This ATPase, as well as the other members of the SWI/SNF complexes, are among the most mutated genes in human malignancies [85], clearly establishing a relation between chromatin remodeling complexes and genome instability. Bayona-Feliu et al. [86] depleted the ATPase BRG1 in human cells and observed, as before with depletion of PBRM1 [84], an increase in R-loops and TRCs leading to DSB [86]. Transient transfection of RNase H1-GFP fusion construct reduced the phosphorylation of H2AX, a marker of DSBs, concomitantly with a reduction of RNA–DNA hybrids detected by S9.6. BRG1-depleted cells showed a slowing in replication fork progression due to R-loops creating TRCs, as evidenced by restoration of replication velocity by overexpression of RNase H1. These studies established a relation between R-loop-mediated conflicts and chromatin remodeling. SWI/SNF complexes likely modulate the chromatin structure to allow R-loop resolution and prevent DNA damage. Because defects of chromatin remodeling complexes are very common in cancer cells, their role in R-loop resolution suggests a link between R-loops and cancer propensity [86], as also observed in other systems.

9 Detection and Quantification of R-Loops by RNase H1

By immunocytochemistry, Ginno et al. [48] determined that an N-terminal HA-tagged RNase H1 colocalized with the S9.6 antibody, concluding that in vivo there is an abundance of R-loops in the genome that are bound and recognized by RNase H1. To identify these R-loops, Ginno et al. [48] developed two techniques for genome-wide R-loop profiling. DRIP (DNA–RNA Immunoprecipitation) uses the S9.6 antibody to specifically bind and isolate R-loops from purified cellular DNAs, which then can be identified and quantified by qPCR or sequencing. This technique has been improved and modified by numerous labs and adapted to many different cell types [14, 48, 58, 87–90]. The second method, called DRIVE (DNA–RNA In Vitro Enrichment) utilizes a catalytically deficient, but binding competent human RNase H1 mutant protein to precipitate R-loops, followed by qPCR or sequencing. Both techniques helped to determine R-loop accumulation at thousands of human promoters, specifically those that contained CpG islands.

Although the S9.6 antibody has been widely used to immunoprecipitate R-loops and characterize the presence and abundance of RNA–DNA hybrids in different systems and conditions [14, 48, 58, 87–90], it has some drawbacks. The S9.6 antibody also recognizes dsRNA [91–93], although with lower affinity. Removing RNA before S9.6 immunoprecipitation helped increase specificity

[92–94] but it does not entirely solve the problem. The use of catalytically inactive RNase H1 to pull down hybrids circumvents dsRNA interference but is a less sensitive method.

Another approach that captures R-loops in vivo, called R-loop Chromatin ImmunoPrecipitation (R-ChIP), involves the use of catalytically inactive tagged human RNase H1 expressed in cells to bind and immunoprecipitate RNA–DNA hybrids [36]. They made a quadruple RNase H1 mutant which has the commonly used D210N mutation plus three specific mutations in the HBD (W43A, K59A, K60A) that abolished binding to RNA–DNA hybrids [95, 96]. This quadruple mutant has no binding or catalytic activity and constitutes a good negative control. To the single (D210N) and quadruple mutant (WKKD) they added a nuclear localization signal (NLS) and a V5 tag at the C-terminus for immunoprecipitation. Cells expressing single or quadruple mutant were cross-linked and immunoprecipitated with an anti-V5 antibody. The DNA part of the RNA–DNA hybrid was converted to dsDNA, amplified and deep sequenced [36, 97]. R-ChIP using inactive RNase H1 confirmed an R-loop distribution relatively similar to those previously described by using other methods [48–50, 54], although with several significant differences. They found an abundance of R-loops at gene promoters with G/C skew and G-quadruplexes in the nontemplate DNA strand and at transcriptional pauses. In contrast to DRIP-seq techniques that showed a more even distribution of R-loop peaks across promoter, gene body and 3-terminal regions, R-ChIP peaks concentrated within promoter and intergenic enhancer regions. This work also proposed a mechanism for R-loop initiation and elongation, according to which the free RNA end emerging from the elongating RNAPII search for regions with nearby G/C-skewed sequence which would favor the formation of a stable RNA–DNA hybrid. This new model solves the topological issue posed by the fact that a hybrid propagating in the same direction as the elongating RNAPII will create helical torsion that would hinder the advance of the hybrid, altering its stability [36].

A recent alternative approach for mapping R-loops in vivo, called MapR [98], makes use of the catalytically inactive RNase H1 to bind RNA–DNA hybrids in fixed and permeabilized cells thereby eliminating the need for a tag for affinity purification. RNase H1 is fused to micrococcal nuclease for direct and efficient release of the hybrids after binding, a general technique known as "CUT & RUN" [99]. It has the advantage of capturing R-loops inside cells without the need to create stable cell lines expressing RNase H1, which can be difficult, time-consuming, and even impossible when using primary cell lines. It is also a gentle procedure that eliminates the harsh conditions used for extracting DNA that might affect stability and integrity of the RNA–DNA hybrid. R-loops mapped by this method closely resemble those described by the other in vivo method, R-ChIP [36], and differs significantly from the ex vivo methods using S9.6 antibody. To further explore these differences, Wang et al. [100] utilized the CUT & TAG technology with a similar approach to in vivo MapR but using, instead of inactive RNase H1, a construct that contains only two copies of the HBD attached to GST and His_6 tags. When using this fusion protein in permeabilized cells, they obtained RNA–DNA hybrid peaks similar to the ones from other in vivo methods, although the double HBD had higher affinity for hybrids and gave higher signal to noise ratios. However, when this construct was utilized ex vivo after DNA extraction and purification, peaks were more comparable to those obtained with the S9.6 antibody. They concluded that the discrepancies in R-loop mapping are due mostly to the procedure utilized for isolating and purifying the RNA–DNA hybrid. R-loops associated with promoter proximal pause regions may be short and unstable under harsh purification procedures but may be more prevalent and relevant in vivo than longer more stable but less abundant R-loops found along gene bodies. However, this method assumes that all RNA–DNA hybrids are captured by the added RNase H1.

10 Is Overexpression of RNase H1 a Valid Tool to Study R-Loops? Benefits and Limitations

From the studies outlined here using overexpression of RNase H1 and from other works using mostly the S9.6 antibody to detect R-loops, the existence of different classes of R-loops is emerging (Fig. 3) [101]. Some R-loops seem to be intrinsic parts of physiological processes and RNase H1 overexpression alters the outcome resulting in detrimental effects (Fig. 3a). R-loops that form at promoter-proximal RNAPII pause sites are short [102] and depending on the sequence context and associated factors could activate [49–51] or inhibit [47, 103] transcription. R-loops at the 3′-end of genes are associated with transcription termination and chromatin remodeling [5, 39]. Harmful R-loops arise from defects in cellular processes, particularly those related to mRNA biogenesis, and their unscheduled formation might lead to TRC and genomic instability (Fig. 3b) [9, 17, 22, 29, 30, 35, 52–54, 56–59, 65, 66, 104]. In these situations, overexpression of RNase H1 has a positive outcome by partially reversing R-loops adverse effects. It is quite remarkable that overexpression of RNase H1 affects many R-loops forming from disparate defects. How these defects avoid complete reversal of phenotypes remains unclear. Nevertheless, overexpression of RNase H1 is a powerful tool to study R-loops formation and resolution and the consequences of R-loop accumulation under pathological conditions. However, several

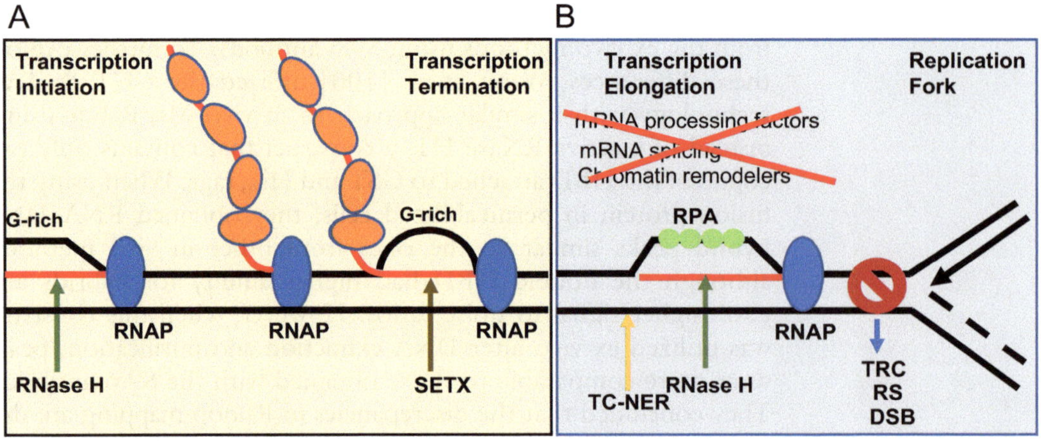

Fig. 3 R-loops arise from physiological and pathological conditions. (**a**) Physiological R-loops are enriched at both ends of transcribed genes. In mammalian active genes, R-loops form at G-rich and GC-skewed promoter-proximal regions associated to RNAPII pausing. R-loop formation at promoter regions prevents the methylation and silencing of genes, acting as transcription activators. These R-loops might be processed by RNase H1. R-loops forming at G-rich termination regions facilitate transcription termination after RNAPII pausing. These R-loops are unwound by Senataxin to promote poly(A) addition. (**b**) Harmful R-loops arise form transcription elongation deficiencies due to defects in mRNA processing factors, pre-mRNA splicing or chromatin remodelers and might lead to transcription replication conflicts (TRC), replicative stress (RS) and dsDNA breaks (DSB). The ssDNA binding protein RPA can bind the ssDNA portion of the R-loop activating RNase H1 and signaling the DNA damage response. In addition to RNase H1, helicases (not shown), the Fanconi Anemia pathway (not shown) and the transcription-coupled nucleotide excision repair (TC-NER) might process unscheduled R-loops. RNA polymerase (RNAP) is represented by a blue oval, mRNA binding complexes by orange ovals and RPA by green ovals. DNA is black and RNA is red. The red cross signifies defects in these processes. The transcription–replication conflicts are represented by a red prohibition sign

considerations should be made before reaching conclusions based solely on this method:

- Mammalian RNase H1 has a mitochondrial localization signal that directs the enzyme to mitochondria [26], where is essential for mtDNA replication [4]. Overexpression of the mitochondrial form of the enzyme is toxic for cells and a balance between the cellular concentrations of the mitochondrial and the nuclear forms is essential for cellular fitness [27]. The nuclear form of RNase H1 is more suitable for overexpression and should be used for complementation studies, although increased amount of this isoform could eliminate some native R-loops affecting genome-wide gene expression and indirectly modulating the phenotypes obtained.
- Overexpression of RNase H1 in most cases, partially suppresses or ameliorates genomic instability defects associated to R-loop

accumulation/stabilization. These defects are caused by depletion, inhibition or deletion of factors involved in mRNA biogenesis, which alter transcription initiation, elongation, or termination, mRNA splicing or packaging and transport. These alterations create an artificial increase of R-loops, which might not be processed by physiological levels of RNase H1. These R-loops might be natural substrates for RNase H1, but an increased expression of the enzyme could be required to have enough of it to process the hybrids excess. Alternatively, these R-loops might not be natural substrates for RNase H1, but an overload of the enzyme could force the artificial degradation of them. The fact that additional amounts of RNase H1 are needed to eliminate certain R-loops and to produce a phenotype might signal that under physiological conditions RNase H1 does not process these RNA–DNA hybrids, falsely associating RNase H1 with a cellular process.

- In most cases, the association between RNase H1 overexpression and the reduction of genome instability has not been demonstrated to be due to R-loop processing. What constitutes an "harmful" R-loop and how its elimination by RNase H1 resolves a defect has not been clearly stablished.
- The use of inactive RNase H1 and the HBD, constructs that bind but do not cleave RNA–DNA hybrids, have improved genome-wide methods to isolate and quantify R-loops in cells. These modified RNases H1 constitute a more sensitive and specific tool than the previously developed and extensively used S9.6 antibody, but not without some important drawbacks. The fact that RNase H1 does not binds some RNA–DNA hybrids identified by the S9.6 antibody might indicate that not all R-loops are accessible and/or substrates of RNase H1, always or in certain conditions.
- Overexpression of catalytically inactive RNase H1 may compete with the endogenous enzyme, preventing the processing of natural R-loops and affecting the observed phenotypes.

The general notion that accumulation of R-loops creates genome instability by inducing TRCs, RS and DSB remains elusive to probe. In most cases the spatial-temporal relation between R-loops and DNA damage has not been clearly established. New tools and more sensitive techniques, in addition to overexpression of RNase H1 need to be developed to improve our understanding of R-loop formation, distribution, and processing as well as what makes some R-loops "good" and physiologically relevant and what are the characteristic of "harmful" R-loops and why and how they induce genome instability. In the last 10 years our knowledge about R-loops has grown enormously, going from being considered an irrelevant accident resulting from transcription to become major regulators of many cellular processes. Expanding the use of RNase H1 as a reporter should help to further our understanding of the role of R-loops in health and diseases.

References

1. Itoh T, Tomizawa J (1980) Formation of an RNA primer for initiation of replication of ColE1 DNA by ribonuclease H. Proc Natl Acad Sci U S A 77(5):2450. https://doi.org/10.1073/pnas.77.5.2450
2. Masukata H, Tomizawa J-i (1990) A mechanism of formation of a persistent hybrid between elongating RNA and template DNA. Cell 62(2):331–338. https://doi.org/10.1016/0092-8674(90)90370-T
3. Yu K, Chedin F, Hsieh CL, Wilson TE, Lieber MR (2003) R-loops at immunoglobulin class switch regions in the chromosomes of stimulated B cells. Nat Immunol 4(5):442–451. https://doi.org/10.1038/ni919
4. Holmes JB, Akman G, Wood SR, Sakhuja K, Cerritelli SM, Moss C, Bowmaker MR, Jacobs HT, Crouch RJ, Holt IJ (2015) Primer retention owing to the absence of RNase H1 is catastrophic for mitochondrial DNA replication. Proc Natl Acad Sci U S A 112(30): 9334–9339. https://doi.org/10.1073/pnas.1503653112
5. Skourti-Stathaki K, Proudfoot Nicholas J, Gromak N (2011) Human senataxin resolves RNA/DNA hybrids formed at transcriptional pause sites to promote Xrn2-dependent termination. Mol Cell 42(6):794–805. https://doi.org/10.1016/j.molcel.2011.04.026
6. Ohle C, Tesorero R, Schermann G, Dobrev N, Sinning I, Fischer T (2016) Transient RNA-DNA hybrids are required for efficient double-strand break repair. Cell 167(4): 1001–1013.e1007. https://doi.org/10.1016/j.cell.2016.10.001
7. Cohen S, Puget N, Lin Y-L, Clouaire T, Aguirrebengoa M, Rocher V, Pasero P, Canitrot Y, Legube G (2018) Senataxin resolves RNA:DNA hybrids forming at DNA double-strand breaks to prevent translocations. Nat Commun 9(1):533. https://doi.org/10.1038/s41467-018-02894-w
8. Aguilera A, García-Muse T (2012) R Loops: from transcription byproducts to threats to genome stability. Mol Cell 46(2):115–124. https://doi.org/10.1016/j.molcel.2012.04.009
9. Li X, Manley JL (2005) Inactivation of the SR protein splicing factor ASF/SF2 results in genomic instability. Cell 122(3):365–378. https://doi.org/10.1016/j.cell.2005.06.008
10. Liu B, Alberts BM (1995) Head-on collision between a DNA replication apparatus and RNA polymerase transcription complex. Science 267(5201):1131. https://doi.org/10.1126/science.7855590
11. Hamperl S, Bocek MJ, Saldivar JC, Swigut T, Cimprich KA (2017) Transcription-replication conflict orientation modulates R-loop levels and activates distinct DNA damage responses. Cell 170(4):774–786.e719. https://doi.org/10.1016/j.cell.2017.07.043
12. Hamperl S, Cimprich KA (2014) The contribution of co-transcriptional RNA:DNA hybrid structures to DNA damage and genome instability. DNA Repair 19:84–94. https://doi.org/10.1016/j.dnarep.2014.03.023
13. Hamperl S, Cimprich KA (2016) Conflict resolution in the genome: how transcription and replication make it work. Cell 167(6): 1455–1467. https://doi.org/10.1016/j.cell.2016.09.053
14. Barroso S, Herrera-Moyano E, Muñoz S, García-Rubio M, Gómez-González B, Aguilera A (2019) The DNA damage response acts as a safeguard against harmful DNA–RNA hybrids of different origins. EMBO Rep 20(9):e47250. https://doi.org/10.15252/embr.201847250
15. Morales JC, Richard P, Patidar PL, Motea EA, Dang TT, Manley JL, Boothman DA (2016) XRN2 links transcription termination to DNA damage and replication stress. PLoS Genet 12(7):e1006107. https://doi.org/10.1371/journal.pgen.1006107
16. Sollier J, Cimprich KA (2015) Breaking bad: R-loops and genome integrity. Trends Cell Biol 25(9):514–522. https://doi.org/10.1016/j.tcb.2015.05.003
17. Sollier J, Stork CT, Garcia-Rubio ML, Paulsen RD, Aguilera A, Cimprich KA (2014) Transcription-coupled nucleotide excision repair factors promote R-loop-induced genome instability. Mol Cell 56(6):777–785. https://doi.org/10.1016/j.molcel.2014.10.020

18. Lindahl T (1993) Instability and decay of the primary structure of DNA. Nature 362(6422):709–715. https://doi.org/10.1038/362709a0
19. Cerritelli SM, Crouch RJ (2009) Ribonuclease H: the enzymes in eukaryotes. FEBS J 276(6):1494–1505. https://doi.org/10.1111/j.1742-4658.2009.06908.x
20. Drolet M, Phoenix P, Menzel R, Massé E, Liu LF, Crouch RJ (1995) Overexpression of RNase H partially complements the growth defect of an Escherichia coli delta topA mutant: R-loop formation is a major problem in the absence of DNA topoisomerase I. Proc Natl Acad Sci U S A 92(8):3526–3530. https://doi.org/10.1073/pnas.92.8.3526
21. Huertas P, Aguilera A (2003) Cotranscriptionally formed DNA:RNA hybrids mediate transcription elongation impairment and transcription-associated recombination. Mol Cell 12(3):711–721. https://doi.org/10.1016/j.molcel.2003.08.010
22. Paulsen RD, Soni DV, Wollman R, Hahn AT, Yee M-C, Guan A, Hesley JA, Miller SC, Cromwell EF, Solow-Cordero DE, Meyer T, Cimprich KA (2009) A genome-wide siRNA screen reveals diverse cellular processes and pathways that mediate genome stability. Mol Cell 35(2):228–239. https://doi.org/10.1016/j.molcel.2009.06.021
23. Sparks JL, Chon H, Cerritelli SM, Kunkel TA, Johansson E, Crouch RJ, Burgers PM (2012) RNase H2-initiated ribonucleotide excision repair. Mol Cell 47(6):980–986. https://doi.org/10.1016/j.molcel.2012.06.035
24. Hwang JK, Alt FW, Yeap L-S (2015) Related mechanisms of antibody somatic hypermutation and class switch recombination. Microbiol Spectr 3(1):3.1.08. https://doi.org/10.1128/microbiolspec.MDNA3-0037-2014
25. Maul RW, Chon H, Sakhuja K, Cerritelli SM, Gugliotti LA, Gearhart PJ, Crouch RJ (2017) R-loop depletion by over-expressed RNase H1 in mouse B cells increases activation-induced deaminase access to the transcribed strand without altering frequency of isotype switching. J Mol Biol 429(21):3255–3263. https://doi.org/10.1016/j.jmb.2016.12.020
26. Cerritelli SM, Frolova EG, Feng C, Grinberg A, Love PE, Crouch RJ (2003) Failure to produce mitochondrial DNA results in embryonic lethality in Rnaseh1 null mice. Mol Cell 11(3):807–815. https://doi.org/10.1016/S1097-2765(03)00088-1
27. Suzuki Y, Holmes JB, Cerritelli SM, Sakhuja K, Minczuk M, Holt IJ, Crouch RJ (2010) An upstream open reading frame and the context of the two AUG codons affect the abundance of mitochondrial and nuclear RNase H1. Mol Cell Biol 30(21):5123–5134. https://doi.org/10.1128/mcb.00619-10
28. Banaszynski LA, Chen L-c, Maynard-Smith LA, Ooi AGL, Wandless TJ (2006) A rapid, reversible, and tunable method to regulate protein function in living cells using synthetic small molecules. Cell 126(5):995–1004. https://doi.org/10.1016/j.cell.2006.07.025
29. Sordet O, Redon CE, Guirouilh-Barbat J, Smith S, Solier S, Douarre C, Conti C, Nakamura AJ, Das BB, Nicolas E, Kohn KW, Bonner WM, Pommier Y (2009) Ataxia telangiectasia mutated activation by transcription- and topoisomerase I-induced DNA double-strand breaks. EMBO Rep 10(8):887–893. https://doi.org/10.1038/embor.2009.97
30. Tresini M, Warmerdam DO, Kolovos P, Snijder L, Vrouwe MG, Demmers JA, van IJcken Wilfred FJ, Grosveld FG, Medema RH, Hoeijmakers JH, Mullenders LH, Vermeulen W, Marteijn JA (2015) The core spliceosome as target and effector of non-canonical ATM signalling. Nature 523(7558):53–58. https://doi.org/10.1038/nature14512
31. Shiloh Y, Ziv Y (2013) The ATM protein kinase: regulating the cellular response to genotoxic stress, and more. Nat Rev Mol Cell Biol 14(4):197–210. https://doi.org/10.1038/nrm3546
32. Dvinge H, Kim E, Abdel-Wahab O, Bradley RK (2016) RNA splicing factors as oncoproteins and tumour suppressors. Nat Rev Cancer 16(7):413–430. https://doi.org/10.1038/nrc.2016.51
33. Pellagatti A, Armstrong RN, Steeples V, Sharma E, Repapi E, Singh S, Sanchi A, Radujkovic A, Horn P, Dolatshad H, Roy S, Broxholme J, Lockstone H, Taylor S, Giagounidis A, Vyas P, Schuh A, Hamblin A, Papaemmanuil E, Killick S, Malcovati L, Hennrich ML, Gavin A-C, Ho AD, Luft T, Hellström-Lindberg E, Cazzola M, Smith CWJ, Smith S, Boultwood J (2018) Impact of spliceosome mutations on RNA splicing in myelodysplasia: dysregulated genes/pathways and clinical associations. Blood 132(12):1225–1240. https://doi.org/10.1182/blood-2018-04-843771
34. Yoshida K, Sanada M, Shiraishi Y, Nowak D, Nagata Y, Yamamoto R, Sato Y, Sato-Otsubo-A, Kon A, Nagasaki M, Chalkidis G, Suzuki Y, Shiosaka M, Kawahata R, Yamaguchi T,

Otsu M, Obara N, Sakata-Yanagimoto M, Ishiyama K, Mori H, Nolte F, Hofmann W-K, Miyawaki S, Sugano S, Haferlach C, Koeffler HP, Shih L-Y, Haferlach T, Chiba S, Nakauchi H, Miyano S, Ogawa S (2011) Frequent pathway mutations of splicing machinery in myelodysplasia. Nature 478(7367): 64–69. https://doi.org/10.1038/nature10496

35. Chen L, Chen J-Y, Huang Y-J, Gu Y, Qiu J, Qian H, Shao C, Zhang X, Hu J, Li H, He S, Zhou Y, Abdel-Wahab O, Zhang D-E, Fu X-D (2018) The augmented R-loop is a unifying mechanism for myelodysplastic syndromes induced by high-risk splicing factor mutations. Mol Cell 69(3):412–425.e416. https://doi.org/10.1016/j.molcel.2017.12.029

36. Chen L, Chen J-Y, Zhang X, Gu Y, Xiao R, Shao C, Tang P, Qian H, Luo D, Li H, Zhou Y, Zhang D-E, Fu X-D (2017) R-ChIP using inactive RNase H reveals dynamic coupling of r-loops with transcriptional pausing at gene promoters. Mol Cell 68(4):745–757.e745. https://doi.org/10.1016/j.molcel.2017.10.008

37. Kim E, Ilagan Janine O, Liang Y, Daubner Gerrit M, Lee Stanley CW, Ramakrishnan A, Li Y, Chung Young R, Micol J-B, Murphy ME, Cho H, Kim M-K, Zebari Ahmad S, Aumann S, Park Christopher Y, Buonamici S, Smith Peter G, Deeg HJ, Lobry C, Aifantis I, Modis Y, Allain Frederic HT, Halene S, Bradley Robert K, Abdel-Wahab O (2015) SRSF2 mutations contribute to myelodysplasia by mutant-specific effects on exon recognition. Cancer Cell 27(5):617–630. https://doi.org/10.1016/j.ccell.2015.04.006

38. Lee SC-W, Dvinge H, Kim E, Cho H, Micol J-B, Chung YR, Durham BH, Yoshimi A, Kim YJ, Thomas M, Lobry C, Chen C-W, Pastore A, Taylor J, Wang X, Krivtsov A, Armstrong SA, Palacino J, Buonamici S, Smith PG, Bradley RK, Abdel-Wahab O (2016) Modulation of splicing catalysis for therapeutic targeting of leukemia with mutations in genes encoding spliceosomal proteins. Nat Med 22(6):672–678. https://doi.org/10.1038/nm.4097

39. Skourti-Stathaki K, Kamieniarz-Gdula K, Proudfoot NJ (2014) R-loops induce repressive chromatin marks over mammalian gene terminators. Nature 516(7531):436–439. https://doi.org/10.1038/nature13787

40. Jenuwein T (2006) The epigenetic magic of histone lysine methylation. FEBS J 273(14): 3121–3135. https://doi.org/10.1111/j.1742-4658.2006.05343.x

41. Margueron R, Reinberg D (2011) The Polycomb complex PRC2 and its mark in life. Nature 469(7330):343–349. https://doi.org/10.1038/nature09784

42. Di Croce L, Helin K (2013) Transcriptional regulation by Polycomb group proteins. Nat Struct Mol Biol 20(10):1147–1155. https://doi.org/10.1038/nsmb.2669

43. Guenther MG, Levine SS, Boyer LA, Jaenisch R, Young RA (2007) A chromatin landmark and transcription initiation at most promoters in human cells. Cell 130(1):77–88. https://doi.org/10.1016/j.cell.2007.05.042

44. Stock JK, Giadrossi S, Casanova M, Brookes E, Vidal M, Koseki H, Brockdorff N, Fisher AG, Pombo A (2007) Ring1-mediated ubiquitination of H2A restrains poised RNA polymerase II at bivalent genes in mouse ES cells. Nat Cell Biol 9(12):1428–1435. https://doi.org/10.1038/ncb1663

45. Kanhere A, Viiri K, Araújo CC, Rasaiyaah J, Bouwman RD, Whyte WA, Pereira CF, Brookes E, Walker K, Bell GW, Pombo A, Fisher AG, Young RA, Jenner RG (2010) Short RNAs are transcribed from repressed Polycomb target genes and interact with Polycomb repressive complex-2. Mol Cell 38(5): 675–688. https://doi.org/10.1016/j.molcel.2010.03.019

46. Mikkelsen TS, Ku M, Jaffe DB, Issac B, Lieberman E, Giannoukos G, Alvarez P, Brockman W, Kim T-K, Koche RP, Lee W, Mendenhall E, O'Donovan A, Presser A, Russ C, Xie X, Meissner A, Wernig M, Jaenisch R, Nusbaum C, Lander ES, Bernstein BE (2007) Genome-wide maps of chromatin state in pluripotent and lineage-committed cells. Nature 448(7153): 553–560. https://doi.org/10.1038/nature06008

47. Skourti-Stathaki K, Torlai Triglia E, Warburton M, Voigt P, Bird A, Pombo A (2019) R-loops enhance Polycomb repression at a subset of developmental regulator genes. Mol Cell 73(5):930–945.e934. https://doi.org/10.1016/j.molcel.2018.12.016

48. Ginno PA, Lott PL, Christensen HC, Korf I, Chedin F (2012) R-loop formation is a distinctive characteristic of unmethylated human CpG island promoters. Mol Cell 45(6): 814–825. https://doi.org/10.1016/j.molcel.2012.01.017

49. Ginno PA, Lim YW, Lott PL, Korf I, Chédin F (2013) GC skew at the 5′ and 3′ ends of

human genes links R-loop formation to epigenetic regulation and transcription termination. Genome Res 23(10):1590–1600. https://doi.org/10.1101/gr.158436.113

50. Sanz LA, Hartono SR, Lim YW, Steyaert S, Rajpurkar A, Ginno PA, Xu X, Chedin F (2016) Prevalent, dynamic, and conserved R-loop structures associate with specific epigenomic signatures in mammals. Mol Cell 63(1):167–178. https://doi.org/10.1016/j.molcel.2016.05.032

51. Tan-Wong SM, Dhir S, Proudfoot NJ (2019) R-loops promote antisense transcription across the mammalian genome. Mol Cell 76(4):600–616.e606. https://doi.org/10.1016/j.molcel.2019.10.002

52. Kotsantis P, Silva LM, Irmscher S, Jones RM, Folkes L, Gromak N, Petermann E (2016) Increased global transcription activity as a mechanism of replication stress in cancer. Nat Commun 7(1):13087. https://doi.org/10.1038/ncomms13087

53. Gorthi A, Romero JC, Loranc E, Cao L, Lawrence LA, Goodale E, Iniguez AB, Bernard X, Masamsetti VP, Roston S, Lawlor ER, Toretsky JA, Stegmaier K, Lessnick SL, Chen Y, Bishop AJR (2018) EWS–FLI1 increases transcription to cause R-loops and block BRCA1 repair in Ewing sarcoma. Nature 555(7696):387–391. https://doi.org/10.1038/nature25748

54. Stork CT, Bocek M, Crossley MP, Sollier J, Sanz LA, Chédin F, Swigut T, Cimprich KA (2016) Co-transcriptional R-loops are the main cause of estrogen-induced DNA damage. elife 5:e17548. https://doi.org/10.7554/eLife.17548

55. Tuduri S, Crabbé L, Conti C, Tourrière H, Holtgreve-Grez H, Jauch A, Pantesco V, De Vos J, Thomas A, Theillet C, Pommier Y, Tazi J, Coquelle A, Pasero P (2009) Topoisomerase I suppresses genomic instability by preventing interference between replication and transcription. Nat Cell Biol 11(11):1315–1324. https://doi.org/10.1038/ncb1984

56. Promonet A, Padioleau I, Liu Y, Sanz L, Biernacka A, Schmitz A-L, Skrzypczak M, Sarrazin A, Mettling C, Rowicka M, Ginalski K, Chedin F, Chen C-L, Lin Y-L, Pasero P (2020) Topoisomerase 1 prevents replication stress at R-loop-enriched transcription termination sites. Nat Commun 11(1):3940. https://doi.org/10.1038/s41467-020-17858-2

57. Bhatia V, Barroso SI, García-Rubio ML, Tumini E, Herrera-Moyano E, Aguilera A (2014) BRCA2 prevents R-loop accumulation and associates with TREX-2 mRNA export factor PCID2. Nature 511(7509):362–365. https://doi.org/10.1038/nature13374

58. García-Rubio ML, Pérez-Calero C, Barroso SI, Tumini E, Herrera-Moyano E, Rosado IV, Aguilera A (2015) The Fanconi anemia pathway protects genome integrity from R-loops. PLoS Genet 11(11):e1005674. https://doi.org/10.1371/journal.pgen.1005674

59. Schwab Rebekka A, Nieminuszczy J, Shah F, Langton J, Lopez Martinez D, Liang C-C, Cohn Martin A, Gibbons Richard J, Deans Andrew J, Niedzwiedz W (2015) The Fanconi anemia pathway maintains genome stability by coordinating replication and transcription. Mol Cell 60(3):351–361. https://doi.org/10.1016/j.molcel.2015.09.012

60. Kwak H, Lis JT (2013) Control of transcriptional elongation. Annu Rev Genet 47(1):483–508. https://doi.org/10.1146/annurev-genet-110711-155440

61. Delmore Jake E, Issa Ghayas C, Lemieux Madeleine E, Rahl Peter B, Shi J, Jacobs Hannah M, Kastritis E, Gilpatrick T, Paranal Ronald M, Qi J, Chesi M, Schinzel Anna C, McKeown Michael R, Heffernan Timothy P, Vakoc Christopher R, Bergsagel PL, Ghobrial Irene M, Richardson Paul G, Young Richard A, Hahn William C, Anderson Kenneth C, Kung Andrew L, Bradner James E, Mitsiades Constantine S (2011) BET bromodomain inhibition as a therapeutic strategy to target c-Myc. Cell 146(6):904–917. https://doi.org/10.1016/j.cell.2011.08.017

62. Filippakopoulos P, Qi J, Picaud S, Shen Y, Smith WB, Fedorov O, Morse EM, Keates T, Hickman TT, Felletar I, Philpott M, Munro S, McKeown MR, Wang Y, Christie AL, West N, Cameron MJ, Schwartz B, Heightman TD, La Thangue N, French CA, Wiest O, Kung AL, Knapp S, Bradner JE (2010) Selective inhibition of BET bromodomains. Nature 468(7327):1067–1073. https://doi.org/10.1038/nature09504

63. Muhar M, Ebert A, Neumann T, Umkehrer C, Jude J, Wieshofer C, Rescheneder P, Lipp JJ, Herzog VA, Reichholf B, Cisneros DA, Hoffmann T, Schlapansky MF, Bhat P, von Haeseler A, Köcher T, Obenauf AC, Popow J, Ameres SL, Zuber J (2018) SLAM-seq defines direct gene-regulatory functions of the BRD4-MYC

axis. Science 360(6390):800–805. https://doi.org/10.1126/science.aao2793

64. Winter GE, Mayer A, Buckley DL, Erb MA, Roderick JE, Vittori S, Reyes JM, di Iulio J, Souza A, Ott CJ, Roberts JM, Zeid R, Scott TG, Paulk J, Lachance K, Olson CM, Dastjerdi S, Bauer S, Lin CY, Gray NS, Kelliher MA, Churchman LS, Bradner JE (2017) BET bromodomain proteins function as master transcription elongation factors independent of CDK9 recruitment. Mol Cell 67(1): 5–18.e19. https://doi.org/10.1016/j.molcel.2017.06.004
65. Edwards DS, Maganti R, Tanksley JP, Luo J, Park JJH, Balkanska-Sinclair E, Ling J, Floyd SR (2020) BRD4 prevents R-loop formation and transcription-replication conflicts by ensuring efficient transcription elongation. Cell Rep 32(12):108166. https://doi.org/10.1016/j.celrep.2020.108166
66. Lam FC, Kong YW, Huang Q, Vu Han T-L, Maffa AD, Kasper EM, Yaffe MB (2020) BRD4 prevents the accumulation of R-loops and protects against transcription–replication collision events and DNA damage. Nat Commun 11(1):4083. https://doi.org/10.1038/s41467-020-17503-y
67. Giraud-Panis M-J, Pisano S, Benarroch-Popivker D, Pei B, Le Du M-H, Gilson E (2013) One identity or more for telomeres? Front Oncol 3:48–48. https://doi.org/10.3389/fonc.2013.00048
68. Arora R, Lee Y, Wischnewski H, Brun CM, Schwarz T, Azzalin CM (2014) RNaseH1 regulates TERRA-telomeric DNA hybrids and telomere maintenance in ALT tumour cells. Nat Commun 5(1):5220. https://doi.org/10.1038/ncomms6220
69. Yehezkel S, Segev Y, Viegas-Péquignot E, Skorecki K, Selig S (2008) Hypomethylation of subtelomeric regions in ICF syndrome is associated with abnormally short telomeres and enhanced transcription from telomeric regions. Hum Mol Genet 17(18): 2776–2789. https://doi.org/10.1093/hmg/ddn177
70. Yehezkel S, Shaked R, Sagie S, Berkovitz R, Shachar-Bener H, Segev Y, Selig S (2013) Characterization and rescue of telomeric abnormalities in ICF syndrome type I fibroblasts. Front Oncol 3:35–35. https://doi.org/10.3389/fonc.2013.00035
71. Sagie S, Toubiana S, Hartono SR, Katzir H, Tzur-Gilat A, Havazelet S, Francastel C, Velasco G, Chédin F, Selig S (2017) Telomeres in ICF syndrome cells are vulnerable to DNA damage due to elevated DNA:RNA hybrids. Nat Commun 8(1):14015. https://doi.org/10.1038/ncomms14015
72. Maréchal A, Zou L (2015) RPA-coated single-stranded DNA as a platform for post-translational modifications in the DNA damage response. Cell Res 25(1):9–23. https://doi.org/10.1038/cr.2014.147
73. Sikorski TW, Ficarro SB, Holik J, Kim T, Rando OJ, Marto JA, Buratowski S (2011) Sub1 and RPA associate with RNA polymerase II at different stages of transcription. Mol Cell 44(3):397–409. https://doi.org/10.1016/j.molcel.2011.09.013
74. Nguyen HD, Yadav T, Giri S, Saez B, Graubert TA, Zou L (2017) Functions of replication protein A as a sensor of R loops and a regulator of RNaseH1. Mol Cell 65(5): 832–847.e834. https://doi.org/10.1016/j.molcel.2017.01.029
75. Wu H, Lima WF, Crooke ST (2001) Investigating the structure of human RNase H1 by site-directed mutagenesis*. J Biol Chem 276(26):23547–23553. https://doi.org/10.1074/jbc.M009676200
76. Boguslawski SJ, Smith DE, Michalak MA, Mickelson KE, Yehle CO, Patterson WL, Carrico RJ (1986) Characterization of monoclonal antibody to DNA · RNA and its application to immunodetection of hybrids. J Immunol Methods 89(1):123–130. https://doi.org/10.1016/0022-1759(86)90040-2
77. Gaidamakov SA, Gorshkova II, Schuck P, Steinbach PJ, Yamada H, Crouch RJ, Cerritelli SM (2005) Eukaryotic RNases H1 act processively by interactions through the duplex RNA-binding domain. Nucleic Acids Res 33(7):2166–2175. https://doi.org/10.1093/nar/gki510
78. Teloni F, Michelena J, Lezaja A, Kilic S, Ambrosi C, Menon S, Dobrovolna J, Imhof R, Janscak P, Baubec T, Altmeyer M (2019) Efficient pre-mRNA cleavage prevents replication-stress-associated genome instability. Mol Cell 73(4):670–683.e612. https://doi.org/10.1016/j.molcel.2018.11.036
79. Kabeche L, Nguyen Hai D, Buisson R, Zou L (2018) A mitosis-specific and R loop–driven ATR pathway promotes faithful chromosome segregation. Science 359(6371):108–114. https://doi.org/10.1126/science.aan6490
80. Chan FL, Marshall OJ, Saffery R, Won Kim B, Earle E, Choo KHA, Wong LH (2012) Active transcription and essential role of RNA polymerase II at the centromere during mitosis. Proc Natl Acad Sci 109(6):1979. https://doi.org/10.1073/pnas.1108705109

81. Gent JI, Dawe RK (2012) RNA as a structural and regulatory component of the centromere. Annu Rev Genet 46(1):443–453. https://doi.org/10.1146/annurev-genet-110711-155419
82. Mashtalir N, D'Avino AR, Michel BC, Luo J, Pan J, Otto JE, Zullow HJ, McKenzie ZM, Kubiak RL, St. Pierre R, Valencia AM, Poynter SJ, Cassel SH, Ranish JA, Kadoch C (2018) Modular organization and assembly of SWI/SNF family chromatin remodeling complexes. Cell 175(5):1272–1288.e1220. https://doi.org/10.1016/j.cell.2018.09.032
83. Gerlinger M, Horswell S, Larkin J, Rowan AJ, Salm MP, Varela I, Fisher R, McGranahan N, Matthews N, Santos CR, Martinez P, Phillimore B, Begum S, Rabinowitz A, Spencer-Dene B, Gulati S, Bates PA, Stamp G, Pickering L, Gore M, Nicol DL, Hazell S, Futreal PA, Stewart A, Swanton C (2014) Genomic architecture and evolution of clear cell renal cell carcinomas defined by multiregion sequencing. Nat Genet 46(3): 225–233. https://doi.org/10.1038/ng.2891
84. Chabanon RM, Morel D, Eychenne T, Colmet-Daage L, Bajrami I, Dorvault N, Garrido M, Meisenberg C, Lamb A, Ngo C, Hopkins SR, Roumeliotis TI, Jouny S, Hénon C, Kawai-Kawachi A, Astier C, Konde A, Del Nery E, Massard C, Pettitt SJ, Margueron R, Choudhary JS, Almouzni G, Soria J-C, Deutsch E, Downs JA, Lord CJ, Postel-Vinay S (2021) PBRM1 deficiency confers synthetic lethality to DNA repair inhibitors in cancer. Cancer Res 81(11):2888. https://doi.org/10.1158/0008-5472.CAN-21-0628
85. Kadoch C, Hargreaves DC, Hodges C, Elias L, Ho L, Ranish J, Crabtree GR (2013) Proteomic and bioinformatic analysis of mammalian SWI/SNF complexes identifies extensive roles in human malignancy. Nat Genet 45(6):592–601. https://doi.org/10.1038/ng.2628
86. Bayona-Feliu A, Barroso S, Muñoz S, Aguilera A (2021) The SWI/SNF chromatin remodeling complex helps resolve R-loop-mediated transcription–replication conflicts. Nat Genet 53(7):1050–1063. https://doi.org/10.1038/s41588-021-00867-2
87. Wahba L, Costantino L, Tan FJ, Zimmer A, Koshland D (2016) S1-DRIP-seq identifies high expression and polyA tracts as major contributors to R-loop formation. Genes Dev 30(11):1327–1338. https://doi.org/10.1101/gad.280834.116
88. El Hage A, Webb S, Kerr A, Tollervey D (2014) Genome-wide distribution of RNA-DNA hybrids identifies RNase H targets in tRNA genes, retrotransposons and mitochondria. PLoS Genet 10(10): e1004716. https://doi.org/10.1371/journal.pgen.1004716
89. Castellano-Pozo M, García-Muse T, Aguilera A (2012) R-loops cause replication impairment and genome instability during meiosis. EMBO Rep 13(10):923–929. https://doi.org/10.1038/embor.2012.119
90. Marabitti V, Lillo G, Malacaria E, Palermo V, Sanchez M, Pichierri P, Franchitto A (2019) ATM pathway activation limits R-loop-associated genomic instability in Werner syndrome cells. Nucleic Acids Res 47(7): 3485–3502. https://doi.org/10.1093/nar/gkz025
91. Kinney JSVR, Vonderfecht SL, Eiden JJ, Yolken RH (1989) Monoclonal antibody assay for detection of double-stranded RNA and application for detection of group A and non-group A rotaviruses. J Clin Microbiol 27: 6–12
92. Hartono SR, Malapert A, Legros P, Bernard P, Chédin F, Vanoosthuyse V (2018) The affinity of the S9.6 antibody for double-stranded RNAs impacts the accurate mapping of R-loops in fission yeast. J Mol Biol 430(3):272–284. https://doi.org/10.1016/j.jmb.2017.12.016
93. Smolka JA, Sanz LA, Hartono SR, Chédin F (2021) Recognition of RNA by the S9.6 antibody creates pervasive artifacts when imaging RNA:DNA hybrids. J Cell Biol 220(6):e202004079. https://doi.org/10.1083/jcb.202004079
94. Zhang ZZ, Pannunzio NR, Hsieh C-L, Yu K, Lieber MR (2015) Complexities due to single-stranded RNA during antibody detection of genomic rna:dna hybrids. BMC Res Notes 8(1):127. https://doi.org/10.1186/s13104-015-1092-1
95. Nowotny M, Gaidamakov SA, Ghirlando R, Cerritelli SM, Crouch RJ, Yang W (2007) Structure of human RNase H1 complexed with an RNA/DNA hybrid: insight into HIV reverse transcription. Mol Cell 28(2): 264–276. https://doi.org/10.1016/j.molcel.2007.08.015
96. Nowotny M, Cerritelli SM, Ghirlando R, Gaidamakov SA, Crouch RJ, Yang W (2008) Specific recognition of RNA/DNA hybrid and enhancement of human RNase H1 activity by HBD. EMBO J 27(7):1172–1181. https://doi.org/10.1038/emboj.2008.44

97. Chen J-Y, Zhang X, Fu X-D, Chen L (2019) R-ChIP for genome-wide mapping of R-loops by using catalytically inactive RNA-SEH1. Nat Protoc 14(5):1661–1685. https://doi.org/10.1038/s41596-019-0154-6
98. Yan Q, Shields EJ, Bonasio R, Sarma K (2019) Mapping native R-loops genome-wide using a targeted nuclease approach. Cell Rep 29(5):1369–1380.e1365. https://doi.org/10.1016/j.celrep.2019.09.052
99. Thakur J, Henikoff S (2018) Unexpected conformational variations of the human centromeric chromatin complex. Genes Dev 32(1):20–25
100. Wang K, Wang H, Li C, Yin Z, Xiao R, Li Q, Xiang Y, Wang W, Huang J, Chen L, Fang P, Liang K (2021) Genomic profiling of native R loops with a DNA-RNA hybrid recognition sensor. Sci Adv 7(8):eabe3516. https://doi.org/10.1126/sciadv.abe3516
101. Castillo-Guzman D, Chédin F (2021) Defining R-loop classes and their contributions to genome instability. DNA Repair 106:103182. https://doi.org/10.1016/j.dnarep.2021.103182
102. Adelman K, Lis JT (2012) Promoter-proximal pausing of RNA polymerase II: emerging roles in metazoans. Nat Rev Genet 13(10):720–731. https://doi.org/10.1038/nrg3293
103. Skourti-Stathaki K, Proudfoot NJ (2014) A double-edged sword: R loops as threats to genome integrity and powerful regulators of gene expression. Genes Dev 28(13):1384–1396. https://doi.org/10.1101/gad.242990.114
104. García-Muse T, Aguilera A (2016) Transcription–replication conflicts: how they occur and how they are resolved. Nat Rev Mol Cell Biol 17:553. https://doi.org/10.1038/nrm.2016.88

Chapter 8

RNase H1 Hybrid-Binding Domain-Based Tools for Cellular Biology Studies of DNA–RNA Hybrids in Mammalian Cells

Sónia Silva, Cristina Guillén-Mendoza, and Andrés Aguilera

Abstract

R loops are abundant noncanonical DNA–RNA hybrid structures that can occur during DNA-based processes, such as transcription, replication and DNA damage, and can lead both to physiologically favorable and pathological outcomes. With an increasing body of work feeding the field of R loop biology, our understanding of the processes in which R loops intervene and the consequences of meddling with R loop formation and dissolution has greatly increased but it has also led to new questions and sometimes opposing possibilities. Proper detection of these structures is a crucial factor to advance our knowledge about R loops and factors associated with their formation and removal. Here, we describe the use of fluorescently tagged HBD, the hybrid-binding domain of RNase H1, as a tool for analyzing DNA–RNA hybrids in different contexts using live-cell microscopy and immunofluorescence experiments.

Key words DNA–RNA hybrids, R loops, RNase H1, Hybrid-binding domain (HBD), DNA double-strand breaks (DSB), GFP

1 Introduction

R loops are three-stranded DNA–RNA hybrid structures that form when an RNA molecule invades the DNA double helix, annealing with its complementary strand and leaving a displaced single-stranded stretch of DNA. It has long been known that R loops play important roles in both prokaryotic and eukaryotic physiological processes that can be both beneficial, such as regulation of transcription, replication and chromatin states; or detrimental, leading to increased genomic instability and ultimately genetic disorders and cancer [1, 2]. More recently, it has been shown that R loops accumulate around DNA double strand breaks (DSBs), particularly in actively transcribed regions of the genome, through there are still many questions regarding the contribution of these structures to the different stages and outcomes of DNA repair [3].

Andrés Aguilera and Alexey Ruzov (eds.), *R-Loops: Methods and Protocols*, Methods in Molecular Biology, vol. 2528,
https://doi.org/10.1007/978-1-0716-2477-7_8,

Particularly in the last decade, R loops have become an intense matter of study propelling our current knowledge about the frequency and genomic location of R-loops with great specificity, as well as generating a growing list of surveillance factors influencing their formation or removal, and ultimately the consequences of R-loop formation for the cell in different physiological processes. Nevertheless, it is still debated and sometimes difficult to precise the physiological relevance of the vast distribution of naturally occurring R loops in genomes, even though occurring at a very low frequency, and the ever-growing number of the factors identified as intervening in R-loop metabolism.

The detection of R loops has relied heavily on the use of the monoclonal S9.6 antibody [4], which recognizes DNA–RNA hybrids with great affinity but has been shown to have several limitations derived from its ability to bind other nucleic acid species, particularly dsRNA. The other classical tool in the R-loop biology field are the DNA–RNA hybrid-specific nucleases of the RNase H family [5, 6], particularly relevant in validating R-loop associated phenotypes by over-expression in vivo and used in vitro for verifying the nature of hybrid detection in molecular biology techniques, such as DNA–RNA immunoprecipitation (DRIP) [7, 8]. Thus, the interference of dsRNA with the analysis of DNA–RNA hybrid levels using the S9.6 antibody is minor in molecular biology methodologies, but in immunofluorescence-based cell biology studies it can produce significant interference [9–11]. The use of RNase III to degrade dsRNA-associated signal can eliminate such an interference and, in addition, the fidelity of the S9.6 signal should be confirmed by treating with RNase H and showing suppression of the signal. Alternatively, to avoid detection artifacts due to S9.6 interferences and as means of corroborating findings on R-loop levels and distribution, other tools have been developed [10]. Particularly relevant for their broad applicability, are the tools established from engineered human RNase H1, exploiting its ability to specifically recognize DNA–RNA hybrids through its hybrid-binding domain (HBD) and allowing for R-loop visualization and mapping, overcoming many of the caveats of S9.6 antibody-based techniques for in vivo cell biology [12–15]. One such tool is the use of the human RNase H1 HBD expressed with a tag, fluorescent or not, that allows easy visualization and pull-down of DNA–RNA hybrids [13, 16]. In particular, the 52-residue DNA–RNA hybrid-binding domain (HBD) of RNase H1 fused to green fluorescent protein (GFP) has been successfully used for the quantitative analysis of DNA–RNA hybrids at defined genomic loci by molecular biology performing GFP ChIP and was also shown to be a valid alternative for immunofluorescence [13]. Here, we describe in detail the applications of the RNase H1 HBD-GFP for the detection of DNA–RNA hybrids in human cells by immunofluorescence for the global appreciation of hybrid accumulation in a cell population and in the context of localized DNA damage induction by live-cell microscopy.

2 Materials

2.1 Immuno-fluorescence Analysis of DNA–RNA Hybrids Using HBD-GFP

1. Cell culture 6-well plates.
2. 24-well plates.
3. Polylysine-coated round coverslips (*see* **Note 1**).
4. Adherent cell line (e.g., U2OS).
5. Lentiviral particles for expression of GFP (or GFP variant)-tagged HBD (*see* **Notes 2** and **3**).
6. Transduction reagent (e.g., Polybrene).
7. Appropriate culture medium (e.g., DMEM supplemented with 10% FBS, glutamine and antibiotics (e.g., 100 U/mL penicillin and 100 U/mL streptomycin)).
8. PBS.
9. Preextraction (PE) buffer (0.5% Triton X-100, 20 mM HEPES-KOH (pH 7.9), 50 mM NaCl, 3 mM $MgCl_2$, 300 mM sucrose) (*see* **Note 4**).
10. Fixative solution (4% formaldehyde, 2% sucrose in PBS).
11. Permeabilization solution (0.5% Triton X-100 in PBS).
12. Blocking solution (2% bovine serum albumin (BSA), 0.1% Triton X-100 in PBS).
13. Primary antibody anti-GFP.
14. Alexa-conjugated secondary antibody.
15. DAPI solution (1 μg/mL in PBS).
16. Distilled H_2O.
17. Sharp-angled forceps to handle coverslips.
18. Microscope slides.
19. Mounting medium (e.g., Invitrogen™ Prolong Gold Antifade Mountant, Thermo Fisher Scientific).
20. Rocking shaker.
21. Fluorescence microscope.

2.2 Live Cell Microscopy Analysis of HBD-GFP Recruitment to Microirradiation-Induced DNA Damage

1. Live cell microscopy compatible Petri dish or chamber slides (e.g., Ibidi μ-Slide 4 Well Glass Bottom).
2. Adherent cell line (e.g., U2OS).
3. Standard cell culture incubator.
4. Appropriate culture medium (e.g., DMEM-Dulbecco's Modified Eagle Medium supplemented with 10% FBS, glutamine,

and antibiotics (e.g., 100 U/mL penicillin and 100 U/mL streptomycin).

5. PBS.
6. Lentiviral particles or transient expression plasmid encoding for GFP (or GFP variant)-tagged HBD (*see* **Notes 2** and **3**).
7. Transduction reagent (e.g., Polybrene) or transfection reagent of choice.
8. Imaging medium without phenol red for live cell microscopy (e.g., Leibovitz's L-15 medium), supplemented with 10% FBS.
9. Confocal spinning disk microscope equipped with a photoablation laser of 355 nm (*see* **Note 5**), incubator with temperature and CO_2 control, and laser lines for 488 nm.
10. Microscope software (e.g., MetaMorph® Microscopy Automation and Image Analysis Software, Molecular Devices).
11. FIJI (https://imagej.net/software/fiji/).
12. Data analysis and representation software (e.g., Microsoft Excel, Microsoft; and Prism, GraphPad) (*see* **Note 6**).

3 Methods

3.1 Immunofluorescence Analysis of DNA–RNA Hybrids Using HBD-GFP

1. Seed U2OS cells at a density of 1.2×10^5 cells/well over polylysine-coated coverslips in a 6-well plate and incubate in standard conditions for 24 h (*see* **Note 7**).
2. Prepare the appropriate volume of transduction mix for the number of wells to use in the experiment. Per well prepare 2 mL of warm culture medium, polybrene at a final concentration of 10 μg/mL and the volume of lentiviral suspension to yield a MOI of ≥1 (*see* **Note 3**). Remove old culture medium, add the transduction mix to each well and incubate for 24 h.
3. Remove the medium from viral transduction with the vacuum system and wash once with PBS (*see* **Note 8**).
4. Transfer each coverslip, cells facing up, to a well in a 24-well plate containing 1 mL cold PBS.
5. Place the plate flat over ice and remove PBS with the vacuum system.
6. Add 500 μL of preextraction (PE) solution per well and incubate for 1 min on ice (*see* **Note 9**).
7. Carefully remove the PE solution using a P1000 pipette and add 1 mL of cold PBS, gently dripping down the wall of the well (*see* **Note 10**).

8. Carefully remove PBS and add 500 μL of fixative solution per well. Carefully place the plate on the bench, and incubate for 15 min at room temperature (RT) (*see* **Note 11**).
9. Wash twice with 1 mL PBS (*see* **Note 12**).
10. Remove PBS and add 500 μL of permeabilization solution per well. Incubate at RT for 10 min.
11. Wash twice with 1 mL PBS and add 500 μL of blocking solution per well. Incubate at RT for 1 h with gentle rocking.
12. Remove the blocking solution and add the primary anti-GFP antibody (1:500) in 250 μL of blocking solution per well. Incubate overnight at 4 °C with gentle rocking.
13. Wash three times with 1 mL PBS (5 min each wash, with gentle rocking) and add the secondary anti-mouse Alexa Flour conjugated antibody (1:1000 in blocking solution). Incubate for 1 h at RT with gentle rocking.
14. Wash twice with 1 mL PBS (5 min each, with gentle rocking).
15. Add 500 μL of DAPI solution and incubate 5 min at RT.
16. Wash once with distilled H_2O.
17. Pick the coverslip from a well using sharp-angled forceps, drain the excess liquid by tipping the coverslip on filter paper and then place it on a microscope slide over a drop of mounting medium, cells facing down (*see* **Note 13**).
18. Place slides in a flat surface at RT and protected from light overnight to allow mounting medium to cure before acquiring images with a fluorescence microscope (Fig. 1). Slides can be kept at RT or at 4 °C for long-term storage.

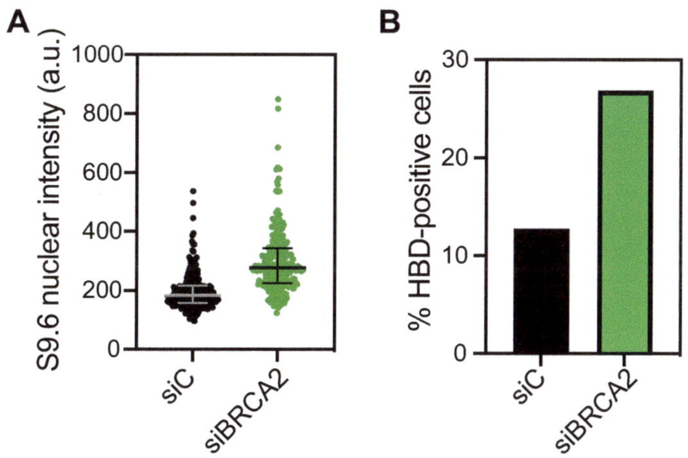

Fig. 1 Detection of DNA–RNA hybrids using the HBD-GFP in human cells. (**a**) Immunofluorescence (IF) using the S9.6 antibody or anti-GFP in control and BRCA2-depleted U2OS cells, known to accumulate R loops. (**b**) Quantification of HBD-positive cells in control and BRCA2-depleted U2OS cells

3.2 Live Cell Microscopy Analysis of HBD-GFP Recruitment to Microirradiation-Induced DNA Damage

1. Seed U2OS cells at a density of 40,000 cells/well (one well per condition to assay) in Ibidi μ-Slide 4 Well Glass Bottom, in 500 μL of medium and allow cells to attach to the plate for 24 h in a standard cell culture incubator.
2. Remove the culture medium and add viral transduction mix for expression of HBD-GFP or GFP as a negative control. Per chamber prepare 500 μL of warm culture medium, add polybrene at a final concentration of 10 μg/mL and the volume of lentiviral suspension to yield a multiplicity of infection (MOI) = 1 (*see* **Note 3**).
3. After 24 h, aspirate the medium with the vacuum system and wash once with PBS (*see* **Note 8**). Add 500 μL per well of prewarmed imaging medium.
4. Turn on the spinning disk confocal microscope and associated computer and control systems with enough time in advance to allow for the incubation chamber and stage to equilibrate to 37 °C and 5% CO_2.
5. Apply immersion oil along the chambers to be imaged of the Ibidi slide and transfer it to the microscope stage over the 100× objective (*see* **Note 14**). Make sure to securely place the slide on the sample holder of the spinning disk microscope to avoid displacements due to stage motion through the duration of the experiment.
6. Focus the cells with transmitted light through the ocular.
7. Once the focal plane is set, select an imaging field containing fluorescent cells showing appropriate expression level of HBD-GFP using fluorescence excitation (*see* **Note 15**).
8. Take a snapshot on the 488 channel and use this image to draw a line (region of interest—ROI) inside every nuclei of interest in the field, where the photoablation laser will irradiate to generate DNA damage (*see* **Note 16**).
9. Save the ROIs in your experiment folder, which you will later use in the analysis.
10. Upload all set ROIs into the Laser Module. Set up the output of the 355 laser to 30% of the maximum power, 1 repetition and thickness of 1 (*see* **Note 17**).
11. Select the number of preirradiation images to acquire prior to laser microirradiation and click *Setup MDA* to transfer all the information to the *Acquisition Module* (*see* **Note 18**).
12. Set the total number of images to acquire, the time intervals for image acquisition and the duration of the experiment in the Acquisition module (e.g., take sequential images in intervals of 5 s for the course of 3 min, with a total of 37 images containing one preirradiation image).

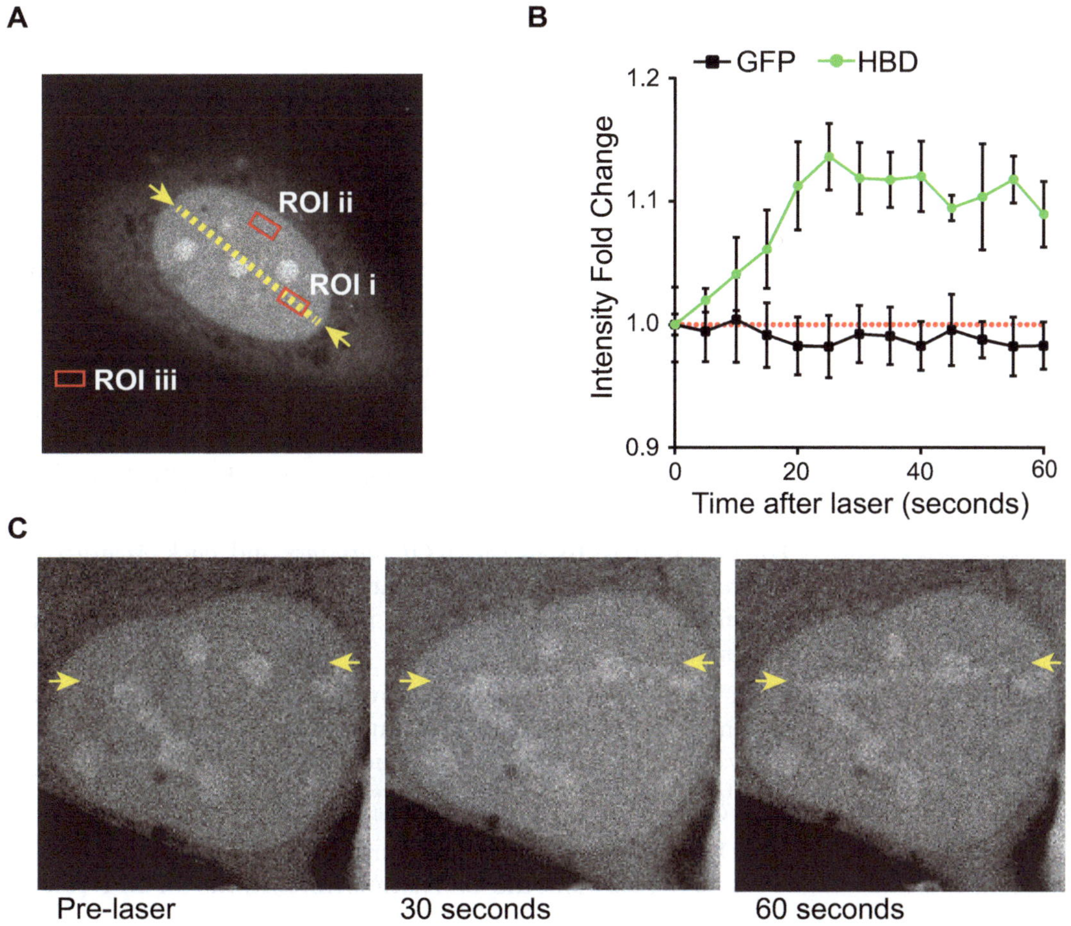

Fig. 2 Quantification of HBD-GFP accumulation at sites of laser microirradiation-induced DNA damage in U2OS cells. (**a**) Schematic representation showing the laser incision site (yellow dotted line) and selection of the ROIs: (i) Laser ROI; (ii) Cell ROI; iii) Background ROI (in red boxes). (**b**) Quantification of HBD-GFP accumulation at laser-induced DNA damage sites, represented as mean fold change (±SEM) in fluorescence intensity relative to preirradiation over time. (**c**) Accumulation of HBD-GFP at sites of laser-induced DNA damage over time in U2OS cells. The mean intensity of the three ROIs is measured in preirradiation images (left panel), defining the baseline of the relative fluorescence intensity, and at the same sites through the duration of the time-lapse after irradiation (middle and right panels)

13. Once acquisition is finished, shut down the system in the correct manner, turning off all microscope devices and clean the lens and stage with Whatman lens cleaning paper.
14. Export the microscopy images and proceed to analysis.
15. Using FIJI, import the complete set of images from the experiment and build a *Stack* containing all the timepoints of the experiment (*see* **Note 19**).

16. Import the ROI file generated in MetaMorph (*see* **Note 20**).
17. For each cell to analyze, define three regions of interest (ROIs) with the same area (e.g., Area = 135) (*see* Fig. 2a):
    (a) Laser ROI—coinciding with the laser incision region (defined by the MetaMorph ROI imported).
    (b) Cell ROI—located inside the nucleus of the cell of interest and outside the laser incision region. This will allow to calculate the relative enrichment at the laser microirradiation region compared to a random nonirradiated region of the nucleus and correct for decay of fluorescence resulting from the acquisition process.
    (c) Background ROI—placed in an area of the field without fluorescence to correct for background noise. These ROIs should automatically appear on the *ROI Manager* window.
18. Select all ROIs in the *ROI manager* and click *Measure* and repeat it for all the time points. A *Results* window will automatically popup displaying all measurement values, which can be directly exported to Excel.
19. Use the *Mean Gray Value*, which corresponds to the GFP intensity, to calculate the normalized *Corrected Mean Intensity* value for each analyzed nucleus at each time point as $\frac{(\text{Laser ROI Mean} - \text{Background ROI Mean})}{(\text{Cell ROI Mean} - \text{Background ROI Mean})}$ (*see* **Note 21**).
20. Data can be represented as bar charts or kinetics curves, plotting the time and the *Corrected Mean intensity* for GFP (Fig. 2) (*see* **Note 22**).

4 Notes

1. To reduce the risk of cell detachment, especially during preextraction, we use coverslips previously coated with poly-L-lysine.
2. The 52-residue DNA–RNA hybrid-binding domain (HBD) of RNase H1 was originally cloned into pEGFPC1 vector (Clontech) generating the pHBD-GFP transient expression vector, subsequently used to generate the stable HeLa cell lines used for ChIP experiments [13]. To achieve more consistent numbers of HBD-GFP expressing cells and even expression levels in hard-to-transfect cell lines, lentiviral vectors carrying the HBD-GFP cassette were engineered. The corresponding fragment was PCR-amplified from the original pHBD-GFP vector [13] and cloned into pSIN-DUAL-GFP1 (kind gift from Dr. JA Pintor-Toro, CABIMER, Sevilla, Spain) [17] using the enzymes *Bam*HI and *Not*I.

3. Handling of recombinant Lentiviruses requires the use of a Class II biological safety cabinet. For lentiviral production, the pSIN-DUAL-GFP1 and pSIN-DUAL-GFP-HBD plasmid were transfected into 293 T cells together with pCMVDR8.91 (Addgene plasmid # 8455) and pCMV-VSV-G (Addgene plasmid # 8454) [18]. Virus particles for GFP and HBD-GFP were collected from the culture medium after 48–72 h post-transfection by ultra-centrifugation, tittered for yield calculation and stored at −80 °C for coming experiments. MOI is calculated as the number of infective particles per cell and should be calculated for each cell type used. The number of infective particles is obtained in the viral titration as the number of positive cells (GFP+) seen by the cytometer and extrapolated to the total number of cells contained in the well.
4. We recommend that the PE buffer should be prepared fresh for every experiment and cooled at 4 °C prior to use.
5. Here we use a 355 nm photoablation laser, but other lasers could be used, moreover, a sensitization with an agent such as Hoechst could be used if necessary (for details *see* ref. 19).
6. In case there is no access to Prism software, data analysis and representation can be done entirely on Microsoft Excel.
7. Up to four coverslips can be placed in each well of a 6-well plate in case several factors need to be assayed in parallel by IF under the same condition applied in each well.
8. After washing off the medium used for viral transduction, lentiviral particles have been removed and further incubations and processing of the cell cultures can be performed in normal tissue culture settings.
9. Add the PE solution slowly, allowing it to drip along the wall of the well using a P1000 micropipette to avoid cell rupture and detachment because of the detergent present in this solution. Incubation time with the PE solution should be optimized for different cell lines.
10. Do not shake the plates during this time nor apply the PBS directly over the cells, as cells are extremely sensitive and this might alter cell shape or lead to cell detachment or bursting.
11. Formaldehyde is a hazardous chemical. Wear gloves and lab coat and always work in a chemical fume hood.
12. At this point you can proceed directly to **step 10** or store samples in PBS at 4 °C for a few days, after which the quality of IFs might be affected.
13. Invitrogen™ Prolong Gold Antifade Mountant is designed to protect fluorescent dyes from fading during fluorescence microscopy experiments and cures for longer-term storage.

14. In our imaging system, the photoablation laser has been set up for use with the 100× objective. Consult with your imaging facility manager before setting up the photoablation conditions in your experiment as conditions including which objective to use may vary.
15. Here we describe the steps for laser irradiating and imaging at a single stage position. It is possible to irradiate and image several stage positions in one single experiment. Consult with your imaging facility manager to correctly setup your experiment.
16. Using the snapshot to define ROIs will minimize the exposure time with fluorescence excitation. Be mindful that during cell selection exposure time with fluorescence excitation should be kept to the necessary minimum to avoid any phototoxicity and photobleaching prior to the start of the experiment.
17. These laser conditions should be optimized for each imaging system and cell line.
18. Acquire at least one preirradiation image of each position.
19. In case there is an observable drift of the cell/nucleus through the course of the experiment, a logarithmical correction should be applied to maintain the same location of the ROI(s) (e.g., Image J pluging StackReg).
20. MetaMorph software exports ROIs in *.rgn* format and to import it into the ROI Manager in FIJI, a suitable plugin can be used (e.g., *mm_nd_and_roi_pack.jar*).
21. This calculation eliminates noisy signal from the data and normalizes the intensity levels from the *Laser region* to the *Cell region*, yielding the value for relative enrichment at the DNA damage sites compared to nonirradiated regions.
22. For meaningful quantitative results of laser microirradiation experiments, for each condition studied, at least ten nuclei per experiment from two independent experiments should be analyzed.

Acknowledgments

We thank Dr. Paloma Domínguez Giménez from the Microscopy Unit at CABIMER for her valuable help setting up the confocal spinning disk microscope experiments. The Aguilera lab is funded by the Spanish Ministry of Economy and Competitiveness, Junta de Andalucía, European Research Council, and the European Union (FEDER).

References

1. García-Muse T, Aguilera A (2019) R loops: from physiological to pathological roles. Cell 179:604–618. https://doi.org/10.1016/j.cell.2019.08.055
2. Niehrs C, Luke B (2020) Regulatory R-loops as facilitators of gene expression and genome stability. Nat Rev Mol Cell Biol 21:167–178. https://doi.org/10.1038/s41580-019-0206-3
3. Marnef A, Legube G (2021) R-loops as Janus-faced modulators of DNA repair. Nat Cell Biol 23:305–313. https://doi.org/10.1038/s41556-021-00663-4
4. Boguslawski SJ, Smith DE, Michalak MA et al (1986) Characterization of monoclonal antibody to DNA · RNA and its application to immunodetection of hybrids. J Immunol Methods 89:123–130. https://doi.org/10.1016/0022-1759(86)90040-2
5. Nowotny M, Gaidamakov SA, Ghirlando R et al (2007) Structure of human RNase H1 complexed with an RNA/DNA hybrid: insight into HIV reverse transcription. Mol Cell 28: 264–276. https://doi.org/10.1016/j.molcel.2007.08.015
6. Cerritelli SM, Crouch RJ (1998) Cloning, expression, and mapping of ribonucleases H of human and mouse related to bacterial RNase HI. Genomics 53:300–307. https://doi.org/10.1006/geno.1998.5497
7. Vanoosthuyse V (2018) Strengths and weaknesses of the current strategies to map and characterize R-loops. Non-coding RNA 4:9. https://doi.org/10.3390/ncrna4020009
8. García-Rubio M, Barroso SI, Aguilera A (2018) Detection of DNA-RNA hybrids in vivo BT—genome instability: methods and protocols. In: Muzi-Falconi M, Brown GW (eds) Genome instability. Methods in molecular biology, vol 1672. Springer, New York, pp 347–361. https://doi.org/10.1007/978-1-4939-7306-4_24
9. Hartono SR, Malapert A, Legros P et al (2018) The affinity of the S9.6 antibody for double-stranded RNAs impacts the accurate mapping of R-loops in fission yeast. J Mol Biol 430: 272–284. https://doi.org/10.1016/j.jmb.2017.12.016
10. Chédin F, Hartono SR, Sanz LA, Vanoosthuyse V (2021) Best practices for the visualization, mapping, and manipulation of R-loops. EMBO J 40:1–13. https://doi.org/10.15252/embj.2020106394
11. Silva S, Camino LP, Aguilera A (2018) Human mitochondrial degradosome prevents harmful mitochondrial R loops and mitochondrial genome instability. Proc Natl Acad Sci U S A 115:11024–11029. https://doi.org/10.1073/pnas.1807258115
12. Wu H, Lima WF, Crooke ST (2001) Investigating the structure of human RNase H1 by site-directed mutagenesis. J Biol Chem 276: 23547–23553. https://doi.org/10.1074/jbc.M009676200
13. Bhatia V, Barroso SI, García-Rubio ML et al (2014) BRCA2 prevents R-loop accumulation and associates with TREX-2 mRNA export factor PCID2. Nature 511:362–365. https://doi.org/10.1038/nature13374
14. Chen L, Chen JY, Zhang X et al (2017) R-ChIP using inactive RNase H reveals dynamic coupling of R-loops with transcriptional pausing at gene promoters. Mol Cell 68:745–757.e5. https://doi.org/10.1016/j.molcel.2017.10.008
15. Crossley MP, Brickner JR, Song C et al (2021) Catalytically inactive, purified RNase H1: a specific and sensitive probe for RNA–DNA hybrid imaging. J Cell Biol 220:e202101092. https://doi.org/10.1083/jcb.202101092
16. Wang K, Wang H, Li C et al (2021) Genomic profiling of native R loops with a DNA-RNA hybrid recognition sensor. Sci Adv 7:1–17. https://doi.org/10.1126/sciadv.abe3516
17. Bernal JA, Roche M, Méndez-Vidal C et al (2008) Proliferative potential after DNA damage and non-homologous end joining are affected by loss of securin. Cell Death Differ 15:202–212. https://doi.org/10.1038/sj.cdd.4402254
18. Stewart SA, Dykxhoorn DM, Palliser D et al (2003) Lentivirus-delivered stable gene silencing by RNAi in primary cells. RNA 9:493–501. https://doi.org/10.1261/rna.2192803
19. Tampere M, Mortusewicz O (2016) DNA damage induction by laser microirradiation. Bio-Protocol 6:1–7. https://doi.org/10.21769/bioprotoc.2039

Chapter 9

Detection and Quantification of RNA Modifications on RNA–DNA Hybrids Using SID-UPLC-MS/MS

Marta Starczak, Abdulkadir Abakir, Alexey Ruzov, and Daniel Gackowski

Abstract

R-loops are three-stranded nucleic acid structures consisting of an RNA–DNA hybrid and an unpaired strand of nontemplate DNA that represent a major source of genomic instability and are involved in regulation of several important biological processes in eukaryotic cells. A growing body of experimental evidence suggests that RNA moieties of RNA–DNA hybrids may convey RNA modifications influencing various aspects of R-loop biology. Here we present a protocol for quantitative analysis of RNA modifications on RNA–DNA hybrids using stable-isotope dilution ultraperformance liquid chromatography coupled with tandem mass spectrometry (SID-UPLC-MS/MS). Supplemented by other techniques, this method can be instrumental in deciphering the roles of RNA modifications in R-loop metabolism.

Key words R-loops, RNA–DNA hybrids, RNA modifications, N6-methyladenosine, Tandem mass spectrometry, RNase H

1 Introduction

R-loops are non-B DNA structures that form by annealing of nascent RNA to template DNA during transcription and, thus, consist of an RNA–DNA hybrid and a nontemplate DNA strand [1, 2]. These structures frequently occur in the genome [3] retaining a dual biological significance: they are known as a critical source of replication stress, DNA damage, transcription-elongation defects and genomic instability [1, 4] and, at the same time, specific classes of R-loops may also contribute to gene regulation via promoting transcriptional termination, recruiting chromatin modifiers and determining the profiles of promoter DNA methylation [5].

Given the importance of R-loop regulation for a wide range of pathological and physiological processes, eukaryotic cells developed specific mechanisms for R-loop resolution that involve DNA/RNA helicases, nucleases and chromatin remodelling complexes [6–10]. In addition to these mechanisms, several recent studies imply

Andrés Aguilera and Alexey Ruzov (eds.), *R-Loops: Methods and Protocols*, Methods in Molecular Biology, vol. 2528, https://doi.org/10.1007/978-1-0716-2477-7_9,

that RNA components of RNA–DNA hybrids possess such RNA modifications as N6-methyladenosine (m^6A) [11–14], RNA 5-methylcytosine (m^5C) [15] and possibly 3-methylcytidine (m^3C) [16] that may regulate R-loop stability [11] or promote homologous recombination and DNA repair [15, 17], consequently, affecting different aspects of R-loop metabolism. It is worth noticing that most of these reports heavily rely on the use of anti-m^6A/m^5C antibodies, and, therefore, critically depend on their specificity and sensitivity. Moreover, the fact that the antibodies against RNA modifications readily detect their DNA counterparts (6mA, 5mC) in denaturing conditions, represents an additional major challenge for such studies [11]. This, together with the anticipated importance of RNA modifications for R-loop biology, warrants development of alternative techniques for analysis of RNA modifications in the R-loop context.

Here we describe a protocol for direct mass spectrometry-based analysis of RNA modifications that we have previously employed for detection and quantification of the m^6A present on RNA–DNA hybrids in human cancer and pluripotent stem cells [11]. This technique is based on the treatment of the total nucleic acids with RNase H, an enzyme that specifically digests the RNA moieties of RNA–DNA hybrids, followed by ultrafiltration of the samples and subsequent quantification of the modified RNA nucleosides released upon RNase H treatment (Fig. 1). This method may be instrumental for both quantification of the levels of RNA modifications on RNA–DNA hybrids and detection of novel modified RNA nucleosides on these structures in different cell lines and tissues. In combination with antibody-based mapping techniques and functional assays, this technique should contribute to elucidating the roles of different RNA modifications in R-loop biology.

2 Materials

Prepare all solutions using ultrapure water and analytical grade reagents. Prepare and store all reagents at room temperature (unless indicated otherwise).

2.1 Cell Culture

1. Essential 8TM medium.
2. Corning® Matrigel®.
3. Phosphate-buffered saline (PBS).
4. TrypLE™ Select Enzyme (ThermoFisher).
5. ROCK inhibitor (Tocris Bioscience).
6. Nonessential amino acids.
7. 100 μM β-mercaptoethanol
8. GlutaMAX (Gibco™).

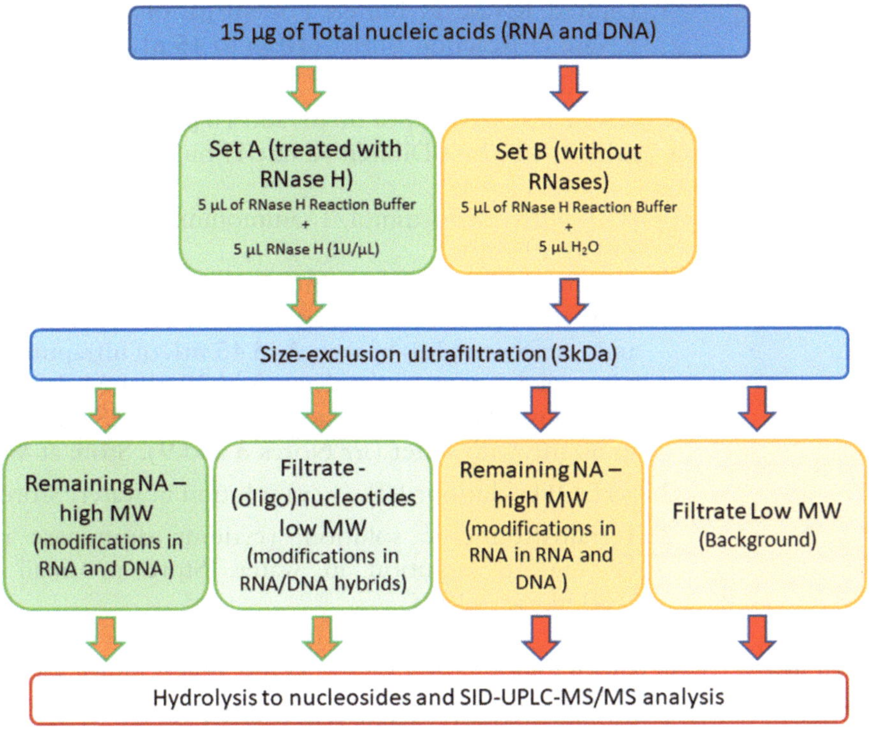

Fig. 1 Schematic illustration of the protocol for direct SID-UPLC-MS/MS analysis of RNA modifications on RNA–DNA hybrids

9. DMEM-F12.
10. Knockout Serum Replacement (Gibco™).

2.2 Total Nucleic Acid Extraction

1. Buffer B: 5 mmol/L Na_2EDTA, 0.15 mmol/L deferoxamine, 10 mmol/L Tris-HCl, pH 8.0.

   For 500 mL of buffer, add 930.6 mg Na_2EDTA, 605.7 mg Trizma base, and 49.26 mg deferoxamine to glass beaker containing about 450 mL water. Stir until completely dissolved and adjust the pH to 8.0 with HCl (*see* **Note 1**). Make up to 500 mL with water (*see* **Note 2**). Store at 4 °C (*see* **Note 3**).
2. SDS solution: 10% (m/v) solution in water (*see* **Note 4**).
3. Proteinase K solution: 20 mg/mL solution in water (*see* **Note 5**).
4. Chloroform–isoamyl alcohol mixture (24:1): For 500 mL of mixture, mix 20 mL of isoamyl alcohol and 480 mL chloroform in a tightly closed glass bottle (*see* **Note 6**).

2.3 RNase H Treatment of the Samples

The protocol is based on the use of New England Biolabs RNase H (M0297L) and RNase H Reaction Buffer.

1. Diluted RNase H buffer: For 50 μL of buffer add 5 μL of RNase H Reaction Buffer (10×) to 45 μL ultrapure water (*see* **Note 7**).
2. RNase H 1 U/μL: For 50 μL add 10 μL of RNase H (5000 U/mL) to 40 μL of Diluted RNase H buffer (*see* **Note 8**).

2.4 Nucleic Acids Hydrolysis to Nucleosides

1. NP1 buffer: 200 mmol/L ammonium acetate, 0.2 mmol/L $ZnCl_2$, pH 4.6.

   Prepare 1 mL of 1 M $ZnCl_2$ solution in water (136.28 mg/1 mL). Weight 770.8 mg ammonium acetate (LC/MS grade) and transfer to glass beaker. Add 45 mL of ultrapure water and stir until completely dissolved. Add 20 μL 1 M $ZnCl_2$ solution and adjust pH to 4.6 with acetic acid (LC/MS grade). Make up to 50 mL with water (*see* **Notes 3** and **9**). Store at 4 °C.
2. NH_4OH solution: 10% (v/v) solution in water. Store at 4 °C.
3. Tetrahydrouridine solution: (cytidine deaminase inhibitor) 10 mg/mL solution in water. Store in small portions (~30 μL) at −20 °C.

2.5 SID-UPLC-MS/MS Analysis

All reagents should be LC/MS grade, unless indicated otherwise.

1. Mobile phase A1: acetic acid 0.05% (v/v).

   For 2 L of mobile phase add 1000 μL of acetic acid to 2 L ultrapure filtered water, sonicate 15 min to mix and degas (*see* **Note 10**).
2. Mobile phase B1: acetonitrile (*see* **Note 11**).
3. Washing solutions—as necessary, depends on system configuration—see your system manual for reference.

   Chromatographic system components necessary to perform SID-UPLC-MS/MS analyses:
4. Gradient pumps (preferably high-pressure gradient) min. Operating pressure 15,000 psi.
5. Programmable six-port two-position switching valve min. Operating pressure 15,000 psi.
6. Column thermostat.
7. High-pressure tee.
8. PDA detector, preferably equipped with short path-length flow cell.
9. Ultra-sensitive tandem quadrupole mass spectrometer with electrospray or unispray ion source.
10. Column 1: CORTECS UPLC T3 Column, 120 Å, 1.6 μm, 3 mm × 150 mm with CORTECS T3 VanGuard Pre-column, 120 Å, 1.6 μm, 2.1 mm × 5 mm (Waters).
11. Tubes, connectors, steering workstation with software.

2.6 Preparation of the Mix of Internal Standards for MS Analysis

1. $[D_3]$ -N6-methyladenosine (m^6A).
2. $[^{15}N_5]$ -adenosine (A).
3. $[^{13}CD_3]$ -5-methylcytidine (m^5C).
4. $[^{13}CD_2]$ -5-hydroxymethylcytidine (hm^5C).
5. Spectrophotometer.

3 Methods

Here, we describe a protocol for the detection and quantification of RNA modifications on RNA–DNA hybrids in human pluripotent stem cells (hPSCs); however, this technique has also been validated on other cell lines (Exemple results of the analysis of RNA modification on RNA–DNA hybrids in human cancer and pluripotent stem cells are presented in Table 1) and can potentially be used for the detection and quantification of chemical modifications on RNA–DNA hybrids in various animal models.

Carry out all procedures at room temperature unless otherwise specified.

3.1 Cell Culture: Preparing Matrigel™ Coated Plates and Culturing hPSCs

Carry out all cell culture in a mycoplasma free Class II safety cabinet.

1. Thaw a stock of Matrigel™ solution in the fridge (5 °C) overnight.
2. Aseptically, transfer the thawed Matrigel™ solution to a 50 mL conical tube containing sterile DMEM/F-12 (*see* **Note 12**).
3. Add 6 mL of Matrigel™ solution to a sterile T75 cell culture flask.
4. Incubate the Matrigel™ coated flask for a minimum of 30 min at room temperature. Alternatively, the coated flask can also be incubated for 30 min at 37 °C/5% CO_2 to accelerate polymerisation (*see* **Note 13**).
5. Immediately before use, carefully aspirate the excess Matrigel™ solution, without dislodging the Matrigel® coating from the flask.
6. Add 10 mL of freshly prepared E8 medium supplemented with 10 μM rock inhibitor.
7. Plate the cell culture flask with 25,000 hPSCs/cm^2.
8. Rock the flask gently to ensure even distribution of cells across the flask and incubate at 37 °C with 5% CO_2.
9. The E8 medium is replaced every 24 h without the rock inhibitor supplementation until the cells are ready to passage, collect for DNA extraction or freeze (*see* **Note 14**).

Table 1
Exemplary results of the analysis of RNA modifications on RNA–DNA hybrids in human pluripotent stem cells and in leukocytes from patients with chronic lymphocytic leukemia (CLL)

	m^6A in ultrafiltrate [fmol/μg input DNA]		*m^5C in ultrafiltrate [fmol/μg input DNA]*		*m^6A in remaining NA [rN/1000 rA]*		*m^5C in remaining NA [rN/10^6dN]*		*hm^5C in remaining NA [rN/1000 rA]*		*References*
	RNase H	*Without RNases*	*RNase H*	*Without RNases*	*RNase H*	*Without RNases*	*RNase H*	*Without RNases*	*RNase H*	*Without RNases*	
ReBI PAT hiPSC	6.916	2.700	2.902	1.928	2.549	2.555	106.141	88.799	ND	ND	Abakir et al. [11] and unpublished data
HUES7 hESC	18.894	1.754	0.603	0.772	3.935	3.999	238.436	229.245	ND	ND	
	m^6A in ultrafiltrate [fmol/μg input DNA]		*m^5C in ultrafiltrate [fmol/μg input DNA]*		*m^6A in remaining NA [rN/1000 rA]*		*m^5C in remaining NA [rN/1000 rA]*		*hm^5C in remaining NA [rN/1000 rA]*		*References*
	RNase H	*Without RNases*	*RNase H*	*Without RNases*	*RNase H*	*Without RNases*	*RNase H*	*Without RNases*	*RNase H*	*Without RNases*	
CLL989	1.140	0.239	3.348	0.238	8.384	10.042	25.910	32.933	ND	0.088	Unpublished data
CLL1907	6.105	1.871	61.658	1.112	9.128	10.485	23.518	27.318	0.028	0.041	

3.2 Preparing the Cell Culture for DNA Extraction

1. Once the cells reach 70–80% confluency (approximately on the third day after plating), aspirate the spent E8 medium.
2. Add 10 mL of PBS to wash the cells.
3. Aspirate excess PBS, add 3 mL of TrypLE Select Enzyme and incubate the cells for 4 min at 37 °C in a humidified incubator with 5% CO_2 to dissociate the cells.
4. Inactivate TryPLE by adding 7 mL of BGK solution (DMEM-F12 medium supplemented with 100 μM β-mercaptoethanol, 1% GlutaMAX, 1% nonessential amino acids, 1% GlutaMAX, and 15% Knockout Serum Replacement).
5. Pellet the dissociated cells in 15 mL conical tube by centrifugation for 4 min at 160 × *g*.
6. Without removing the pellet, aspirate the supernatant (containing excess BGK solution) and add 10 mL of PBS.
7. Pellet the resuspended cells in 15 mL conical tube by centrifugation for 4 min at 160 × *g*.
8. The cells are ready for proteinase K digestion and DNA extraction.

3.3 Nucleic Acid Extraction from Cells

1. Disperse pellet of the cells in 2 mL of ice-cold buffer B (*see* **Notes 15** and **16**).
2. Add 100 μL 10% SDS (to final concentration of 0.5%) and gently mix the mixture using a polypropylene Pasteur pipette (*see* **Note 17**).
3. Incubate for 30 min at 37 °C.
4. Add 400 μL Proteinase K (to final concentration 4 mg/mL) and incubate the mixture for 1.5 h at 37 °C (*see* **Note 18**).
5. Cool the mixture and add 1 volume (~2.5 mL) of phenol–chloroform–isoamyl alcohol (25:24:1), and shake vigorously (for at least 30 s) (*see* **Note 19**).
6. Centrifuge for 15 min at 2800 × *g* at 4 °C.
7. Transfer aqueous layer (top) to new centrifuge tube (*see* **Note 20**).
8. Add 1 volume (~2 mL) of chloroform–isoamyl alcohol (24:1) mixture and shake vigorously (for at least 30 s) (*see* **Note 19**).
9. Centrifuge for 10 min at 2800 × *g* at 4 °C.
10. Transfer the aqueous (top) layer containing nucleic acids to new centrifuge tube as before (*see* **Note 20**).
11. Add 3 volumes of cold 96% ethanol to the supernatant in order to precipitate high molecular weight nucleic acids. Mix very gently (NA precipitation can be visible).

12. Remove the precipitate with a plastic spatula and wash with ethanol (*see* **Note 21**).
13. Place the NA in 0.5 mL microcentrifuge tube and add 150 μL water (*see* **Note 22**).

3.4 Digestion to Determine the Percentage of DNA

1. Prepare a NP1 reaction mixture containing 10 μL of NP1 buffer, 0.2 μL of nuclease P1 (100,000 U/mL), and 0.2 μL of tetrahydrouridine (10 mg/mL) per sample (*see* **Note 23**).
2. Mix 10 μL of the hydrated nucleic acid sample with 10.4 μL of NP1 reaction mixture. Mix thoroughly.
3. Incubate at 37 °C for 3 h.
4. Prepare rSAP reaction mixture containing 2.6 μL of 10% NH_4OH and 1.2 μL of shrimp alkaline phosphatase (1000 U/mL) per sample.
5. Add 3.8 μL of rSAP reaction mixture per sample. Mix thoroughly (*see* **Note 24**).
6. Incubate at 37 °C for 1 h.
7. Perform a chromatographic separation to find out the ratio of guanosine to 2′-deoxyguanosine (the ratio of RNA to DNA in the sample) (*see* **Note 25**).

3.5 RNase H Treatment of the Samples

1. Check the concentrations of the compounds using spectrophotometer.
2. For RNase H treatment use 15 μg of DNA per sample (*see* **Note 26**).
3. Prepare two identical sets of samples: Set A (treated with RNase H) and Set B (without RNase) (*see* **Note 27**).
4. Make up each sample with ultrapure water to a final volume of 45 μL.
5. To all samples add 5 μL RNase H Reaction Buffer (10×) (*see* **Note 28**).
6. To samples from Set A add 5 μL of freshly prepared RNase H (1 U/μL), to samples from Set B add 5 μL ultrapure water.
7. Incubate at 37 °C overnight (*see* **Note 29**).

3.6 Ultrafiltration

In order to separate individual fractions (Fig. 1), we use ultrafiltration on Amicon Ultra-0.5 MWCO 3 kDa Centrifugal Filter (Merck) (*see* **Note 30**).

1. Prepare the filtration device, insert the filter into the provided microcentrifuge tube.
2. Apply 150 μL of ultrapure water on the membrane and spin for 10 min at 14,000 × *g* at 22 °C (*see* **Note 31**).
3. Transfer the filter to the new microcentrifuge tube, add ~50 μL of sample to filtration device and cap it (*see* **Notes 32** and **33**).

4. Centrifuge for 15 min at 14,000 × *g* at 22 °C.
5. Add 50 μL of ultrapure water and centrifuge for 25 min at 14,000 × *g* at 22 °C. Repeat this step twice.
6. Separate filter from the microcentrifuge tube. Save tubes containing ultrafiltrate for further analysis.
7. To recover the nucleic acids place the filter upside down in a clean microcentrifuge tube and centrifuge at 1000 × *g* for 5 min at 22 °C (*see* **Note 34**).

3.7 Hydrolysis to Nucleosides

1. Prepare a NP1 reaction mixture 1 (for the ultrafiltrates): containing 100 μL of NP1 buffer, 0.5 μL of nuclease P1 (100,000 U/mL) and 0.5 μL of tetrahydrouridine (10 mg/mL) per sample (*see* **Notes 23** and **35**).
2. Prepare a NP1 reaction mixture 2 (for the remaining NA): containing 50 μL of NP1 buffer, 1 μL of nuclease P1 (100,000 U/mL), and 1 μL of tetrahydrouridine (10 mg/mL) per sample (*see* **Notes 23** and **35**).
3. Mix the ultrafiltrate (~100 μL) with 101 μL of NP1 reaction mixture 1. Mix thoroughly (*see* **Note 36**).
4. Mix the remaining NA sample (~50 μL) with 52 μL of NP1 reaction mixture 2. Mix thoroughly (*see* **Note 36**).
5. Incubate at 37 °C overnight.
6. Prepare rSAP reaction mixture 1 (for the ultrafiltrates): containing 15 μL of 10% NH_4OH and 3 μL of shrimp alkaline phosphatase (1000 U/mL) per sample.
7. Prepare rSAP reaction mixture 2 (for the remaining NA): containing 13 μL of 10% NH_4OH and 6 μL of shrimp alkaline phosphatase (1000 U/mL) per sample.
8. Add 18 μL of rSAP reaction mixture 1 to ultrafiltrates and 18 μL of rSAP reaction mixture 2 to remaining NA samples. Mix thoroughly (*see* **Note 37**).
9. Incubate at 37 °C for 1.5 h.
10. Samples should be ultrafiltered prior to injection (*see* **Note 38**).
11. Concentrate the remaining NA samples in the vacuum concentrator (to final volume ~ 25 μL) (*see* **Note 39**).
12. Completely dry the ultrafiltrates in the vacuum concentrator (*see* **Note 39**).
13. Spike remaining NA hydrolysates with a solution of internal standard in volumetric ratio 4:1 (20 μL remaining NA hydrolysate +5 μL internal standard solution) (*see* **Note 40**).
14. Resuspend the ultrafiltrates in 20 μL of ultrapure water and spike with a solution of internal standard in volumetric ratio 4: 1 (20 μL remaining NA hydrolysate +5 μL internal standard solution) (*see* **Note 40**).

Table 2
Wavelengths, ε values, and concentrations of the internal standards used for SID-UPLC-MS/MS

Compound	Stock solution concentration [mmol/L] (fmol/μL)	Concentration in IS mixture [fmol/μL]	Concentration in injection [fmol/μL]	Wavelength (nm); ε
$[D_3]$ -m^{6}A	0.05 (50000)	1	0.2	266; 11,490
$[^{15}N_5]$ -A		25	5	259; 15,400
$[^{13}CD_3]$ -m^{5}C				278; 7283
$[^{13}CD_2]$ -hm^{5}C				273; 7640

3.8 Preparation of the Mix of Internal Standards for MS Analysis

1. Prepare solutions of standards: $[^{15}N_5]$ -A, $[D_3]$ -m^{6}A, $[^{13}CD_3]$ -m^{5}C, and $[^{13}CD_2]$ - hm^{5}C in ultrapure water.
2. Check the concentrations of the compounds using spectrophotometer (the corresponding wavelengths and ε values are listed in Table 2. Calculate the concentration using the following formula.

$$c = \frac{A}{\varepsilon \times l}$$

A—measured absorbance (relative to pure solvent), *ε*—molar coefficient of extinction $[M^{-1}\ cm^{-1}]$, *c*—concentration of the substance [M], *l*—layer thickness [cm].

3. Prepare the stock solutions of the compounds (0.05 mM) by their diluting with ultrapure water (*see* **Note 41**).
4. Prepare the internal standard mixture (550 μL per 100 samples) with 1 fmol/μL of $[D_3]$ -m^{6}A and 25 fmol/μL of other compounds. The compounds and their concentrations are listed in Table 2 (*see* **Note 42**).

3.9 SID-UPLC-MS/MS Analysis

1. Tune and optimize mass spectrometer signal using solutions of analyzed compounds.
2. Prepare LC gradient separation methods using data in Table 3. Optimize gradient shape and concentration of acetic acid in mobile phase 1 to assure baseline separation of the peaks of interest.
3. Prepare MRM acquisition method using data in Table 4 or own optimized values.
4. Keep column in 50 °C.
5. Prepare calibration samples sets, separate for compound analysed in UV and MS/MS next perform 5 injection per calibration point plot calibration curves, and calculate calibration equations using internal standard mode for MS traces and external standard mode for UV traces.

Table 3
Conditions of chromatographic separation

Time	Flow (mL/min)	%A	%B	Curve
1st dimension chromatography				
Initial	0.5	99.3	0.7	Initial
0.5	0.5	99.3	0.7	6
5	0.5	50	50	7
7.2	0.5	50	50	6
7.21	0.5	99.3	0.7	6
9	0.5	99.3	0.7	6

6. To assess recovery and system imprecision spike samples of pooled DNA, extracted from desired material, with ca. one, two, and four times of expected analyte concentration and analyze in triplicate, together with unspiked samples. Subsequently calculate recovery for each spiked concentration and imprecision of the method. Plot observed vs expected curves; accept method accuracy when extension of curve is within value of 90–110% of the adduct concentration without adding the analyte.
7. Prepare sample batch and perform analyses of in three to five technical replicates. Adjust injection volume to current sample volume and concentration (this method works well in a broad range of injection volumes, from 0.2 μL to 4 μL).
8. Calculate concentrations of native deoxynucleosides (d*N*) in every sample. To determine amount of deoxynucleosides in samples derived from dsDNA, quantify d*G* and d*T* and calculate d$N = 2 \times (\mathrm{d}G + \mathrm{d}T)$. In ssDNA samples quantification of all deoxynucleosides is necessary to calculate total d*N* level in the sample. Calculate mean, SD and RSD from all technical replicates. Check system performance and reanalyze sample and when RSD $> 2.5\%$.
9. Calculate concentrations of all modified nucleosides and rA in the sample, using proper blank subtraction if necessary (*see* **Note 43**). Calculate mean, SD and RSD from all technical replicates. Check system performance and reanalyze sample and when RSD $> 15\%$.
10. Calculate relative amounts of modified nucleosides in remaining material dividing its concentration by total concentration of rA.
11. Calculate total content of RNA–DNA hybrids dividing amount of RNase H released rA fraction by amount of input DNA, applying corrections for nonspecific rN release and DNA purity in input sample.

Table 4
Transition patterns-specific detector settings for the analyzed compounds

Compound name	Ionization mode	Nominal molecular mass (Da)	Pseudomolecular ion formulation	Nominal parent ion (Da)	Nominal daughter ion (Da)	ESI capillary (kV)	ESI cone (V)	US (kV)	US cone (V)	Collision energy (eV)
5-Hydroxymethyl-cytidine	Positive	273	[M + H]+	274	142	1.2	15	0.4	18	12
[$^{13}CD_2$] - 5-Hydroxymethyl-cytidine	Positive	260	[(M + 3) + H]+	277	145	1.2	15	0.4	18	12
5-Methylcytidine	Positive	257	[M + H]+	258	126	1.2	18	0.4	18	14
[$^{13}CD_3$] - 5-Methyl-cytidine	Positive	261	[(M + 4) + H]+	262	130	1.2	18	0.4	18	14
N6-Methyladenosine	Positive	281	[M + H]+	282	150	1.2	9	0.4	10	18
[D_3] - N6-Methyladenosine	Positive	284	[(M + 3) + H]+	285	153	1.2	9	0.4	10	18
Adenosine	Positive	267	[M + H]+	268	136	1.2	6	0.4	6	16
[$^{15}N_5$] - Adenosine	Positive	272	[(M + 5) + H]+	273	141	1.2	6	0.4	6	16

4 Notes

1. We recommend using diluted HCl (3–4% v/v) to avoid a sudden drop in pH below the required value.
2. Check the pH of the buffer again before first use. If necessary, adjust the pH to 8 with diluted HCl.
3. Buffers can be stored at 4 °C for up to 1 month unless you notice signs of bacterial/fungal/other contamination.
4. If possible, all work should be completed under fume hood in a sealed tube. Leave the solution on a stirrer overnight (it foams and is hardly soluble). Heating to ~37 °C will facilitate dissolution. SDS can precipitate even at room temperature, therefore, the solution may need to be heated before use.
5. Proteinase K solution should be prepared fresh each time. When preparing the solution, make an excess of it (at least 2 additional volumes).
6. All the work should be completed in a fume hood and the mixture should be stored under it.
7. Diluted RNase H buffer should be prepared fresh each time. When preparing the solution, make an excess of it (at least 2 additional volumes).
8. RNase H (1 U/μL) cannot be frozen because it loses its activity. Prepare fresh enzyme solution each time.
9. Check the pH of the buffer again before the first use. If necessary, adjust the pH to 4.6 with diluted acetic acid (LC/MS grade).
10. To prepare mobile phases, use clean, borosilicate glass bottles. Do not wash the glass with detergents or using dish-washer. Flush bottle with 5 to 10 ca.100 mL portions of LC/MS grade water.
11. Do not acidify organic solvents, it may cause high background noise in some MRM transitions.
12. Gently pipette up and down (without generating bubbles) to obtain a homogenous solution.
13. Unless used immediately, the coated flaks can be stored at 4 °C for a maximum of 1 week. Ensure stored vessels are equilibrated to room temperature 30 min prior to use.
14. For consistent and accurate results, daily monitoring of the cells in culture under a light microscope (e.g., using a phase contrast microscopy) to check for abnormal changes in morphology, growth rate, presence of differentiated cells, and so on, is critical. Immediately discard cultures in case of any abnormal changes and restart with a fresh vial of hPSCs with a relatively low passage, in aseptic cell culture conditions.

15. The buffer volume must be adjusted to the amount of cells. Remember that both too small and too large volumes of buffer can cause DNA isolation to fail.
16. Do not use any RNase when isolating genetic material. Treatment with RNase A–RNase T1 mixture substantially depleted fraction of RNA released by RNAse H.
17. After adding SDS, the sample should become denser/more viscous. Some samples become viscous after incubation at 37 °C. Sometimes the lysate does not thicken (especially when very little material is used) which does not usually lead to the failure of the DNA extraction. If the sample is extremely viscous (cannot be mixed with a Pasteur pipet), dilute it with buffer B.
18. After 45 min incubation with proteinase K, mix each sample with a Pasteur pipette. After this time, mixture should be less viscous (there should not be any problems with pipetting). If the samples are still thick, consider using one of the following solutions: (a) add buffer B to dilute the sample, (b) add another portion of proteinase K solution, (c) extend the incubation time. When the samples are extremely viscous, we suggest using a combination of all three approaches listed above.
19. Phenol causes severe burns and chloroform is a carcinogen: wear gloves, goggles, and lab coat. Work under the fume hood if possible. Check if your tubes are tight when the phenol–chloroform is centrifuged. We know from our experience that some test tubes start to leak when they come in contact with phenol or chloroform. We strongly recommend experimentally checking the tightness of the used tubes. Keep tubes capped tightly.
20. Be careful to avoid transferring the unwanted proteins (concentrated in the interface between the aqueous and organic layers).
21. Carefully wind the genetic material into a plastic spatula and transfer to a small container with 70% ethanol. Wash for 30–40 s, then gently dry on filter paper. Repeat this process 2–3 times. If DNA sticks to the paper, place it in the ethanol for longer time. If the material adheres to the filter paper, it may be difficult to recover it.
22. For further analysis, the obtained genetic material must be well hydrated. For this reason, let it rehydrate overnight.
23. The presented DNA hydrolysis protocol is based on the use of New England Biolabs enzymes [Nuclease P1 M0660S, shrimp alkaline phosphatase (rSAP) M0371L]. The amounts of enzymes used and the incubation times described were experimentally adjusted. If you use other enzymes, perform a test hydrolysis to select the optimal conditions.

24. Check the pH of the sample using a universal pH indicator. The pH should be at least 8.0.
25. Another method can be used to evaluate the percentage of DNA. However, it is very important that each sample for RNase H treatment has the same DNA input.
26. The minimum amount of DNA for reliable quantification of m^6A and m^5C is 5 μg; however, we recommend using 15 μg of DNA.
27. Remember that the final sample volume is 45 μL. For this reason, the volume of collected DNA must be ≤45 μL.
28. The presented protocol is based on the use of New England Biolabs RNase H (M0297L) and RNase H Reaction Buffer (10×). The amounts of enzyme used and the incubation time described were experimentally adjusted. If you use other enzymes, perform a test hydrolysis to select the optimal conditions for each.
29. Try to reduce sample evaporation and condensation on the lid of the tube.
30. Attention is required, that defective filters may occur, which allow fragments with a molecular weight greater than 3 kDa to pass through.
31. The purpose of this step is to wash out production impurities (e.g., glycerin) from the membrane, that could otherwise interfere with further analysis. Do not let the filter membrane dry out.
32. As the dead volume of the filtering device is ~40 μL, remember to remove any residual water before adding the sample.
33. Spin the samples before applying on filter to remove any drops from the lid.
34. After the procedure, approximately 100–110 μL of ultrafiltrate (containing released (oligo) nucleotides with molecular weight less than 3 kDa) and ~50 μL of remaining nucleic acids should be recovered.
35. The volume of NP1 buffer must be equal to the volume of the DNA sample. The amount of the enzymes should be matched proportionally.
36. We suggest creating a blank sample (100/50 μL water + 101/52 μL NP1 reaction mixture1/2) and handling it in parallel with the other samples. It allows to eliminate errors in the quantitative analysis of nucleosides via estimation of the background levels of nucleosides derived from the enzymes used for nucleic acids digestion.
37. Check the pH of the sample using a universal pH indicator. The pH should be at least 8.0. If it is lower, add 1 μL of 10% NH_4OH, mix the sample and check the pH again.

38. In our laboratory, we use filter plates with modified polyethersulfone membranes (10K) and centrifuge them for 1 h 30 min at 1800 × *g* at 4 °C. Filter plates should be tested for the filtration conditions and for background MS noise.
39. The drying time depends on the number of the vials in vacuum concentrator and on the system itself.
40. The crucial step here is to assure that the method accuracy is precise by volumetric mixing of the sample and mixture of internal standards.
41. The standards stock solutions can be stored at −20 °C for several months.
42. The working solutions of internal standards can be aliquoted and stored at −20 °C for several months.
43. In every batch of samples prepare "blank" sample containing internal standards which is processed as the sample (RNase H treated, hydrolyzed, and ultrafiltered) to monitor system performance and purity of used chemicals. This also allows to monitor matrix effect in every batch, not only during method validation.

References

1. Garcia-Muse T, Aguilera A (2019) R loops: from physiological to pathological roles. Cell 179(3):604–618. https://doi.org/10.1016/j.cell.2019.08.055
2. Chedin F, Benham CJ (2020) Emerging roles for R-loop structures in the management of topological stress. J Biol Chem 295(14): 4684–4695. https://doi.org/10.1074/jbc.REV119.006364
3. Nadel J, Athanasiadou R, Lemetre C et al (2015) RNA:DNA hybrids in the human genome have distinctive nucleotide characteristics, chromatin composition, and transcriptional relationships. Epigenetics Chromatin 8: 46. https://doi.org/10.1186/s13072-015-0040-6
4. Gomez-Gonzalez B, Aguilera A (2019) Transcription-mediated replication hindrance: a major driver of genome instability. Genes Dev 33(15–16):1008–1026. https://doi.org/10.1101/gad.324517.119
5. Niehrs C, Luke B (2020) Regulatory R-loops as facilitators of gene expression and genome stability. Nat Rev Mol Cell Biol 21(3): 167–178. https://doi.org/10.1038/s41580-019-0206-3
6. San Martin-Alonso M, Soler-Oliva ME, Garcia-Rubio M et al (2021) Harmful R-loops are prevented via different cell cycle-specific mechanisms. Nat Commun 12(1): 4451. https://doi.org/10.1038/s41467-021-24737-x
7. Yuan W, Al-Hadid Q, Wang Z et al (2021) TDRD3 promotes DHX9 chromatin recruitment and R-loop resolution. Nucleic Acids Res 49(15):8573–8591. https://doi.org/10.1093/nar/gkab642
8. Liu Z, Gao X, Zhou Z et al (2021) San1 deficiency leads to cardiomyopathy due to excessive R-loop-associated DNA damage and cardiomyocyte hypoplasia. Biochim Biophys Acta Mol basis Dis 1867(11):166237. https://doi.org/10.1016/j.bbadis.2021.166237
9. Bayona-Feliu A, Barroso S, Munoz S et al (2021) The SWI/SNF chromatin remodeling complex helps resolve R-loop-mediated transcription-replication conflicts. Nat Genet 53(7):1050–1063. https://doi.org/10.1038/s41588-021-00867-2
10. Abakir A, Ruzov A (2021) SWI/SNF complexes as determinants of R-loop metabolism. Nat Genet 53(7):940–941. https://doi.org/10.1038/s41588-021-00884-1
11. Abakir A, Giles TC, Cristini A et al (2020) N (6)-methyladenosine regulates the stability of RNA:DNA hybrids in human cells. Nat

Genet 52(1):48–55. https://doi.org/10.1038/s41588-019-0549-x

12. Yang X, Liu QL, Xu W et al (2019) m(6)A promotes R-loop formation to facilitate transcription termination. Cell Res 29(12):1035–1038. https://doi.org/10.1038/s41422-019-0235-7
13. Marnef A, Legube G (2020) m(6)A RNA modification as a new player in R-loop regulation. Nat Genet 52(1):27–28. https://doi.org/10.1038/s41588-019-0563-z
14. Kang HJ, Cheon NY, Park H et al (2021) TonEBP recognizes R-loops and initiates m6A RNA methylation for R-loop resolution. Nucleic Acids Res 49(1):269–284. https://doi.org/10.1093/nar/gkaa1162
15. Chen H, Yang H, Zhu X et al (2020) m(5)C modification of mRNA serves a DNA damage code to promote homologous recombination. Nat Commun 11(1):2834. https://doi.org/10.1038/s41467-020-16722-7
16. Zhang LH, Zhang XY, Hu T et al (2020) The SUMOylated METTL8 Induces R-loop and tumorigenesis via m3C. iScience 23(3):100968. https://doi.org/10.1016/j.isci.2020.100968
17. Zhang C, Chen L, Peng D et al (2020) METTL3 and N6-methyladenosine promote homologous recombination-mediated repair of DSBs by modulating DNA-RNA hybrid accumulation. Mol Cell 79(3):425–442.e427. https://doi.org/10.1016/j.molcel.2020.06.017

Chapter 10

DNA–RNA Hybrids at Telomeres in Budding Yeast

Carolin B. Wagner and Brian Luke

Abstract

It has recently been demonstrated that budding yeast telomeres are transcribed into TERRA, a long noncoding RNA. Due to the G-rich nature of the coding strand, TERRA has a tendency to form DNA–RNA hybrids and potentially R-loops, which in turn, promote repair at short telomeres. Here, we report two methods to detect DNA–RNA hybrids at yeast telomeres, namely, DRIP, which employs the S9.6 hybrid-recognizing antibody, and R-ChIP, which takes advantage of a catalytic dead form of RNase H1 (Rnh1-cd). We use cross-linked material for both protocols as we have found that this does not negatively affect recovered material, and furthermore allows the precipitation of other proteins from the identical cross-linked material. Although both methods are successful in terms of detecting DNA–RNA hybrids at telomeres, the R-ChIP method yields an approximately ten-fold increased enrichment.

Key words DNA–RNA hybrids, R-loops, DRIP, R-ChIP, S9.6 antibody, Telomeres, Yeast

1 Introduction

R-loops are three stranded DNA–RNA hybrid-containing structures, in which one DNA strand is displaced by an RNA strand. They arise dynamically in a genome-wide manner [1]. The majority of R-loops are thought to form cotranscriptionally, although it has been suggested that they may act *in trans* in some instances [2–4]. R-loops are involved in the regulation of DNA transactions including transcription regulation, chromatin state establishment and DNA repair [5]. R-loops can be removed by nucleolytic degradation of the RNA strand by the DNA–RNA hybrid specific ribonucleases RNase H1 or RNase H2 [6]. If they are not removed in a timely manner, they can lead to DNA damage and genome instability following an encounter with the DNA replication machinery [7]. For insights into the mechanistic and regulatory functions of R-loops, it is essential to be able to reproducibly measure R-loop levels, either genome-wide or at specific loci of interest. We use budding yeast, *Saccharomyces cerevisiae,* to study the noncoding RNA, TERRA (Telomere repeat-containing RNA) and its

Andrés Aguilera and Alexey Ruzov (eds.), *R-Loops: Methods and Protocols*, Methods in Molecular Biology, vol. 2528,
https://doi.org/10.1007/978-1-0716-2477-7_10,

functions at telomeres [8]. We have previously demonstrated that TERRA forms DNA–RNA hybrids at short telomeres, which promotes homology-directed repair. The repair of short telomeres is critical to prevent the rapid onset of replicative senescence [9].

DNA–RNA hybrid immunoprecipitation (DRIP) is a well-established method for detecting R-loops at specific loci [10, 11]. In this technique, R-loops are immunoprecipitated through the use of the S9.6 antibody, which recognizes DNA–RNA hybrids. In the context of TERRA, the immunoprecipitated DNA is analyzed by qPCR using subtelomere-specific oligonucleotides (Fig. 1) to determine the relative R-loop levels. Technical variations of DRIP methodology are possible, but our method relies on cross-linking and sonication to fragment the chromatin. We prefer this method for telomeres because (a) we can use the same samples to perform ChIP on other proteins of interest, (b) the yield is comparable to restriction digested DNA, and (c) it provides the most reproducible results in our hands. We do not use S1 nuclease treatment prior to sonication, as has previously been suggested [12], as this did not improve DNA–RNA hybrid recovery at the telomere. The combination of DRIP qPCR with the overexpression of RNase H1 or the catalytically dead version of RNase H1 (Rnh1-cd)(D193N) [13] enables studies of R-loop levels in hybrid destabilizing or stabilizing conditions, respectively. Furthermore, the use of Rnh1 and Rnh1-cd can serve as experimental controls. Epitope tagging of overexpressed Rnh1 and Rnh1-cd allow ChIP to capture DNA–RNA hybrids bound by Rnh1 or Rnh1-cd (R-ChIP), hence giving an alternative to the S9.6 antibody approach [14]. In the R-ChIP protocol, an inducible system is used for overexpression of Rnh1 and Rnh1-cd, which allows studies of the short term effects of overexpression, and prevents damage accumulation due to persisting R-loop stabilization by Rnh1-cd. Here, we outline both protocols, DRIP and R-ChIP (Fig. 1) and show that R-ChIP may be more suited to detect DNA–RNA hybrids at yeast telomeres.

2 Materials

All used chemicals should be of high quality. For the DRIP procedure, the pH of the buffers must be very precise. We have observed more reproducible results when we use commercial buffer stocks for buffer preparation.

2.1 Yeast Strains

The experiments are carried out in an *rnh1Δ rnh201Δ* (BY4741, MATa, *his3Δ1, leu2Δ0, ura3Δ0, met15Δ0, rnh1::KANMX, rnh201::NATMX*) background. The strains are transformed with an empty vector and the plasmids for overexpression of *RNH1-HA* and *RNH1-cd-HA* (Table 1).

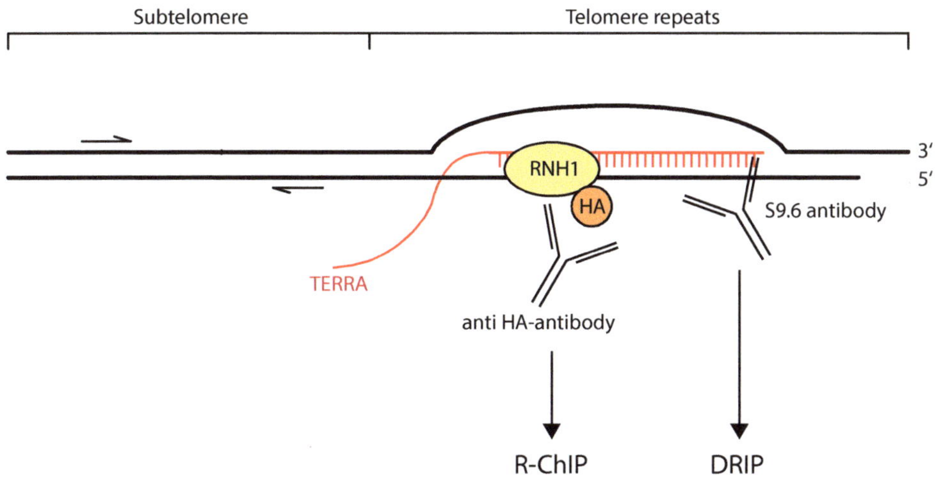

Fig. 1 To detect telomere TERRA DNA–RNA hybrids at telomeres we employ 2 approaches. Either the S9.6 (hybrid recognizing) antibody is used for DRIP, or an epitope (HA)-tagged version of RNase H1 (RNH1) is used for R-loop ChIP (R-ChIP). Specific PCR primers in the subtelomere can be used to amplify individual telomeres (e.g., 1 L = left arm of chromosome 1)

Table 1
Overexpression plasmids

Plasmid name	Strain description	Systematic name	Selectivity marker
pBL211	Empty vector (EV)	pRS425 pGal	*LEU2*
pBL352	*RNH1* overexpression	pRS425 pGal-*RNH1*-HA	*LEU2*
pBL633	*RNH1*-cd overexpression	pRS425 pGal-*RNH1*(D193N)-HA	*LEU2*

2.2 Culture Growth

1. Agar plates (Synthetic dropout (SC) minus leucine: 1.66 g SC-Leu powder, 6.7 g Difco yeast nitrogen base without amino acids (BD) and 24 g agar, dissolved in ddH_2O to 900 ml final and autoclaved for 20 min. Before casting the plates, 100 ml 20% glucose (w/v) is added).
2. SC-Leucine liquid medium containing 2% raffinose: 1.66 g SC-Leu powder and 6.7 g Difco yeast nitrogen base without amino acids (BD), dissolved in ddH_2O to 900 ml final and autoclaved for 20 min. 100 ml 20% (w/v) raffinose is added to the cooled media (*see* **Note 1**).
3. SC-Leucine liquid medium containing 1% raffinose and 2% galactose (w/v): 1.66 g SC-Leu powder and 6.7 g Difco yeast nitrogen base without amino acids (BD), dissolved in ddH_2O to 950 ml final and autoclaved for 20 min. 100 ml 20% galactose (w/v) and 50 ml 20% raffinose (w/v) are added to the cooled media.

4. 1 l Erlenmeyer flasks
5. 50 ml bioreactor tubes.

2.3 Cross-Linking of Cell Cultures

1. 37% formaldehyde (w/v)
2. 2.5 M glycine
3. 1× phosphate buffered saline (PBS, pH 7.4)
4. Ice.
5. 15 ml centrifuge tubes.

2.4 Cell Lysis

1. Homogenizer (MP Biomedicals Fastprep-24).
2. Lysing Matrix C tubes (MP Biomedicals).
3. FA lysis buffer minus SOD: 50 mM HEPES-KOH (pH 7.5), 140 mM NaCl, 1 mM ethylenediaminetetraacetic acid (EDTA), 1% Triton X-100 (v/v).
4. FA lysis buffer plus SOD: 50 mM HEPES-KOH (pH 7.5), 140 mM NaCl, 1 mM EDTA, 1% Triton X-100 (v/v), 0.1% sodium deoxycholate (SOD, w/v).
5. cOmplete™ Mini EDTA-free protease inhibitor cocktail tablets (Roche),
6. Refrigerated tabletop centrifuge.
7. 20% sodium dodecyl sulfate (w/v) (SDS).

2.5 Sonication

1. 15 ml Bioruptor Pico tubes (Diagenode).
2. Bioruptor sonication beads (Diagenode).
3. Bioruptor Pico (Diagenode).

2.6 Immunoprecipitation

1. Bradford reagent.
2. Spectrophotometer.
3. Dynabeads™ Protein G (Invitrogen).
4. Bovine serum albumin (BSA) 20 mg/ml, molecular biology grade.
5. Magnetic racks for 1.5 ml centrifugation tubes and 15 ml centrifugation tubes.
6. Rotating wheel for 1.5 ml centrifugation tubes.
7. Heating block for 1.5 ml centrifugation tubes.
8. FA 500/lysis buffer: 50 mM HEPES-KOH (pH 7.5), 500 mM NaCl, 1 mM EDTA, 1% Triton X-100 (v/v), 0.1% sodium deoxycholate (w/v).
9. Buffer III: 10 mM Tris–HCl (pH 8.0), 1 mM EDTA, 250 mM LiCl, 1% Tergitol-type NP-40 (v/v), 1% sodium deoxycholate (w/v).
10. 1× TE-buffer: 0.1 M Tris–HCl (pH 8.0), 0.05 M EDTA.

11. Elution buffer B: 50 mM Tris–HCl (pH 7.5), 1% SDS (w/v), 10 mM EDTA.
12. Rat anti-HA high affinity antibody clone 3F10 (Roche).
13. S9.6 antibody.
14. Proteinase K, 20 mg/ml.
15. QIAquick PCR purification kit (Qiagen).

2.7 Verification of Sonication

1. 100% EtOH
2. 10 mg/ml RNase A, DNase and Proteinase free.
3. 1× TBE buffer: 89 mM Tris base (pH 8.0), 89 mM boric acid, 2 mM EDTA.
4. Agarose.
5. SYBR safe DNA Gel stain (Invitrogen).
6. DNA loading dye: 10 mM EDTA (pH 8.0), 15% Ficoll Type 400 (w/v), Orange G.
7. Agarose electrophoresis chamber.
8. Power supply.

2.8 qPCR

1. qPCR cycler.
2. qPCR well plate and sealing tape.
3. PCR Primers (*see* Table 2).
4. DYNAMO FLASH SYBR GREEN qPCR master mix (Thermo Scientific).
5. Bio-Rad CFX Manager (Software for qPCR analysis).

Table 2
Primers for qPCR

Targeted telomere	Forward	Reverse
1L	CGGTGGGTGAGTGGTAGTAAG TAGA	ACCCTGTCCCATTCAACCATAC
6 Y′	GGCTTGGAGGAGACGTACATG	CTCGCTGTCACTCCTTACCCG
13R	ACGGTTATGGTGCACGATGGG	GATTTAGCATGTCTCTTAACTTACCC TCC
15L	GGGTAACGAGTGGGGAGGTAA	CAACACTACCCTAATCTAACCCTGT

3 Methods

3.1 Culture Growth

1. Strains are streaked out from cryo stocks (−80 °C) onto a SC-Leucine agar plate and incubated at an appropriate temperature (usually 30 °C) for 3 days.
2. Single colonies are patched onto a fresh agar plate. The patches should be at least 1 cm^2 in size. The plate is incubated overnight at an appropriate temperature.
3. Overnight cultures of the strains are inoculated in 10 ml of SC-Leu + 2% Raffinose using the entire patch to ensure sufficient cell density on the following day. The overnight cultures are incubated shaking at 200 rpm at an appropriate temperature.
4. The overnight cultures are diluted to OD_{600} 0.2 in 200 ml of SC-Leu medium containing 1% Raffinose and 2% Galactose (w/v) in 1 L Erlenmeyer flasks. The cultures are incubated at an appropriate temperature until they reach at least OD_{600} 0.8. This takes about 8 h (*see* **Note 2**).

3.2 Cross-Linking of Cell Cultures

Make sure everything needed for cross-linking is prepared to maintain precise timing, which prevents differences in the cross-linking efficiencies. Formaldehyde and glycine should be aliquoted in 15 ml centrifugation tubes for every sample in advance. Big ice buckets that can hold all used flasks should be ready when starting to cross-link.

1. The OD_{600} is adjusted to precisely 0.8 in 200 ml total volume with fresh medium for all cultures (*see* **Note 3**).
2. The cross-linking is started by adding 6.5 ml 37% formaldehyde (0.84% final volume) to each culture at room temperature (RT). For precision, formaldehyde should be added with exactly 30 s intervals between different cultures.
3. The cross-linking reaction is incubated for 10 min at RT. The flasks are shaken every few minutes by hand.
4. After 10 min, 10 ml 2.5 M glycine are added. The quenching reaction is carried out for exactly 5 min at RT; again, shaking the flasks periodically.
5. Following the quenching reaction, flasks are placed on ice, for between 5 min and 2 h (*see* **Note 4**).
6. The cells from each culture are collected in a single 50 ml tube in four repeated steps of centrifugation at 4 °C and 3000 × *g* for 3 min to collect all 200 ml of the cell culture.

7. The collected cell pellets are washed twice in 20 ml ice cold 1× PBS.
8. The cell pellets are transferred to 1.5 ml centrifugation tubes and spun down again at 4 °C and 3000 × *g* for 3 min. All supernatant is removed, and the cell pellets are stored at −80 °C until further processing.

3.3 Cell Lysis

All steps until the elution step should be carried out on ice. The working bench and pipets should be cleaned in advance to prevent RNase contamination.

1. The cell pellets are thawed on ice.
2. In the meantime, 10 ml FA lysis buffer *minus* SOD is supplemented with one protease inhibitor tablet, and for each sample 5 ml FA lysis buffer *plus* SOD is supplemented with 0.5 protease inhibitor tablet (*see* **Note 5**).
3. Each cell pellet is resuspended in 200 μl FA lysis buffer minus SOD with protease inhibitor.
4. The cell suspensions are transferred to lysing matrix C tubes.
5. The cells are lysed using a bead mill at the speed of 6 m/s for 30 s. Place on ice for 1 min and repeat once more with the same settings.
6. The lysate is recovered by adding 800 μl of the FA lysis buffer plus SOD with protease inhibitor and vortexing briefly.
7. The lysate is transferred to fresh 2 ml centrifugation tubes. Centrifuge 17,000 × *g* at 4 °C for 7 min (*see* **Notes 6** and 7).
8. The supernatant is discarded (*see* **Note 8**).
9. The chromatin pellet is resuspended in 750 μl of the FA lysis buffer plus SOD with protease inhibitor. 20 μl of 20% SDS is added, followed by another 750 μl of the FA lysis buffer plus SOD with protease inhibitor.

3.4 Preparation of the Magnetic Beads (See Note 9)

1. For each sample, 300 μl of the Invitrogen Dynabeads Protein G is transferred into a 15 ml centrifugation tube.
2. The buffer level is marked on the tube (*see* **Note 10**).
3. 10 ml of ice cold 1× PBS is added, and the beads are resuspended by shaking.
4. The tube is placed in a magnetic rack, and 6 ml of the buffer is removed (*see* **Note 11**).
5. For blocking, 5% BSA (final) is added and the beads are incubated for 1 h on a rotating wheel at 4 °C.
6. The tube is placed in the magnetic rack and all buffer is removed. The beads are washed with 10 ml ice cold 1× PBS.

7. The tube is placed in the magnetic rack again, and all buffer is removed. The beads are washed with 10 ml FA lysis buffer plus SOD.
8. In the magnetic rack, the buffer is removed to initial volume mark (*see* **Note 12**).

3.5 Sonication

1. Between 0.8 g to 1 g sonication beads are weighed into 15 ml sonication tubes.
2. The samples are transferred into the sonication tubes.
3. The sonication is carried out with 10 cycles (30 s on, 30 s off) twice with 15 min on ice in between. These settings are optimized for the Bioruptor Pico from Diagenode to reach fragment sizes between 100 bp and 500 bp.
4. The samples are transferred to fresh 2 ml (*see* **Note 6**) centrifugation tubes. Centrifuge 17,000 × *g* at 4 °C for 15 min.
5. The supernatant is transferred to fresh 2 ml centrifugation tubes. This is the ChIP extract (*see* **Note 13**). It is split up for the following samples.
   - 100 μl for the sonication control
   - 2 μl for the Bradford assay
   - 3.2 ml of ChIP extract adjusted to 1 mg/ml protein for the IP samples (1 ml each for S9.6 antibody, anti-HA antibody, and minus antibody samples) and the Input sample (50 μl).
6. The protein concentration of the samples is measured by Bradford assay.
7. The samples are diluted to 1 mg/ml in 3.2 ml in FA lysis buffer plus SOD with protease inhibitor. The solution is split 3 times into 1 ml for the IP (directly processed, *see* Subheading 3.6) and 50 μl for the input sample which is stored at −20 °C until further processing.
8. The samples for the sonication control are mixed with 100 μl of Elution buffer B and 7.5 μl Proteinase K (20 mg/ml). The mixture is incubated at 65 °C with 450 rpm shaking on a heating block overnight.
9. 1 ml of 100% Ethanol is added to each sample, mixed well and incubated for 15–20 min at −80 °C.
10. The samples are centrifuged at top speed for 45 min at 4 °C.
11. The supernatant is discarded, and the pellet is air dried.
12. The DNA is resuspended in 20 μl ddH_2O, then 1 μl of RNase A (10 mg/ml) is added and the reaction is incubated for 30 min at 37 °C.

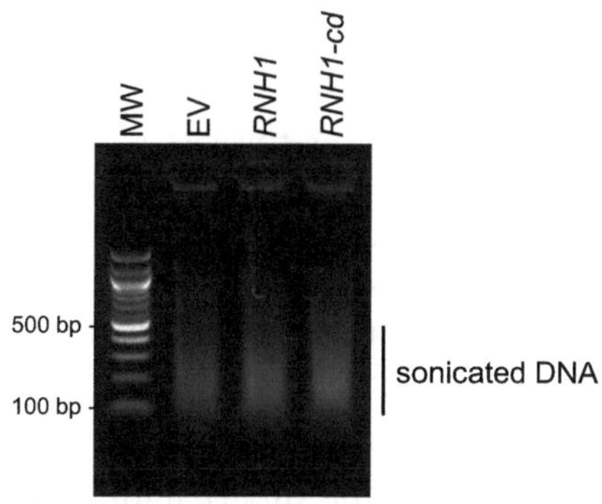

Fig. 2 Sonication control. The size of the sonicated DNA is controlled on an 1.5% agarose gel. The majority of the fragment smear should have a size between 100 bp and 500 bp

13. The samples are supplemented with a DNA loading dye of your choice and loaded to an 1.5% agarose gel using a 100 bp ladder. Run the gel 35 min at 100 V.
14. The size of the fragment smear is controlled. It should range from about 100 bp to 500 bp (Fig. 2). If this is not the case, the sonication was not successful.

3.6 Immunoprecipitation

1. For reduction of background noise, the ChIP extracts are precleared. For this reason, add 30 μl of the prepared magnetic beads (from Subheading 3.4) to each tube with a cut tip. Make sure the beads are well resuspended (*see* **Notes 14** and **15**).
2. The samples are incubated for 1 h at 4 °C on a rotating wheel.
3. The tubes are placed in a magnetic rack and the supernatant is transferred to a fresh tube. The different antibodies are added to the samples (2 μg per sample for the anti-HA antibody and 4 μg per sample for the S9.6 sample).
4. The samples are incubated for 30 min at 4 °C on a rotating wheel.
5. 50 μl of the prepared magnetic beads are added.
6. The samples are incubated overnight on a rotating wheel at 4 °C.
7. The samples are washed in four steps, each increasing the stringency of the binding. For each wash, the samples are placed on a magnetic rack, the buffer is replaced by 1 ml of the next buffer and the samples are incubated for 5 min at a rotating wheel at 4 °C. The buffers should be precooled at 4 °C. The order of the buffers is as follows.
   - FA cell lysis buffer plus SOD.
   - FA 500/lysis buffer.

- Buffer III.
- 1× TE buffer

8. After the last washing step, the buffer is completely removed and 100 μl of Elution buffer B is added immediately. The input samples are thawed at this point and 150 μl of Elution buffer B is added. The input samples are taken through the following steps together with the IP samples.
9. For elution from the beads, it is important to be precise. The tubes are kept for exactly 8 min at 65 °C shaking at 850 rpm.
10. The supernatant is transferred to a fresh tube and replaced by fresh 100 μl of Elution buffer B. The elution is repeated, and the supernatant is collected again together with the supernatant from the first elution.
11. 7.5 μl of Proteinase K (20 mg/ml) is added to each sample and the mixture is incubated overnight at 65 °C shaking at 450 rpm.
12. The DNA is purified with a PCR purification kit and eluted in 50 μl ddH_2O.

3.7 DRIP qPCR

1. The qPCR is carried out in a 384-well plate. The reaction mix per well is the following.
   - 1 μl DNA
   - 2 μl ddH_2O
   - 1 μl of each primer (5 μM)
   - 5 μl 2× DYNAMO FLASH SYBR GREEN qPCR Master mix
2. PCR condition:
   - 10 min at 95 °C
   - 15 s at 95 °C — 40 cycles
   - 1 min at 60 °C
   - Melting curve (65.0 °C till 96.5 °C in 0.5 °C steps).

3.8 Computational Analysis

1. The qPCR results are analyzed using the Bio-Rad RFX Manager software. The C_q determination method is set to "regression" and the data is exported.
2. The mean C_q values of the technical replicates is calculated.
3. The C_q value of the input is corrected, as the input sample is diluted 1:20 for the qPCR.

$$\log_2(20) = 4.322$$

$$C_q(\text{corr.input}) = C_q(\text{input}) - 4.322$$

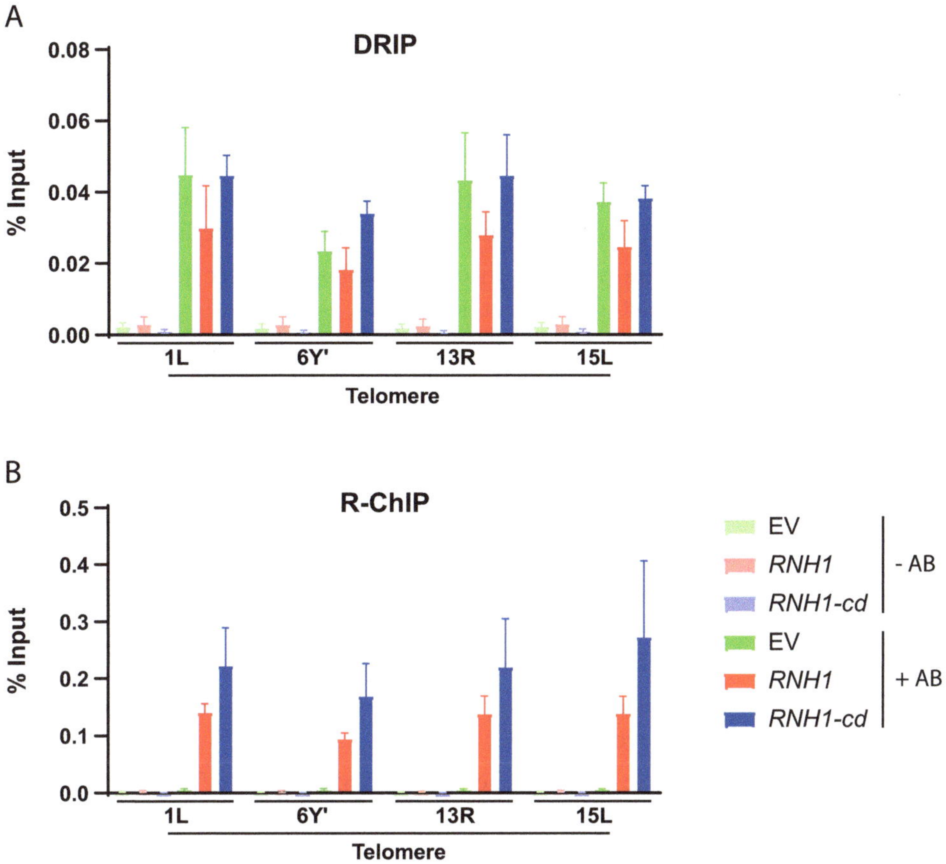

Fig. 3 Exemplary DRIP and R-ChIP results. Figure A shows the DRIP results, comparing R-loop levels in strains transformed either with empty vector (EV), RNH1 or RNH1-cd following 8 h induction with galactose. Figure B shows the corresponding R-ChIP results. Note that the % input is approximately 10× higher for the R-ChIP method, suggesting that it may be the method of choice for detecting DNA–RNA hybrids at yeast telomeres. All experiments were carried out in a strain without RNase H activity (*rnh1Δ rnh201Δ*). Note that A and B were identical samples, hence the legend in (B) also applies to (A)

4. Calculate percent input of the samples with and without antibody.

$$\Delta C_q = C_q(\text{corr.input}) - C_q(\text{sample} \pm \text{antibody})$$

$$\text{Percent input} = 100 \times 2^{\Delta C_q}$$

5. The resulting percent input values are plotted (Fig. 3).

4 Notes

1. Do not use glucose in the overnight culture. The shift from glucose to galactose containing medium leads to an extreme slowdown of culture growth and very slow induction.

2. The galactose induced protein expression is carried out over two population doublings. Depending on the research question, the growth conditions can be changed.
3. It is important that all compared samples have the same OD while cross-linking to ensure the same cross-linking efficiencies.
4. After quenching the cross-linking reaction, the flasks should be well covered with ice, at least above the liquid level.
5. The protease inhibitor tablets dissolve faster if the tubes are shaken/tilted at 4 °C.
6. Tubes larger than the actual sample volume are recommended due to the high tendency of the buffers to form foam.
7. In the meantime, prepare the sonication beads.
8. This step prevents competition with soluble protein.
9. The preparation of the beads can be carried out whenever time allows.
10. This step is important to maintain the same bead concentration after blocking and washing of the beads.
11. Never centrifuge magnetic beads.
12. Beads that are prepared as described here can be stored up to 1 week at 4 °C.
13. The ChIP extract can be frozen at −80 °C, but this reduces the IP efficiency.
14. Use the same cut pipette tip for every sample to ensure equal bead pipetting.
15. To be sure the beads do not sediment during pipetting, first pipet from the bottom of the tube, remove the pipette from the liquid and then let the suspension out on top of the liquid. Pipet the needed beads from the middle of the liquid.

Acknowledgments

We thank Olga Vydzhak and Kristi Jensen for comments on the manuscript. BL's lab is supported by the Heisenberg Program of the DFG-LU1709/2-1.

References

1. Crossley MP, Bocek M, Cimprich KA (2019) R-loops as cellular regulators and genomic threats. Mol Cell 73:398–411. https://doi.org/10.1016/j.molcel.2019.01.0242
2. Lafuente-Barquero J, García-Rubio ML, Martin-Alonso MS et al (2020) Harmful DNA–RNA hybrids are formed in cis and in a rad51-independent manner. elife 9:1–19
3. Wahba L, Gore SK, Koshland D (2013) The homologous recombination machinery modulates the formation of RNA-DNA hybrids and associated chromosome instability. elife 2: e00505

4. Marnef A, Legube G (2021) R-loops as Janus-faced modulators of DNA repair. Nat Cell Biol 23:305–313. https://doi.org/10.1038/s41556-021-00663-4
5. Niehrs C, Luke B (2020) Regulatory R-loops as facilitators of gene expression and genome stability. Nat Rev Mol Cell Biol 21:167–178
6. Cerritelli SM, El Hage A (2020) RNases H1 and H2: guardians of the stability of the nuclear genome when supply of dNTPs is limiting for DNA synthesis. Curr Genet 66:1073–1084. https://doi.org/10.1007/s00294-020-01086-8
7. Gaillard H, Aguilera A (2016) Transcription as a threat to genome integrity. Annu Rev Biochem 85:291–317
8. Rippe K, Luke B (2015) TERRA and the state of the telomere. Nat Struct Mol Biol 22:853–858. https://doi.org/10.1038/nsmb.3078
9. Graf M, Bonetti D, Lockhart A et al (2017) Telomere length determines TERRA and R-loop regulation through the cell cycle. Cell 170:72–85.e14
10. Mischo HE, Gómez-González B, Grzechnik P et al (2011) Yeast Sen1 helicase protects the genome from transcription-associated instability. Mol Cell 41:21–32. https://doi.org/10.1016/j.molcel.2010.12.007
11. Chédin F, Hartono SR, Sanz LA et al (2021) Best practices for the visualization, mapping, and manipulation of R-loops. EMBO J 40:e106394. https://doi.org/10.15252/embj.2020106394
12. Wahba L, Costantino L, Tan FJ et al (2016) S1-DRIP-seq identifies high expression and polyA tracts as major contributors to R-loop formation. Genes Dev 30:1327–1338. https://doi.org/10.1101/gad.280834.116
13. Zimmer AD, Koshland D (2016) Differential roles of the RNases H in preventing chromosome instability. Proc Natl Acad Sci U S A 113:12220–12225. https://doi.org/10.1073/pnas.1613448113
14. Chen L, Chen JY, Zhang X et al (2017) R-ChIP using inactive RNase H reveals dynamic coupling of R-loops with transcriptional pausing at gene promoters. Mol Cell 68:745–757.e5

Chapter 11

Detection of TERRA R-Loops at Human Telomeres

Galina Glousker, Rita Valador Fernandes, Marianna Feretzaki, and Joachim Lingner

Abstract

R-loops are three-stranded nucleic acid structures composed of a DNA–RNA hybrid and a displaced DNA strand. The long noncoding RNA TERRA forms R-loops at telomeres influencing the telomeric chromatin composition and impacting on telomere maintenance mechanisms by semiconservative DNA replication, homology directed DNA repair and telomerase. Here, we describe a method to detect R-loops at telomeres, which involves immunoprecipitation with the R-loop recognizing S9.6 antibody, followed by detection of telomeric DNA by either dot-blot hybridization with a radiolabeled telomeric probe, or qPCR using DNA primers that are specific for subtelomeric sequences.

Key words Telomeres, TERRA long noncoding RNA, DNA–RNA hybrids, R-loops, DRIP, Immunoprecipitation, S9.6 monoclonal antibody

1 Introduction

The detection of R-loops at telomeres is crucial to study how these structures are being formed and how they impinge on telomere biology. Of note, TERRA R-loops can form posttranscription *in trans* in dependency of the RAD51 DNA recombinase [1]. Furthermore, they are critical to promote telomere maintenance by homology directed repair which is particularly important in ALT cancer cells [2].

The detection of telomeric R-loops involves the isolation of total nucleic acids (DNA and RNA) from human cells under conditions which maintain nucleic acid double-stranded helical structures (Fig. 1a). The nucleic acid fraction is then sonicated to obtain smaller fragments with a length of approximately 100–500 base pairs (Fig. 1b). From the sonicated fraction, the DNA–RNA hybrids are immunoprecipitated using the commercially available

Galina Glousker, Rita Valador Fernandes, and Marianna Feretzaki are equally contributed.

Andrés Aguilera and Alexey Ruzov (eds.), *R-Loops: Methods and Protocols*, Methods in Molecular Biology, vol. 2528,
https://doi.org/10.1007/978-1-0716-2477-7_11,

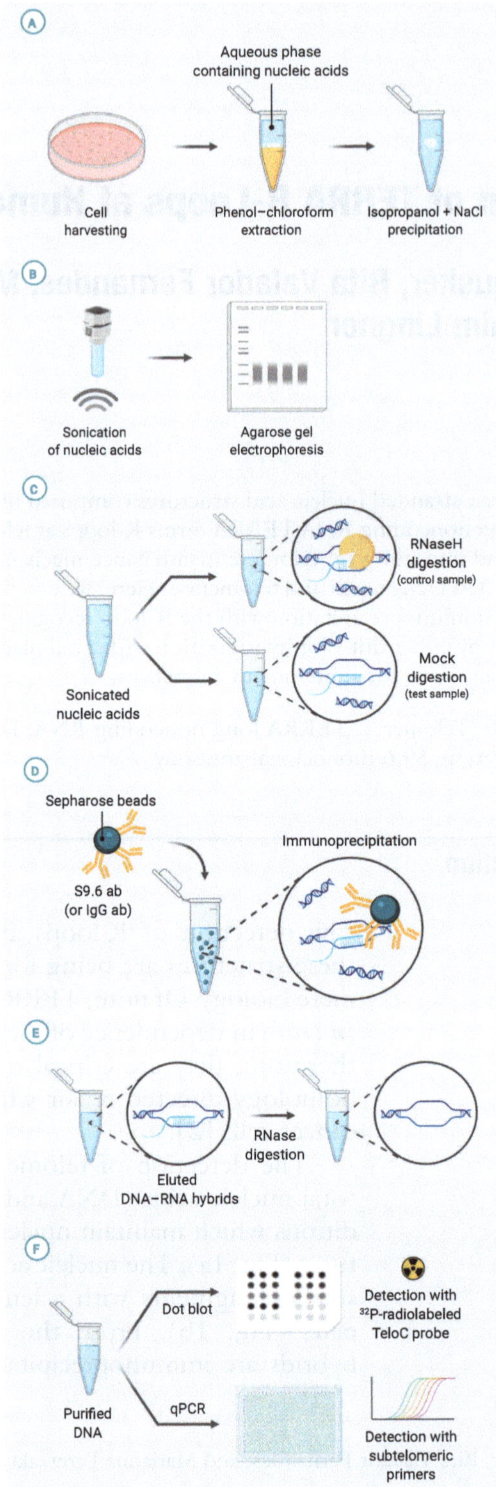

Fig. 1 Schematic depiction of a DNA–RNA immunoprecipitation, followed by detection of telomeric R-loops by dot blot or qPCR. Illustration created with BioRender.com. (**a**) Isolation of nucleic acids. (**b**) Fragmentation of nucleic acids. (**c**) RNaseH negative control. (**d**) S9.6 immunoprecipitation of DNA:RNA hybrids. (**e**) Elution and RNase digestion. (**f**) DNA analysis

S9.6 monoclonal antibody [3] which binds to R-loops (Fig. 1d). This step is often referred to as DRIP (for DNA–RNA immunoprecipitation). As a control for specificity, the nucleic acid fraction is treated in parallel prior to immunoprecipitation with the RNase H enzyme, which destroys the RNA moiety in DNA–RNA hybrid structures (Fig. 1c). Thus, this sample provides a negative control and the background signal. Finally, immunoprecipitates are eluted and treated with RNase (Fig. 1e). Telomeric DNA is then detected in the precipitated test and control samples in one of two ways (Fig. 1f). Either, the precipitated nucleic acids are spotted on a nylon membrane and telomeric DNA is detected by Southern hybridization with a radioactively labelled telomeric probe (as in Fig. 2a, b). Alternatively, telomeric DNA containing fragments are quantified by qPCR using DNA primers that recognize subtelomeric sequences residing in the vicinity of the chromosome-terminal telomeric 5′-TTAGGG-3′ repeats (as in Fig. 2c).

2 Materials

Work in RNase-free conditions. Clean all the surfaces and pipettes with RNase AWAY (Thermo Fisher Scientific) or an equivalent product.

2.1 Buffers and Solutions

1. RLN buffer: 50 mM Tris–HCl pH 8.0, 140 mM NaCl, 1.5 mM $MgCl_2$, 0.5% Nonidet P-40. Filter and store at 4 °C. Prior to use, add 100 U/mL RNasin Plus (Promega N2615) and DTT to 1 mM.
2. 10× RNase H buffer: 0.2 M HEPES–KOH pH 7.5, 0.5 M NaCl, 0.1 M $MgCl_2$, 10 mM DTT. Prepare fresh, do not filter.
3. DIP-1 buffer: 10 mM HEPES–KOH pH 7.5, 275 mM NaCl, 0.1% SDS, 1% Triton X-100, 0.1% Na-deoxycholate. Filter, store at 4 °C, short term.
4. DIP-2 buffer: 50 mM HEPES–KOH pH 7.5, 140 mM NaCl, 1 mM EDTA pH 8.0, 1% Triton X-100, 0.1% Na-deoxycholate. Filter, store at 4 °C, short term.
5. DIP-3 buffer: 50 mM HEPES–KOH pH 7.5, 500 mM NaCl, 1 mM EDTA pH 8.0, 1% Triton X-100, 0.1% Na-deoxycholate. Filter, store at 4 °C, short term.
6. DIP-4 buffer: 10 mM Tris–HCl pH 8.0, 1 mM EDTA pH 8.0, 250 mM LiCl, 1% NP-40, 1% Na-deoxycholate. Filter, store at 4 °C, short term.
7. TE buffer: 10 mM Tris–HCl pH 8.0, 1 mM EDTA pH 8.0. Filter, store at RT.
8. Elution buffer: 20 mM Tris–HCl pH 8.0, 0.1% SDS, 0.1 M $NaHCO_3$, 0.5 mM EDTA pH 8.0. Filter, store at RT. Prior to use, add RNase-DNase free (Roche) to 10 μg/mL final concentration.

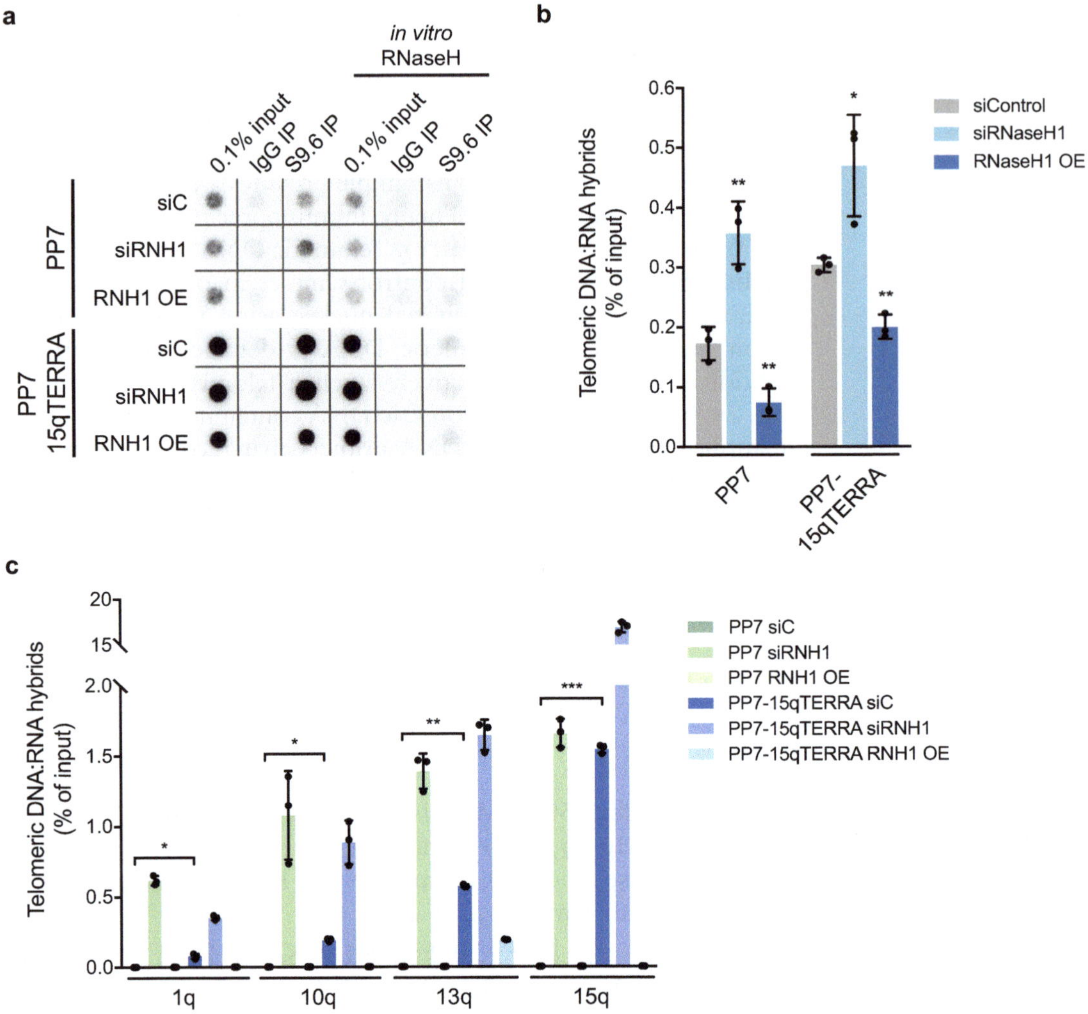

Fig. 2 Detection of telomeric R-loops by DRIP. Nucleic acids used for DRIP were isolated from HeLa cells expressing from transfected plasmids either *Pseudomonas aeruginosa* Phage 7 (PP7) RNA stem-loops or PP7-15qTERRA RNA. The cells were also transfected with nontargeting siRNA (siControl), siRNA targeting RNaseH1 (siRNaseH1) or a plasmid overexpressing RNaseH1 (RNaseH1 OE), as indicated. (**a**) DRIP samples were spotted on a nylon membrane and telomeric DNA was detected with an α-^{32}P-radiolabeled telomeric probe. The telomeric signal from PP7-15qTERRA plasmid containing cells is much stronger due to presence of telomeric repeats in the plasmid and R-loops that are formed at the site of transcription. Depletion of endogenous RNase H1 increased the DRIP signal whereas RNase H1 overexpression reduced it. In vitro RNase H treatment prior to immunoprecipitation with the S9.6 antibody reduced the signal to background levels. (**b**) Quantification of three independent biological replicates as in **a**. (**c**) DRIP samples were analyzed by qPCR and DNA was detected with primers targeting 1q, 10q, 13q, or 15q subtelomeric DNA sequences, as indicated. Three independent biological replicates are shown. Of note, expression of PP7-15qTERRA from plasmids increased the DRIP signal at telomeres 1q, 10q, and 13q which is indicative of R-loop formation at these telomeres posttranscription *in trans*. The very strong signal obtained with 15q stems mostly from the plasmid DNA which contains 15q subtelomeric sequences

9. 2× SSC buffer: 30 mM sodium citrate pH 7.0, 300 mM NaCl.
10. Denaturation buffer: 0.5 M NaOH, 1.5 M NaCl.
11. Neutralization buffer: 0.5 M Tris–HCl pH 7.0, 1.5 M NaCl.
12. Church Buffer: 0.5 M $NaHPO_4$, 1 mM EDTA, pH 8.0, 1% (weight/volume) BSA, 7% SDS.

2.2 Isolation of Nuclei and Sonication

1. 1 mL syringes.
2. 0.9 × 40 mm needles.
3. β-mercaptoethanol.
4. RA1 buffer (NucleoSpin RNA, Macherey-Nagel).
5. Phase Lock Gel Heavy 2 mL tubes (5PRIME).
6. 25:24:1 phenol–chloroform–isoamyl alcohol (pH 7.8–8.2).
7. 15 mL Falcon tubes.
8. Covaris E220 focused-ultrasonicator.
9. AFA intensifier for sonication (PN 500141).
10. Microtubes (AFA fiber Screw Cap 6 × 16 mm (Ref 520,096)).

2.3 DNA–RNA Immunoprecipitation

1. RNase H (Roche).
2. 0.5 M EDTA pH 8.0.
3. Protein G Sepharose 4 Fast Flow (Cytiva).
4. Tube rotator.
5. Table top centrifuge.
6. Anti-DNA-RNA Hybrid [S9.6] antibody (Kerafast 022715).
7. Normal mouse IgG control.
8. RNase (DNase-free).
9. NucleoSpin Gel and PCR Cleanup kit (Macherey-Nagel).
10. QIAquick PCR Purification kit (Qiagen).

2.4 Dot Blot

1. Hybond N+ membrane (GE Healthcare).
2. Bio-dot filter paper.
3. Bio-dot blot apparatus.
4. Stratalinker UV crosslinker.
5. Hybridization oven.
6. Hybridization bottles.
7. TeloC probe (PCR product generated as described [4] and labeled using RadPrime DNA Labeling System [Invitrogen]).
8. Alu probe (5′ end labeled 5′-gtgatccgcccgcctcggcctcccaaagtg-3′).
9. RadPrime DNA Labeling System (Invitrogen).
10. T4 Polynucleotide kinase (NEB).

11. α-^{32}P dCTP (10 μCi/μL; 3000 Ci/mmol).
12. γ-^{32}P ATP (10 μCi/μL; 3000 Ci/mmol).
13. Mini quick spin Oligo Column (Roche).
14. Mini quick spin DNA Column (Roche).
15. Phosphorimager screen.
16. Typhoon phosphorimager (GE).

2.5 qPCR

1. 12-channel 0.5–10 μL pipette.
2. Primers:
   (a) 1q subtelomeric primers (5′-cagcgtcgcaactcaaatg-3′; 5-′-ccctcaccctccatgagtaata-3′);
   (b) 10q subtelomeric primers (5′-gcattcctaatgcacacatgac-3′; 5′-tacccgaacctgaaccctaa-3′);
   (c) 13q subtelomeric primers (5′-gcacttgaaccctgcaatacag-3′; 5′-cctgcgcaccgagattct-3′);
   (d) 15q subtelomeric primers (5′-aaccctaaccacatgagcaacg-3′; 5′-gctgcattaaagggtccagt-3′).
3. Power SYBR Green PCR Master Mix (Applied Biosystems).
4. 384-well reaction plate.
5. Optical adhesive film.
6. QuantStudio 7 Flex Real-Time PCR System (Applied Biosystems) or equivalent.

3 Methods

Clean all surfaces and tools with RNase AWAY.

3.1 Cells Harvesting and Isolation of Nucleic Acids

1. Harvest ten million cells per condition for DRIP assay. Harvest additional one million cells for Western blot analysis if needed. Spin down the cells for 5 min at 300 × *g* at 4 °C. Keep cells on ice.
2. Wash cells with 1–2 mL cold PBS and transfer into Eppendorf microfuge tubes. Spin down again for 5 min at 300 × *g* at 4 °C.
3. Aspirate PBS. Loosen pellet thoroughly by flicking the tube. Resuspend 10 × 10^6 cells in 175 μL of cold RLN buffer containing DTT and RNaseIN, pipet up and down once, and incubate for 5 min on ice for cell lysis (*see* **Note 1**).
4. Spin down at 300 × *g*, for 2 min, 4 °C. Remove the supernatant carefully and discard.
5. Wash the nuclei containing pellet in 500 μL cold PBS. Spin down at 300 × *g*, for 2 min, 4 °C. Remove the supernatant carefully and discard.

6. Warm tubes to RT and continue to work at RT. Move to fume hood. Loosen nuclear pellet thoroughly by flicking the tube.
7. Pellet gel in Phase Lock Gel (PLG) Heavy 2 mL tubes by centrifugation at 13,000 × *g* for 1 min.
8. Prepare nuclear lysis buffer: add 10 μL of β-mercaptoethanol per 1 mL of buffer RA1 (prepare 1 mL of nuclear lysis buffer per condition).
9. Resuspend nuclei in 500 μL of nuclear lysis buffer at RT.
10. Pass the lysate 10 times up and down through a needle.
11. Transfer the lysate into the PLG tubes. Add 250 μL H_2O and 750 μL 25:24:1 phenol/chloroform/isoamyl alcohol. Mix thoroughly, but do not vortex.
12. Centrifuge at 13,000 × *g* for 5 min to separate phases.
13. Transfer the nucleic-acid-containing aqueous upper phase (NA) into a fresh 2 mL Eppendorf microfuge tube on ice.
14. Perform isopropanol precipitation: 750 μL NA + 15 μL 5 M NaCl +750 μL ice-cold isopropanol. Shake vigorously! Incubate on ice for 30 min, then spin down for 30 min at 10,000 × *g* at 4 °C (in 2 mL Eppendorf tubes) (*see* **Note 2**).
15. Wash pellet 2× with 1 mL of ice cold 70% ethanol.
16. Let the pellet air dry (*see* **Note 3**).
17. Add 135 μL H_2O to each pellet from 10 × 10^6 cells, flick to detach the pellet from the wall of the tube, incubate for approximately 1 h shaking in the cold room until nucleic acids are dissolved.
18. Take sample "A" before sonication: 3 μL.

3.2 Sonication with Covaris E220

1. Transfer each condition into a microtube for sonication (130 μL per tube). Perform sonication using the following conditions to achieve fragment length < 500 bp.

Condition	a single condition
Duty factor	10%
Peak incident power (W)	140
Cycles per burst	200
Time (seconds)	150
AFA intensifier	Yes
Temperature	4 °C
Water level	6

2. Take sample “B” after sonication: 3 μL.
3. Test sonication profile:
   (a) Add to 3 μL of sample (A and B) 7 μL of H_2O.
   (b) add 2 μL of 6× loading buffer.
   (c) Analyze on a 1% TBE-agarose gel.

You can freeze the samples in liquid nitrogen and keep them at −80 °C for a few days.

3.3 DNA–RNA Immunoprecipitation

1. Measure the concentration of nucleic acids with Nanodrop (DNA). Use the same amount of nucleic acids for all your samples (15–60 μg depending on the concentration—*see* **Note 4**).
2. Prepare fresh RNase H buffer.
3. Perform the RNase H control +/−RNase H treatment - example:
   (a) nucleic acids,
   (b) 15 μL 10× RNase H buffer
   (c) 10 μL RNase H (1 U/μL; Roche) (control) or H_2O (sample)
   (d) H_2O to 150 μL.
4. Incubate at 37 °C for 90 min.
5. Add 2 μL 0.5 M EDTA (pH 8.0) to stop the reaction.
6. During the RNase H treatment equilibrate beads (*see* **Note 5**). Use 200 μL beads per condition (equals to 400 μL of 50% slurry stock (*see* **Note 6**)).
   (a) Put the 50% Sepharose G bead stock on the wheel in the cold room for 5 min to mix it very well. Take what you need into a new tube and spin them down for 2 min at 300 × *g* at 4 °C. Remove the supernatant.
   (b) Add the same volume of DIP-1 buffer, wash beads on the wheel for 5 min at 4 °C. Spin down for 2 min at 300 × *g* at 4 °C, discard the supernatant. Repeat this step once.
   (c) Resuspend the beads in DIP-1 buffer to have a 50% slurry.
7. Preclear the nucleic acids. Mix in a 2 mL microfuge tube the following.
   (a) 150 μL nucleic acids after RNase H treatment.
   (b) 1270 μL DIP-1 buffer (ten-fold dilution of nucleic acids).
   (c) 80 μL of protein G beads 50% slurry.
8. Incubate for 1 h at 4 °C on the wheel. Spin down 5 min at 300 × *g* and continue with the precleared extract.

9. Put aside 1% input.
10. Assemble the IP reaction.
    (a) 600 μL of precleared extract.
    (b) 3 μg of antibody (spin down before use):
        - Tube 1: S9.6 antibody (Kerafast 022715).
        - Tube 2: mouse IgG (sc2025).
    (c) 80 μL of protein G beads 50% slurry.
11. Incubate overnight on the wheel at 4 °C.
12. Spin down for 2 min at 300 × *g* at 4 °C, discard the supernatant.
13. Wash the beads 5 min on a wheel at 4 °C with ice-cold buffers (spin for 2 min at 300 × *g* at 4 °C).
    (a) 1 mL DIP-2 buffer.
    (b) 1 mL DIP-3 buffer.
    (c) 1 mL DIP-4 buffer.
    (d) 1 mL TE.
14. Spin for 5 min at 300 × *g* at 4 °C.
15. Resuspend beads and inputs in 100 μL Elution buffer containing RNase (DNase-free).
16. Incubate for 1–2 h at 65 °C (or overnight).
17. Extract DNA using:
    (a) *If you continue with dot-blot*: Macherey-Nagel NucleoSpin Gel and PCR Cleanup kit and elute in 100 μL H_2O (Heat up H_2O to 70 °C = > Incubate for 5 min before elution).
    (b) *If you continue with qPCR*: Qiagen PCR Purification kit for qPCR and elute in 100 μl H_2O (Heat up H_2O to 70 °C = > Incubate for 5 min before elution).
18. Store DNA at −20 °C or directly continue with dot-blot or qPCR.

3.4 Dot Blot

1. Clean Dot-Blot apparatus (rinse with 0.02% SDS; spray with 70% ethanol, rinse with H_2O multiple times; dry completely).
2. Prewarm Church buffer and hybridization oven.
3. Denature DNA for 10 min at 95 °C and keep on ice for 5 min or longer.
4. Assemble the dot blot apparatus with N+ Hybond membrane soaked in 2× SSC, Whatman paper soaked in 2×SSC, then wash the slots with 200 μL 2× SSC.
5. Apply the samples. For the input (which is 1%), add 900 μL of H_2O and load 200 μL (0.2%), 100 μL (0.1%), and 50 μL (0.05%), for the IP samples load the entire sample (*see* **Note 7**).

6. Wash again the slots that have been used with 2× SSC.
7. UV-crosslink the DNA onto the membrane (do not let the membrane get completely dry), using autocrosslink as a setting.
8. Denature the membrane in 0.5 M NaOH, 1.5 M NaCl for 15 min on a shaker at room temperature.
9. Neutralize the membrane in 0.5 M Tris–HCl pH 7.0, 1.5 M NaCl for 10 min on a shaker at room temperature.
10. Prehybridize the membrane with Church Buffer for at least 1 h at 65 °C when using the telomeric probe and at 55 °C for the alu probe.
11. Label telomeric probe or alu probe.
    (a) Telomeric probe.
    - μL DNA (200 ng; telomeric PCR product (Porro et al., 2014)).
    - 20 μL H_2O.
    - Denature for 5 min at 95 °C => put on ice immediately, add the following.
    - 1 μL dATP (500 μM), 1 μL dGTP (500 μM), 1 μL dTTP (500 μM).
    - 20 μL 2,5× Random Prim buff.
    - 1 μL Klenow enzyme.
    - 5 μL α-^{32}P dCTP (3000 Ci/mmol; 10 μCi/μL).
    - H_2O to 50 μL.
    - Incubate at 37 °C for at least 10 min.
    - Purify the labeled DNA using the Mini quick spin DNA Column.
    - Denature the radiolabeled probe for 5 min at 95 °C.
    - Place on ice.
    - Add to 20–30 mL of Church buffer.

    (b) Alu probe.
    - 1 μL oligonucleotide (from a 10 μM stock).
    - 27 μL H_2O.
    - 4 μL T4 PNK buffer.
    - 2 μL T4 PNK.
    - 5 μL γ-^{32}P ATP (3000 Ci/mmol; 10 μCi/μL).
    - H_2O to 40 μL.
    - Incubate at least 1 h at 37 °C.
    - Purify using Mini quick spin Oligo Column.
    - Add to 20–30 mL of Church buffer.

12. Hybridize with radiolabeled telomeric probe or alu probe overnight.

    On the following day:

13. Rinse with 4× SSC.
14. Wash three times for 30 min at 65 °C in 1× SSC, 0.5% SDS at 65 °C for the telomeric probe or 55 °C for the alu probe.
15. Rinse in 4× SSC in a tray to remove SDS before sealing the membrane.
16. Place the membrane into a plastic folder, seal properly and expose to a phosphorimager screen (expose overnight, prolong exposure time if needed).

3.5 qPCR

1. Set up a qPCR master mix for each primer pair (all components should be kept on ice).
   (a) 5 μL Power SYBR Green (prolonged direct light exposure should be avoided).
   (b) 1 μL of each primer (forward and reverse) from a 10 μM stock.
   (c) H_2O to a total volume of 9 μL per well.
2. Prepare serial dilutions of the collected 1% input in H_2O (at least two dilutions; for example, dilution factors of 10 and 100—*see* **Note 7**).
3. Transfer all samples (diluted inputs and immunoprecipitates) and master mixes into 0.2 mL PCR tube strips, to facilitate pipetting samples with multichannel pipette.
4. Resorting to a multichannel pipette (in order to minimize pipetting-derived errors), transfer 9 μL of each primer pair master mix into each well of a 384-well reaction plate.
5. Resorting to a multichannel pipette, add 1 μL of each input dilution and each DRIP sample into each well of the reaction plate. At least two technical replicates per input dilution/sample should be carried out. For each primer pair a control reaction with no template should also be included.
6. Seal the reaction plate with optical adhesive film.
7. Briefly centrifuge the plate at 2000 × *g*.
8. Load on the qPCR instrument.
9. Perform the qPCR reaction according to manufacturer's instructions: 95 °C for 10 min, followed by 95 °C for 15 s, and annealing and extension at 60 °C for 1 min for 40 cycles, and finalized with dissociation stage for melting curve analysis.

3.5.1 qPCR Data Analysis

1. Using the Ct values obtained for the dilutions of input samples, perform a regression analysis to determine the equation of the standard curve of each input.
2. Using each input equation, calculate the corresponding S9.6 and IgG IPs as percentage of input.

4 Notes

1. Do not scale up. The efficient lysis is only possible with no more than 10×10^6 cells per 175 μL of RLN buffer. If your signal is too low and you want to increase the starting number of cells to 20×10^6, split your cells into two tubes with 10×10^6 cells in each.
2. The work in the fume hood lasts for at least 1 h.
3. You can load your Western blot during that time. Turn on Covaris in a timely manner to ensure that the water is cold and degassed when you start sonication.
4. Depending on the cell line, basal levels of telomeric RNA–DNA hybrids will vary. Adjust cell number and amount of nucleic acids accordingly (e.g., for HeLa cells 30 μg/IP are required, while for U2OS cells 7.5 μg/IP should suffice).
5. Do not use Falcon tubes to spin down the beads as the beads tend to be lost in these tubes. Use 5 mL or 2 mL tubes.
6. Start with 20% more beads than you need.
7. The input dilutions should be adjusted so that the amount of target DNA in each immunoprecipitated sample lies within the linear dynamic range of each assay.

Acknowledgments

The laboratory was supported by the Swiss National Science Foundation (SNFS grant 310030_184718), the SNFS-funded National Center of Competence in Research (NCCR) RNA and disease network (grant 182880), and the European Union's Horizon 2020 research and innovation programme under grant agreement 812829.

References

1. Feretzaki M, Pospisilova M, Valador Fernandes R et al (2020) RAD51-dependent recruitment of TERRA lncRNA to telomeres through R-loops. Nature 587:303–308
2. Silva B, Arora R, Bione S et al (2021) TERRA transcription destabilizes telomere integrity to initiate break-induced replication in human ALT cells. Nat Commun 12:3760
3. Boguslawski SJ, Smith DE, Michalak MA et al (1986) Characterization of monoclonal antibody to DNA · RNA and its application to immunodetection of hybrids. J Immunol Methods 89:123–130
4. Porro A, Feuerhahn S, Lingner J (2014) TERRA-reinforced association of LSD1 with MRE11 promotes processing of uncapped telomeres. Cell Rep 6:765–776

Chapter 12

R-Loops and Mitochondrial DNA Metabolism

Ian J. Holt

Abstract

R-loops forming inadvertently during transcription can threaten genome stability, but R-loops are also formed intentionally, as a means of regulating transcription and other aspects of DNA metabolism. The study of R-loops in mitochondria is in its infancy, and yet there is already clear evidence that they are predominantly located in the major regulatory region of the mammalian mitochondrial genome. Here, we describe how mitochondrial R-loops have been characterized to date, with the emphasis on the problems of their being extremely labile, and how to minimize their loss during extraction. The oft-overlooked issues of RNA–DNA hybrids *not* being synonymous with R-loops, and adventitious RNA hybridization to DNA, are tackled head on; and possible new approaches are described and placed in the context of future research lines that could reveal the detailed roles of R-loops in the metabolism of mitochondrial DNA.

Key words R-loop, RNA–DNA hybrid, Mitochondrial DNA, Mitochondrial DNA replication, Mitochondrial transcription

1 Introduction

Before detailing the protocols used to identify and characterize mitochondrial R-loops, it is important to review the evidence and the competing arguments, as the issue of RNA–DNA hybrids and DNA replication in mitochondria is mired in controversy. It is also important to recognize that the RNA–DNA hybrids detected can include species other than simple R-loops. Hence, these considerations aim to aid both experimental design and data interpretation.

First, a basic map is essential: mammalian mitochondrial DNA (mtDNA) comprises only 16 kb of double stranded DNA, which is densely packed with 37 genes that are essential for mitochondrial ATP production (Fig. 1). Each DNA strand, heavy (H) and light (L), is transcribed to yield all the coding RNAs in a single polycistronic transcript. In striking contrast to nuclear DNA, there are no introns and only one substantial noncoding region, which contains several *cis*-elements, including the promoters of the polycistronic transcripts, LSP and HSP. Because of its multiple critical roles in

Andrés Aguilera and Alexey Ruzov (eds.), *R-Loops: Methods and Protocols*, Methods in Molecular Biology, vol. 2528,
https://doi.org/10.1007/978-1-0716-2477-7_12,

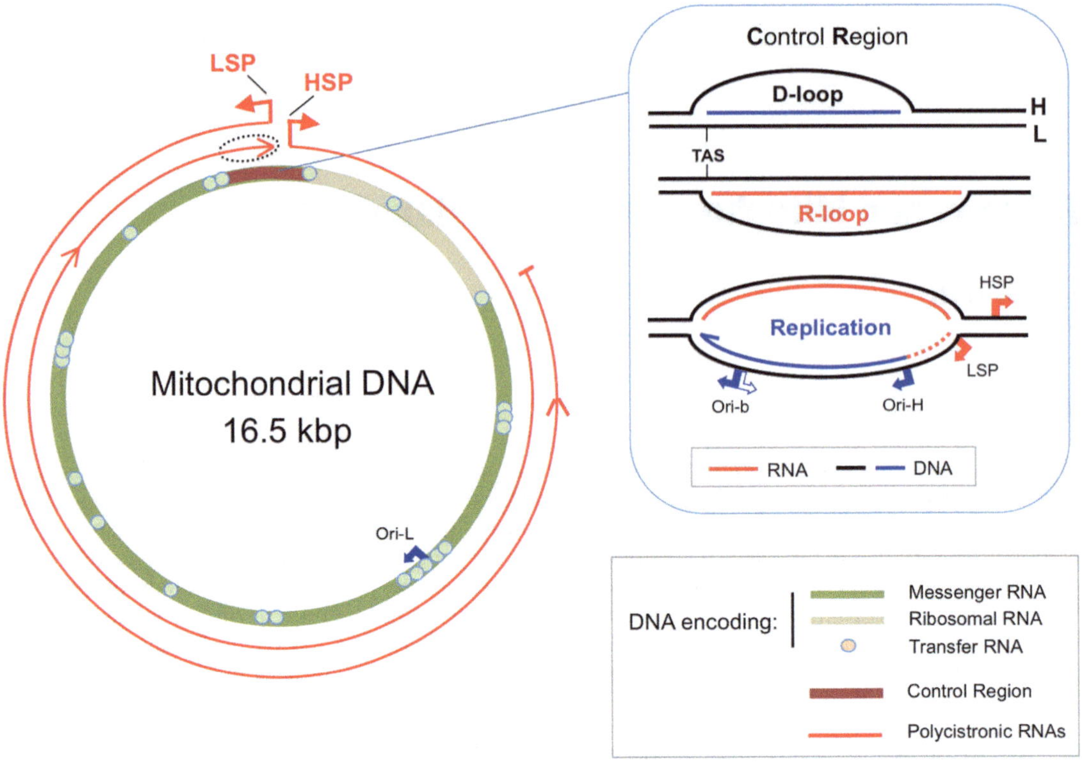

Fig. 1 A schematic map of mammalian mitochondrial DNA. Mammalian mtDNA contains one sizeable noncoding, or control region that contains several critical *cis*-elements for replication and transcription, and is the location of triple-stranded species that form a D-loop or a R-loop (inset). The end of the D-loop and start of the R-loop coincide at TAS (Termination associated sequence). Transcription of polycistronic RNAs from the Heavy and Light Strand Promoters (HSP and LSP) yield all the mitochondrial RNAs required for ATP production, and which in the case of the HSP transcript can include the RNA corresponding to the R-loop (broken ellipse). Synthesis of RNA from LSP is also the prelude to strand-asynchronous mtDNA replication, with transition to DNA synthesis at Ori-H (origin of heavy strand DNA replication) (inset), or Ori-b (origin of bidirectional replication). Second strand DNA synthesis is often delayed until leading strand DNA synthesis reaches the site known as Ori-L

mtDNA replication and expression the major noncoding region is known as *the control region*, or CR (Fig. 1).

There is agreement that the major mechanism of mammalian mtDNA replication in many cells and tissues is strand-asynchronous (as oppose to coupled leading and lagging strand replication in the nucleus), and that the RNA primer at the origin of replication starts from the same promoter (LSP), as the polycistronic transcript derived from the L-strand of mtDNA. DNA synthesis frequently starts downstream of LSP either at a site denoted Ori-H/O_H (the origin of heavy-strand DNA replication) or further downstream at another site called Ori-b [1, 2]. Ten to sixty per cent of mtDNAs possess a third strand of DNA (a displacement/ D-loop) that covers a substantial portion of the CR, called 7S

DNA [3, 4]. The D-loop is widely described as an aborted or stalled replication intermediate; however, it would be both reckless and wasteful to produce such an open structure at high frequency for no good purpose. Moreover, while in mice, Ori-H has a consistent 5′-end that is identical to that of 7S DNA [2, 5]; in humans none of the heterogeneous 5′-ends of 7S DNA (between nucleotide positions (np) 100–200) coincides with the start site of the growing nascent H-strands (i.e., bona fide replication intermediates). Instead, the 5′ ends map to a single site close to np 60 [6, 7]. Therefore, we infer that the D-loop has specific roles in mtDNA metabolism (see below), and does not represent a temporary pause in replication, nor premature termination of replication, although others maintain that one or both of these is true.

A major complication when considering RNA–DNA hybrids in mitochondria is that we have proposed that such species form the basis of a unique mechanism of replication—transcript-dependent mtDNA replication [8]. Despite the fact that the RNAs can be cross-linked to the mtDNA in replication-competent mitochondria, and intact cells [8], this mechanism of mtDNA replication is disputed [9]. The transcript dependent model of mtDNA replication (TDR) proposed that mature transcripts are hybridized to the displaced lagging-strand template in the long interval between the initiation of DNA replication in the CR and the synthesis of the second (lagging) strand of DNA [8]. However, the predicted first transcript required by the TDR model, proposed in 2013 [8], is an L-strand RNA mapping to the CR, which prior to 2016 had not been detected.

Mitochondria have long been known to possess an abundant RNA that is the same (H) strand as 7S DNA, it originates at LSP and terminates just short of the 5′ end of the 7S DNAs, at the second of three conserved sequence elements, CSB1 [10]. Although it is much shorter than 7S DNA (particularly in rodents), it bears the name 7S RNA, and while it is often described as a stable R-loop based on in vitro experiments [11], it is not known what proportion of 7S RNA forms a RNA–DNA hybrid in mitochondria. The small gap between CSB1 and the 5′ ends of the D-loop, and the slightly greater gap to Ori-H (in human mtDNA), suggests that 7S RNAs cannot serve as primers for 7S DNA synthesis or H-strand replication. And although the synthesis of RNA from LSP to Ori-H (and other site[s]) is needed to generate a primer for the initiation of DNA replication, an extended RNA covalently linked to nascent H-strands of DNA or 7S DNA is barely detectable [5], unless RNase H1 is ablated [2, 6]. This led us to propose that the predicted RNA primer from LSP to Ori-H may not normally exist in the full-length form [12], which is logical as elsewhere primers are typically 1–12 nts, rather than hundreds of nucleotides in length.

Modified mitochondrial isolation protocols and pretreatments led, in 2016, to the identification of another abundant RNA mapping to the CR [6]. This RNA forms R-loops on a substantial fraction of the total population of mtDNA, and is called LC-RNA for L-strand, control region RNA. LC-RNA is the complement of the D-loop, and it spans 400–800 nts in murine solid tissues [6] (Fig. 2a). Because many researchers studying mtDNA continue to overlook the mitochondrial R-loop (of LC-RNA), the evidence for this species is detailed below.

1.1 Evidence for the Prominent, Specific, Mitochondrial R-Loop

(a) Fragments of mtDNA including the control region yield one or two small bubble structures (SBs) of similar size on 2D agarose gels. The invariant one is the D-loop form of mtDNA containing 7S DNA, the other unquestionably contains RNA–DNA hybrid, as it is sensitive to RNase H. No such SBs have been found associated with any other region of mtDNA.

(b) Incubation with single-stranded DNA binding protein alters the mobility of the SBs on 2D gels to a similar extent, indicating that they both contain a displaced single strand of DNA of similar length. Hence, the RNase H-sensitive SB is composed of RNA–DNA hybrid in the form of an R-loop.

(c) The RNA associated with the mtDNA could include a single-stranded RNA tail at one or both ends; however, RNase T_1 has no appreciable effect on the mobility of the RNase H-sensitive SB, and so the RNA–DNA hybrid region is substantial and accounts for the difference in mobility compared to linear duplex DNA.

(d) RNA–DNA hybrids on mtDNA can be lost during extraction, and are preserved by inter-strand crosslinking. The SB signal is enhanced by mitochondrial purification procedures on ice and by inter-strand cross-linking (see protocols).

(e) Restriction enzymes cannot cleave RNA–DNA hybrids, and so the SB formed of an R-loop should be refractive to digestion at multiple sites across the CR. Concordantly, only one of the two SBs was cut with restriction enzymes. Denaturation revealed that the cleaved SB contains 7S DNA. All the 7S DNA was cut by the enzymes, indicating that all the D-loops are sensitive to restriction digestion. These results indicate that one SB contains 7S DNA, whereas the other is resistant to restriction digestion in the control region, as predicted for an R-loop.

(f) Gel extracted and denatured SBs derived from highly purified mitochondrial DNA yield a third strand of nucleic acid of two types: one that is DNase-sensitive, RNase H-resistant (7S DNA), while the other is RNase H-sensitive, DNase-

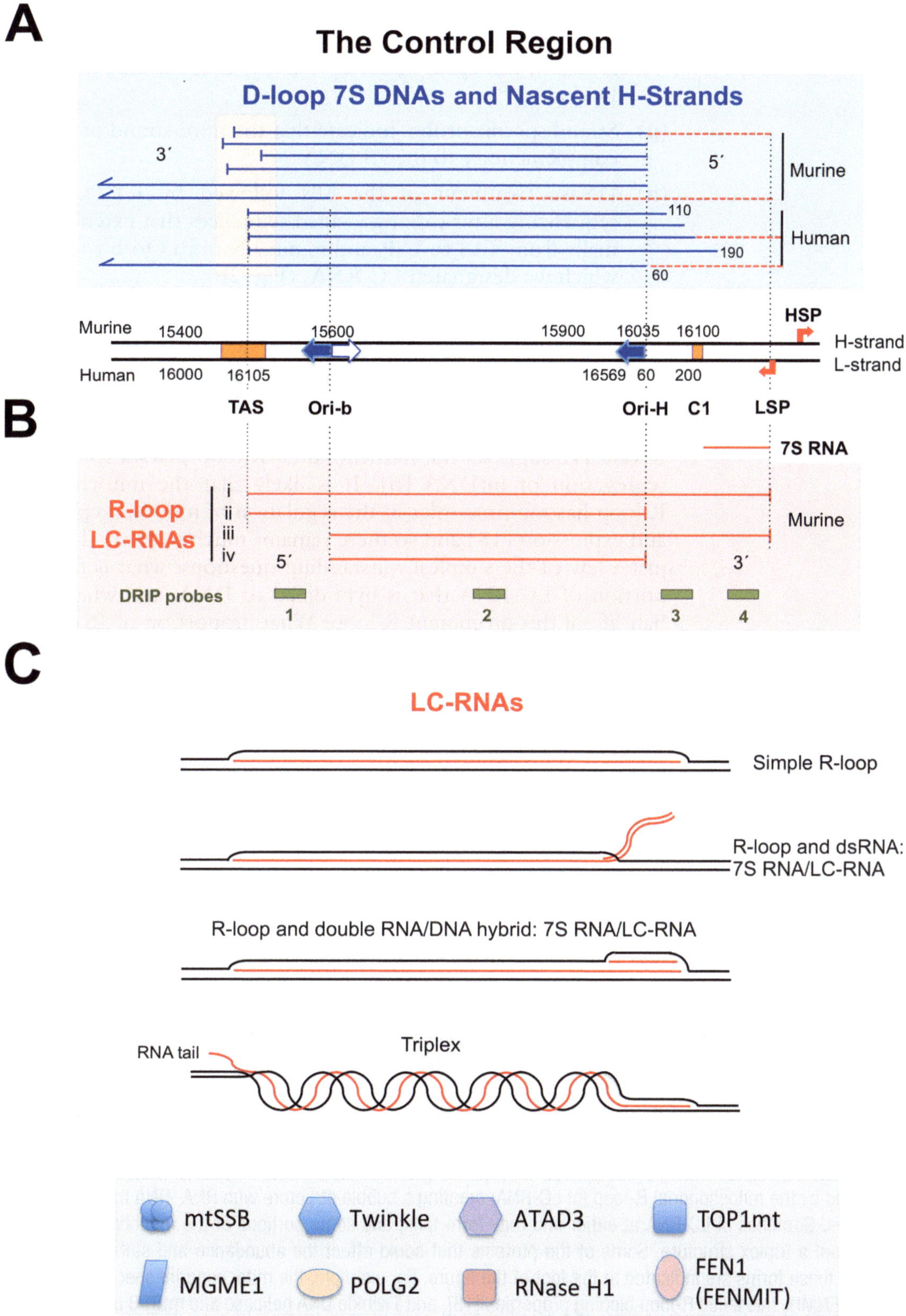

Fig. 2 The mitochondrial R-loop. (**a**) D-loops: triple stranded mtDNA molecules with a third strand of DNA have been recognized since the early 1970s. (**b**) In 2016, abundant mitochondrial R-loops were also found in the

resistant, this again indicates that it is RNA in the form of an R-loop.

(g) Strand specific probes indicate that the third strand of RNA is complementary to the 7S DNA.

(h) DNase treatment of the SBs followed by RT-PCR and sequencing after cloning yielded sequences that extend maximally from TAS to LSP, and minimally from Ori-b to Ori-H, which we designated LC-RNAs (Fig. 2b).

Collectively these data amount to compelling evidence of a major specific mitochondrial R-loop, as detailed in Fig. 2, which we denote *the mitochondrial R-loop* [6]. The extraordinary high abundance of the R-loops formed of LC-RNA is, of itself, evidence that they are formed deliberately, and the study of cells with mutant RNase H1 suggests the mitochondrial R-loop plays a role in the segregation of mtDNA [6]. It is likely that the mitochondrial R-loop has multiple roles in the regulation of mtDNA replication and expression [15] and so there remains much to learn. To name just a few of the simplest outstanding questions: what is the proportion of LC-RNA that is hybridized to DNA, and what is the half-life of this prominent R-loop? What proportion of 7S RNA is associated with the mitochondrial R-loops—either as RNA–DNA hybrid or double-stranded RNA with the 3′ end of LC-RNA (Fig. 2c)? And how are the mitochondrial R-loop and LC-RNA affected in cellular and animal models with deficiencies in a range of mtDNA transacting proteins, particularly those known to bind R-loops and single-stranded RNA or DNA, such as those illustrated in Fig. 2c

Fig. 2 (continued) control region, comprising L-strand RNA spanning much of the same area as 7DNA, often additionally including the region between the origin of replication (Ori-H) and the light strand promoter (LSP) (*see* Box 1 for further details). TAS—Termination Associated Sequence; Ori-b and Ori-H—origins of replication; C1—conserved sequence block 1. Suggested probe positions for assessing RNA–DNA hybrids via DRIP (or other methods) are illustrated in green. To note, probe 3 will detect the longer LC-RNAs that include the region Ori-H-LSP as will probe 4, and the latter will additionally detect R-loops containing 7S RNA. If the samples are denatured then 2 strand-specific riboprobes equivalent to probe 4 can be used, making probe 3 redundant. (**c**) LC-RNA can potentially form a variety of structures: as well as a complete R-loop it may bind to 7S RNA to create double stranded RNA portion spanning C1-LSP; or 7S RNA might hybridize to the L-strand displaced by the mitochondrial R-loop (of LC-RNA) creating a bubble structure with RNA–DNA hybrids on both branches. Stretches of LC-RNA at either end may form tails, and some portions of the mitochondrial R-loop may adopt a triplex structure. Some of the proteins that could affect the abundance and stability of one or other of these forms are indicated at the foot of the figure. For example, the mitochondrial specific isoform of FEN1, FENMIT, has tailed R-loop binding properties [13], and Twinkle DNA helicase and mtSSB both have DNA and RNA binding properties [14]

1.2 Aberrant R-Loops: An Exaggerated Threat to Mitochondrial Genomic Stability?

Because R-loops can form spontaneously during transcription, it tended to be assumed that they were unwanted events that threatened genomic stability. A more balanced view has emerged in recent years, which recognizes that R-loops can be created intentionally to regulate transcription or other aspects of DNA metabolism. However, the pendulum may have to shift still further; that is to say, inadvertent R-loops that disrupt DNA replication and transcription may be negligible, in normal circumstances.

Silva et al. [16] repressed the mitochondrial degradosome (comprising Suv3a and PNPase) to determine what effect this would have on mitochondrial R-loops, with one idea being that those generated during transcription would persist and might cause defects in the mtDNA. However, the principal effect was an increase in the incidence of RNA–DNA hybrids concentrated in the Control Region. Thus, the main conclusion is that the mitochondrial degradosome is required for processing RNA–DNA hybrids in the CR. As it was already known that primer processing is mediated by RNase H1 [4], probably in conjunction with ExoG [17], the inference is that the degradosome instead processes LC-RNAs and/or 7S RNAs, that is, bona fide R-loops. Given that there was little change in RNA–DNA hybrid abundance in other regions of the mtDNA following degradosome repression [16], either other enzymes remove deleterious R-loops elsewhere, or the threat they pose to mitochondrial genome stability is minimal. There are good reasons to believe the latter is the case, as mitochondria have excellent systems in place to avoid these hazards. First, a suite of proteins, many organized in defined structures known as mitochondrial RNA granules [18, 19], have RNA binding properties and are juxtaposed the mtDNA, and so are able to capture the RNA as it is transcribed, which would avoid it binding to the DNA and forming an R-loop behind the advancing transcription machinery. Second, I have proposed that the mitochondrial R-loop (of LC-RNA) functions, in concert with 7S DNA (and possibly 7S RNA), to ensure that replication and transcription are physically and temporally segregated, which would prevent cotranscriptional R-loop formation interfering in replication [20]. This idea remains valid despite a recent paper suggesting RNA and DNA synthesis sometimes occur on the same mtDNA molecule [21], as the study did not assess the possible contribution of 7S DNA synthesis to the 'replication' signal, nor did it consider that recently synthesized transcripts might associate with actively replicating mtDNA as part of the transcript-dependent replication mechanism [8]. In summary, while in disease states the mechanisms that prevent conflicts between the replication and transcription machineries could well go awry, in the normal course of events inadvertent R-loops on mtDNA may be exceedingly rare.

1.3 RNA–DNA Hybrids Are Not Synonymous with R-Loops

A key take-home message from the background information is the recognition that methods targeting RNA–DNA hybrids will not only detect R-loops of mtDNA but also unprocessed primers and intermediates of the transcript dependent mtDNA replication. Moreover, RNAs associated with DNA may adopt more complex forms than simple R-loops (Fig. 2c), and below I also discuss the theoretical and practical reality of adventitious hybridization of RNA creating RNA–DNA hybrids post–nucleic acid extraction, which should be considered for all studies of R-loops, not exclusively those in mitochondria.

1.4 Adventitious RNA Hybridization to DNA During Extraction

Few groups analyzing RNA–DNA hybrids in samples of extracted nucleic acids consider the potential problem of adventitious RNA hybridization (ARH). During any DNA extraction protocol, once the protease and detergent have taken effect, the DNA is in solution with an excess of RNA, and so RNA can hybridize to DNA spontaneously, and is expected to do so more readily where single-stranded DNA is present. From the first observation of RNA–DNA hybrids associated with replicating mtDNA in 2002, the possibility of adventitious hybridization of transcripts to the mtDNA was considered. It was immediately discounted, because in mixing experiments the RNA spontaneously hybridizing to the mtDNA did not reproduce the structures isolated from mitochondria [22]. Much later cross-linking experiments on isolated replication-competent mitochondria, and separately on intact cells, demonstrated that the RNAs are hybridized to replicating mtDNA molecules prior to mitochondrial/cell rupture and protease treatment [8]. Nevertheless, a year later, another study performed mixing experiments and concluded that mixing could create a bubble arc ([9] and Fig. 3). However, the bubble arc was not fully duplex as suggested in the cartoon illustration [9], on the contrary the bubble arc was highly similar to the "truncated" bubble arcs we create when we treat isolated mtDNA from solid tissues with RNase H (Fig. 3a). Thus, a more accurate interpretation would be that short regions of RNA can hybridize to available single-stranded DNA (Fig. 3, [22]), but there is no evidence that long stretches of RNA/DNA form spontaneously under any of the in vitro conditions yet tested. Further confirmation that the truncated bubble arcs comprise only short patches of RNA–DNA hybrid could be produced by treating the material with and without RNase T_1 or RNase A, as this would reduce the transcripts to small pieces of RNA, whereas if they were largely hybridized to the mtDNA the transcript lengths would change little, if at all. There is other evidence that the problem of ARH has been overstated. First, the DRIP analysis on extracted nucleic acids was in agreement with the IF analysis in the study of the mitochondrial degradosome [16]. Second the LC-RNA that forms the abundant mitochondrial R-loop is principally found on molecules that are otherwise

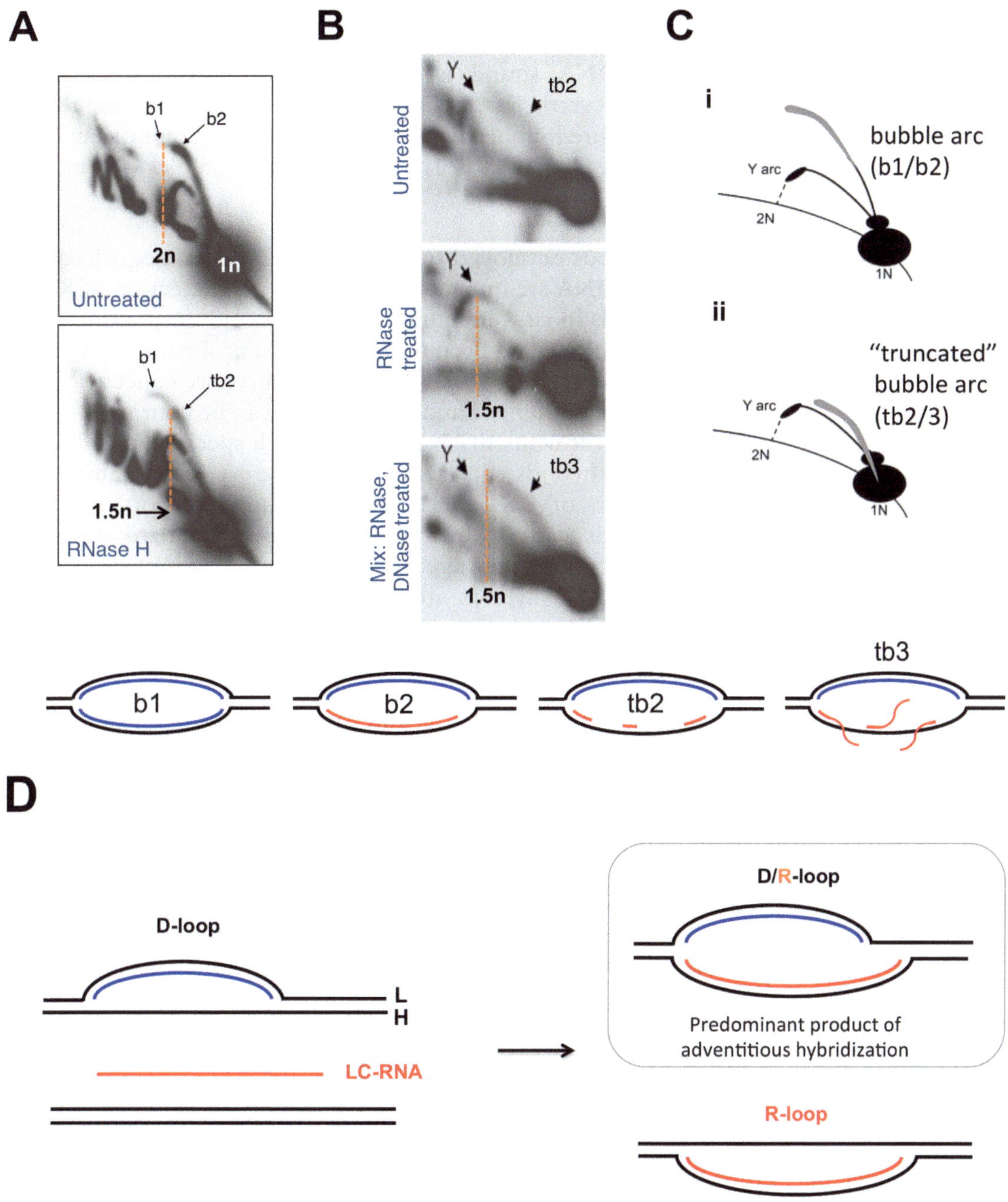

Fig. 3 Adventitious RNA hybridization does not create the intermediates of transcript-dependent replication or the mitochondrial R-loop. (**a**) Substantially intact bubbles with an RNA/DNA branch (b2) have similar mobility to duplex DNA bubbles (b1); both have a maximal size/mass of ~2n (e.g., 8 kb starting from a 4 kb linear duplex DNA molecule), marked by a vertical orange broken line. RNase H treatment alters b2 such that it is closer to the ascending Y arc and the maximal mass is ~1.5 n (also marked by a vertical orange broken line); this we term a truncated bubble arc (tb2), because its length and mass are considerably shorter and less than one comprising duplex DNA (b1) or RNA–DNA hybrid throughout (b2), as illustrated beneath, panels A-C. The b2 arc after RNase H treatment (tb2) has little in the way of RNA–DNA hybrid regions (evidenced by its maximal mass decreasing from ~2n to ~1.5n), and nor do the species formed during the mixing experiment exceed 1.5n (panel **b** - lowest gel image), as the bubble-like arc (tb3) is half the length of the Y arc. Gel images of

dsDNA, whereas ARH would be expected to occur much more readily on molecules with a D-loop, as the displaced (single-stranded) DNA strand is the complement of the LC-RNA (Fig. 3d). Nevertheless, in the methods we describe a straightforward procedure (Subheading 3.11) to assess the extent of any ARH that is applicable to all isolated nucleic acid samples of whatever source.

1.5 General Methods of RNA/DNA Analysis Applicable to Mitochondrial DNA

Three methods developed for nuclear DNA that are equally applicable to mtDNA are immunofluorescence (IF) detection of RNA–DNA hybrids in situ, immunoprecipitation and deep sequencing. The S9.6 antibody has been widely used as a probe for RNA–DNA hybrids, for example, [16, 23], and while there are concerns that it can also detect double stranded RNA, this can be controlled for by RNase III treatments that will remove such species, while leaving RNA–DNA hybrids intact. An alternative tool is a catalytically inactive form of RNase H1 [24]. Immunodetection of RNA–DNA hybrids with S9.6 antibodies follows standard protocols that are described elsewhere in this book. The advantage of IF is that RNA–DNA hybrids are observed directly in cells, thereby avoiding RNA loss during extraction, although it remains possible that mitochondrial R-loops can be disrupted during the fixing process, which typically includes partial denaturation with the detergent Triton X-100. As interstrand cross-linking alters the confirmation of RNA–DNA hybrids, such that they are resistant to RNase H, it is likely that this will affect antibody recognition. Hence, cross-linking cannot be used in combination with immunodetection; other disadvantages are that it is not possible to map the RNA–DNA hybrids, nor determine the strand, via IF. Similarly immunoprecipitation of RNA–DNA hybrids (DRIP) followed by amplification of the RNA is equally applicable to nuclear and mitochondrial DNA, although each may present its own unique problems. Its application to mtDNA is detailed below (*see* **Note 1**).

Protocols have also been established for preparing libraries generated from RNA–DNA hybrids in nuclear DNA, after affinity capture [25] that can be applied directly to mtDNA isolated from cells or tissues. This approach not only gives genome-wide coverage it identifies the RNA strand. However, the technique identifies

Fig. 3 (continued) panel **b**, and illustrative bubble arcs, **c**, are reproduced and adapted from [9]. (**d**) Mixing mtDNA with and without D-loops together with an excess of RNA is expected to greatly favor D/R loops as oppose to R-loops. The D-loop has a displaced H-strand that is complementary to much, or all, of the length of the LC-RNAs (*see* Fig. 2a, b) and so the spontaneous hybridization of LC-RNA to the D-loop should occur readily when the two are in solution, whereas hybridization of LC-RNA to duplex linear DNA is thermodynamically less favorable. However, the reality is the opposite: D/R-loops are barely detectable at steady-state in mouse liver mitochondria or human cultured cells, whereas R-loops of LC-RNA are highly abundant [6]. Hence, the mitochondrial R-loops of LC-RNA are likely formed in an active process

RNA–DNA hybrids and does not indicate which of the identified RNAs was in the form of an R-loop. Nevertheless, it does have considerable potential. Considering once again the increase in RNA–DNA hybrids that follows repression of the mitochondrial degradosome [16], library analysis would reveal the exact sequences of the RNAs that increase in the control region of mtDNA, and any other changes across the rest of the genome, which could be numerous, albeit less marked. Furthermore, by circularizing the nucleic acids after denaturation and treating with an enzyme such as MGME1 that degrades circular single-stranded DNA prior to library preparation, one could subtract nascent DNA strands with a primer. Like all the procedures that utilize extracted nucleic acids, everything hinges on the quality of the material, and the methods section details the current best approaches to producing high quality, artefact-free mtDNA.

2 Materials

All chemicals should be certified RNase- and DNase-free. The entire procedure should be performed on ice and if possible in a cold room, with centrifuges pre-cooled to 4 °C.

2.1 Isolation of Mitochondria from Solid Tissue

1. Phosphate-buffered saline: 10 mM phosphate buffer, pH 7.4, 2.7 mM KCl, 137 mM NaCl.
2. Homogenization buffer (HB): 225 mM Mannitol, 75 mM Sucrose, 2 mM EDTA and 10 mM HEPES-NaOH (pH 7.6) and 1 mg/mL fatty acid–free bovine serum albumin (BSA).
3. Motorized Dounce Homogenizer with tight-fitting Teflon pestle.
4. Six to ten laboratory mice of any standard strain 2–3 months of age.
5. Refrigerated centrifuge with rotor suitable for pelleting material at up to 10,000 × g_{max} in volumes of 10–50 mL.

2.2 Sucrose Gradient Centrifugation

Single-step sucrose gradient buffer (for 50 mL).

10 mM HEPES, pH 7.8	0.5 mL (1 M)
5 mM EDTA	0.5 mL (0.5 M)
2 mM DTT	100 µL (1 M)

Sucrose component (for 10 mL):

1 M sucrose	3.42 g
1.5 M sucrose	5.13 g

2.3 Mitochondrial DNA Extraction

1. Lysis buffer (LB): 20 mM HEPES–NaOH, pH 7.8, 75 mM NaCl, 50 mM EDTA (from a stock of 0.5 M EDTA pH 8.0).
2. 20% (w/v) Sodium lauroyl sarcosinate (sarkosyl).
3. Proteinase K (PK): 20 mg/mL in water.
4. PCIA: equilibrated phenol:chloroform:isoamyl alcohol (25: 24:1).
5. CIA: chloroform: isoamyl alcohol (24:1).
6. TE: 10 mM Tris–HCl, pH 8.0, 1 mM EDTA.
7. Isopropanol.
8. Ethanol.

2.4 Isolation of Mitochondria from Cells

Polyallomer Centrifuge tubes (13 × 51 mm) (Ref: 326819; Beckman)

- **PMSF** (MW = 174.19 g/mol).
  - For 100 mM PMSF: add 174.2 mg PMSF and vol up to 10 mL with 96% absolute ethanol.
- **DTT** (MW = 154.2 g/mol).
  - For 1 M DTT: add 771 mg DTT and vol. up to 5 mL with mQ H_2O.
- **HEPES-NaOH, pH 7.8** (50 mL).
  - dissolve 11.9 g HEPES in 45 mL mQ H_2O, pH with NaOH, adjust the final volume to 50 mL.
- **HYPOTONIC BUFFER (50 mL)**—*sufficient for cells from 40 × 15 cm plates.*

20 mM HEPES, pH 7.8 (pH with NaOH)	1 mL (1 M)
5 mM KCl	0.25 mL (1 M)
1.5 mM $MgCl_2$	75 μL(1 M)
2 mM DTT	100 μL (1 M)
1 mg/mL BSA	50 mg
1 mM PMSF	500 μL (100 mM)
1/50 Roche inhibitor (EDTA free)	1× tablet (dissolve pellet in 1 mL water)

- **2.5× MSH buffer (50 mL)**—*sufficient for 80, 15 cm plates*

525 mM Mannitol	4.78 g
175 mM sucrose	2.99 g
20 mM HEPES, pH 7.8	1 mL (1 M)
5 mM EDTA	0.5 mL (0.5 M)
1 mg/ml BSA	50 mg
2 mM DTT	100 μL (1 M)
1 mM PMSF	500 μL (100 mM)
1/50 Roche inhibitor)	1× tablet (EDTA free)

- **1× MSH buffer (50 mL) prepared without BSA.**

210 mM Mannitol	1.91 g
70 mM sucrose	1.20 g
20 mM HEPES, pH 7.8	1 mL (1 M)
2 mM EDTA	200 μL (0.5 M)
2 mM DTT	100 μL (1 M)
1 mM PMSF	500 μL (100 mM)
1/50 Roche inhibitor.	1× tablet (EDTA free)

- **1× MSH buffer (for 50 mL) without inhibitors**—*sufficient for 40, 15 cm plates*

210 mM Mannitol	1.91 g
70 mM sucrose	1.20 g
20 mM HEPES, pH 7.8	1 mL (1 M)
2 mM EDTA	200 μL (0.5 M)

- **0.5× MSH buffer (for 25 mL) without inhibitors**—*sufficient for 20, 15 cm plates.*

105 mM Mannitol	0.48 g
35 mM sucrose	0.30 g
20 mM HEPES, pH 7.8	500 μL (1 M)
2 mM EDTA	100 μL (0.5 M)

2.5 Nucleic Acid Restriction and Modification, and Agarose Gel Electrophoresis

1. A range of restriction endonucleases (*see* **Note 2**).
2. RNase T_1, RNase H. The former degrades single-stranded RNA, whereas the latter degrades the RNA component of RNA–DNA hybrids.
3. 5 M sodium chloride.
4. 10 mM Tris–HCl, pH 8.0.
5. TE: 10 mM Tris–HCl, pH 8.0, 1 mM EDTA.
6. 0.5 M EDTA, pH 8.0.
7. TBE: 89 mM Tris base, 87 mM boric acid, 2 mM EDTA, pH 8.0.
8. Loading buffer: 40% sucrose, 0.025% (w/vol) xylene cyanol, and bromophenol blue.
9. Low electroendosmotic (LEEO) agarose and ethidium bromide (EB): 10 mg/ml.
10. Gel tank with ports for buffer circulation, peristaltic pump.
11. Power supply and Ultraviolet (UV) light box.

2.6 Transfer of mtDNA to Solid Support & Hybridization (See Note 3)

1. Depurination buffer (DPB): 0.25 M HCl.
2. Denaturing buffer (DNB): 0.5 M NaOH, 1.5 M NaCl.
3. Nylon membrane, 3MM Whatman filter paper & paper towels.
4. Hybridization buffer: 0.25 M sodium phosphate, pH 7.2, 7% (w/vol) sodium dodecyl sulfate.
5. 20× SSC: 3 M NaCl, 0.3 M sodium citrate, pH 7.0.
6. Washing buffer 1 (WB1): 1× SSC.
7. Washing buffer 2 (WB2): 1× SSC/0.1% (w/vol) sodium dodecyl sulfate.
8. X-ray film or phosphorimager cassettes.

3 Methods

3.1 Cotranscriptional R-Loop Formation Assessed by Nascent RNA Labeling and S9.6 Antibody Detection (See Note 4)

1. Plate cells 1–2 days in advance on glass coverslips or chambers suitable for inverted microscope analysis, such that they are 50–70% confluent on the day of the experiment.
2. Prepare a stock solution of 250 mM bromouridine (BrU) in water, add to the culture medium to a final concentration of 2.5–5 mM BrU and return the cells to the CO_2 incubator for 30 min at 37 °C. (*see* **Note 5**).
3. Without washing add paraformaldehyde to 4%, and incubate for a further 15 min at 37 °C; wash 3 times with PBS. Note: For 300 μL of medium add 32.4 μL of 37% PFA stock).

4. Incubate the cells for 30 min in IF buffer (permeabilization + blocking). IF BUFFER: 0.3% Triton X-100, 3% BSA, or 5% normal goat serum.
5. Dilute the primary antibodies against BrU and RNA–DNA hybrids (e.g., S9.6) with IF and incubate overnight at 4 °C on a rocking or rotating platform at low speed (*see* **Note 6**).
6. Recover primary antibodies, wash coverslips/chambers 3 times with PBS and add secondary antibodies and DAPI for 1 h in the dark, at room temperature.
7. View labeled antibodies via fluorescence or confocal microscopy.
8. Quantify signals using ImageJ or other software package.

3.2 Extraction of Mitochondria from Mouse Liver (See Note 7)

All solutions should be precooled and all operations carried out on ice in a cold room.

1. Remove the livers from the mice, separate and discard any associated fat, blood and blood vessels.
2. Weigh the liver. Four grams of liver provides enough material for ≥20, 2D gel panels or at least twice as many 1D gels.
3. Mince the tissue finely with scissors and rinse repeatedly with ice cold 1:10 diluted HB until there is no residual blood or fat.
4. After the final rinse, add 4 mL HB/g minced liver and homogenize using 3 strokes of a motorized tight-fitting Dounce homogenizer set at speed 5.
5. Centrifuge the homogenate at 600 × g_{max} for 10 min at 4 °C to pellet nuclei and large debris. Recentrifuge supernatant at 7000 × g_{max} for 10 min to obtain a crude mitochondrial pellet. Resuspend the mitochondrial pellet in 5 vol HB/g liver, centrifuge at 7000 × g_{max} for 10 min, and repeat this step with 2.5 vol HB/g liver.
6. Centrifuge the supernatant at 7000 × g_{max} for 10 min at 4 °C to pellet the mitochondria.
7. Discard the supernatant and suspend the mitochondrial pellet in 5 vol of HB/g, that is, 20 mL for 4 g of liver, with a loose fitting glass homogenizer to ensure a homogeneous solution.
8. Repeat **steps 5** and **6**.
9. Resuspend the second mitochondrial pellet in 0.5 vol HB/g liver.
10. Remove 10 μL of mitochondrial suspension and estimate the protein concentration.
11. Prepare tubes with a single-step sucrose gradient. For a 35 mL tube, add 16 mL of 1.5 M sucrose, 2 mM EDTA, 10 mM HEPES-NaOH (pH 7.8) and an equal volume of 1.0 M

sucrose, 2 mM EDTA, 10 mM HEPES-NaOH (pH 7.8), gently and slowly adding the first few milliliters of the 1 M sucrose to minimize mixing of the 2 solutions.

12. Add 1–3 mL of the resuspended mitochondria (5 mg/mL protein) to the top of the gradient and adjust the weights of the tubes to within 0.01 g before ultra-centrifugation. Centrifuge the samples at 40–60,000 × *g*, 30 min at 4 °C in a SW60Ti rotor or equivalent. The mitochondria sediment at the boundary between the layers of 1 M and 1.5 M sucrose.
13. Remove and discard most of the gradient above the mitochondrial band. Using a pipette with a wide aperture, or a 1 mL Gilson-type pipette tip cut with a sterile razor blade, aspirate the mitochondria.
14. Dilute the mitochondria to 250 mM sucrose, 2 mM EDTA and 10 mM HEPES pH 7.8 and sediment by centrifugation at 10,000 × g_{max} for 10 min, at 4 °C.

3.3 Mitochondrial DNA Extraction from Isolated Mitochondria (See Note 8)

1. Resuspend the mitochondria in 75 mM NaCl, 50 mM EDTA, 20 mM HEPES pH 7.2 (*without detergent*). Add Proteinase K (stock 20 mg/mL in water) to a final concentration of 0.1 mg/mL and incubate on ice for 45 min.
2. Add 20% sodium sarkosyl to a final concentration of 1% and mix gently.
3. Add an equal volume of 1:1 phenol: chloroform/isoamyl alcohol, mix gently yet thoroughly.
4. Centrifuge 13,000 × g_{max} for 5 min at 4 °C.
5. Transfer the upper aqueous phase to a clean tube, mix gently with 1 vol of CIA, and spin at 16,000 × g_{max} for 5 min at 4 °C.
6. Recover the upper, aqueous phase and add 1 vol of isopropanol and NaCl to 100 mM, mix gently, and incubate at −20 °C for 1 h or overnight.
7. Centrifuge at 20,000 × g_{max} for 20 min at 4 °C.
8. Wash the pellet with 70% ethanol and briefly air-dry.
9. Resuspend the pellet in TE, and determine the nucleic acid concentration, for example, by UV spectrometry.

3.4 Isolation of Mitochondria and mtDNA from Cultured Cells (See Note 9)

Before collecting cells, gather one 50 mL falcon for every 10 plates, and record the weight of one of them.

1. Harvest cells from 20–40 × 150 mm plates, at 80–90% confluent.
2. Discard the medium and bang to dislodge the cells (for HEK293T cells) or treat with trypsin if required, for example, fibroblasts.

3. Wash the cells in 5 mL PBS (with or without trypsin) per 15 cm plate and place in 50 mL Falcon tubes. Use another 5 mL to rinse and recover the majority of the residual cells. This second solution can be recycled to a second plate and batches of 8–10 plates can be treated in this way, such that the cells can be collected from the plates without pausing, even when trypsin is required.
4. Pellet the cells by centrifugation at 1100 × *g* for 5 min.
5. Aspirate the media, and resuspend and combine the pellets in at least 40 mL PBS, in the preweighed 50 mL falcon tube. Centrifuge again at 1100 × *g* for 5 min. Aspirate and discard the PBS, calculate the volume of the cell pellet (1 g of cells = 0.8 mL wet volume).
6. Resuspend cells in 10 vol of hypotonic buffer with a pipette. Stand on ice for 10 min in the cold room. From this point on, the entire purification process is performed on ice in the cold room.
7. Homogenize using a tight-fitting 15 mL Dounce homogenizer with 15–20 strokes (check cell disruption with light microscope using a 1:10 dilution mainly). 80–90% cells should be broken, whereas nuclei should be mainly intact.
8. Pipette homogenate into 50 mL tube and immediately mix with 2 mL 2.5× MSH for every 3 mL of hypotonic buffer added. Mix thoroughly by pipetting up and down several times with a serological pipette.
9. Spin 700 × *g* × 1 min at 4 °C in to remove cell debris and nuclei. Keep the pellet if you wish to check enrichment.
10. Recentrifuge supernatant, 700 × *g*, 10 min at 4 °C to remove residual nuclei and cell debris.
11. Centrifuge the second supernatant: 9000 × *g* for 10 min at 4 °C.
12. Resuspend the pellet in 5 mL 1× MSH.
13. Centrifuge 8500 × *g* × 10 min at 4 °C to pellet the mitochondria. Discard the supernatant.
14. Further purification of mitochondria on a single-step sucrose gradient is as per solid tissue, except that the volumes are smaller. Thus, for 5 mL ultracentrifuge tubes, add 2.2 mL of 1.5 M sucrose, followed by 2.2 mL of 1 M sucrose.
15. Resuspend the mitochondria in 1/8 vol of the original cell pellet, in 1× MSH. For ~2 g of cells, resuspend in 250 μL, ~500 μL total with the mitochondria, and divide between 2 gradients.
16. mtDNA extraction from mitochondria isolated from cells is as per those of solid tissues (*see* Subheading 3.2).

3.5 Total Cellular DNA Isolation from Mammalian Cultured Cells After Crosslinking

1. After cross-linking has been applied (*see* Subheading 3.7), TMP that is classified as toxic must be discarded according to the local biosafety protocols.
2. Wash the cells with PBS, treating the first wash as toxic waste.
3. Remove PBS and lyse cells on the 90 mm dish with 4 mL 20 mM HEPES–NaOH, pH 7.8, 75 mM NaCl, 50 mM EDTA, 1% sodium sarkosyl on ice.
4. Transfer the solution to a 15-mL tube containing 4 mL phenol immediately.
5. Mix gently but thoroughly by rotation for 15 min at room temperature.
6. Centrifuge the sample at 6000 × g_{max} for 15 min at room temperature.
7. Transfer the aqueous (upper) phase to a clean 15-mL tube (Add 1 vol CIA and repeat **step 5**.
8. Centrifuge the mixture at 6000 × g_{max} for 15 min at 4 °C.
9. Recover the aqueous (upper) phase and aliquot in 500 μL in 1.5 mL tubes.
10. Add 5 M NaCl to a final concentration of 175 mM and 1 vol isopropanol. Mix gently and keep on ice for 15 min.
11. Centrifuge the sample at 20,000 × g_{max} for 20 min at 4 °C.
12. Discard the supernatant and wash the pellet with 70% ethanol.
13. Air-dry the pellet briefly, suspend in 10 mM Tris–HCl at pH 8.0 and determine the concentration by UV spectrometry.

3.6 Total Cellular DNA Isolation from Tissue Homogenates

1. If cross-linking has been applied (*see* Subheading 3.7), centrifuge the homogenate at 9000 × g_{max} for 10 min at 4 °C and discard the TMP safely.
2. Resuspend the homogenate in 4–10 mL 20 mM HEPES–NaOH, pH 7.8, 75 mM NaCl, 50 mM EDTA, 1% sodium sarkosyl, 100 μg/mL proteinase K and incubate on ice for 20–60 min.
3. Transfer the solution to a 15 mL or 50 mL polyurethane tube containing an equal vol of phenol.
4. Mix gently but thoroughly.
5. Centrifuge the sample at 6000 × g_{max} for 15 min at room temperature.
6. Transfer the aqueous (upper) phase to another tube containing an equal volume of phenol, and repeat **step 5**, once, or twice if protein is visible as a white band at the boundary of the aqueous and phenol phases.
7. Transfer the aqueous (upper) phase to a tube containing 1 vol of CIA and repeat **step 5**.

8. Recover the aqueous (upper) phase and aliquot in 500 μL in 1.5 mL tubes.
9. Follow **steps 10–13** of the protocol for recovering DNA from cells.

3.7 Inter-Strand Cross-Linking with Psoralen and UV Light

(For an explanation of how cross-linked material can be analyzed to infer the presence of R-loops, *see* **Notes 10** and **11**),

1. For nucleic acids cross-linking of cells, wash cell monolayers with DMEM without serum.
2. Incubate cells in DMEM supplemented with 10 μM 4,5′,8-trimethylpsoralen (TMP) for 10 min at 37 °C.
3. Expose the cells to 365 nm UV light for 10 min at room temperature. Cells grown as a monolayer are ideal as the cross-linking can be performed without detaching the cells, and every cell will receive a very similar dose of UV and thus a similar number of cross-links.
4. Incubate 5–50 mg/mL isolated mitochondria in HB, lacking BSA, with 10 μM TMP for 10 min on ice.
5. Spread the mitochondria in a shallow layer, ~ 2 mL, in a 90–100 mm tissue culture plate, expose to 365 nm UV light for 10 min.
6. Homogenize the tissue of interest with a motorized Dounce homogenizer with a Teflon pestle in the case of liver and kidney, and with an Ultra-turrax for skeletal and cardiac muscle.
7. Suspend the tissue for homogenization in 0.2 × HB without BSA after homogenization determine the protein concentration.
8. Incubate between 5–50 mg/mL of the homogenate with 10 μM TMP for 10 min on ice.
9. Spread the homogenate in a shallow layer, ~ 2 mL, in a 90–100 mm tissue culture plate, expose to 365 nm UV light for 10 min.
10. Extract the nucleic acids according to the protocols described below.

3.8 Nucleic Acid Modifications and 1D and 2D-AGE

1. Digest 3 μg lots of total mitochondrial nucleic acid with 10 U of the appropriate restriction endonuclease(s) for 30–180 min at 37 °C in a total volume of 200 μL.
2. To precipitate the digestion products add NaCl to 100 mM (maximally 4 μL of 5 M NaCl for a vol of 200 μL), 5 μL glycogen (10 mg/ml), and 2 vol of 100% EtOH (420 μL). Invert the tubes several times. Precipitate at −20 °C 1 h, or overnight.
3. Centrifuge the sample at 20,000 × g_{max} for 20 min at 4 °C.

4. Discard the supernatant and wash the pellet with 70% ethanol.
5. Air-dry the pellet, suspend in TE or in 10 mM Tris–HCl, pH 8.0 if it is to be treated with modifying reagents. If no additional treatment is required then the sample is ready for AGE.
6. Our standard conditions for the modifying enzymes are 1 U of RNase H for 15 min at 37 °C in 20 mM HEPES–KOH (pH 7.8), 50 mM KCl, 10 mM $MgCl_2$, 1 mM DTT; 100 U of RNase T_1 nuclease in 10 mM Tris–HCl (pH 7.4), 40 mM NaCl, 1 mM $MgCl_2$ for 15 min at 37 °C (*see* **Note 12**).
7. Stop the reaction by flash-freezing the sample on dry ice/ethanol.
8. For low-resolution 1D gels we typically use 1% Tris borate or tris acetate gels without circulating the buffer. Dissolve 1 g of agarose in 100 mL TBE buffer and cast in a 110 × 140 mm. Electrophoresis can be anywhere between 3 and 16 h at room temperature dependent on the fragment sizes; for fragments of 10–16 kb the latter is appropriate.
9. To analyze nascent DNA strands, 7S DNA and the RNA components of R-loops samples are denatured prior to gel loading and electrophoresis by heating to 85 °C for 15 min in 80% formamide. This can be applied to specific species gel-extracted from 2D agarose gels, such as small bubbles [6].
10. To resolve and map nascent strands to a resolution of 5–10 nts on fragments of a few hundred nucleotides a sodium borate (tris-free) buffer is used [6, 26] (*see* **Note 13**).
11. Circulate the buffer with a peristaltic pump and monitor the temperature periodically, as toward the end of the run the resistance increases and thus the temperature, and the separation of the fragments accelerates and the gel can melt if left unattended.
12. For 2D-AGE, prepare a 100 mL 0.4% agarose gel in TBE buffer and cast in a 110 × 140 mm gel tray for the first electrophoresis step, that is, 1D-AGE.
13. Submerge gel in a tank containing TBE.
14. Load samples and appropriate size markers (e.g., kilobase ladder).
15. Run the first dimension at 0.7 V/cm for 20 h at room temperature.
16. After 20 h, cut out the lanes containing the samples with the aid of a razor blade and a ruler.
17. Prepare 400 mL of 1% molten agarose gel in TBE containing 500 ng/mL EB.

18. Rotate each 1D gel slice 90° counter-clockwise and place it in a 200 × 200 mm gel-casting tray.
19. Remove excess TBE buffer around the gel slice with 3MM filter paper and adhere the gel slice to the tray with 1% molten agarose.
20. Cool the 1% molten agarose to 50 °C in a water bath and add to the gel tray from the opposite side of the gel slice and wait until it has solidified.
21. Place the gel in a tank containing cold TBE with 500 ng/mL EB.
22. Using a peristaltic pump, circulate the buffer from the positive to the negative electrode to avoid the EB accumulating at the cathode.
23. Run the second dimension at 6 V/cm for 6 h at 4 °C.
24. The gel can be dried using a standard gel dryer but the local regulations concerning radioactive condensate must be consulted and adhered to.

3.9 Transfer to Solid Support

1. After 2D-AGE, remove the gel from the tank and invert into a glass dish.
2. Add 500 mL DPB for a dish measuring 300 × 400 mm, rock gently for 30 min, and discard solution (*acid turns xylene cyanol green and BPB yellow*).
3. Add 500 mL DNB and rock gently for 5 min. Pour off the solution and repeat the treatment for 25 min (*xylene cyanol and BPB return to their original colors*).
4. Add 500 mL NB and rock gently for 10 min. Discard the solution and repeat the treatment for 10–20 min.
5. Cut a piece of nylon membrane to the dimensions of the gel, wet in water and place face down on the gel removing any bubbles with a roller or glass rod.
6. Soak two sheets of 3MM Whatman filter paper in water and place them on top of the membrane, again avoiding bubbles.
7. Stack 10–12 cm of paper towel on top of the filter paper.
8. Place a glass plate or a tray on top and a weight to create enough pressure to maintain uniform contact across the gel. Occasionally a novice will select a weight of 1 Kg or more in the mistaken belief that this will more rapidly and effectively "squeeze" the nucleic acids out of the gel.
9. Blot overnight on the bench.
10. Remove and discard paper towels and 3MM Whatman filter paper. Place the membrane face up on 3MM Whatman filter paper and let stand for a few min.

11. Covalently link the DNA to the membrane in a UV cross-linker: total energy 1200 × 100 mJ/cm^2.
12. Expose the membrane to an X-ray film or phosphorimager cassette for 30 min to several days.
13. Because of the sensitivity of RNA to extremes of pH, DPB and DNB are not appropriate after UTP labeling of mitochondria. Instead the gel should be equilibrated in 6X SSC buffer, 20 mM NaOH for 30 min.
14. Rinse the membrane in 6× SSC, place the gel on a bed of three layers of 3MM Whatman filter paper soaked in 6× SSC and follow **steps 5–12** above.

3.10 Hybridization to Determine the Steady-State Level of R-Loops

1. Place the membrane DNA side up in a hybridization tube and add 15 mL prewarmed hybridization buffer at 60 °C. Hybridization solution for riboprobes is 2× SSPE, 2% sodium dodecyl sulfate, 5× Denhardt's Reagent, 5% Dextran sulfate buffer.
2. Incubate in a hybridization oven for at least 30 min at 60 °C.
3. Pour off the solution and repeat the incubation with fresh 15 mL prewarmed hybridization buffer at 60 °C.
4. Label a mtDNA specific probe. Riboprobes are prepared using T7-maxiscript kit (Ambion) as per the manufacturer's instructions (*see* **Note 14**).
5. Note dsDNA probes must be denatured at 95 °C for 5 min and chilled on ice for 2 min before incubating with the membrane.
6. Incubate overnight at 65 °C for dsDNA probes, 60 °C for riboprobes and oligonucleotides >80 mers Wash temperatures may need to be adjusted down for these last two types of probe. In all cases there should be next to no detectable probe in the wash solution at the end of the washing process.
7. In the case of riboprobes, discard the hybridization solution containing the probe and wash the membrane four to six times with 0.1–1× SSPE, 0.5% SDS, at 60 °C for 20 min.
8. Remove the membrane from the tube, air-dry briefly, and wrap in cling film (Saran Wrap).
9. Expose to X-ray film for 0.5–7 days at −80 °C, or to a phosphor cassette for 1–24 h.

3.11 Evaluating Adventitious Hybridization

1. To assess the potential contribution of ARH to the RNA–DNA hybrids analyzed by DRIP or 1D and 2D-AGE isolate nucleic acids from cells or tissues or the mitochondria thereof as described above.
2. Treat one fraction with either: 5 Units of RNase A per microgram of nucleic acid in 50 mM NaCl, 10 mM Tris pH 7.6, at

37 °C for 60 min. (Many laboratories have RNase A in abundance from plasmid preparation kits).

3. Treat another fraction with 1–5 units of DNase per microgram of nucleic acid at 37 °C for 60 min, after restriction digestion.
4. Phenol-chloroform extract the samples from **steps 1** and **2**; after ethanol or isopropanol precipitation resuspend in the same volume of 0.1 mM EDTA, 10 mM Tris pH 8.0. Mix the RNA (DNase-treated material) with the DNA (RNase A–treated) in equal proportions (i.e., maintain the RNA–DNA ratio found in the cells/mitochondria).
5. Heat to 37 °C for 30 min then apply DRIP, 1D or 2D-AGE, as per the other fractions that have not been treated with RNase or DNase.

4 Notes

1. Detailed DRIP protocols are described in other chapters. There follows some considerations pertaining to its application to mtDNA analysis. DRIP analysis comprises immunoprecipitation to capture RNA–DNA hybrids followed by RT-PCR to amplify captured RNA, which include R-loops [27]. As such, it requires sets of primers for each target region. A variety of protocols are available, most of which have been applied to cultured cells, with few studies of solid tissues. The basics steps are universal, cells are lysed with a strong detergent (SDS), and the proteins digested prior to phenol-chloroform extraction of nucleic acids. The nucleic acids are then incubated with an antibody that binds to RNA–DNA hybrids, typically S9.6. As controls, fractions of the samples are treated with RNase H, RNase A or RNase T_1, and RNase III. The last is to remove double-stranded RNA, which is also detected by the S9.6 antibody. For laboratories on a tight budget, RNase A can be applied at high- and low-salt concentrations; in the latter case (0–100 mM NaCl), it degrades RNA–DNA hybrids, as well as double stranded RNA, and so in theory neither RNase H nor RNase III is required. Quantitative PCR is then performed targeting the regions of interest. Changes in the abundance of RNA–DNA hybrids in each region are determined as the percentage of input recovered for each immunoprecipitated sample.

   Minimally probes are needed to identify/distinguish LC-RNA and 7S RNA in the CR; preferably 3 probes to distinguish the short medium and long forms of LC-RNA, with a fourth that will detect both the long LC-RNAs and any RNA–DNA hybrids formed of 7S RNA (Fig. 2b). Elsewhere a scattering of 3–4 probes around the mitochondrial genome would be better than 1 or 2. Investigators may also wish to include a probe to ND6, as it is the only mRNA encoded on the H-strand, and its expression often differs from the other 10 mRNAs [19]. ND6 maps close to the center of a broad initiation zone for the replication of mtDNA and so ND6 mRNA might be important for initiation of replication outside the CR [7].

Advantages: Multiple RNA–DNA hybrids from around the genome can be mapped with material generated in a single experiment. The method is likely to be more sensitive than detecting small bubble structures by 2D-AGE.

Limitations: It may not be compatible with cross-linked material. Adventitious hybridization could contribute to the signals (*see* Subheading 3.11). Conversely, loss of RNA–DNA hybrids during extraction can diminish the signals, so the quality of the results will hinge on the intactness of the RNA–DNA hybrids of the isolated mtDNA. Interpretation of the results needs to consider, that the RNA–DNA hybrids will not be exclusively R-loops, but will also include replication intermediates, according to the bootlace model of mtDNA replication [8], and nascent strands that retain the RNA primer. Although RNA primers are barely detectable at steady state in control cells, they can achieve high levels when primer processing is impaired [2].

Future prospects: by including an additional control—RNA–DNA hybrids formed adventitiously by mixing purified mitochondrial RNA and DNA, and treated with and without RNases (see introduction, Subheading 3.11 and Fig. 3)—the contribution of adventitious hybridization during extraction can be assessed and subtracted from the total. Combining it with library preparation, see next section can achieve an unprecedented depth of coverage.

2. Most restriction enzymes cut exclusively double-stranded DNA; a handful, such as *Acc*I and *Bsp*HI, also cut single-stranded DNA, and we have shown that these also cut the triple stranded DNA of the mitochondrial D-loops. In our experience no restriction enzyme cuts (both strands of) RNA–DNA hybrids.

3. This protocol is equally suitable for 1D and 2D agarose gels.

4. BrU labeling is likely to be the most reliable method for assessing whether mitochondrial R-loop formation is cotranscriptional, as it gives robust labeling within 1 h. Two alternative approaches used previously are ethidium bromide (EB) disruption of mitochondrial transcription for 5 h, or a 2 h incubation with the RNA chain terminator cordycepin. The first of these did produce a substantial decrease in RNA–DNA hybrids [16]; however, mitochondrial transcripts are markedly

depleted by EB within 3–6 h [28] (and our unpublished data). A 2 h incubation with cordycepin, that is converted to an RNA chain terminator, produced a more modest decrease in RNA–DNA hybrids in mitochondria when the degradosome was repressed [16]; so there remains some doubt as to whether the majority of the R-loops formed cotranscriptionally. BrU labeling could help resolve the issue, as it has the advantage that it is a direct reporter of newly synthesized RNA and requires incubations as short as 30 min. BrU labeling can also be used in pulse-chase experiments, which could answer the question of the half-life of the LC-RNA and the mitochondrial R-loop.

5. Prewarm the BrU solution to 37 °C until all the precipitate disappears. Incubation time may need to be optimized for the cell type under study. For HeLa cells 30 or 60 min work similarly well. BrU is rapidly converted to Br-UTP and thence incorporated into nascent transcripts.
6. Triton X100 can be reduced from 0.3% to 0.03% for the incubation with the primary antibodies but this is not an essential adjustment.
7. In our experience, RNA–DNA hybrids in mitochondria are extremely labile, and so their preservation is a critical prerequirement for assessing mitochondrial R-loops. In 2002, we showed that purifying mitochondria on a sucrose-step gradient greatly improved the yield of mitochondrial RNA–DNA hybrids compared to mitochondria isolated by differential centrifugation [21]. Including a sucrose gradient step markedly decreases the amount of contaminating nuclear DNA, which is sometimes visible as a pellet at the base of the tube. Conversely, purification of mtDNA via cesium chloride based isopycnic density centrifugation should be avoided, as it results in the loss of RNA from mtDNA [29].
8. A broad range of tests can be applied to mtDNA extracted from tissues, to determine comprehensive details of its composition and properties. It can be treated with a battery of nucleic acid modifying enzymes and binding proteins, and resolved on 1D and 2D gels, either in its native form, or denatured to reveal the individual component parts [6, 7]. The size and map positions of any R-loops can be determined to great accuracy, and its precise sequence can be determined. Despite all the modifications to the protocols we have devised, some RNA–DNA hybrids may be depleted or lost during extraction, and oppositely adventitious hybridization of RNA could create forms that were not present in vivo (although as noted elsewhere the latter is at most a minor problem).
9. As with mtDNA isolated from solid tissues, a broad range of tests can be applied to isolated mtDNA from cells, which will

reveal its composition and properties. Cultured cells have the major benefit that it is possible to modify aspects of mtDNA replication readily, including that of factors essential for mtDNA replication such as RNase H1 [2]. Of course, it is equally possible to create tissue-specific knockouts, if necessary where gene excision occurs only after birth, but this requires a much greater investment of resources than manipulation of the genes in cells. In our experience, the major limitation of nucleic acid extraction from whole cells is that it invariably results in a lower yield of RNA–DNA hybrids of mtDNA than from purified mitochondria of solid tissues. Moreover, the yield of replication intermediates from human fibroblasts is much lower than from HEK cells, which might also be true of R-loop abundance. Two modifications that singularly failed to give good results were (a) whole cell lysis on ice with 1% SDS/Sarkosyl and any amount of Proteinase K, and (b) high concentrations of EDTA (5 or 10 mM). As with solid tissue, yields of RNA/DNA are better when the mtDNA is prepared from isolated mitochondria purified on a sucrose-gradient. Hence, the clear implication is that extra-mitochondrial RNase, or RNA helicase activities result in the loss of the hybrids once the cell contents mix. This inference is further corroborated by the fact that inter-strand, cross-linked RNA–DNA hybrids are resistant to RNase H, and cross-linking increases R-loop yields in cultured cells as well as tissues [6, 8]. Notwithstanding the limitations above, the relative ease of altering gene expression by gene silencing or Crispr-CAS gene editing means that cultured cells will continue to be used heavily to study many aspects of mtDNA metabolism, including R-loop formation and function, aided by cross-linking to reduce their loss during extraction. Additionally there is a vast array of cell disease models of mtDNA dysfunction available in laboratories around the world, and the detailed study of a pathological RNase H1 mutant indicates how these can reveal insights into R-loop metabolism and human disease [6].

10. Introducing interstrand cross-links into mtDNA via psoralen and ultraviolet (UV) light has been to shown to be a highly effective means of preserving RNA–DNA hybrids of mtDNA [6, 8]. After isolation of cross-linked mtDNA from whole cells or tissues, or mitochondria thereof, the material can be restriction digested and resolved by 1D and 2D-AGE [6]. The improved preservation of RNA–DNA hybrids can be spectacular [8]. In the case of the small bubbles formed by the mitochondrial R-loop, and resolved by 2D-AGE, they were undetectable in mtDNA isolates from rat liver homogenates, but represented a substantial fraction of all the mtDNAs when cross-linking was applied prior to extraction. To note the

markedly higher yield of mitochondrial R-loops when cross-linking is performed in intact cells and mitochondria represents further good evidence that it does not form by adventitious RNA hybridization. Cross-linking is versatile as it can be applied to cells prior to extraction and tissue homogenates, as well as isolated mitochondria. Notably the isolated mitochondria are replication and transcription competent [8], and so are expected to reflect mtDNA metabolism in vivo. The major limitation of UV/psoralen cross-linking is that it is difficult to reverse, and the current method involves strong alkali treatment, which will of course simultaneously degrade the RNA component of any RNA–DNA hybrids. Nor can one easily determine which RNA strand is forming the RNA loop. Hence, this approach has to be used in conjunction with analysis of material isolated without cross-linking. For example, the latter material can be used to show that a R-loop maps to a particular part of the mitochondrial genome, and the strand, whereas cross-linking should be used to gain a much more accurate idea of its true abundance. Although it has not yet been demonstrated empirically, it is expected that the change in structure (B to Z), which renders cross-linked RNA–DNA hybrids resistant to RNase H, will also affect protein binding, in which case DRIP analysis is also excluded. Nevertheless, cross-linking is expected to be used in the future to clarify just how abundant mitochondrial R-loops are in a variety of situations, as it greatly enhances the yield, and can be demonstrated by a simple ′blocked-site′ assay, which is described below.

11. Gel electrophoresis of isolated mtDNA from cells or tissues after treating with restriction and other nucleic acid modifying enzymes can reveal a wealth of information about the nature and properties of different forms of mtDNA, and can distinguish whether RNA–DNA hybrids are R-loops or other species. One indirect yet simple and effective method to interrogate specific regions and sites for the presence of RNA–DNA hybrids exploits the fact that RNA–DNA hybrids are refractive to restriction endonuclease digestion. We call this the "Blocked-Site Assay" for RNA–DNA hybrids. In contrast, all three strands of D-loops can be cleaved by several enzymes that cut duplex, but not single-stranded, DNA [6]. Hence, isolated mtDNA can be digested with restriction enzymes that target sites within the region covered by a known or suspected R-loop and the products separated by 1D-AGE; longer fragments than the predicted digestion products are obtained wherever an RNA–DNA hybrid bridges a restriction site. If the material has not been cross-linked then this supposition can be confirmed by treating a fraction of the material

with RNase H prior to restriction digestion and the removal of the RNA will instead result in complete digestion of the mtDNA. The assay does not indicate which strand is RNA.

12. If the mtDNA has not been cross-linked, RNase H treatment prior to denaturation will fully degrade RNAs that form an RNA–DNA hybrid (R-loops) but also the transcripts hybridized to replication intermediates [8]. RNase H will modify nascent DNA strands and 7S DNA that retain a RNA primer. However, this procedure is not sufficiently sensitive to detect dispersed replication initiation sites spread across an initiation zone [30].

13. To map H-strand DNA ends (initiation or termination sites) in the CR and distinguish them from 7S DNA, we had to produce fragments slightly longer than 7S DNA itself [7]. Now that it is established that nascent strands of human mtDNA invariably start downstream of 7S DNA (~ np 60 [6, 7], there is now no reason why the mtDNA cannot be cut very close to the 5′ end (e.g., ~ np 10–30 of human mtDNA). Combined with PAGE this could give single-nucleotide resolution, which together with RNase H2 treatment would reveal even the shortest primers.

14. For many years, we used exclusively DNA probes, latterly we have chiefly used riboprobes because with denatured nucleic acids one obtains information as to the specific strand detected; moreover, riboprobes label to high specific activity (up to 90% incorporation), denaturation is not required and they are easy to remove for reprobing.

5 Concluding Remarks

On the one hand, the small size of the mitochondrial genome and the presence of one prominent R-loop greatly simplify the task of determining the contribution of R-loops to mitochondrial DNA metabolism. On the other hand, mtDNA is refractive to genetic manipulation and R-loop preservation is a challenge. Nevertheless, the protocols and contextual framework presented here can serve as a platform to advance our understanding of the function of mitochondrial R-loops. Pressing questions to resolve are the roles of the prominent R-loop, and its sister the D-loop, in the regulation of transcription and replication, as well as in mtDNA membrane attachment and segregation. The methods laid out here can aid in these investigations, and an inescapable conclusion is that multiple approaches will be required. One improved technique that would help considerably is a method of reversible interstrand cross-linking, so that well preserved RNA–DNA hybrids can be sequenced after extraction.

Acknowledgments

The author would like to thank all those past members of his laboratory who contributed to the application of these procedures to the study of mammalian mitochondrial DNA replication; and for their help in understanding the often-perplexing data produced, especially Aurelio Reyes and Takehiro Yasukawa.

References

1. Yasukawa T et al (2005) A bidirectional origin of replication maps to the major noncoding region of human mitochondrial DNA. Mol Cell 18:651–662
2. Holmes JB et al (2015) Primer retention owing to the absence of RNase H1 is catastrophic for mitochondrial DNA replication. Proc Natl Acad Sci U S A 112:9334–9339
3. Kasamatsu H, Robberson DL, Vinograd J (1971) A novel closed-circular mitochondrial DNA with properties of a replicating intermediate. Proc Natl Acad Sci U S A 68:2252–2257
4. Annex BH, Williams RS (1990) Mitochondrial DNA structure and expression in specialized subtypes of mammalian striated muscle. Mol Cell Biol 10:5671–5678
5. Chang DD, Hauswirth WW, Clayton DA (1985) Replication priming and transcription initiate from precisely the same site in mouse mitochondrial DNA. EMBO J 4:1559–1567
6. Akman G et al (2016) Pathological ribonuclease H1 causes R-loop depletion and aberrant DNA segregation in mitochondria. Proc Natl Acad Sci U S A 113:E4276–E4285
7. Cluett TJ et al (2018) Transcript availability dictates the balance between strand-asynchronous and strand-coupled mitochondrial DNA replication. Nucleic Acids Res 46: 10771–10781
8. Reyes A et al (2013) Mitochondrial DNA replication proceeds via a 'bootlace' mechanism involving the incorporation of processed transcripts. Nucleic Acids Res 41:5837–5850
9. Miralles Fusté J et al (2014) In vivo occupancy of mitochondrial single-stranded DNA binding protein supports the Strand displacement mode of DNA replication. PLoS Genet 10: e1004832
10. Jemt E et al (2015) Regulation of DNA replication at the end of the mitochondrial D-loop involves the helicase TWINKLE and a conserved sequence element. Nucleic Acids Res 43:9262–9275
11. Wanrooij PH et al (2012) A hybrid G-quadruplex structure formed between RNA and DNA explains the extraordinary stability of the mitochondrial R-loop. Nucleic Acids Res 40:10334–10344
12. Holt IJ (2019) The Jekyll and Hyde character of RNase H1 and its multiple roles in mitochondrial DNA metabolism. DNA Repair 84: 102630
13. Kazak L et al (2013) A cryptic targeting signal creates a mitochondrial FEN1 isoform with tailed R-loop binding properties. PLoS One 8:e62340
14. Hensen F et al (2019) Mitochondrial RNA granules are critically dependent on mtDNA replication factors Twinkle and mtSSB. Nucleic Acids Res 47:3680–3698
15. Holt IJ (2019) Survey and summary: the mitochondrial R-loop. Nucleic Acids Res 47:5480–5489
16. Silva S, Camino LP, Aguilera A (2018) Human mitochondrial degradosome prevents harmful mitochondrial R loops and mitochondrial genome instability. Proc Natl Acad Sci U S A 115:11024–11029
17. Wu CC et al (2019) A unique exonuclease ExoG cleaves between RNA and DNA in mitochondrial DNA replication. Nucleic Acids Res 47:5405–5419
18. Antonicka H et al (2013) The mitochondrial RNA-binding protein GRSF1 localizes to RNA granules and is required for posttranscriptional mitochondrial gene expression. Cell Metab 17: 386–398
19. Jourdain AA et al (2013) GRSF1 regulates RNA processing in mitochondrial RNA granules. Cell Metab 17:399–410
20. Ojala D et al (1981) A small polyadenylated RNA (7 S RNA), containing a putative ribosome attachment site, maps near the origin of human mitochondrial DNA replication. J Mol Biol 150:303–314

21. Brüser C, Keller-Findeisen J, Jakobs S (2021) The TFAM-to-mtDNA ratio defines inner-cellular nucleoid populations with distinct activity levels. Cell Rep 37(8):110000. https://doi.org/10.1016/j.celrep.2021.110000. PMID: 34818548.
22. Yang MY et al (2002) Biased incorporation of ribonucleotides on the mitochondrial L-strand accounts for apparent strand-asymmetric DNA replication. Cell 111:495–505
23. Renaudin X et al (2021) BRCA2 deficiency reveals that oxidative stress impairs RNaseH1 function to cripple mitochondrial DNA maintenance. Cell Rep 36:109478
24. Crossley MP et al (2021) Catalytically inactive, purified RNase H1: a specific and sensitive probe for RNA-DNA hybrid imaging. J Cell Biol 220:e202101092
25. Sanz LA, Chédin F (2019) High-resolution, strand-specific R-loop mapping via S9.6-based DNA–RNA immunoprecipitation and high-throughput sequencing. Nat Protoc 14:1734–1755
26. Brody JR, Kern SE (2004) Sodium boric acid: a Tris-free, cooler conductive medium for DNA electrophoresis. BioTechniques 36:214–216
27. Sanz LA, Castillo-Guzman D, Chédin F (2021) Mapping R-loops and RNA:DNA hybrids with S9.6-based immunoprecipitation methods. J Vis Exp (174). https://doi.org/10.3791/62455
28. Maniura-Weber K et al (2004) Transient overexpression of mitochondrial transcription factor A (TFAM) is sufficient to stimulate mitochondrial DNA transcription, but not sufficient to increase mtDNA copy number in cultured cells. Nucleic Acids Res 32:6015–6027
29. Pohjoismäki JLO et al (2010) Mammalian mitochondrial DNA replication intermediates are essentially duplex but contain extensive tracts of RNA–DNA hybrid. J Mol Biol 397:1144–1155
30. Bowmaker M et al (2003) Mammalian mitochondrial DNA replicates bidirectionally from an initiation zone. J Biol Chem 278:50961–50969

Chapter 13

Highly Purified Top1-Bound DNA Fragments

Jessica Marinello and Giovanni Capranico

Abstract

Topoisomerase 1–DNA cleavage complexes (Top1ccs) form at DNA sites of Top1 activity and are increased by a highly specific anticancer drug (camptothecin, CPT) in living cells. Various methods are available to detect Top1ccs in cultured cells, including protocols based on the use of specific antibodies. Here, we describe a protocol to isolate Top1ccs at high purity, which does not depend on antibodies.

Key words Topoisomerase I, Top1cc (Topoisomerase 1–DNA cleavage complex), Cesium chloride gradient, K^+-SDS precipitation

1 Introduction

DNA topoisomerase I (Top1) reduces the torsional tension of a DNA duplex with a DNA cleavage-rejoining activity. The reaction proceeds via a complex intermediate wherein Top1 is covalently bound to a 3′ end of a DNA strand (Top1–DNA cleavage complex, Top1cc). Thus, Top1ccs form at DNA sites of Top1 activity. Top1ccs can easily be detected in cultured cells using a highly specific anticancer poison (camptothecin, CPT), or related derivatives, as Top1 poisons act by inhibiting the DNA rejoining reaction of the enzyme, thus increasing Top1ccs. When using short times of CPT treatment (2–5 min), the method can basically reveal the physiological genomic sites of Top1 activity in living cells. Moreover, Top1 deletion or poisoning was shown to alter genomic R-loop levels in cultured cells [1–5]; therefore, mapping Top1cc sites are relevant to define the mechanism of Top1 poison-induced R-loop modulation.

The detection of Top1cc can be achieved with specific antibody during precipitation steps of cellular lysates or in immunofluorescence microscopy [6, 7]. Different protocols are variations (ICE or RADAR assays) of an original method developed by the MT Muller laboratory previously [8], which used an ultracentrifugation step to

Andrés Aguilera and Alexey Ruzov (eds.), *R-Loops: Methods and Protocols*, Methods in Molecular Biology, vol. 2528, https://doi.org/10.1007/978-1-0716-2477-7_13,

isolate DNA-bound Top1 from free Top1. Here, we have made some modifications of the ICE bioassay [8] and included steps to map Top1ccs along the cellular genome.

Genomic Top1 sites can be mapped with ChIP techniques in mammalian and yeast cells [9–11]; however, ChIP protocols detect DNA sites bound to Top1 either in a covalently and noncovalently manner. Levels of chromatin-bound Top1 (measured with ChIP) are prominent at promoter regions of active genes [10, 11], whereas levels of active Top1 (detected as Top1ccs) are higher in the body and termination regions of active genes [10] and telomeres of human and yeast cells [9, 12]. Therefore, ChIP and ICE assays are different as only ICE assays can detect genomic sites of active enzymes and CPT-induced reversible strand breaks.

Our protocol is based on the high efficiency of both CsCl gradient purification of DNA and K^+-SDS precipitation of proteins to detect genomic sites of Top1ccs in cultured cells independently from antibodies. However, it depends on the high specificity of CPT for its unique target. This method can be extended to DNA topoisomerase II when using a compound, such as etoposide or teniposide, which specifically target this enzyme.

2 Materials

General Considerations.

For the following protocol, make sure to use always ultrapure DNase/RNase free distilled water and molecular biology grade reagents (tested against DNase and RNase). Prepare and store all solution at room temperature unless otherwise specified. Use DNA/RNA-low-binding-1.5 mL tubes for each step that requires this volume.

2.1 DNA Purification by Cesium Chloride Density Gradient

1. Cell culture medium.
2. 75 cm^2 flasks.
3. Sarkosyl 1% solution in DNase/RNase-free distilled water.
4. Protease inhibitor cocktail.
5. Cell scraper.
6. 15 mL falcon tubes.
7. Cesium chloride (CsCl).
8. 0.22-μm pore-size sterile syringe filters.
9. 2 mL and 50 mL syringes (22 G needle).
10. TE (Tris–HCl 10 mM, pH 8.0, EDTA 1 mM) buffer.
11. 5 mL Polyallomer tubes for ultracentrifugation (Beckman Coulter Ref. 344,057, Open-Top Thinwall Ultra-Clear Tube).

12. Ultracentrifuge (Beckman Coulter Optima L-90 K, swing rotor SW50.1).
13. Micro volume spectrophotometer for DNA quantification (NanoDrop, ThermoFisher Scientific).
14. Centrifugal filter units for protein and nucleic acid purification, concentration and desalting, 3 kDa cutoff (Merck Millipore Amicon Ultra-0.5 Centrifugal Filter Unit UFC500396).

2.2 Top1cc Purity Evaluation by Western Blot in Purified DNA Samples

1. DNase I, RNase-free, 1 U/μL.
2. Precast Mini Protein Gel.
3. Tris–Glycine SDS Running Buffer 10×: 250 mM Tris–HCl pH 8.3, 1,93 M glycine, 1% (w/v) SDS.
4. Laemmli Sample Buffer 4×: 250 mM Tris–HCl pH 6.8, 8% (w/v) SDS, 40% glycerol, 0.04% (w/v) Bromophenol Blue.
5. Transfer Buffer 1×: 25 mM Tris, 192 mM glycine, 20% (v/v) Methanol, pH 8.3.
6. TBS 10×: 200 mM Tris, 1.5 M NaCl, pH 7.6.
7. TBS-T: TBS 1×, Tween 20 0.1% (v/v).
8. TBS-T-milk: TBS 1×, Tween 20 0.1% (v/v), milk 5% (w/v).
9. Actin Antibody (Actin I-19, sc-1616, Santa Cruz Biotechnology, dilution 1:500).
10. Top1 Antibody (Top1 C-15, sc-5342, Santa Cruz Biotechnology, dilution 1:200).
11. Anti-goat HRP-conjugated (Santa Cruz Biotechnology sc-2922, dilution 1:2000).
12. Nitrocellulose blotting membrane.
13. Ponceau S.
14. Pierce ECL Plus Western Blotting Substrate.
15. Imaging system for chemiluminescence detection.

2.3 DNA Fragmentation and K^+-SDS Precipitation of Top1cc

1. Lysis Buffer 2×: SDS 2% (w/v), 20 mM Tris–HCl pH 8, 20 mM EDTA.
2. Ultrapure BSA 50 mg/mL DNAse/RNase/Protease free.
3. Sonicator (Bioruptor Diagenode or any other bath sonicator).
4. SDS solution 0.5% (w/v) in DNase/RNase-free distilled water.
5. Spectrophotometer and quartz cuvette.
6. Proteinase K solution 20 mg/mL.
7. Phenol–chloroform–isoamyl alcohol, 25:24:1.
8. Ethanol.
9. 3 M sodium acetate solution, pH 5.2.
10. Agarose.

11. TBE Electrophoresis Buffer 1×: 89 mM Tris, 89 mM Boric Acid, 2 mM EDTA.
12. KCl 2.5 M.

2.4 Real-Time PCR Analyses of Recovered DNA Fragments (Top1cc)

1. SYBR Green mix for real-time PCR.
2. 96-well plate for real-time qPCR.
3. Primers.
4. Real time PCR.

3 Methods

3.1 DNA Purification by Cesium Chloride Density Gradient

1. Treat exponentially growing cells (HCT116 cancer cells) cultured in flasks with Top1 poisons (*see* **Note 1**).
2. At the end of treatment, remove cell medium and immediately add 1% Sarkosyl (1 mL of solution is used in 75 cm^2 flasks but the volume must be adapted to cell growing surface). The Sarkosyl solution must be pre-heated at 37 °C and added of protease inhibitors (final dilution 1×, *see* **Note 2**). Then, collect lysed cells at room temperature (RT) using a cell scraper into falcon tubes. Note that no PBS washing is performed before cell lysis to prevent reversal of Top1ccs.
3. Shear cell lysate with a syringe (20 G needle) and two/three strokes.
4. Shortly before cell lysis, a ~1.88 g/mL CsCl stock solution is prepared in TE buffer, filtered and maintained at RT (*see* **Note 3**).
5. Prepare a step gradient of 4 layers of CsCl solution at decreasing density from the bottom to the top, using the step solutions reported in Table 1. First, add protease inhibitors to the CsCl solution (Halt Protease Inhibitor Cocktail 1× final, Thermo-Fisher Scientific). Then, put 1 mL of the A solution at the bottom of a polyallomer tube, then gently stratify 1 mL of B solution on top of A, and then proceed with C and finally D (1 mL each) in the same manner (*see* **Note 4**).
6. Then, gently stratify 1 mL of cell lysate on the top of the gradient and top off with mineral oil.
7. Gently, put the tube in a swing-out rotor of a ultracentrifuge and carefully balance it. Then, centrifuge at 115,000 × *g* for 18 h at 25 °C with the break off (*see* **Note 5**). We use a Beckman Coulter ultracentrifuge with the swing-out SW 50.1 rotor at 115,282 × *g*.
8. After the rotor has come to a complete stop, carefully collect 200-μL fractions from the bottom of the gradients with a sterile syringe avoiding mixing the gradient. At the end, you have collected around 20 fractions (*see* **Note 6**).

Table 1
Combinations of stock CsCl solution and TE buffer to prepare the CsCl step solutions

	A	B	C	D
TE buffer	0.07 mL	0.2 mL	0.45 mL	0.55 mL
CsCl stock solution	0.93 mL	0.8 mL	0.55 mL	0.45 mL

9. Determine the DNA containing fractions reading absorbance at 260 nm with a Nanodrop spectrophotometer. Determine CsCl density of each DNA containing fraction with analytical balance (*see* **Note** 7).
10. Pool together fractions with higher DNA concentrations, those that correspond to DNA peak. DNA containing fractions are in this way separated from DNA-free fractions.
11. Remove CsCl and concentrate DNA using Amicon Ultra-0.5 Centrifugal Filter Units with Ultracel-3 membrane (Merck Millipore) to a final volume of about 50 μL.
12. Conserve each DNA sample from **step 11** at 4 °C for evaluation of purity by Western Blotting (Subheading 3.2) or next steps of Top1ccs mapping (Subheading 3.3).

3.2 Evaluation of Top1cc Content of Purified DNA with Western Blotting

1. To check Top1cc purity and obtain a good signal by Western Blotting, the total amount of purified DNA from a 75 cm^2 flask is necessary. Therefore, digest a 50 μL fraction from Subheading 3.1, **step 11** with 2 U of DNase I for 2 h at 37 °C.
2. Add samples of Laemmli Sample buffer (final concentration 1×), 5% β-mercaptoethanol (v/v) and boil for 10 min before loading on precast mini protein gel (*see* **Note 8**).
3. Run the gel at 120 V for about 1 h at RT in Tris–Glycine SDS Running Buffer (final concentration 1×) (*see* **Note 8**).
4. Transfer protein on nitrocellulose membrane for 2 h at 4 °C, 70 V.
5. Color the membrane with ponceau S solution to verify proteins transfer.
6. Block membrane in TBS-T-milk for 1 h at RT under gentle agitation.
7. Incubate membrane overnight with primary antibody (for actin the used dilution is 1:500 and for Top1 the used dilution is 1: 200) in TBS-T-milk at 4 °C (*see* **Note 9**).
8. Wash membrane three times in TBS-T for 10 min at RT under gentle agitation.

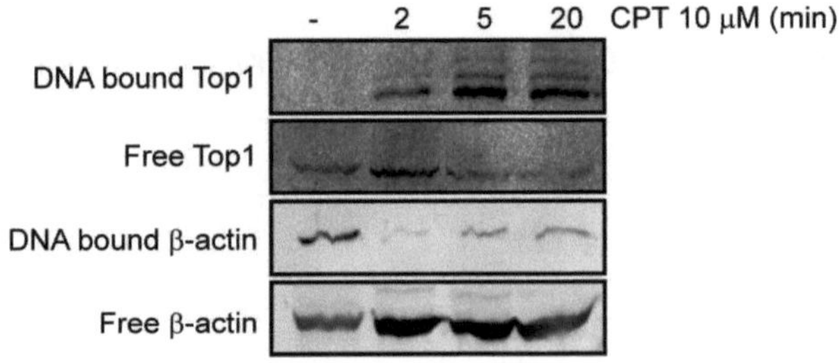

Fig. 1 CPT determines a rapid and transient burst of Top1ccs. ICE bioassay was performed on HCT116 cells after 2, 5, and 20 min of 10 μM CPT treatment. The gel shows both CsCl gradient fractions containing DNA bound proteins and free proteins. β-Actin was used as loading control. (Adapted from [3] with permission by Oxford University Press (license number 5135941252333))

9. Incubate membrane with secondary antibody (used dilution 1: 2000) diluted in TBS-T-milk for 1 h at RT under gentle agitation.
10. Wash membrane three times in TBS-T for 10 m at RT under gentle agitation.
11. Detect protein signal with Pierce ECL Plus Western Blotting Substrate and ChemiDoc XRS (Bio-Rad). Representative image of the result is reported in Fig. 1.

3.3 Fragmentation and K^+-SDS Precipitation of Top1cc

1. To map Top1ccs, pool together from 4 to 5 samples of gradient-recovered DNA from a 75 cm^2 flask each (Subheading 3.1, **step 11**). Then, add an appropriate volume of Lysis Buffer 2× to obtain a final 1× concentration. Add 100 μg (2 μL) of BSA 50 mg/mL to each 50 μL of recovered DNA sample. For example, if 5 samples from Subheading 3.1, **step 11** are pooled together, add 10 μL of BSA 50 mg/mL. Make aliquots of 300 μL in 1.5 mL tube each (*see* **Note 10**).
2. Sonicate one aliquots of 300 μL of the collected samples using Bioruptor (Diagenode) for 15 min on ice at medium intensity: 30 s ON, 30 s OFF.
3. As sonication efficiency depends on DNA concentration, quantify DNA reading absorbance at 260 nm in quartz cuvette (15 μL of DNA solution +735 μL of SDS 0.5%, 10 min at RT and then reading with Spectrophotometer).
4. Dilute DNA samples to 15 U Abs/mL adding Lysis Buffer 1×. Perform three additional cycles of sonication with Bioruptor (Diagenode) at same conditions described at **step 2**. All aliquots of same sample are mixed together to homogenize before storing.

   PAUSE POINT. Shared DNA can be stored at −80 °C if immediate processing is not possible. It is preferably to prepare aliquots of about 250 μL before stocking samples to avoid freeze–thaw cycles.

5. Use 50 μL of sonicated DNA to determine the fragment size by agarose gel. Add 50 μL TE to 50 μL of sonicated DNA, to dilute SDS to 0.5%. Add proteinase K to samples (final concentration 0.5 mg/mL) and incubate at 52 °C for 2 h. Extract DNA with phenol–chloroform–isoamyl alcohol (25:24:1) and precipitate with ethanol before loading on 1.2% agarose gel (*see* **Note 11**).

   CRITICAL STEP. Properly shared DNA should have a size range between 200 and 500 bp. Since sonication is one of the most critical steps for generating a high-quality resolution experiment, optimizing the sonication conditions by pilot experiments is highly recommended when working with a new cell type or a different sonicator.

6. For K^+-SDS precipitation, add KCl 2.5 M to 250 μL of extracted DNA from **step 4** (1/10 of the total volume, final concentration 0.25 M KCl) and incubate for 15 min in ice.
7. Centrifuge samples at 1200 × *g* for 3 min, 4 °C.
8. Wash pellets 5 times with KCl 0.25 M. In particular, add three volumes of KCl and keep 5 min in ice. Then centrifuge samples at 1200 × *g* for 3 min, 4 °C, and discard supernatant.
9. Resuspend DNA pellet in 500 μL of TE buffer (twice the initial volume of precipitated DNA), added of Proteinase K (final concentration 0.5 mg/mL) and incubate for 3 h at 52 °C.
10. Prepare INPUT, that represent about 20% of the total used DNA, similarly to **step 9**. In particular, add 50 μL of TE Buffer to 50 μL of DNA from **step 4** and then add Proteinase K (final concentration 0.5 mg/mL), incubating for 3 h at 52 °C.
11. Purify precipitated DNA and INPUT by two rounds of phenol–chloroform–isoamyl alcohol (25:24:1) extractions and precipitate with ice cold ethanol in presence of sodium acetate (*see* **Note 11**).
12. Resuspend dried pellets in 80–100 μL of ultrapure RNase–DNase free water and quantify by real time qPCR or use for library preparation and sequencing.

    PAUSE POINT. The recovered DNA can be stored at −80 °C if immediate processing by real-time qPCR is not possible.

3.4 Real-Time PCR Analyses of Recovered DNA Fragments (Top1cc)

1. To evaluate the quality of Top1-bound DNA purification, perform real-time qPCR on specific DNA loci as reported in Table 2. Analysis can then be extended to other DNA regions of interest.
2. Perform qPCR with 2 μL of K^+-SDS precipitated DNA from Subheading 3.3, **step 12** in a 15-μL reaction mixture with SSoAdvanced Universal SYBR Green Supermix in one well of

Table 2
Primers for K + SDS Precipitation Assay Validation. Positive (+) and negative (–) regions are reported

Gene		Primer forward	Primer reverse
GAPDH	+	TAGCTCAGGCCTCAAGACC TT	AAGAAGATGCGGCTGACTGTC
POLR2A	+	GGGGAGAGACAAACTGCCG TA	GGGGAAATAAGGAGCGAAAGG
MYC	+	TCGGGGCTTTATCTAACTCG	GCTATGGGCAAAGTTTCGTG
VEGFA	+	TTTAAAAGTCGGCTGG TAGCG	CGGTGCTGGAATTTGATATTC
SP2	+	CGCCTCTGGTTGATTGGTTA	CACCGCTCATTACGACATCTT
TDG	+	GATTGGGCTGTCCTACATCA	CGGTAAGACAACCCCAGAGA
ATF1	+	CCCCCAGGCTTGTGTAGAT	CCGAATTCACTACCCACTTCC
LMNB2 Replication ORI	+	AGATGCATGCCTAGCGTG TTC	TAGCTACACTAGCCAGTGACC TTTTTCCT
α -SAT	–	AGACACAAGCATTC TCAGCAA	CTTTTTCATCATAGGCCTCAA
MYOD1	–	CGCTTTCCTTAACCACAAA TC	AACACGGTCGTCATAGAAGT
LOR	–	TTGCCAGCATCTTCTCTCCT	GCCAGAAGACCAGAAACACC
FOXA2	–	GGTCCCTGTGTGCCTACTGT	CCCGGTTTTATCCCTTGAAT

Table 3
Real time qPCR program

Step	Denature	Anneal and extend	Melt curve	Hold
1	95 °C, 2 min			
2–41	95 °C, 15 s	60 °C, 1 min		
42			From 60 °C to 95 °C, 0.5 °C/min	
43				4 °C

a 96-well plate, using the qPCR program listed in Table 3. For each DNA sample, perform qPCR in three technical replicates. Routinely control specificity of PCR products by melting curve analysis and agarose gel electrophoresis. The results are calculated as percentage of input, and relative enrichment is evaluated by comparing positive vs negative regions (Fig. 2).

NGS and Final Considerations.

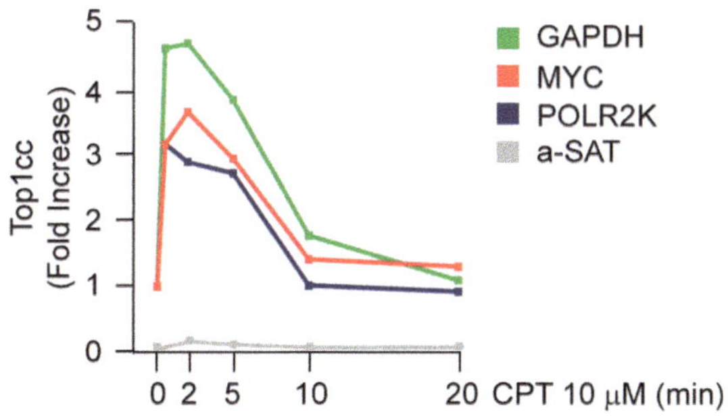

Fig. 2 K^+SDS precipitation assay on HCT116 cells. Cells were treated for 2, 5, 10, and 20 min with 10 μM CPT. The recovery of Top1ccs at indicated gene promoters is reported as fold increase or percentage on input on not treated cells against treatment time. Three transcribed genes (GAPDH, MYC, and POLR2K) and one not transcribed region (α-SAT) are reported as representative. (Adapted from [3] with permission by Oxford University Press (license number 5135941252333))

Levels of Top1ccs can then be measured with qPCR at selected loci as described in this section. Moreover, recovered Top1-bound DNA fragments can be sequenced to map Top1ccs genome-wide following standard NGS methodologies.

4 Notes

1. To obtain an efficient recovery of Top1cc, we treated about 10^7 exponentially growing HCT116 cells, but the correct amount of cells must be optimize for each cell line.
2. Protease Inhibitor Cocktails are usually furnished at 100× concentration and require a dilution 1:100. Cocktails could be replaced using single inhibitors as aprotinin (used final concentration 1 μg/mL), leupeptin (used final concentration 1 μg/mL), pepstatin (used final concentration 1 μg/mL), and PMSF (used final concentration 1 mM).
3. A ~7 M stock CsCl solution (density of ~1.88 g/mL) is prepared solubilizing 16.1 g of CsCl in about 9 mL of TE buffer in 50 mL falcon tubes. To favor solubilization, heat solution up to 50 °C under occasional stirring. Once solution is clear, adjust the volume with TE to exactly 13.5 mL and filter with a 0.22-μm pore-size sterile syringe filters into a new tube.
4. To optimize DNA separation on gradient, the gradient itself must be prepared carefully. In particular each gradient solution has to be gently stratified over the other, without allowing solution to mix up.

5. Do not centrifuge at 4 °C since cesium chloride could precipitate.
6. Collecting fractions from the bottom of the gradient avoids contamination of DNA containing fractions with free proteins.
7. After gradient separation, DNA locate at the bottom of the gradient (fractions with higher density) and Top1ccs colocalize with DNA peak. Free proteins and topoisomerases are retained at the top of the gradient.
8. Any specific buffer used for Bolt 4–12% Bis-Tris Plus (NW04120BOX, ThermoFisher Scientific) can be substituted with any other appropriate buffer.
9. The primary and secondary antibody concentration which give the best staining with minimum background must be optimized dependently on each brand and lot antibody.
10. The isolation of Top1ccs can also be performed avoiding CsCl gradient steps. In this case, the protocol may start by performing treatments with Top1 poisons on exponentially growing cells. At the end of treatment, remove cell medium and add directly 1.2 mL of Lysis Buffer 1× for 10 min to the flasks placed on ice. Note that no PBS washing is performed before cell lysis. Then, collect cells using cell scraper and transfer in falcon tubes, maintaining cells on ice during the procedure. Continue with protocol from Subheading 3.3, **step 2**.
11. Perform each extraction adding 1 volume of phenol–chloroform–isoamyl alcohol to DNA samples. Vortex briefly and centrifuge at room temperature for 5 min at 10,000 × *g*. Carefully remove the upper aqueous phase and transfer the layer to a new tube. Be sure not to carry over any phenol during pipetting. If necessary, proceed again with an additional extraction or add to DNA solution 1/10 volume of Sodium Acetate 3 M, pH 5.2, and 2.5 volumes of ice cold 100% ethanol. Put samples at −20 °C overnight or at −80 °C for at least 1 h and then centrifuge at maximum speed for 30 min at 4 °C. Wash pellets once with ice cold 70% ethanol, centrifuging at 4 °C maximum speed for 20 min. Finally air dry the pellets and resuspend in appropriate volume of ultrapure DNase/RNase free water.

Acknowledgments

Our research projects are supported by AIRC under IG 2019—ID 23032 project—PI Capranico Giovanni, and from Ministry of University and Research under PRIN 2017-KSZZJW_004, PI Marinello Jessica.

References

1. El Hage A, French SL, Beyer AL, Tollervey D (2010) Loss of topoisomerase I leads to R-loop-mediated transcriptional blocks during ribosomal RNA synthesis. Genes Dev 24: 1546–1558
2. Tuduri S, Crabbé L, Conti C, Tourrière H, Holtgreve-Grez H, Jauch A, Pantesco V, De Vos J, Thomas A, Theillet C et al (2009) Topoisomerase I suppresses genomic instability by preventing interference between replication and transcription. Nat Cell Biol 11:1315–1324
3. Marinello J, Chillemi G, Bueno S, Manzo SG, Capranico G (2013) Antisense transcripts enhanced by camptothecin at divergent CpG-island promoters associated with bursts of topoisomerase I-DNA cleavage complex and R-loop formation. Nucleic Acids Res 41: 10110–10123
4. Manzo SG, Hartono SR, Sanz LA, Marinello J, De Biasi S, Cossarizza A, Capranico G, Chedin F (2018) DNA topoisomerase I differentially modulates R-loops across the human genome. Genome Biol 19:100
5. Promonet A, Padioleau I, Liu Y, Sanz L, Biernacka A, Schmitz AL, Skrzypczak M, Sarrazin A, Mettling C, Rowicka M et al (2020) Topoisomerase 1 prevents replication stress at R-loop-enriched transcription termination sites. Nat Commun 11:1–12
6. Kiianitsa K, Maizels N (2013) A rapid and sensitive assay for DNA-protein covalent complexes in living cells. Nucleic Acids Res 41: e104
7. Patel AG, Flatten KS, Peterson KL, Beito TG, Schneider PA, Perkins AL, Harki DA, Kaufmann SH (2016) Immunodetection of human topoisomerase I-DNA covalent complexes. Nucleic Acids Res 44:2816–2826
8. Trask DK, DiDonato JA, Muller MT (1984) Rapid detection and isolation of covalent DNA/protein complexes: application to topoisomerase I and II. EMBO J 3:671–676
9. Lotito L, Russo A, Chillemi G, Bueno S, Cavalieri D, Capranico G (2008) Global transcription regulation by DNA topoisomerase I in exponentially growing Saccharomyces cerevisiae cells: activation of telomere-proximal genes by TOP1 deletion. J Mol Biol 377: 311–322
10. Baranello L, Wojtowicz D, Cui K, Devaiah BNN, Chung H-J, Chan-Salis KY, Guha R, Wilson K, Zhang X, Zhang H et al (2016) RNA polymerase II regulates topoisomerase 1 activity to favor efficient transcription. Cell 165:357–371
11. Khobta A, Ferri F, Lotito L, Montecucco A, Rossi R, Capranico G (2006) Early effects of topoisomerase I inhibition on RNA polymerase II along transcribed genes in human cells. J Mol Biol 357:127–138
12. Kang M, Muller M, Chung I (2004) Telomeric DNA damage by topoisomerase I. a possible mechanism for cell killing by Camptothecin. J Biol Chem 279:12535–12541

Chapter 14

R-Loop Immunoprecipitation: A Method to Detect R-Loop Interacting Factors

Chiara Beghè and Natalia Gromak

Abstract

R-loops are non-B-DNA structures consisting of an RNA/DNA hybrid and a displaced single-stranded DNA. They arise during transcription and play important biological roles. However, perturbation of R-loop levels represents a source of DNA damage and genome instability resulting in human diseases, including cancer and neurodegeneration. In this chapter, we describe a protocol which allows detection of R-loop interactors using affinity purification with S9.6 antibody, specific for RNA/DNA hybrids, followed by Western blotting or mass spectrometry. Multiple specificity controls including addition of synthetic competitors and RNase H treatment are described to verify the specificity of identified R-loop-binding factors. The identification of new R-loop interacting factors and the characterization of their involvement in R-loop biology provides a powerful resource to study the role of these important structures in health and disease.

Key words RNA/DNA hybrids, R-loop, Immunoprecipitation, S9.6 antibody, R-loop interactors, Affinity purification, Slot Blot, DHX9, APOBEC3B

1 Introduction

R-loops are three-stranded structures which occur when nascent RNA re-anneals to the template DNA strand. This leads to the formation of an RNA/DNA hybrid and a displaced non-transcribed ssDNA [1]. R-loops arise naturally during transcription and play a critical role in many biological processes, including immunoglobulin class-switch recombination, gene expression, epigenetics, DNA replication and repair [2]. Mis-regulation of R-loops can result in increased DNA damage and genome instability, and it has been recently associated with neurodegenerative diseases and cancer [3, 4]. Therefore, an understanding of molecular mechanisms underlying R-loop regulation is crucial for uncovering their cellular functions.

Over the last few years, many proteins associated with R-loops have been identified reflecting the diversity of R-loop processes.

Andrés Aguilera and Alexey Ruzov (eds.), *R-Loops: Methods and Protocols*, Methods in Molecular Biology, vol. 2528,
https://doi.org/10.1007/978-1-0716-2477-7_14,

These proteins can regulate cellular R-loop levels either directly or indirectly, mostly by preventing the nascent RNA from hybridizing to DNA, thus reducing excessive R-loop accumulation. Among them, there are proteins required for efficient transcriptional elongation, termination, polyadenylation, RNA splicing, packaging, and export [5–9]. R-loop levels can be directly modulated by dedicated enzymes, such as members of the RNase H family that specifically degrade the RNA within the RNA/DNA hybrids [10]. Furthermore, RNA/DNA helicases, including senataxin (SETX) [9], Aquarius (AQR) [11], and DHX9 [12], can unwind RNA/DNA hybrids, while topoisomerase I (TOP 1) can relax negative supercoils by cutting one of the DNA strands, preventing R-loop formation [13]. However, despite the multiple fundamental functions of R-loops in cellular processes and human disease, our current knowledge about R-loop regulation and factors involved in these processes is still lagging behind.

To close this gap and identify R-loop-interacting proteins in human cells in a comprehensive way, we recently employed an unbiased affinity purification approach with S9.6 antibody, followed by Western blotting (WB) or mass spectrometry (MS) [12]. This protocol has been validated to allow the purification of RNA/DNA hybrid-binding factors from the chromatin fraction of non-cross-linked HeLa cells. RNA/DNA hybrid immunoprecipitation (IP) relies on the use of the monoclonal S9.6 antibody, which has 0.6 nM affinity for RNA/DNA hybrids, with no sequence specificity [14, 15] (Figs. 1 and 2). However, S9.6 antibody is also known to bind double-stranded RNA (dsRNA), especially AU-rich dsRNA, with weaker 2.7 nM affinity [15]. Considering this, we performed the R-loop IP with multiple specificity controls. First, to reduce the detection of unspecific RNA-mediated interactions and avoid undesired binding of S9.6 to dsRNA, IP with S9.6 antibody is carried out in the presence of RNase A (Figs. 1 and 2a, b). The RNase A quantity has been optimized to ensure efficient RNA trimming without affecting the integrity of RNA/DNA hybrids [12]. Second, R-loop IP reactions were carried out under competitive conditions with the addition of short synthetic double-stranded oligonucleotides (Figs. 2c and 4c). Addition of synthetic RNA/DNA hybrids resulted in a loss of immunoprecipitated proteins at higher competitor concentrations. Importantly, the IP was unaffected when it was carried out with dsDNA or dsRNA competitors, confirming the specificity of this approach.

Finally, to confirm the specificity of the RNA/DNA hybrid interactors detected, a modified version of the RNA/DNA hybrid IP with RNase H treatment in vitro was designed (Fig. 3). This was required, since the RNase H treatment cannot be performed directly in the R-loop IP buffers due to their composition. In this protocol, genomic DNA containing endogenous RNA/DNA hybrids was isolated from cells and treated (or not) with

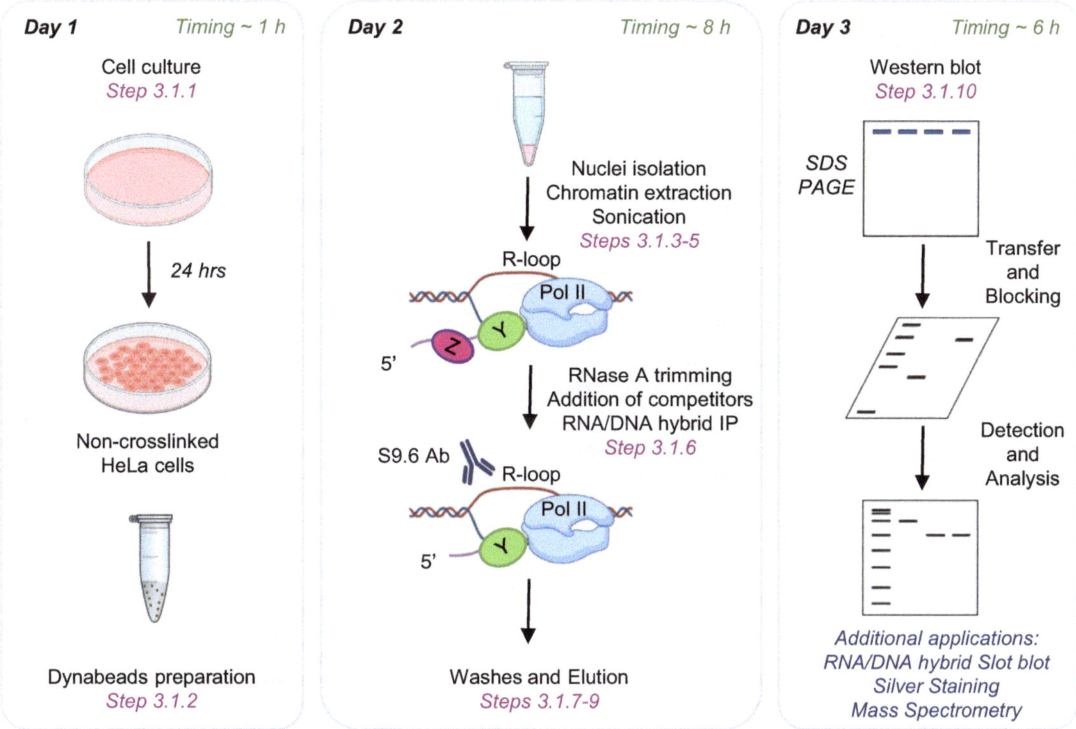

Fig. 1 Overview of the RNA/DNA hybrid IP workflow in HeLa cells. Cells are plated 24 h before the experiment. After gentle nuclei isolation followed by chromatin extraction, sonication, and RNase A trimming, RNA/DNA hybrids are immuno-precipitated with the S9.6 antibody. IP samples are analyzed by SDS-PAGE followed by WB. Alternative analysis can be performed using RNA/DNA hybrid slot blot (SB), silver staining, and mass spectrometry as demonstrated in [12]. Figure is created with BioRender.com

RNase H, prior to enrichment for RNA/DNA hybrids with the S9.6 antibody (Fig. 3). Endogenous RNA/DNA hybrids were then incubated with HeLa nuclear extracts, depleted for RNA/DNA hybrids by treatment with high concentrations of RNase A (Fig. 3). IP with S9.6 antibody resulted in the identification of R-loop interacting factors, however, their pull-down was significantly decreased following RNase H digestion (Fig. 4a, b). Addition of synthetic RNA/DNA hybrid competitors also led to a loss of immuno-precipitated proteins at higher competitor concentrations (Fig. 4c), providing an additional support to the specificity of RNA/DNA hybrid-protein interaction.

R-loop IP protocol is performed in non-cross-linked cells, as crosslinking reagents could induce R-loops [18], preventing the identification of bona fide R-loop interactors. Therefore, this technique allows the detection of endogenous R-loop-mediated interactions in vivo, without perturbing the cellular environment. However, it may be limiting in the detection of transient/weak interactions with RNA/DNA hybrids. Recently Wang et al. analyzed the interaction of proteins with synthetic over-expressed

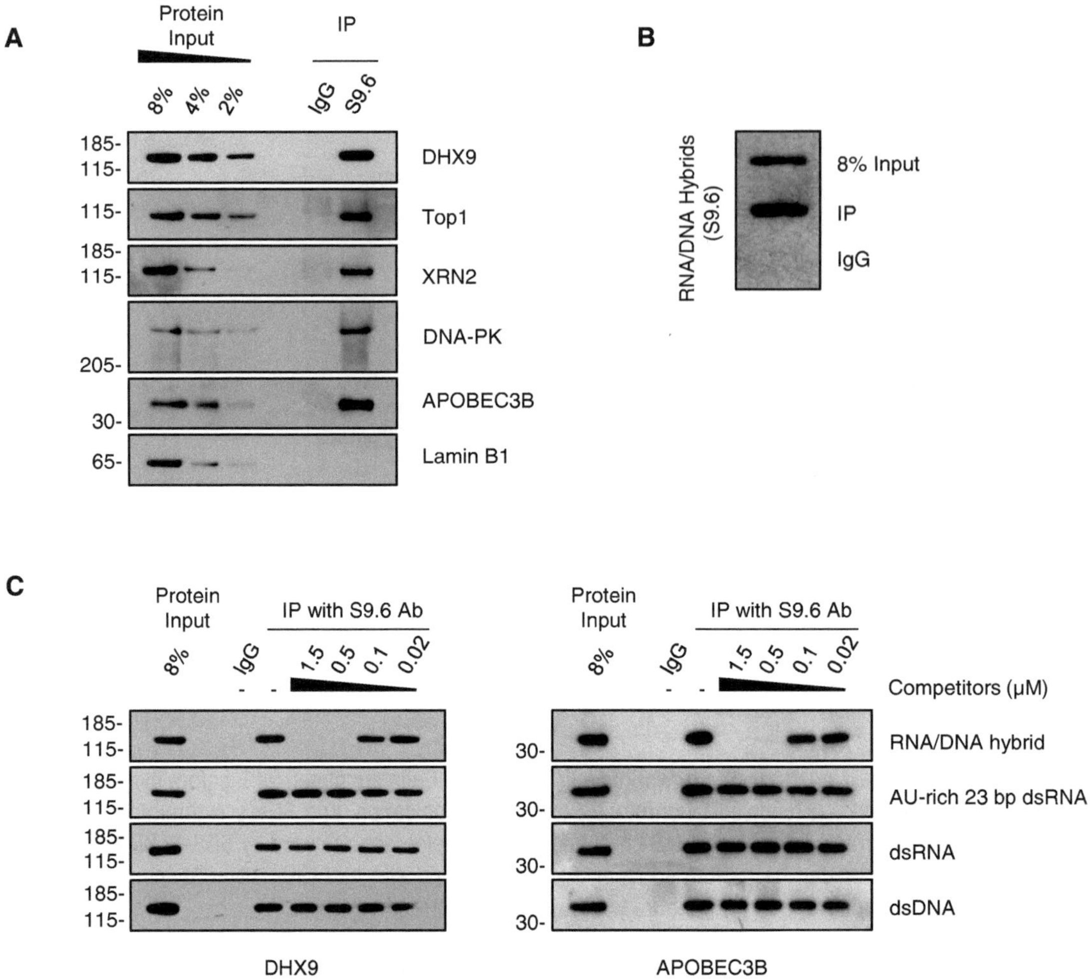

Fig. 2 Validation of the RNA/DNA Hybrid IP method. (**a**) Western blot (WB) analysis of RNA/DNA hybrid IP experiment in HeLa cells probed with antibodies recognizing different R-loop factors, including previously characterized R-loop interactors (DHX9 [12], Top1 [13], and XRN2 [9]) and some more recently identified factors (APOBEC3B [16] and DNA-PK [17]). Lamin B1 is used as a negative control. Inputs are shown on the left and the S9.6-immuno-precipitated samples on the right. IgG lane corresponds to control IP with IgG2a antibody. Protein markers (in kDa) are shown on the left and the primary antibodies used for detection are shown on the right. 8%, 4%, and 2% of input was loaded. (**b**) RNA/DNA hybrid Slot Blot (SB) probed with S9.6 antibody shows enrichment of RNA/DNA hybrids in S9.6 IP sample. 8% of the Input was loaded. (**c**) WB analysis of RNA/DNA hybrid IP experiment, performed with addition of increasing amounts of the indicated synthetic competitors, and probed with antibodies for DHX9 (left panel) and APOBEC3B (right panel). Addition of synthetic RNA/DNA hybrid competitor leads to efficient competition, as demonstrated by a loss of co-purified DHX9 and APOBEC3B proteins. In contrast, addition of dsDNA, dsRNA, and 23 bp AU-rich dsRNA did not affect DHX9 and APOBEC3B IP efficiency

RNA/DNA hybrids from *BAMBI* and *DPP9* genes in human B cells, resulting in detection of ~800 R-loop-binding factors [19]. This raises a question about the specificity and biological significance of these interactions at these specific genomic loci. It

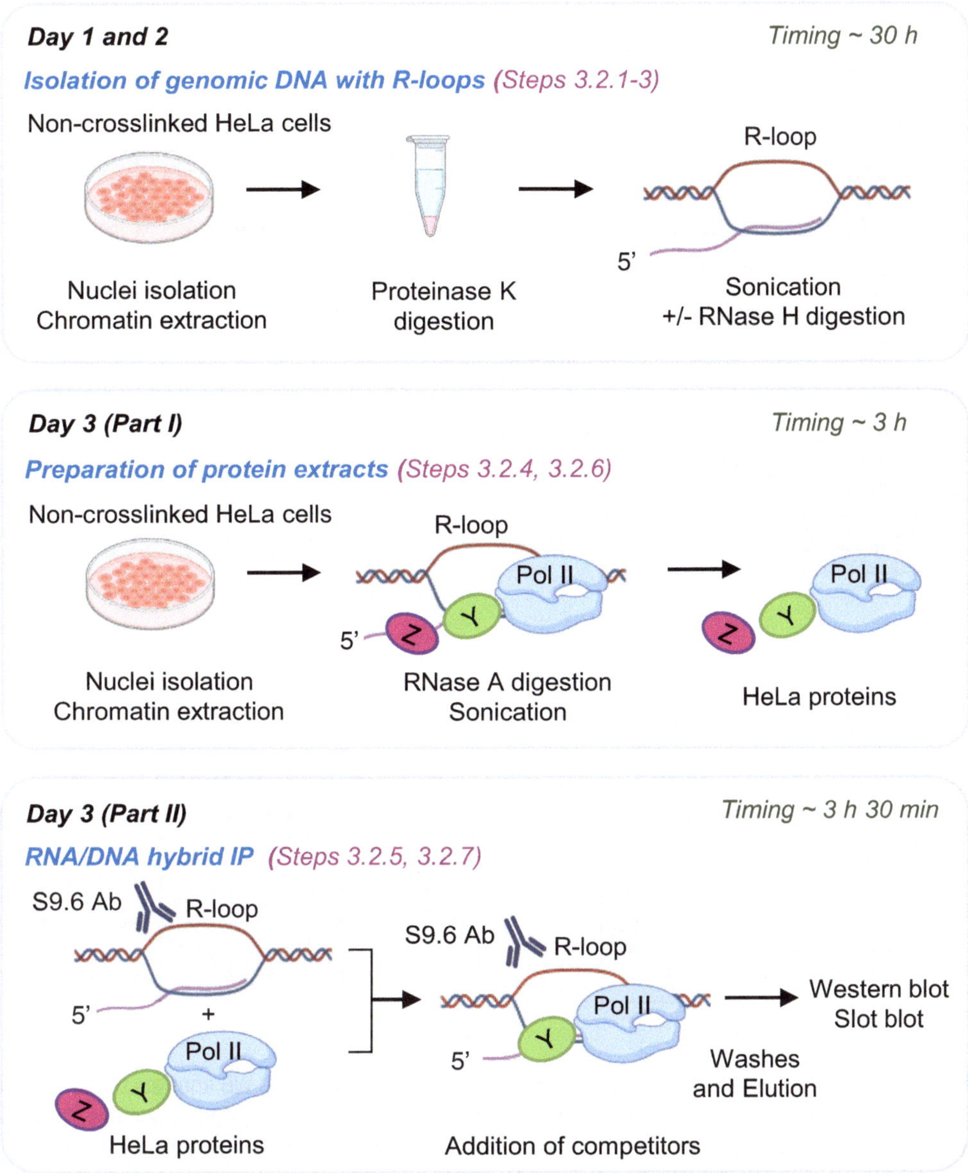

Fig. 3 Overview of the RNA/DNA hybrid IP with RNase H treatment in HeLa cells. The genomic DNA containing R-loops was extracted from non-crosslinked HeLa cells and treated or not with RNase H (+/-RNase H) before enrichment for RNA/DNA hybrids with the S9.6 antibody (Day 1 and 2). Endogenous RNA/DNA hybrids were then incubated with HeLa nuclear extracts, depleted for RNA/DNA hybrids by treatment with high concentrations of RNase A, followed by S9.6 IP (Day 3). Figure is created with BioRender.com

should be noted that an interaction between a particular protein and endogenous or synthetic RNA/DNA hybrids, through the R-loop IP approach, should be interpreted with caution as it does not always reflect a direct involvement of a protein in R-loop regulation in vivo. Furthermore, R-loop regulation is a kinetic

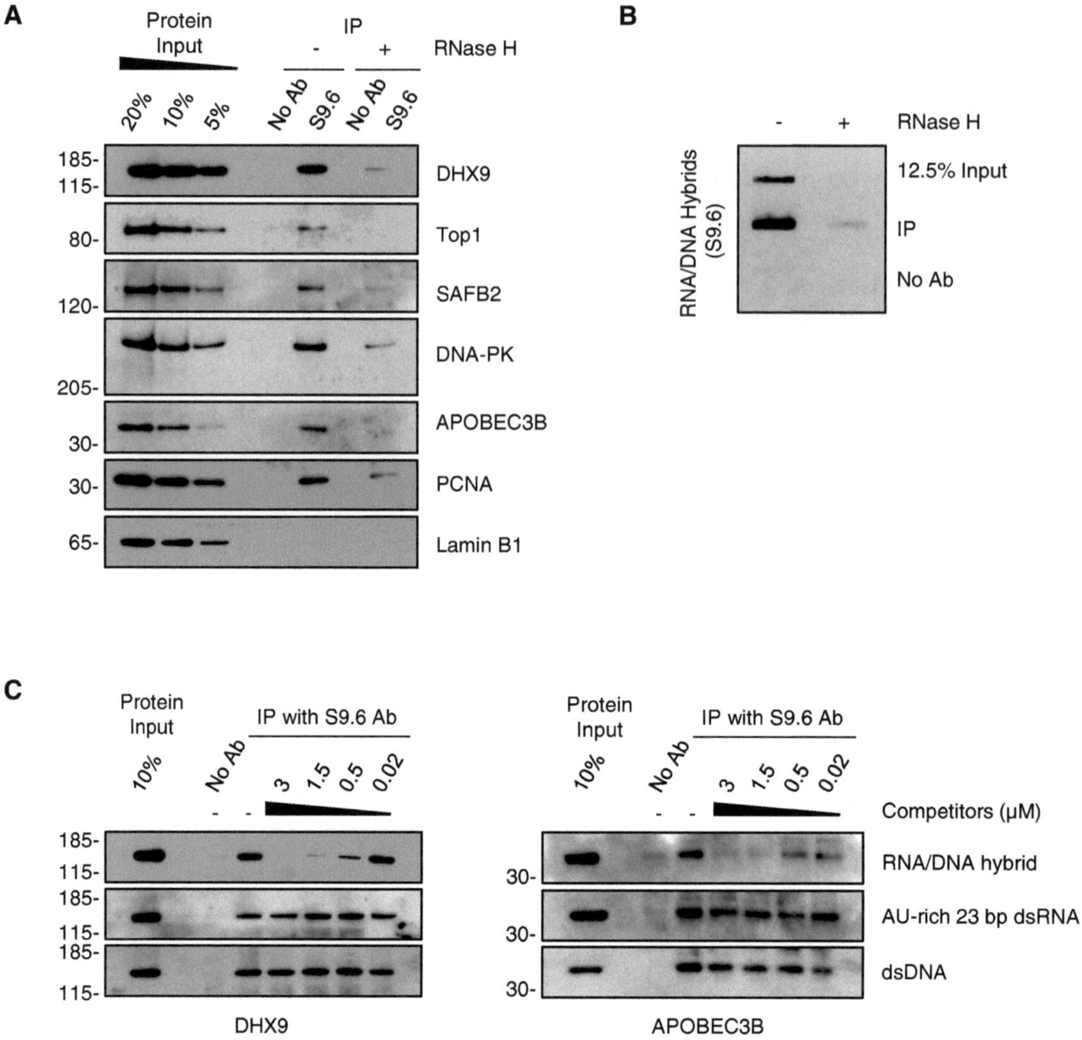

Fig. 4 Validation of RNA/DNA hybrid IP performed with RNase H treatment. (**a**) WB analysis of RNA/DNA hybrid IP experiment, performed with RNase H treatment, and probed with the indicated antibodies. Treatment of genomic DNA with RNase H (+), which decreases the levels of RNA/DNA hybrids (figure **b**), correlates with a decrease in the amount of co-purified proteins. 20%, 10%, and 5% of Input was loaded. (**b**) SB analysis of RNA/DNA hybrid IP carried out with genomic DNA treated with RNase H. SB is probed with the S9.6 antibody. 12.5% of the Input was loaded. (**c**) WB analysis of RNA/DNA hybrid IP experiment, following the protocol in Fig. 3, performed with addition of increasing amounts of the indicated synthetic competitors, and probed with antibodies for DHX9 (left panel) and APOBEC3B (right panel). This experiment is performed without RNase H digestion step (*see* **Note 26**). Addition of synthetic RNA/DNA competitor leads to efficient competition, as demonstrated by a loss of co-purified DHX9 and APOBEC3B proteins. In contrast, addition of dsDNA and 23 bp AU-rich dsRNA did not affect DHX9 and APOBEC3B IP efficiency

process which heavily relies on transcriptional parameters and multiple co-transcriptional RNA-processing events in vivo, aspects which may not necessarily be assessed in the pull-down assays. Thus, additional molecular, cellular, and biochemical approaches are required to validate the biological function of detected R-loop–protein interactions.

Here we provide a detailed RNA/DNA hybrid IP protocol which allows the identification of proteins binding endogenous RNA/DNA hybrids in vivo, using the S9.6 antibody. Mass spectrometry analysis of the R-loop IP resulted in the identification of 469 proteins, which were significantly enriched in the RNA/DNA hybrid IP, a dataset named the "R-loop interactome" [12]. This interactome constitutes proteins of different classes, including helicases, RNA/DNA-binding proteins, chromatin factors, and factors implicated in DNA damage. Among these proteins, factors previously implicated in R-loop biology, including SRSF1, Top1, and DHX9 were uncovered alongside multiple novel R-loop interactors [12]. As a consequence, this R-loop IP technique has provided a foundation for multiple studies to uncover the biology of R-loops.

R-loop IP has been successfully used to reveal novel R-loop-associated functions of multiple factors including APOBEC3B, YTHDF1/2, hnRNPA2B1, METTL3, XPG, XPF, GADD45A, SOX2 [16, 20–25]. The R-loop interactome dataset served as an investigation tool to assess a potential role of candidate proteins in R-loop biology [26–31]. R-loop interactome has identified multiple DEAD/H helicases, helping to reveal their functions in R-loop regulation [12, 32–36]. R-loops have been recently implicated in cancer, and R-loop interactome has played an important role in uncovering new factors involved in this process [16, 17, 37, 38]. Finally, recently the R-loop IP MS dataset was used to design machine-learning algorithms to reveal the features of R-loop binding factors [39, 40].

In conclusion, R-loop IP can be a powerful tool to assess the binding of a candidate protein to RNA/DNA hybrids, helping to uncover its function in R-loop biology in health and disease. Future discovery and characterization of new biological reagents to detect R-loops in cells and advances in proteomics are likely to improve R-loop IP protocol and enhance our understanding of R-loop regulation.

2 Materials

2.1 Biological Materials

Mammalian cells should be routinely checked for mycoplasma contamination. The basic R-loop IP protocol has been tested in several cell lines, including HeLa [12, 16, 20], HEK293T [23], MCF10A [16], primary human lung embryonic WI38 fibroblasts [22], mouse embryonic fibroblasts (MEFs) [21], mouse lung epithelial cells (MLE-12) [24], and mouse embryonic stem cells (mESCs) [25].

2.2 Buffers

Prepare all solutions fresh, using RNase- and DNase-free water. Store all solutions at 4 °C unless otherwise stated.

1. Annealing buffer for competitors: 60 mM KCL, 50 mM Tris-HCl pH 8.
2. Cell Lysis buffer: 85 mM KCL, 5 mM PIPES pH 8.0, 0.5% NP-40.
3. Nuclear Lysis buffer: 10 mM Tris-HCl pH 7.5, 200 mM NaCl, 2.5 mM $MgCl_2$, 0.5% Triton X-100, 0.2% sodium deoxycholate, 0.1% SDS, 0.05% sodium lauroyl sarcosinate.
4. RSB buffer: 10 mM Tris-HCl pH 7.5, 200 mM NaCl, 2.5 mM $MgCl_2$.
5. RSB + T buffer: 10 mM Tris-HCl pH 7.5, 200 mM NaCl, 2.5 mM $MgCl_2$, 0.5% Triton X-100.
6. R-loop IP buffer: 10 mM Tris-HCl pH 7.5, 200 mM NaCl, 2.5 mM $MgCl_2$, 0.5% Triton X-100, 0.05% sodium deoxycholate, 0.025% SDS, 0.0125% sodium lauroyl sarcosinate.
7. Nuclear Lysis Buffer for DNA extraction: 50 mM Tris–HCl pH 8.0, 5 mM EDTA pH 8, 1% SDS.
8. Elution buffer: 1% SDS, 100 mM $NaHCO_3$.
9. Digest buffer: 400 mM Tris-HCl pH 6.5, 100 mM EDTA pH 8.

2.3 Commercial Reagents

1. DMEM supplemented with 10% FBS, 1× penicillin-streptomycin (*see* **Note 1**).
2. PBS.
3. Dynabeads Protein A (Invitrogen, 10002D).
4. PureLink RNase A 20 mg/mL (Life Technologies, 12091-021).
5. Proteinase K (Roche AG, 03115828001).
6. Recombinant RNase H enzyme (NEB, M0297).
7. RNasin Plus Ribonuclease Inhibitor (Promega, N2611).
8. cOmplete™, EDTA-free Protease Inhibitor Cocktail (Roche, 11873580001). For 100× stock solution, dissolve one tablet in 500 μL of RNase- and DNase-free water. The solution can be stored at 4 °C up to 1 month. Working dilution is 1×.
9. PhosSTOP™ (Sigma-Aldrich, 4906845001). To prepare stock solution, dissolve one tablet in 200 μL of RNase- and DNase-free water. The solution can be stored at 4 °C up to 1 month. Working dilution is 1/50. PhosSTOP™ inhibitors can be omitted unless research is focused on proteins which are heavily phosphorylated.

10. Phenylmethylsulfonyl fluoride (PMSF). Prepare 0.1 M PMSF stock solution in isopropanol. Working dilution is 1/200 (*see* **Note 2**).
11. Ethanol 100%.
12. Propan-2-ol, > = 99.8%.
13. BSA. To prepare 0.5% BSA solution, dissolve 0.25 g of BSA in 50 mL of PBS. Sterilization by filtration through a 0.22 μM filter is recommended. Use only freshly-made BSA solutions.
14. DTT. To prepare 1 M DTT solution, dissolve 1.55 g of DTT powder in 10 mL of RNase- and DNase-free water and store the aliquotes at −20 °C.
15. Pierce BCA protein Assay Kit or equivalent.
16. Pierce ECL Western Blotting Substrate or equivalent.
17. Running Buffer, MOPS, or Tris-Acetate (Life Technologies).
18. NuPAGE Transfer Buffer (Life technologies), containing 10% v/v MeOH.
19. NuPAGE LDS Sample buffer (Invitrogen).
20. NuPAGE 4–12% Bis-Tris gel (Life Technologies).
21. NuPAGE 3–8% Tris-Acetate gel (Invitrogen).
22. Hybond-N+ membrane, used for Slot blot (SB) (GE Healthcare).
23. Amersham Hybond-P PVDF membrane, used for WB (GE Healthcare).
24. Synthetic nucleic acid competitors (see sequences in Table 1) were synthesized by Integrated DNA Technologies.

Table 1
Sequences of nucleic acid competitors

Double-stranded competitors	*Sequence (5′ → 3′)*
dsDNA	S: CGG TGT GAA TCA GAC AS: GTC TGA TTC ACA CCG
dsRNA	S: CGG UGU GAA UCA GAC AS: GUC UGA UUC ACA CCG
ds AU-rich RNA	S: AAU UAC AUU GAU AGA AUU AUU AG AS: CUA AUA AUU CUA UCA AUG UAA UU
RNA/ DNA hybrid	S: CGG TGT GAA TCA GAC AS: GUC UGA UUC ACA CCG

Note: *S* sense strand, *AS* anti-sense strand

25. Antibodies used for the IP: αIgG2a (Sigma-Aldrich, M5409), αRNA/DNA Hybrid S9.6 (purified from hybridoma cells, Gromak Lab). Alternatively commercial S9.6 antibody can be used (Kerafast, ENH001). Antibodies used for WB: αDHX9 (Abcam, ab26271, 1:2000), αTopoisomerase I (Abcam, ab109374, 1:500), αLamin B1 (Abcam, ab16048, 1:2000), αAPOBEC3B 5210-87-13 (ARP-12397 (NIH HIV reagent programme), provided by R.Harris lab [41], 1:1000), αDNA-PK (Abcam, ab1832, 1:500), αXRN2 (Proteintech, 11267-1-AP, 1:500), αSAFB2 (Bethyl Laboratories, A301-112A, 1: 300), αPCNA (Abcam, ab92552, 1:500). Antibody used for SB: αRNA/DNA Hybrid S9.6 (purified from hybridoma cells, Gromak Lab, 1:1000). Secondary antibodies used for WB and SB: anti-Mouse IgG (whole molecule)-Peroxidase antibody (Sigma-Aldrich, A8924) and anti-Rabbit IgG (whole molecule)-Peroxidase antibody (Sigma-Aldrich, A0545).

2.4 Equipment

1. Petri dishes, polystyrene, size 100 mm × 15 mm.
2. 11 mL Nunc centrifuge tubes, with conical bottom.
3. Safe-lock 1.5 mL microcentrifuge tubes (e.g. Eppendorf, EP0030123611).
4. 1.5 mL microfuge tubes.
5. 10 μL, 200 μL, 1000 μL Tips.
6. Cell lifter.
7. Syringe.
8. 0.22 μM syringe filter.
9. Rotating wheel.
10. Magnetic rack.
11. Block Heater.
12. ThermoMixer.
13. Vortex Mixer.
14. Bench-top refrigerated centrifuge.
15. Large volume refrigerated centrifuge with adaptors for 15/50 mL tubes.
16. Bioruptor Plus Sonicator System (Diagenode).
17. Nanodrop.
18. SDS-PAGE apparatus.
19. Slot blot apparatus.
20. Carestream Biomax Light Film.

3 Methods

3.1 RNA/DNA Hybrid IP

Here, we describe a protocol for IP of RNA/DNA hybrids and detection of R-loop-interacting proteins from non-crosslinked HeLa cells. For each experimental condition, we recommend to generate three samples which include "Protein Input," the "S9.6 IP," and the "No antibody Control" (or "IgG Control"). A schematic diagram of the experimental workflow is shown in Fig. 1.

Day 1.

3.1.1 Cell Culture (Timing ~30 min)

1. Plate 1.8×10^6 HeLa cells per 100 mm × 15 mm Petri dish 24 h before the experiment. We recommend to use one Petri dish per experimental condition.

3.1.2 Protein A Dynabeads Preparation (Timing ~15 min plus O/N Incubation)

1. Pre-block 40 μL of Protein A Dynabeads for each condition in 1 mL of PBS supplemented with 0.5% BSA in a fresh safe-lock 1.5 mL tube (*see* **Note 3**).
2. Incubate the Protein A Dynabeads on a rotating wheel at 16 rpm at 4 °C o/n.

Day 2.

Perform everything in a 4 °C room or on ice, with ice-cold buffers (*see* **Note 4**).

3.1.3 Preparation of Double-Stranded Competitors (Timing ~2 h 15 min)

1. Mix equimolar amounts of single-stranded oligonucleotides (e.g. 40 μM), resuspended in Annealing buffer, as described previously [42]. *See* Table 1 for competitor sequences.
2. Heat at 95 °C for 5 min in a Block Heater. Cool down the dry block by taking it out from the Block Heater at room temperature for ~2 h (*see* **Note 5**). Store the annealed competitors on ice. Meanwhile proceed with the protocol.

3.1.4 Conjugation of S9.6 Antibody to the Protein A Dynabeads (Timing ~2 h 15 min)

1. Incubate Protein A Dynabeads from Subheading 3.1.2, **step 2** on a magnetic rack for 1 min. Using Gilson P1000 pipette, carefully remove the supernatant.
2. Wash Protein A Dynabeads in 1 mL of RSB + T buffer three times (*see* **Note 6**).
3. Resuspend the Protein A Dynabeads in 1 mL of R-loop IP buffer supplemented with inhibitors.
4. Add 2 μg of S9.6 antibody to the IP sample. Add 2 μg of IgG2a matched isotope mouse antibody (or no antibody) to the Control sample.
5. Incubate antibody-Protein A Dynabeads conjugates on a rotating wheel at 16 rpm at 4 °C for at least 2 h. If necessary, the antibody-beads conjugates can be incubated for longer.

3.1.5 Preparation of Nuclear Extracts (Timing ~2 h)

Process the plates corresponding to different experimental conditions in parallel. The volumes indicated below refer to one Petri dish.

1. Prior the experiment, check the confluence of the cells under the microscope. Ideally, cells should reach ~85% confluence on the day of experiment.
2. Wash the cells on the plate with 10 mL of ice-cold PBS 2–3 times. Using cell lifter scrape non-cross-linked cells into 5 mL of PBS supplemented with inhibitors and collect the cells in 11 mL Nunc centrifuge tubes (*see* **Note** 7).
3. Pellet cells at 500 × *g* at 4 °C for 5 min in a large volume refrigerated centrifuge. Using Gilson P1000 pipette, carefully remove the supernatant.
4. Resuspend the cells in 0.6 mL of Cell Lysis buffer supplemented with protease inhibitors. Transfer the cell lysate into a fresh safe-lock 1.5 mL tube. Incubate the lysate for 10 min on ice.
5. Pellet the nuclei at 500 × *g* at 4 °C for 5 min in a bench-top refrigerated centrifuge. Using Gilson P1000 pipette, carefully remove the supernatant.
6. Repeat the Cell lysis step by resuspending the pellet in 0.6 mL of Cell Lysis buffer supplemented with inhibitors. This ensures efficient cell lysis to pellet the nuclei.
7. Pellet the nuclei at 500 × *g* at 4 °C for 5 min. Using Gilson P1000 pipette, carefully remove the supernatant.
8. Resuspend cell nuclei in 0.3 mL of the Nuclei Lysis buffer supplemented with inhibitors. Incubate the lysate for 10 min on ice.
9. Sonicate for 10 min, using medium power setting and cycle condition 30 s ON 30 s OFF with Bioruptor Plus Sonicator System (Diagenode) to generate DNA fragments of ~500–700 bp (*see* **Note 8**).
10. Measure protein concentration with a BCA protein Assay Kit.
11. Dilute lysate with RSB + T buffer, supplemented with protease inhibitors, to a total volume of 1.2 mL and store it on ice until IP (Subheading 3.1.6). Keep 100 μL fraction of diluted protein lysate as "Protein Input" sample and store it at −20 °C for WB analysis (Subheading 3.1.10).

3.1.6 Immunoprecipitation (Timing ~2.5 h)

1. Incubate antibody-Protein A Dynabeads conjugates from Subheading 3.1.4, **step 5** on a magnetic rack for 1 min (*see* **Note 9**). Using Gilson P1000 pipette, carefully remove the supernatant.
2. Wash Protein A Dynabeads three times in 1 mL of RSB + T buffer (*see* **Note 6**).

3. Add diluted quantified protein lysate (from Subheading 3.1.5, **step 11**) to the antibody conjugated to Protein A Dynabeads in the R-loop IP buffer, supplemented with inhibitors. We recommend to use 50–100 μg of protein lysate for each IP reaction. Adjust the sample to a total volume of 400 μL with R-loop IP buffer. The amount of protein extract can be optimized according to each experimental setting.
4. Optional step: Add synthetic competitors (*see* **Note 10**).
5. Incubate the antibody-Protein A Dynabeads conjugates and the protein lysate on a rotating wheel at 16 rpm at 4 °C for 30 min.
6. Add 12 μL of 0.1 ng/μL freshly prepared RNase A dilution. Prepare RNase A working dilutions in PBS from the working stock (20 mg/mL) freshly before each experiment (*see* **Note 11**).
7. Incubate the samples on the rotating wheel at 16 rpm at 4 °C for 1 h 30 min.

3.1.7 Washes (Timing ~1 h)

1. Incubate Protein A Dynabeads on a magnetic rack for 1 min. Using Gilson P1000 pipette, carefully remove the supernatant. If necessary, keep the supernatant as a "flow-through" fraction.
2. Wash the Protein A Dynabeads with 1 mL of RSB + T buffer four times (*see* **Note 12**).
3. Wash the Protein A Dynabeads with 1 mL of RSB buffer twice.

3.1.8 Elution for Western Blot (Timing ~15 min)

1. Incubate the Protein A Dynabeads in the last RSB wash (Subheading 3.1.7, **step 3**) on a magnetic rack for 1 min. Using Gilson P1000 pipette, carefully remove the supernatant. Beads can be frozen at −20 °C.
2. Resuspend the beads in 40 μL of 1× LDS.
3. Add 5 μL of 1 M DTT, vortex gently.
4. Boil the Protein A Dynabeads for 10 min at 70 °C, vortex them gently and touch spin them briefly for ~10–20 s (less than 500 × *g*). Carefully transfer the eluted proteins to a fresh tube without touching the beads. Eluates can be frozen at −20 °C.

3.1.9 Elution for Slot Blot (Timing ~45 min)

1. Incubate the Protein A Dynabeads from the last RSB wash (Subheading 3.1.7, **step 3**) on a magnetic rack for 1 min. Using Gilson P1000 pipette, carefully remove the supernatant. Beads can be frozen at −20 °C.
2. Add 250 μL of freshly prepared ChIP Elution buffer to the beads and rotate at 16 rpm on a rotating wheel for 15 min at room temperature (*see* **Note 13**).

3. Spin the beads at 16,000 × *g* at room temperature for 1 min and carefully transfer the supernatant into a fresh 1.5 mL tube. Carry out another round of elution by repeating the **steps 2** and **3**.
4. Eluates can be frozen at –20 °C. Add 30 μL of 5 M NaCl and 53 μL of the Digest buffer to eluates to perform Proteinase K digestion and subsequent DNA precipitation for the SB.

Day 3.

3.1.10 Western Blot (WB) to Detect RNA/DNA-Binding Proteins (Timing ~6 h)

Follow a standard WB protocol (*see* **Note 14**), (Figs. 2a, c and 4a, c).

3.1.11 Slot Blot (SB) to Assess the Efficiency of S9.6 Pull-Down (Timing ~4 h)

Follow a standard SB protocol (*see* **Note 15**), (Figs. 2b and 4b).

3.2 Specificity of R-Loop Interactors: RNA/DNA hybrid IP with RNase H Control

A schematic diagram of the experimental workflow is shown in Fig. 3.

Day 1.

3.2.1 Cell Culture for Endogenous Hybrids Extraction (Timing ~30 min)

1. Plate 1.8 × 10^6 HeLa cells per 100 mm × 15 mm Petri dish 24 h before the experiment. We recommend to use four Petri dishes to prepare at least 100 μg of genomic DNA containing R-loops for optimal sonication outcomes in Subheading 3.2.2, **step 18**. The excess of genomic DNA can be stored at –20 °C.

Day 2.

3.2.2 Preparation of Endogenous Hybrids (Timing ~5 h)

Perform everything at 4 °C or on ice, with ice-cold buffers. Process the plates corresponding to different experimental conditions in parallel. The volumes indicated below refer to one Petri dish.

1. Prior the experiment, check the confluence of the cells under the microscope. Ideally cells should reach ~85% confluence on the day of experiment.
2. Wash cells on the plate with 10 mL of ice-cold PBS 2–3 times. Using cell lifter, harvest non-cross-linked cells into 5 mL of PBS and collect the cells in 11 mL Nunc centrifuge tubes (*see* **Note 7**).
3. Pellet cells at 500 × *g* at 4 °C for 5 min in a large volume refrigerated centrifuge.
4. Remove supernatant and gently resuspend the cells in 800 μL of Cell Lysis buffer. Transfer the cellular lysate into a fresh safe-lock 1.5 mL tube. Incubate for 10 min on ice.

5. Pellet the nuclei at 500 × *g* at 4 °C for 5 min in a bench-top refrigerated centrifuge. Using Gilson P1000 pipette, carefully remove the supernatant.
6. Resuspend the nuclei in 800 μL of Nuclear Lysis buffer for DNA extraction (*see* **Note 16**). Incubate for 10 min on ice.
7. Digest the proteins by incubation with 5 μL of Proteinase K (19.1 mg/mL) for 1 h at 55 °C in a Block Heater.
8. Add additional 3 μL of Proteinase K and incubate for 1 h at 55 °C, and add another 2 μL of Proteinase K and incubate for another 1 h at 55 °C. Every time, pipette the sample up and down with Gilson P1000 pipette (*see* **Note 17**).
9. Add 200 μL of 5 M KAc, mix by inversion ~10–20 times (*see* **Note 18**).
10. Spin the samples for 5 min at 16,000 × *g* at 4 °C.
11. Transfer the supernatant containing genomic DNA and R-loops into a new 1.5 mL tube (*see* **Note 19**). Add 800 μL of isopropanol. Mix by inversion (*see* **Note 20**).
12. Pellet DNA for 15 min at 16,000 × *g* at 4 °C. Remove the supernatant (*see* **Note 21**).
13. Wash the pellet with 1 mL of 75% EtOH.
14. Pellet DNA for 10 min at 16,000 × *g* at 4 °C. Remove the supernatant.
15. Air dry the DNA pellet (*see* **Note 22**).
16. Dissolve the DNA pellet in 75 μL of RNase- and DNase-free water at room temperature for ~3 min, transfer to a fresh 1.5 mL tube. Pull together the genomic DNA obtained from the four 100 mm × 15 mm Petri dishes to a final volume of 300 μL.
17. Measure the DNA concentration using Nanodrop.
18. Sonicate genomic DNA. For 100 μg of DNA, sonicate for 10 min, using medium power setting and cycle condition 30 s ON 30 s OFF with Bioruptor Plus Sonicator System (Diagenode), to generate DNA fragments of ~500–700 bp. The total sonication volume and conditions require optimization depending on sonicator models, according to manufacturer's instructions.
19. Set up RNase H treatment reactions as described in the Table 2 (*see* **Note 23**).
20. Incubate samples in a ThermoMixer at 500 rpm at 37 °C o/n.

3.2.3 Preparation of BSA-Blocked and Antibody-Bound Protein A Dynabeads (Timing ~4 h 15 min plus O/N Incubation)

1. Use 40 μL of Dynabeads Protein A for each condition.
2. Pre-block 40 μL of Protein A Dynabeads for each condition (depending on your experimental plan) in 1 mL of PBS supplemented with 0.5% BSA in a fresh safe-lock 1.5 mL tube. Incubate beads on a rotating wheel at 16 rpm at 4 °C for 4 h.

Table 2
Experimental set up for RNase H treatment reaction

Reagent	+ RNase H reaction	– RNase H reaction
Genomic DNA containing R-loops	4 μg	4 μg
10× RNase H buffer	10 μL	10 μL
RNase H enzyme (22.5 U), i.e. 5.625 U of RNase H per μg of DNA	4.5 μL (*see* **Note 23**)	–
RNase-free H20	Up to 100 μL	Up to 100 μL

3. Incubate Protein A Dynabeads on a magnetic rack for 1 min. Using Gilson P1000 pipette, carefully remove the supernatant.
4. Wash Protein A Dynabeads in 1 mL of RSB + T buffer three times (*see* **Note 6**).
5. Resuspend the Protein A Dynabeads in 1 mL of R-loop IP buffer supplemented with freshly prepared inhibitors.
6. Add 2 μg of S9.6 antibody to the IP samples. Add 2 μg of IgG2a matched isotope mouse antibody (or no antibody) to the Control samples.
7. Incubate the samples on a rotating wheel at 16 rpm at 4 °C o/n.

3.2.4 Cell Culture for R-Loop IP with RNase H Digestion (Timing ~1 h)

1. Plate 1.8 × 10^6 HeLa cells per 100 mm × 15 mm Petri dish 24 h before the experiment. We recommend to use one 100 mm × 15 mm Petri dish per experimental condition.

Day 3.

3.2.5 Binding of Genomic RNA/DNA Hybrids to Conjugated Antibody-Protein A Dynabeads (Timing ~2 h 30 min)

1. Stop RNase H reaction from Subheading 3.2.2, **step 20** and store digested and undigested genomic DNA containing R-loops on ice until further usage.
2. Incubate antibody-Protein A Dynabeads conjugates from Subheading 3.2.3, **step 7** on a magnetic rack for 1 min. Using Gilson P1000 pipette, carefully remove the supernatant.
3. Wash the Protein A Dynabeads in 1 mL of RSB + T buffer three times (*see* **Note 6**).
4. Resuspend the Protein A Dynabeads in 1 mL of R-loop IP buffer supplemented with inhibitors.
5. Add 100 μL of genomic DNA containing R-loops, treated and untreated with RNase H (−/+RNase H).
6. Incubate the samples on a rotating wheel at 16 rpm at 4 °C for at least 2 h.

3.2.6 Preparation of Protein Extracts (Timing ~3 h)

1. Follow protocol Subheading 3.1.5, **steps 1–8**.
2. Add 1.5 μL of RNase A (20 mg/mL) to Nuclei Lysis buffer samples to a final concentration of 0.1 mg/mL and incubate for 2 h at 37 °C (*see* **Note 24**).
3. Sonicate for 10 min, using medium power setting and cycle condition 30 s ON 30 s OFF with Bioruptor Plus Sonicator System (Diagenode) (*see* **Note 8).**
4. Measure protein concentration with a BCA protein Assay Kit.
5. Dilute the lysate with RSB + T buffer supplemented with protease inhibitors to a total of 1.2 mL and store on ice until IP (Subheading 3.2.7). Keep 100 μL fraction of diluted and RNase A pre-treated lysate as "Protein Input" and store it at −20 °C for WB (Subheading 3.1.10).

3.2.7 Immunoprecipitation (Timing ~3 h 30 min)

1. Incubate the antibody-Protein A Dynabeads conjugated to genomic R-loops (+/−RNase H) or control beads from Subheading 3.2.5, **step 6** on a magnetic rack 1 min. Using Gilson P1000 pipette, carefully remove the supernatant (*see* **Note 9**).
2. Wash Protein A Dynabeads in 1 mL of RSB + T buffer three times (*see* **Note 6**).
3. Resuspend the Protein A Dynabeads in 1.1 mL of R-loop IP Buffer supplemented with inhibitors.
4. Add 15–20 μg of the RNase A-treated, diluted, and quantified protein lysate from Subheading 3.2.6, **step 5**.
5. Add 5 μL of RNasin Plus Ribonuclease Inhibitor (200 U) to IP samples to block the RNase A activity (*see* **Note 25**).
6. Optional step: competitors can be added at this stage to test the specific enrichment of proteins associated with RNA/DNA hybrids (*see* **Note 26**).
7. Incubate the IP samples on the rotating wheel at 16 rpm at 4 °C for 2 h.
8. Follow protocol Subheadings 3.1.7–3.1.11.

4 Notes

1. The medium used for cell culture needs to be adapted according to the cell line used for the IP procedure.
2. Precaution should be taken when handling PMSF as it poses a serious health hazard.
3. Protein A Dynabeads can easily precipitate. Carefully mix Protein A Dynabeads by gentle swirling before aliquoting 40 μL into each tube to avoid uneven distribution between the samples.

4. We recommend to prepare fresh solutions supplied with freshly prepared inhibitors each time. All solutions should be prepared with DNase/RNase-free water, and stored on ice until usage unless otherwise stated.
5. Heating followed by slow cooling facilitates efficient hybridization. We suggest to cool down the annealed competitors in the dry block removed from the Block Heater at room temperature for at least 2 h. Annealed oligonucleotides can be re-used if properly stored at −20 °C, however, we recommend to repeat the annealing/cooling process prior to their usage.
6. We suggest that for each wash step, the beads should be resuspended in 1 mL of RSB + T buffer and washed by a gentle inversion of the tubes ~5–10 times; there is no need to incubate the samples on a rotating wheel. Then, the beads should be incubated on a magnetic rack for 1 min and the supernatant should be carefully removed with Gilson P1000 pipette.
7. We use non-cross-linked cells because cross-linking reagents can induce the formation of RNA/DNA hybrids, preventing an identification of bona fide R-loop interactors [18]. To spin the cells, we recommend to use 11 mL Nunc centrifuge tubes (e.g. VWR, 391-8506) because their conical bottom helps to avoid the loss of material.
8. We perform sonication before IP to minimize co-purification of unspecific proteins. Conditions of sonication should be optimized according to different sonicator machines and manufacturer's instructions to generate DNA fragments of ~500 bp.
9. If necessary, the beads can be collected using a touch spin for a couple of seconds in a benchtop centrifuge at 4 °C. The spin should not exceed 500 × *g*. Please note that it is important to take extra care when pipetting to avoid the loss of beads.
10. When performing IP experiment with addition of competitors, 1 IP sample for each competitors' concentration used will be required.
11. We carry out IP with S9.6 antibody in the presence of RNase A to reduce unspecific RNA-mediated interactions and avoid S9.6 recognition of double-stranded RNA (dsRNA). The working RNase A concentration was optimised in the lab previously based on the nucleic acid amount in the IP reaction [12], however change in the protocol (e.g different experimental conditions, cell lines or amounts of protein extracts) will require further optimization steps. Prepare fresh RNase A dilutions each time using cold PBS. Prepare the dilutions prior to the start of the experimental procedure and store them on ice until required.

12. We suggest that for each wash step (initially with RSB + T buffer and later with RBS buffer), the beads should be resuspended in 1 mL of wash buffer and incubated on a rotating wheel at 16 rpm for 5 min at 4 °C. Then the beads should be incubated on a magnetic rack for 1 min and the supernatant should be carefully removed with Gilson P1000 pipette.
13. Elution Buffer contains 1% SDS. It needs to be stored at room temperature to avoid SDS precipitation.
14. When performing the WB analysis to detect RNA/DNA-binding proteins, consider the molecular weight of the protein you are trying to detect and choose the appropriate percentage of the gel. 4–12% Bis-Tris gels are the best for proteins ranging between 10 and 200 kDa. 3–12% Tris-Acetate gels are used for proteins >200 kDa. Loading of ½ of the IP should be enough to detect most of the proteins. For very abundant proteins (such as DHX9), loading ¼ of the IP is sufficient. If the protein is detected in the Input lanes but it is not detected in the IP lane, this probably indicates that the protein does not bind RNA/DNA hybrids.
15. We suggest performing S9.6 SB to assess the efficiency of the S9.6 pull-down, especially, when comparing different experimental settings. We suggest to load an equivalent of one IP and ~0.5–1 μg of DNA as "DNA Input" 3.2.5. We recommend to perform Proteinase K digestion of the eluted samples followed by the DNA precipitation to eliminate all proteins in the samples and successfully detect RNA/DNA hybrids on the SB.
16. Extracts will become very viscous and difficult to resuspend. Use Gilson P1000 pipette to resuspend the extracts up and down gently and avoid making bubbles.
17. The samples need to lose all viscosity during the Proteinase K digestion. Gentle mechanical resuspension will make this process more efficient. We suggest to do ~10–20 gentle resuspensions up and down using Gilson P1000 pipette every 30 min to speed up the process. Do not vortex the tubes.
18. The solution will become white (milky).
19. Set the pipette at 900 μL and carefully transfer the supernatant into a fresh 1.5 mL tube avoiding the white precipitate (proteins).
20. The DNA should become visible and form white "hair-like bundles." If needed, DNA can be frozen at −20 °C in isopropanol. This is a safe stop point.
21. The DNA pellet should become visible. Avoid touching the DNA pellet with the pipette when discarding the supernatant.
22. The DNA should become transparent. We suggest to air dry the pellets for ~10 min, however, sometimes longer incubation

periods may be required to allow complete evaporation of ethanol.

23. Initially add 2.5 μL (12.5 U) of RNase H enzyme to the samples and incubate them for 4 h in a ThermoMixer shaking at 500 rpm at 37 °C. Supply with another 2 μL (10 U) of RNase H enzyme after the initial 4 h treatment and incubate o/n. Adjust the number of RNase H digestion reactions according to the experimental set up.
24. The extracts are treated with a high concentration of RNase A prior to sonication to remove all RNA and RNA/DNA hybrids.
25. We use RNasin Plus Ribonuclease Inhibitor to prevent the action of the RNase A on the RNA/DNA hybrids bound to Protein A Dynabeads.
26. RNase H digestion step can be omitted if the experiment is carried out in the presence of RNA/DNA hybrid competitors which provide the specificity control.

Acknowledgments

We thank Maria Sole Giordano for testing this protocol and Prof. R.Harris for providing APOBEC3B antibody. We thank Holly Roach for editing this paper. This work was supported by the Royal Society University Research fellowship (BVD07340), Royal Society Enhancement Award (RGF\EA\180023), EPA Research Fund (Sir William Dunn School of Pathology, University of Oxford), Ataxia UK (ZGROMAK) and Rosetrees Fund (A2196) to NG. CB is supported by the Royal Society Enhancement Award (RGF\EA\180023) to NG and the Sir William Dunn School of Pathology, University of Oxford.

References

1. Garcia-Muse T, Aguilera A (2019) R loops: from physiological to pathological roles. Cell 179(3):604–618. https://doi.org/10.1016/j.cell.2019.08.055
2. Niehrs C, Luke B (2020) Regulatory R-loops as facilitators of gene expression and genome stability. Nat Rev Mol Cell Biol 21(3): 167–178. https://doi.org/10.1038/s41580-019-0206-3
3. Santos-Pereira JM, Aguilera A (2015) R loops: new modulators of genome dynamics and function. Nat Rev Genet 16(10):583–597. https://doi.org/10.1038/nrg3961
4. Groh M, Gromak N (2014) Out of balance: R-loops in human disease. PLoS Genet 10(9): e1004630. https://doi.org/10.1371/journal.pgen.1004630
5. Bhatia V, Barroso SI, Garcia-Rubio ML, Tumini E, Herrera-Moyano E, Aguilera A (2014) BRCA2 prevents R-loop accumulation and associates with TREX-2 mRNA export factor PCID2. Nature 511(7509):362–365. https://doi.org/10.1038/nature13374
6. Huertas P, Aguilera A (2003) Cotranscriptionally formed DNA:RNA hybrids mediate transcription elongation impairment and transcription-associated recombination. Mol Cell 12(3):711–721. https://doi.org/10.1016/j.molcel.2003.08.010

7. Li X, Manley JL (2005) Inactivation of the SR protein splicing factor ASF/SF2 results in genomic instability. Cell 122(3):365–378. https://doi.org/10.1016/j.cell.2005.06.008
8. Santos-Pereira JM, Herrero AB, Garcia-Rubio ML, Marin A, Moreno S, Aguilera A (2013) The Npl3 hnRNP prevents R-loop-mediated transcription-replication conflicts and genome instability. Genes Dev 27(22):2445–2458. https://doi.org/10.1101/gad.229880.113
9. Skourti-Stathaki K, Proudfoot NJ, Gromak N (2011) Human senataxin resolves RNA/DNA hybrids formed at transcriptional pause sites to promote Xrn2-dependent termination. Mol Cell 42(6):794–805. https://doi.org/10.1016/j.molcel.2011.04.026
10. Cerritelli SM, Crouch RJ (2009) Ribonuclease H: the enzymes in eukaryotes. FEBS J 276(6):1494–1505. https://doi.org/10.1111/j.1742-4658.2009.06908.x
11. Sollier J, Stork CT, Garcia-Rubio ML, Paulsen RD, Aguilera A, Cimprich KA (2014) Transcription-coupled nucleotide excision repair factors promote R-loop-induced genome instability. Mol Cell 56(6):777–785. https://doi.org/10.1016/j.molcel.2014.10.020
12. Cristini A, Groh M, Kristiansen MS, Gromak N (2018) RNA/DNA hybrid interactome identifies DXH9 as a molecular player in transcriptional termination and R-loop-associated DNA damage. Cell Rep 23(6):1891–1905. https://doi.org/10.1016/j.celrep.2018.04.025
13. Tuduri S, Crabbe L, Conti C, Tourriere H, Holtgreve-Grez H, Jauch A, Pantesco V, De Vos J, Thomas A, Theillet C, Pommier Y, Tazi J, Coquelle A, Pasero P (2009) Topoisomerase I suppresses genomic instability by preventing interference between replication and transcription. Nat Cell Biol 11(11):1315–1324. https://doi.org/10.1038/ncb1984
14. Boguslawski SJ, Smith DE, Michalak MA, Mickelson KE, Yehle CO, Patterson WL, Carrico RJ (1986) Characterization of monoclonal antibody to DNA.RNA and its application to immunodetection of hybrids. J Immunol Methods 89(1):123–130. https://doi.org/10.1016/0022-1759(86)90040-2
15. Phillips DD, Garboczi DN, Singh K, Hu Z, Leppla SH, Leysath CE (2013) The sub-nanomolar binding of DNA-RNA hybrids by the single-chain Fv fragment of antibody S9.6. J Mol Recognit 26(8):376–381. https://doi.org/10.1002/jmr.2284
16. McCann JL, Cristini A, Law EK, Lee SY, Tellier M, Carpenter MA, Beghè C, Kim JJ, Jarvis MC, Stefanovska B, Temiz NA, Bergstrom EN, Salamango DJ, Brown MR, Murphy S, Alexandrov LB, Miller KM, Gromak N, Harris RS (2021) R-loop homeostasis and cancer mutagenesis promoted by the DNA cytosine deaminase APOBEC3B. bioRxiv:2021.2008.2030.458235. https://doi.org/10.1101/2021.08.30.458235
17. Tummala H, Dokal AD, Walne A, Ellison A, Cardoso S, Amirthasigamanipillai S, Kirwan M, Browne I, Sidhu JK, Rajeeve V, Rio-Machin A, Seraihi AA, Duncombe AS, Jenner M, Smith OP, Enright H, Norton A, Aksu T, Ozbek NY, Pontikos N, Cutillas P, Dokal I, Vulliamy T (2018) Genome instability is a consequence of transcription deficiency in patients with bone marrow failure harboring biallelic ERCC6L2 variants. Proc Natl Acad Sci U S A 115(30):7777–7782. https://doi.org/10.1073/pnas.1803275115
18. Schwab RA, Nieminuszczy J, Shah F, Langton J, Lopez Martinez D, Liang CC, Cohn MA, Gibbons RJ, Deans AJ, Niedzwiedz W (2015) The Fanconi anemia pathway maintains genome stability by coordinating replication and transcription. Mol Cell 60(3):351–361. https://doi.org/10.1016/j.molcel.2015.09.012
19. Wang IX, Grunseich C, Fox J, Burdick J, Zhu Z, Ravazian N, Hafner M, Cheung VG (2018) Human proteins that interact with RNA/DNA hybrids. Genome Res 28(9):1405–1414. https://doi.org/10.1101/gr.237362.118
20. Abakir A, Giles TC, Cristini A, Foster JM, Dai N, Starczak M, Rubio-Roldan A, Li M, Eleftheriou M, Crutchley J, Flatt L, Young L, Gaffney DJ, Denning C, Dalhus B, Emes RD, Gackowski D, Correa IR Jr, Garcia-Perez JL, Klungland A, Gromak N, Ruzov A (2020) N(6)-methyladenosine regulates the stability of RNA:DNA hybrids in human cells. Nat Genet 52(1):48–55. https://doi.org/10.1038/s41588-019-0549-x
21. Goulielmaki E, Tsekrekou M, Batsiotos N, Ascensao-Ferreira M, Ledaki E, Stratigi K, Chatzinikolaou G, Topalis P, Kosteas T, Altmuller J, Demmers JA, Barbosa-Morais NL, Garinis GA (2021) The splicing factor XAB2 interacts with ERCC1-XPF and XPG for R-loop processing. Nat Commun 12(1):3153. https://doi.org/10.1038/s41467-021-23505-1
22. Cristini A, Ricci G, Britton S, Salimbeni S, Huang SN, Marinello J, Calsou P, Pommier Y, Favre G, Capranico G, Gromak N, Sordet O (2019) Dual processing of R-loops and topoisomerase I induces transcription-dependent DNA double-Strand

breaks. Cell Rep 28(12):3167–3181.e6. https://doi.org/10.1016/j.celrep.2019.08.041

23. Volkening K, Farhan SMK, Kao J, Leystra-Lantz C, Ang LC, McIntyre A, Wang J, Hegele RA, Strong MJ (2021) Evidence of synergism among three genetic variants in a patient with LMNA-related lipodystrophy and amyotrophic lateral sclerosis leading to a remarkable nuclear phenotype. Mol Cell Biochem 476(7):2633–2650. https://doi.org/10.1007/s11010-021-04103-7
24. Dobersch S, Rubio K, Singh I, Gunther S, Graumann J, Cordero J, Castillo-Negrete R, Huynh MB, Mehta A, Braubach P, Cabrera-Fuentes H, Bernhagen J, Chao CM, Bellusci S, Gunther A, Preissner KT, Kugel S, Dobreva G, Wygrecka M, Braun T, Papy-Garcia D, Barreto G (2021) Positioning of nucleosomes containing gamma-H2AX precedes active DNA demethylation and transcription initiation. Nat Commun 12(1):1072. https://doi.org/10.1038/s41467-021-21227-y
25. Li Y, Song Y, Xu W, Li Q, Wang X, Li K, Wang J, Liu Z, Velychko S, Ye R, Xia Q, Wang L, Guo R, Dong X, Zheng Z, Dai Y, Li H, Yao M, Xue Y, Scholer HR, Sun Q, Yao H (2020) R-loops coordinate with SOX2 in regulating reprogramming to pluripotency. Sci Adv 6(24):eaba0777. https://doi.org/10.1126/sciadv.aba0777
26. Shiromoto Y, Sakurai M, Minakuchi M, Ariyoshi K, Nishikura K (2021) ADAR1 RNA editing enzyme regulates R-loop formation and genome stability at telomeres in cancer cells. Nat Commun 12(1):1654. https://doi.org/10.1038/s41467-021-21921-x
27. Jimeno S, Prados-Carvajal R, Fernandez-Avila MJ, Silva S, Silvestris DA, Endara-Coll M, Rodriguez-Real G, Domingo-Prim J, Mejias-Navarro F, Romero-Franco A, Jimeno-Gonzalez S, Barroso S, Cesarini V, Aguilera A, Gallo A, Visa N, Huertas P (2021) ADAR-mediated RNA editing of DNA:RNA hybrids is required for DNA double strand break repair. Nat Commun 12(1):5512. https://doi.org/10.1038/s41467-021-25790-2
28. Kim S, Kang N, Park SH, Wells J, Hwang T, Ryu E, Kim BG, Hwang S, Kim SJ, Kang S, Lee S, Stirling P, Myung K, Lee KY (2020) ATAD5 restricts R-loop formation through PCNA unloading and RNA helicase maintenance at the replication fork. Nucleic Acids Res 48(13):7218–7238. https://doi.org/10.1093/nar/gkaa501
29. Alfano L, Caporaso A, Altieri A, Dell'Aquila M, Landi C, Bini L, Pentimalli F, Giordano A (2019) Depletion of the RNA binding protein HNRNPD impairs homologous recombination by inhibiting DNA-end resection and inducing R-loop accumulation. Nucleic Acids Res 47(8):4068–4085. https://doi.org/10.1093/nar/gkz076
30. Villarreal OD, Mersaoui SY, Yu Z, Masson JY, Richard S (2020) Genome-wide R-loop analysis defines unique roles for DDX5, XRN2, and PRMT5 in DNA/RNA hybrid resolution. Life Sci Alliance 3(10):e202000762. https://doi.org/10.26508/lsa.202000762
31. Bayona-Feliu A, Barroso S, Munoz S, Aguilera A (2021) The SWI/SNF chromatin remodeling complex helps resolve R-loop-mediated transcription-replication conflicts. Nat Genet 53(7):1050–1063. https://doi.org/10.1038/s41588-021-00867-2
32. Hao H, Han T, Xuan B, Sun Y, Tang S, Yue N, Qian Z (2019) Dissecting the role of DDX21 in regulating human cytomegalovirus replication. J Virol 93(24):e01222-19. https://doi.org/10.1128/JVI.01222-19
33. Mersaoui SY, Yu Z, Coulombe Y, Karam M, Busatto FF, Masson JY, Richard S (2019) Arginine methylation of the DDX5 helicase RGG/RG motif by PRMT5 regulates resolution of RNA:DNA hybrids. EMBO J 38(15):e100986. https://doi.org/10.15252/embj.2018100986
34. Jiang X, Prabhakar A, Van der Voorn SM, Ghatpande P, Celona B, Venkataramanan S, Calviello L, Lin C, Wang W, Black BL, Floor SN, Lagna G, Hata A (2021) Control of ribosomal protein synthesis by the microprocessor complex. Sci Signal 14(671):eabd2639. https://doi.org/10.1126/scisignal.abd2639
35. Perez-Calero C, Bayona-Feliu A, Xue X, Barroso SI, Munoz S, Gonzalez-Basallote VM, Sung P, Aguilera A (2020) UAP56/DDX39B is a major cotranscriptional RNA-DNA helicase that unwinds harmful R loops genome-wide. Genes Dev 34(13–14):898–912. https://doi.org/10.1101/gad.336024.119
36. Ngo GHP, Grimstead JW, Baird DM (2021) UPF1 promotes the formation of R loops to stimulate DNA double-strand break repair. Nat Commun 12(1):3849. https://doi.org/10.1038/s41467-021-24201-w
37. Bauer M, Nascakova Z, Mihai AI, Cheng PF, Levesque MP, Lampart S, Hurwitz R, Pfannkuch L, Dobrovolna J, Jacobs M, Bartfeld S, Dohlman A, Shen X, Gall AA, Salama NR, Topfer A, Weber A, Meyer TF, Janscak P, Muller A (2020) The ALPK1/TIFA/NF-kappaB axis links a bacterial carcinogen to R-loop-induced replication stress. Nat

Commun 11(1):5117. https://doi.org/10.1038/s41467-020-18857-z

38. Patel PS, Abraham KJ, Guturi KKN, Halaby MJ, Khan Z, Palomero L, Ho B, Duan S, St-Germain J, Algouneh A, Mateo F, El Ghamrasni S, Barbour H, Barnes DR, Beesley J, Sanchez O, Berman HK, Brown GW, Affar EB, Chenevix-Trench G, Antoniou AC, Arrowsmith CH, Raught B, Pujana MA, Mekhail K, Hakem A, Hakem R (2021) RNF168 regulates R-loop resolution and genomic stability in BRCA1/2-deficient tumors. J Clin Invest 131(3):e140105. https://doi.org/10.1172/JCI140105
39. Fournier L-A, Kumar A, Smith T, Su E, Moksa M, Hirst M, Stirling PC (2021) Global prediction of candidate R-loop binding and R-loop regulatory proteins. bioRxiv:2021.2008.2009.454968. https://doi.org/10.1101/2021.08.09.454968
40. Dettori LG, Torrejon D, Chakraborty A, Dutta A, Mohamed M, Papp C, Kuznetsov VA, Sung P, Feng W, Bah A (2021) A tale of loops and tails: the role of intrinsically disordered protein regions in R-loop recognition and phase separation. Front Mol Biosci 8: 691694. https://doi.org/10.3389/fmolb.2021.691694
41. Brown WL, Law EK, Argyris PP, Carpenter MA, Levin-Klein R, Ranum AN, Molan AM, Forster CL, Anderson BD, Lackey L, Harris RS (2019) A rabbit monoclonal antibody against the antiviral and cancer genomic DNA mutating enzyme APOBEC3B. Antibodies (Basel) 8(3):47. https://doi.org/10.3390/antib8030047
42. Rigby RE, Webb LM, Mackenzie KJ, Li Y, Leitch A, Reijns MA, Lundie RJ, Revuelta A, Davidson DJ, Diebold S, Modis Y, MacDonald AS, Jackson AP (2014) RNA:DNA hybrids are a novel molecular pattern sensed by TLR9. EMBO J 33(6):542–558. https://doi.org/10.1002/embj.201386117

Chapter 15

Studies on Protein–RNA:DNA Hybrid Interactions by Microscale Thermophoresis (MST)

Miaomiao Li, Arne Klungland, and Bjørn Dalhus

Abstract

Microscale thermophoresis (MST) is a technology that allows for quantitative analysis of interactions between biomolecules with low sample consumption. MST uses localized temperature fields to measure the diffusion rates of the free and bound states of a fluorescently labeled protein, and to determine the dissociation constant K_D by fitting of the binding isotherm with a 1:1 binding model. Here, we describe the use of MST for quantitative analysis of the interaction of the N-terminal his-tagged 6-methyladenine (m^6A) reader protein YTHDF2 with m^6A modified and unmodified RNA, in single-strand configuration or with RNA:DNA hybrid substrates. The described protocol is also suitable for studies of interactions with proteins binding to double-stranded RNA or DNA substrates.

Key words Microscale thermophoresis (MST), RNA:DNA hybrids (R-loop), RNA modification, 6-methyladenine (m^6A), RNA-binding, YTHDF2

1 Introduction

Microscale thermophoresis (MST) that allows for analyses of proteins interacting with small molecules and with nucleic acids [1, 2] is a quite recent discovery. MST uses a localized temperature field created by an infrared laser source to assess biomolecular interaction [3]. Since the biological matrix can easily be modified, the method allows for reliable conclusions on the function of the proteins and the precise dynamics introduced when interacting with its biological substrate [1]. Interaction of proteins with a range of nucleic acids substrates and complex formation is key to cell viability and processes including replication, transcription, and translation. An overview and assessment of various techniques for analyzing these interactions, including EMSA (electrophoretic mobility shift assays), SPR (surface plasmon resonance), and MST, has been published previously [4]. Here, we present the protocol to study the interaction of YTHDF2 with m^6A-modified

Andrés Aguilera and Alexey Ruzov (eds.), *R-Loops: Methods and Protocols*, Methods in Molecular Biology, vol. 2528,
https://doi.org/10.1007/978-1-0716-2477-7_15,

or unmodified RNA and RNA:DNA hybrid substrates [5]. We find that, in line with EMSA studies, MST analysis demonstrates that YTHDF2 binds preferentially to m^6A-modified RNA and that YTHDF2 shows similar dissociation constants for its interaction with m^6A-modified single-stranded RNA and m^6A-RNA:DNA duplexes [5].

While the frequent modification of adenosine to 6-methyladenosine (m^6A) in messenger RNA was discovered many decades ago [6], the reversible nature of these RNA modifications was only recently discovered [7, 8]. The dynamic potential of m^6A inspired a tremendous effort for understanding the role of m^6A in mRNA and have included its writers, readers, and erasers. One of the key studies is the identification of its reader protein YTHDF2 [9] that also led to the identification of numerous other reader proteins. In recent years, it has been shown that some unique structures required for genome stability involves m^6A-modified RNA and its reader proteins. These include RNA:DNA hybrids (R-loops), G4-quadruplexes and retrotransposons, and m^6A writers, readers and erasers have been shown to engage in the stability and dynamics of such structures [e.g. 5, 10, 11]. While some elegant studies have identified m^6A dependent structural alterations that can regulate RNA–protein interactions [12] the lack of high-resolution structural insight on protein/m^6A-RNA complexes necessitate other approaches. Thus, alternative methods for studying and quantifying the dynamic interactions of protein/m^6A-RNA are very valuable and we here describe a method for studying modified RNA/RNA:DNA hybrid protein interaction by microscale thermophoresis (MST).

MST measures the time-dependent diffusion of a fluorescent molecule moving in a steep temperature gradient. The diffusion rate is determined by a range of factors including particle size, charge, and hydration—parameters that normally differ between a protein in its free state and when bound to a ligand. The detected fluorescence can either stem from the intrinsic fluorescence of the protein itself (known as *label-free MST*) or from a dye bound to the protein (*labeled MST*). Since the intrinsic protein fluorescence, normally from tryptophan residues buried in the core of the protein, is much weaker than modern, high-photon-yield dyes, the label-free version of MST typically requires μM concentrations of the protein, with a corresponding limitation on the range of affinities that can be determined. The instruments for the labeled version of MST, Monolith NT.115 and Monolith Pico, can use protein concentrations down to low nM or even down to pM range, respectively, depending on the sensitivity of the detector. A drawback of labeled MST versus label-free MST is, however, that some proteins can lose their structural integrity or stability upon labeling.

The labeled version of MST uses either a dye covalently attached to the protein or a dye binding to a moiety with very strong affinity to part of the protein, like a terminal tag. This includes nitrilotriacetic acid (NTA)-based dyes binding to a protein's (hexa)histidine tag, which is the method described below. The instrument for labeled MST (Monolith NT.115) comes with different configurations, using two of three possible lasers (red, green, or blue) with corresponding dyes (RED—647 nm; GREEN—557 nm; BLUE—495 nm) for flexibility. The current protocol is identical for all three types of lasers and dye color combinations.

2 Materials

2.1 Materials for Protein Labeling and MST Experiment; Buffer Exchange

1. The NT-647 RED-tris-NTA dye comes as a kit containing 2 ml of 5× PBS-T buffer (PBS buffer with 0.05% Tween-20 when diluted to 1×), 250 pmol dye as a dry powder, and a control peptide to check for binding interference if needed. Store the kit at −20 °C until used (Nanotemper, catalogue no. MO-L008).
2. MST capillaries, standard or premium coated to reduce surface adsorption, are stored at room temperature, according to the manufacturer's instructions (Nanotemper, catalogue no's MO-K022 and MO-K025).
3. Use single- or 8-strips versions of PCR/low-binding tubes for preparation of dye solution, diluted protein, and dilution series of RNA:DNA substrate. The 8-strip PCR tube configuration is highly recommended to prepare the RNA:DNA substrate dilution series.
4. MST experiments using the tris-NTA dye for labeling of proteins with histidine tags requires that the protein solution does not contain compounds that can interfere with the labeling. In particular, the following typical buffer components should not be present above the given concentrations: histidine and imidazole (1 mM), EDTA (5 mM), DTT (5 mM), β-mercaptoethanol (1 mM), and glycerol (10%). See manufacturer's documents for additional information on incompatible compounds. If any of these are present in the given amounts, a buffer exchange step should be included before labeling. The incompatible buffer can be replaced by various means, depending on the amount of available material. Common methods include buffer exchange by centrifugation, size exclusion chromatography or dialysis.

3 Methods

3.1 Purification of Full-Length YTHDF2 with N-Terminal His-Tag

1. A codon-optimized gene sequence corresponding to full-length human YTHDF2 was synthesized and inserted into plasmid pET28b (pET28b-YTHDF2), in frame with the N-terminal His-tag (GeneScript) as described (Abakir et al. 2020).
2. Thaw one tube with BL21(DE3) competent cells on ice and gently mix the competent cells. Add β-mercaptoethanol (β-ME) to the cell suspension to a final concentration of 25 mM. Swirl the tube gently. Incubate the reaction on ice for 10 min, swirling gently every 2 min.
3. Add 1 ng of pET28b-YTHDF2 to the cell suspension and swirl gently. Incubate the transformation reaction on ice for 30 min. Preheat 1 ml SOC medium in a 42 °C water bath or heating block for later use. Heat pulse the transformation reaction in a 42 °C water bath for 45 s. The duration of the heat pulse is critical for optimal transformation efficiency. Incubate the reaction on ice for 2 min. Add 0.9 ml of preheated (42 °C) SOC medium to the transformation reaction and incubate the reaction at 37 °C for 1 h with shaking at 225–250 rpm. Plate the reaction mixture on an LB-agar plate with proper antibiotics (kanamycin for plasmid pET28b), and incubate the plate at 37 °C. On the next day, prepare an overnight culture by inoculating 100 ml LB medium supplemented with the proper antibiotics with a single colony from the plate. Incubate the culture at 37 °C with shaking (180 rpm) for 12–18 h.
4. Before the large scale protein purification, we performed a small scale protein solubility test. BL21(DE3) negative-control and His-YTHDF2 transformed BL21(DE3) colonies are inoculated and cultured in 2 ml LB medium for 4 h, before transferring a volume of each culture into new portions of 5 ml LB medium with proper antibiotics to reach $OD_{600nm} = 0.1$. Incubate at 37 °C and add IPTG to a final concentration of 300 μM when the OD_{600nm} reach 0.6. The culture is further incubated at 18 °C with shaking at 160 rpm until the OD_{600nm} reach 3.
5. The cultures are collected in 1.5 ml microcentrifuge tubes and sonicated with 3× 30 s pulses with amplitude 60. Centrifuge samples and divide into total (T), supernatant (S), and pellet (P). Add SDS loading buffer and boil for 15 min at 95 °C. Load about 0.1 OD of each sample on an SDS-PAGE gel. Use Coomassie blue staining and Western blotting to determine the solubility (Fig. 1).
6. Once the solubility test shows that the target protein is soluble in the supernatant, transfer 5 ml of the overnight culture (from

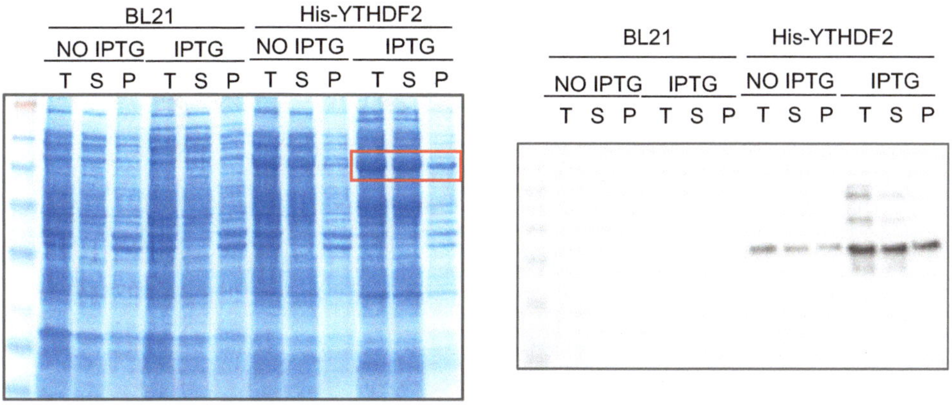

Fig. 1 Analysis of His-tag fusion full-length YTHDF2 protein expression after treatment with 300 μM IPTG. The left panel shows 4–20% Precast Protein Gel stained with colloidal Coomassie. Molecular mass marker is SeeBlue™ Plus2 Pre-stained Protein Standard. The right panel shows the Western blot after detection using His-tag Horseradish Peroxidase-conjugated antibody. *T* total protein, *S* soluble fraction, *P* cell pellet

step 3) to 500 ml of LB medium supplemented with 45.55 g sorbitol, and incubate for about 4 h at 37 °C at 200 rpm, until the OD_{600nm} reach 0.6. IPTG (isopropyl-β-D-1-thiogalactoside) is added to a final concentration of 300 μM, and the culture is further incubated overnight at 18 °C, 160 rpm.

7. Prepare cell pellet by centrifugation at 3000 × *g* for 10 min at 4 °C, and resuspend the pellet in 40 ml lysis buffer (50 mM Tris pH 8.0, 300 mM NaCl, 10 mM β-mercaptoethanol (β-ME), 10 mM imidazole). Sonicate the cell suspension by 3× 30 s pulses with amplitude 60. Keep the cell suspension on ice for 1 min between each sonication.

8. Centrifuge the sonicated samples at 19,000 × *g* for 20 min at 4 °C using a high-speed centrifuge. Decant the supernatant and load it onto 2 ml Protino Ni-NTA agarose, prepared as described by the manufacturer in a 50 ml tube and incubate at 4 °C with rotation for 30 min.

9. After centrifugation at 3000 × *g* for 2 min and discarding the unbound protein extract, the Ni-NT-agarose-bound YTHDF2 is washed twice with 10 ml buffer containing 50 mM Tris pH 8.0, 300 mM NaCl, 10 mM β-ME, and 50 mM imidazole, interspersed with centrifugation at 3000 × *g* for 2 min. Keep wash fractions for SDS-PAGE analysis.

10. Recombinant YTHDF2 is eluted with five 1.5 ml portions of elution buffer (50 mM Tris pH 8.0, 300 mM NaCl, 10 mM β-ME, and 300 mM imidazole). The elutions are analyzed by both SDS-PAGE and Western blotting (Fig. 2).

11. The high concentration of imidazole in the eluted fractions will interfere with protein dye labeling for MST. Hence, it is

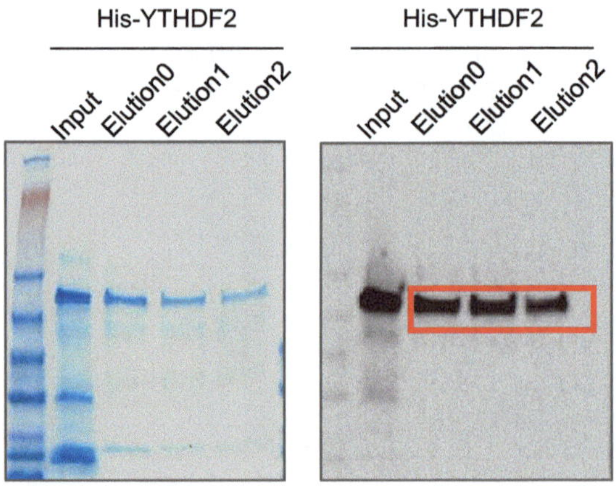

Fig. 2 Analysis of His-tag fusion full-length YTHDF2 protein purified with Protino Ni-NTA agarose. The left panel shows 4–20% Precast Protein Gel stained with colloidal Coomassie. Molecular mass marker is SeeBlue™ Plus2 Pre-stained Protein Standard. The right panel shows the Western blot after detection with His-tag Horseradish Peroxidase-conjugated antibody

necessary to remove imidazole by dialysis. Dialyze the eluted protein at 4 °C overnight using a buffer containing 50 mM Tris pH 8.0, 200 mM NaCl, 10% glycerol, and supplemented with protease inhibitors and DNase.

12. Store the purified full-length YTHDF2 in dialysis buffer at −80 °C until further use. For MST analysis, it is recommended to use freshly purified protein.

3.2 Preparation of Modified RNA and RNA:DNA Hybrid Substrates

1. m^6A-modified RNA oligo is purchased from Midland Certified. (https://oligos.com/custom-oligo-nucleotides/).
2. Non-modified RNA oligo and DNA oligo are purchased from Eurofins. (https://eurofinsgenomics.eu/en/ecom/tools/oligo-analysis/).

RNA oligo	AUGGGCCGUUCAUCUGCUAAAAGGACUGCUUUU GGGGCUUGU
m^6A-RNA oligo	AUGGGCCGUUCAUCUGCUAAAAGGXCUGCUUUU GGGGCUUGU (X is m^6A)
DNA oligo	ACAAGCCCCAAAAGCAGTCCTTTTAGCAGA TGAACGGCCCAT

3. Before MST analysis, the binding activity of His-YTHDF2 to non-modified RNA oligo and m^6A-modified RNA oligo is analyzed using electrophoretic mobility shift assay (Fig. 3). This step is not necessary to perform before the MST assay, but it will provide additional information.

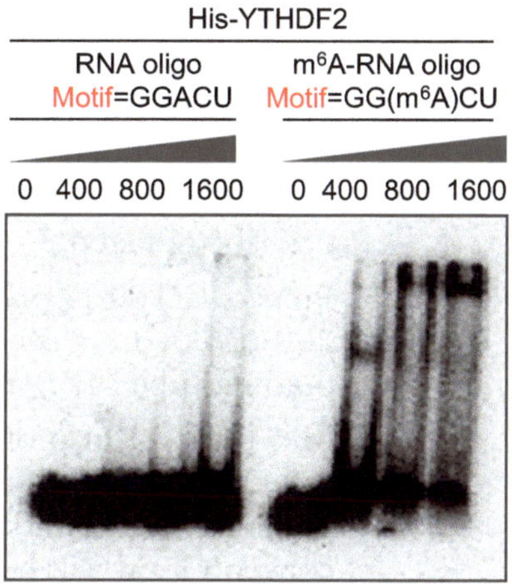

Fig. 3 Electrophoretic mobility shift assay measuring the binding activity of His-YTHDF2 with non-modified RNA oligo and m^{6}A-modified RNA oligo. All the oligos are labeled with ^{32}P-γATP by using T4 PNK, and purified on 20% native PAGE. His-YTHDF2 protein was incubated with the labeled oligos in binding buffer (250 mM MES pH 6.4, 3 M NaCl) on ice for 20 min. The reaction was separated on 7.5% native PAGE in running buffer [40 mM HEPES, 0.114% acetic acid (vol/vol), 1 mM EDTA, pH to 7.0]. The gel was transferred to 3 mm paper and dried at 80 °C for 30 min

4. Spin down the oligo powder and prepare 100 μM stock solutions of the DNA and RNA oligos using UltraPure DNase/RNase-Free Distilled Water.
5. For preparation of the RNA:DNA double-stranded hybrids, mix the relevant RNA and complementary DNA oligo strands in equal molar amounts.
6. For annealing of the RNA:DNA hybrids, heat the oligo mixtures to 94 °C for 2 min and gradually cool the samples on your bench at room temperature.
7. Store the 50 μM annealed oligo substrates at −80 °C. It is recommended to store the substrate in aliquots of 22–23 μl to avoid several thaw-and-freeze cycles, as 20 μl substrate is enough for one MST experiment.

3.3 Preparation of Dye Solution and Labeling of YTHDF2 with NT-647 RED-Tris-NTA Dye

1. Prepare a 1× PBS-T buffer solution by adding 8.0 ml milliQ-filtered or distilled water to the vial containing the 5× PBS-T buffer.
2. Add 50 μl of 1× PBS-T buffer to the vial containing the NT-647 RED-tris-NTA dye (dry powder) to make a 5 μM dye solution. Mix thoroughly to dissolve all the solid material.
3. Dispense the 5 μM dye solution into PCR/low-binding tubes in 2 μl portions for long-time storage. The dye can be stored

for several months at −20 °C in the dark. One portion of 2 μl dye is enough for one single MST experiment.

4. Prepare 100 μl of a 100 nM dye solution in 1× PBS-T by adding 98 μl of 1× PBS-T from **step 1** to a single PCR/low-binding tube containing 2 μl of the 5 μM dye solution described in **step 3**.
5. Prepare a 100 μl protein solution by diluting the protein to 200 nM in 1× PBS-T buffer from **step 1** using a PCR/low-binding tube (*see also* **Note 1**).
6. Mix the 100 μl protein solution (200 nM) with 100 μl dye solution (100 nM) and keep the mixture at room temperature, preferably in the dark, for about 30 min. The final total protein concentration will be 100 nM, while the concentration of the labeled protein is 50 nM, assuming complete binding of the dye with negligible excess free dye.
7. Use a table-top centrifuge to spin the sample at 15,000 × *g* for 10 min at 4 °C. Store on ice, or at the benchtop if protein permits, until further use.

3.4 Preparation of 1:1 Ligand Dilution Series of the RNA:DNA Hybrid Substrate

1. Thaw the RNA:DNA hybrid substrate, stored at −80 °C, on ice and pipet carefully to make the solution homogenous.
2. Place 16 clean and empty PCR/low-binding tubes in a rack and label from 1 through 16. It is highly recommended to use 2 PCR-strips à 8 tubes if available.
3. Fill tubes 2–15 with 10 μl UltraPure DNase/RNase-Free Distilled Water (Thermo Fisher Scientific) which was used to dissolve the oligomers (*see also* **Note 2**).
4. Transfer 20 μl of the 50 μM RNA:DNA hybrid substrate solution (from Subheading 3.2, **step** 7) into tube 1 (Fig. 4a).
5. Transfer 10 μl of the substrate solution from tube 1 into tube 2 and mix thoroughly by careful pipetting up and down multiple times.
6. Repeat **step 5** through the whole series of tubes 2–16 to make a 16-samples 1:1 dilution series starting at 50 μM of the RNA:DNA substrate.
7. Discard the excess 10 μl from the last tube (no. 16) so that each tube contains 10 μl of substrate/ligand.

3.5 Preparing the Complexes and Performing the MST Experiment

1. Add 10 μl of the labeled protein from Subheading 3.3, **step** 7 to each of the 16 tubes with 10 μl of substrate/ligand prepared in Subheading 3.4, **step** 7 (Fig. 4b). Mix carefully by pipetting up and down multiple times. Remember to replace tips between each sample to avoid cross-contamination.

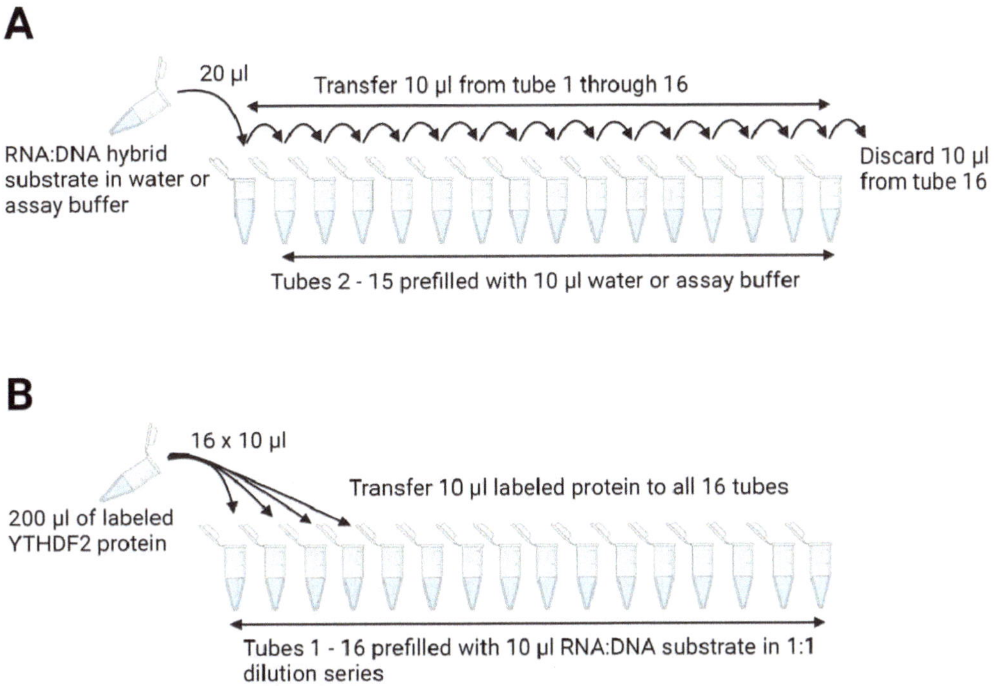

Fig. 4 Preparation of ligand dilution series and protein–ligand complex. (**a**) Dilution of nucleic acid ligand in water or assay buffer by transfer of 10 μl solution throughout the 16-tube series. (**b**) Preparation of the protein-ligand complex by addition of 10 μl labeled protein to each of the 16 nucleic acids solutions from (**a**). Figure created with BioRender.com

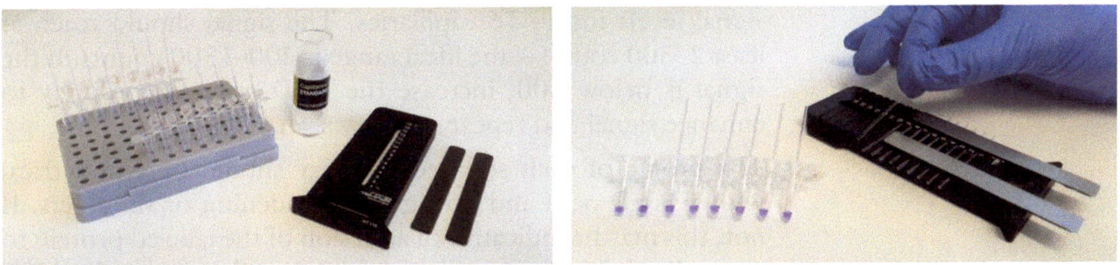

Fig. 5 Filling of capillaries and capillary tray. Left panel: 16 standard MST capillaries are placed in the PCR-strip tubes to soak up the protein-nucleic acid complex solution. Right panel: When filled, each capillary is carefully placed in the capillary tray, starting with the most concentrated sample (tube 1) in position 1. Use the magnetic metal plates to hold capillaries in place while preparing the tray. The authors thank Pernille Blicher for being the hand model

2. Incubate samples at room temperature for several minutes to establish equilibrium conditions.
3. Add one MST capillary (standard or premium coating) to each of the 16 tubes and let the protein-nucleic acid complex fill the capillary (Fig. 5). The tubes with inserted capillaries may have to be tilted horizontally to ease the filling of the capillaries by

capillary forces. The capillary should be almost full, leaving no air bubbles or large gaps of air at either end.

4. Place one capillary in each slot in the capillary tray, starting with tube 1 (the highest ligand concentration) in position 1. Secure the capillaries using the two magnetic, rubber-coated metal strips to ensure fixed positioning of the sample capillaries in the tray (Fig. 5).
5. Once all 16 capillaries are in place, slide the capillary tray into the tray docking port of the Monolith NT.115 instrument.
6. In the MO.control software, select the correct laser (in this case, set *LED color* to RED), the correct number of capillaries to be scanned (in this case from 1 to 16), and set the *LED power* to 60%. Set the concentration of the fluorescent molecule to 50 (in nM, for the total protein concentration). Set the manual temperature controller to 25 °C.
7. Fill in the concentration of the ligand for each of the 16 samples in the *Table of capillaries* table. Select 1:1 dilution factor in the pull-down menu, fill in the correct ligand concentration in the top row for capillary 1, and use click-and-drag to autofill the rest of the column. NB! Use the same unit as for the concentration of the labeled protein in **step 6** above. They should both be either nM or μM, or any other unit, as long as the same unit applies to both.
8. Using the MO.Control software, perform a capillary scan by pressing the ***Start Cap Scan*** button, to check for fluorescence signal levels for all 16 capillaries. The signal should reach at least 2–300 counts—the ideal range is 200–1500 counts. If the signal is below 200, increase the *LED power* accordingly to enhance signal and repeat capillary scan.
9. The profile of each scanned capillary should be symmetrical with a single peak and no sign of broadening or shoulders. If not, this may be indicative of adhesion of the labeled protein to the wall of the glass capillary, which may invalidate the MST analysis. Try changing the type of capillary, buffer composition or add a detergent to identify suitable assay condition.
10. The peak of the fluorescence signal should not vary more than ±10% for all 16 samples, and with no apparent trend in the signal strength along the dilution gradient.
11. A signal that varies randomly more than ±10% throughout the series is indicative of inaccurate pipetting or problems with protein stability/aggregation following mixing of the nucleic acid ligand with the labeled protein. To check for possible protein aggregation, perform a full MST experiment by pressing the ***Start MST Measurement*** button, and observe and inspect the MST curves for signs of problems with protein

aggregation/instability. MST curves of inhomogeneous samples tend to produce bumpy MST profiles with poor shapes. The assay condition must then be optimized before data can be collected and used in MST analysis (replace assay buffer, add detergent or any other stabilizing agent, replace protein with new batch or variant, switch to a covalent dye instead of the NTA-based dye, etc.).

12. If the signal varies *systematically* along the concentration gradient, this may be indicative of *ligand-induced change of fluorescence*, i.e. the ligand affects the fluorescent properties of the NTA-dye directly and does not necessarily reflect binding to the protein of interest. Perform a similar experiment using the accompanying control peptide with your ligand series to exclude the possibility of a direct interaction between the dye and the ligand.
13. If the capillary scan shows all 16 samples to be well within the ±10% range around the average signal, with no particular dose-dependent trend, and with well-defined capillary scan profiles, a full MST data set can be collected.
14. In the *Table of Runs*, a list of several MST experiments can be given, to be run in the given order. Pay attention to the *MST power*—i.e. the power of the infrared laser to produce the temperature gradient. Typically, three series with increasing *MST power* is used; 20%, 40%, and 60%. Use the *LED power* determined in **step 8**.
15. Start the MST experiment by clicking on the ***Start Cap scan + MST measurement*** button to repeat a capillary scan followed by a full MST experiment.
16. Depending on the number of capillaries and number of runs to be executed, the MST experiment normally takes 45 min to complete if using 16 capillaries measured with three different MST powers, and with 5 s *Fluo. Before*, 30 s *MST on*, and 5 s *Fluo. After* for each capillary to be scanned.

3.6 Analyze MST Data for NTA-Labeled YTHDF2 Binding to RNA:DNA Hybrid Substrate

1. After the MST experiment is complete, the data can be analyzed using the MO.Affinity Analysis software.
2. Start the MO.Affinity Analysis software and load the recorded experiment file (.ntp) using the **Load raw data files** button. A preview of all the MST experiments in the data file is shown. Use the **Raw Data Inspection** button to check that the MST traces and the capillary shapes are fine, and that the initial fluorescence is within the ±10% range, for each data set to be used.
3. Select MST as *analysis type*, and select which MST experiments to be analyzed, by clicking on the + sign at the bottom right corner of each respective plot in the preview window on the

left. If several parallel experiments have been run, one can add several MST data for averaging.

4. Once the required selection of data has been added to an analysis set, go to the **Dose Response Fit** tab. Select which set to be analyzed, and tick off the *Kd Model* box to have a 1:1 binding model fitted to the data. Check that the *TargetConc* is equal to the protein concentration (in this case 50 nM). If not, type in the correct number, and tick off the *Fix* box next to it.
5. Inspect the fitted curve and make sure it shows a good fit to the data points. The *Response Amplitude*, i.e., the difference between the bound and free state (high and low concentration of the ligand, respectively), should be larger than the apparent noise level in the data. Typically, a *Response Amplitude* of 3–5 units (F_{norm}, ‰) is indicative of a well-defined fit to the binding isotherm. The dissociation constant K_{D} is reported with confidence interval and in the same units as was used for input of the dilution series.
6. At least three independent experiments using individually prepared dilution series should be used to obtain a reliable K_{D}-value. The three data sets can be merged in the MO.Affinity Analysis software before fitting to the averaged data points.
7. With several data sets and analysis sets loaded into the software, one can compare different data using the **Compare Results** tab. Use *Δ FNorm* or *Fraction Bound* as normalization target to get a better comparison between different data sets (e.g., same protein binding to different substrates).

4 Notes

1. A different buffer can also be used, but take note of pH (should not be too low as binding to the histidine tag is pH-dependent) and any conflicting reagent (*see also* Subheading 2.1).
2. Alternatively, the identical buffer as the RNA:DNA substrates are stored in, if different from pure water; this will then be the assay buffer.

References

1. Wienken CJ, Baaske P, Rothbauer U, Braun D, Duhr S (2010) Protein-binding assays in biological liquids using microscale thermophoresis. Nat Commun 1:100
2. Mueller AM, Breitsprecher D, Duhr S, Baaske P, Schubert T, Längst G (2017) MicroScale thermophoresis: a rapid and precise method to quantify protein-nucleic acid interactions in solution. Methods Mol Biol 1654:151–164
3. Jerabek-Willemsen M, Wienken CJ, Braun D, Baaske P, Duhr S (2011) Molecular interaction studies using microscale thermophoresis. Assay Drug Dev Technol 9:342–353
4. Flores JK, Kariawasam R, Gimenez AX, Helder S, Cubeddu L, Gamsjaeger R, Ataide

SF (2015) Biophysical characterisation and quantification of nucleic acid-protein interactions: EMSA, MST and SPR. Curr Protein Pept Sci 16:727–734

5. Abakir A, Giles TC, Cristini A, Foster JM, Dai N, Starczak M, Rubio-Roldan A, Li M, Eleftheriou M, Crutchley J, Flatt L, Young L, Gaffney DJ, Denning C, Dalhus B, Emes RD, Gackowski D, Corrêa IR Jr, Garcia-Perez JL, Klungland A, Gromak N, Ruzov A (2020) N^6-methyladenosine regulates the stability of RNA:DNA hybrids in human cells. Nat Genet 52:48–55
6. Desrosiers R, Friderici K, Rottman F (1974) Identification of methylated nucleosides in messenger RNA from Novikoff hepatoma cells. Proc Natl Acad Sci U S A 71:3971–3975
7. Jia G, Fu Y, Zhao X, Dai Q, Zheng G, Yang Y, Yi C, Lindahl T, Pan T, Yang YG, He C (2011) N^6-Methyladenosine in nuclear RNA is a major substrate of the obesity-associated FTO. Nat Chem Biol 7:885–887
8. Zheng G, Dahl JA, Niu Y, Fedorcsak P, Huang CM, Li CJ, Vågbø CB, Shi Y, Wang WL, Song SH, Lu Z, Bosmans RP, Dai Q, Hao YJ, Yang X, Zhao WM, Tong WM, Wang XJ, Bogdan F, Furu K, Fu Y, Jia G, Zhao X, Liu J, Krokan HE, Klungland A, Yang YG, He C (2013) ALKBH5 is a mammalian RNA demethylase that impacts RNA metabolism and mouse fertility. Mol Cell 49:18–29
9. Wang X, Lu Z, Gomez A, Hon GC, Yue Y, Han D, Fu Y, Parisien M, Dai Q, Jia G, Ren B, Pan T, He C (2014) N^6-methyladenosine-dependent regulation of messenger RNA stability. Nature 505:117–120
10. Fleming AM, Nguyen NLB, Burrows CJ (2019) Colocalization of m^6A and G-quadruplex-forming sequences in viral RNA (HIV, Zika, hepatitis B, and SV40) suggests topological control of adenosine N^6-methylation. ACS Cent Sci 5:218–228
11. Hwang SY, Jung H, Mun S, Lee S, Park K, Baek SC, Moon HC, Kim H, Kim B, Choi Y, Go YH, Tang W, Choi J, Choi JK, Cha HJ, Park HY, Liang P, Kim VN, Han K, Ahn K (2021) L1 retrotransposons exploit RNA m^6A modification as an evolutionary driving force. Nat Commun 12:880
12. Liu N, Dai Q, Zheng G, He C, Parisien M, Pan T (2015) N^6-methyladenosine-dependent RNA structural switches regulate RNA-protein interactions. Nature 518: 560–564

Chapter 16

Studying R-Loop Recognizing Proteins Using Single-Molecule DNA Curtain Technique and Electrophoretic Mobility Shift Assay

Na Young Cheon and Ja Yil Lee

Abstract

R-loops are nucleic acid structures containing a DNA–RNA hybrid and the associated non-template single-stranded DNA. R-loops are not only involved in many biological processes but also cause genomic instability when they are abnormally regulated. The R-loop regulation pathway consists of multiple steps associated with diverse proteins. The initial and essential step of the pathway is to recognize R-loops in long DNA of human genome. To elucidate the molecular mechanism underlying R-loop recognition by proteins, we utilize a novel high-throughput single-molecule approach called "DNA curtain" as well as electrophoretic mobility shift assays. Here, we describe the detailed protocols for these techniques that both can be used for studying the R-loop recognition mechanisms.

Key words R-loop, DNA curtain, Single-molecule imaging, Electrophoretic mobility shift assay, TonEBP

1 Introduction

The R-loop is a three-stranded nucleic acid structure consisting of a DNA-RNA hybrid and a displaced single-stranded DNA (ssDNA). R-loops play important roles in diverse DNA metabolic events including chromosome segregation, immunoglobulin switching, and signaling for regulation of gene expression [1–4]. However, inappropriate accumulation of R-loops is a serious threat to genomic integrity and stability [5, 6]. R-loops can block replication or transcription, and the displaced ssDNA is a good substrate for nuclease attack, causing DNA damage. It is known that severe diseases, including myelodysplastic syndromes, neurodegenerative diseases, and Ewing's sarcoma, are triggered when an abnormal R-loop accumulation is not eliminated [5, 6].

Tonicity-responsive enhancer binding protein (TonEBP or NFAT5) is a transcription factor that regulates immune responses

Andrés Aguilera and Alexey Ruzov (eds.), *R-Loops: Methods and Protocols*, Methods in Molecular Biology, vol. 2528,
https://doi.org/10.1007/978-1-0716-2477-7_16,

and tonicity controls [7]. Recently, it was reported that TonEBP senses R-loops and then recruits methyltransferase-like 3 (METTL3) and METTL14 for m^{6}A RNA methylation of the R-loops. Then RNase H1 is recruited to the methylated R-loop for eliminating the R-loops [8]. Furthermore, using a high-throughput single-molecule imaging technique called DNA curtain, the R-loop search mechanism by TonEBP has been revealed. TonEBP finds R-loops on DNA by two different molecular mechanisms: 3D collision and 1D diffusion. Such a dual-mode mechanism facilitates rapid search for R-loops through the long human genome. In addition, DNA curtain assay demonstrated that TonEBP diffuses along DNA by sliding without rotating around DNA helix. This finding is consistent with the structure of TonEBP, which encompasses double-stranded DNA and maintains intimate contact with DNA during diffusion [8].

DNA curtain is a hybrid system combining nano-fabrication, microfluidics, and fluorescence imaging that enables the real-time visualization of the physical interaction and kinetics of proteins on individual DNA molecules (Fig. 1) [10, 11]. In a microfluidic flowcell, the surface of which is coated with a lipid bilayer, biotinylated lambda phage DNA (λ-DNA) is anchored on the biotinylated lipid of the bilayer via streptavidin. Under a hydrodynamics flow, DNA molecules move on the lipid bilayer and get stuck at a diffusion barrier, which is formed by the nanometer wide chromium (Cr) or trench barrier (Fig. 1a, b) [11, 12]. Hundreds of DNA molecules are aligned and stretched at the diffusion barrier in the presence of the flow and imaged by total internal reflection florescence microscopy (TIRFM). This type of DNA curtain is called a single-tethered DNA curtain (Fig. 1a). In the single-tethered DNA curtain, DNA molecules are visible only in the presence of the flow because DNA molecules are stretched within the excitation region of the evanescent field. To examine protein–DNA interactions in the absence of flow, another Cr constructs called pedestals, which are wider than the diffusion barrier, are fabricated, and the other end of the λ-DNA is tethered to the pedestal via antigen-antibody bonding (Fig. 1b) [4, 13]. This is called a double-tethered DNA curtain, providing a general platform to investigate protein–DNA interactions.

Here, we introduce the detailed procedures for both single- and double-tethered DNA curtains and electrophoretic mobility shift assay (EMSA) to examine how TonEBP senses R-loops on DNA.

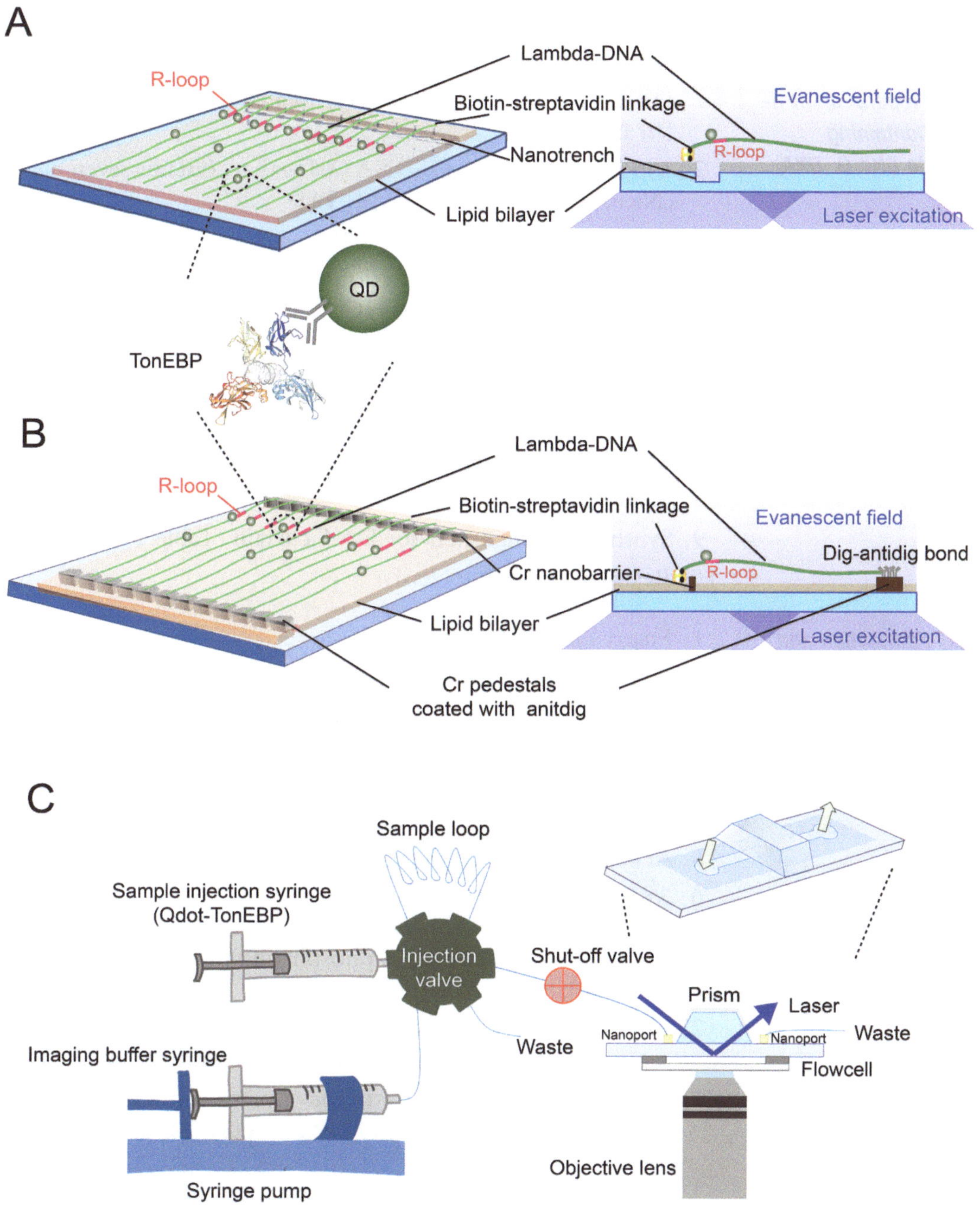

Fig. 1 Schematic of DNA curtain system. (**a**) A schematic of single-tethered DNA curtain assay: slant view (left) and side view (right). In the presence of buffer flow, λ-DNA molecules containing the R-loop (red line) are aligned and stretched at a nano-trench, which serves as a diffusion barrier of the lipid bilayer. The Cy5-labeled R-loop and TonEBP tagged with anti-FLAG conjugated quantum dot (QD) are visualized by TIRFM. The structure of TonEBP is adopted from PDB (1IMH) [9]. (**b**) A schematic of double-tethered DNA curtain assay: slant view (left) and side view (right). The biotinylated end of λ-DNA is anchored on lipid via streptavidin and stalled at a chromium (Cr) nano-barrier, and the other end is tethered to the Cr pedestal via the bonding between digoxigenin and its antibody. In this double-tether configuration, λ-DNA remains stretched even in the absence of flow. (**c**) A schematic of microfluidic system (left) and TIRFM system (right) for DNA curtain. Buffer flows in via a syringe pump, and the protein sample is injected into DNA curtain through the 6-way valve. The flow is transiently switched on and off using the shut-off valve. Fluorophores are excited by a prism-type TIRFM. The evanescent field is built up at the interface between a slide glass and buffer as shown in (**a**) and (**b**)

2 Materials

2.1 Preparation of R-Loop Containing Lambda DNA (λ-DNA)

1. Diethyl pyrocarbonate (DEPC)-treated deionized water (D.I.) (*see* **Note 1**).
2. 1 mg/ml engineered λ-DNA (λ-I3) that contains seven nickase (Nt.BspQI) sites (Fig. 2a) [4, 14].
3. Nickase Nt.BspQI.
4. 10× NEBuffer 3.1 restriction enzyme buffer (New England BioLabs): 500 mM Tris–HCl pH 7.9, 1 M NaCl, 100 mM $MgCl_2$, and 1 mg/ml BSA.
5. T4 DNA Ligase.
6. 100 mM ATP.
7. Microspin™ S-400 HR (GE-27-5140, GE healthcare).
8. Proteinase K.
9. Synthesized oligomers for DNA curtain (Table 1).
10. Restriction enzyme NcoI.

2.2 Fabrication and Cleaning of Slides Containing Nano-Trench Patterns

1. Fused-silica slides.
2. Piranha solution (3:1 mixture of sulfuric acid and hydrogen peroxide).
3. 4% 950 K PMMA (AR-P 671.04, Allresist GmbH).
4. Developing solution (1:3 ratio of methyl isobutyl ketone (MIBK) and isopropanol.
5. Al etchant (AZ-MIF300).
6. 2% Hellmanex III (Z805939, Sigma).

2.3 Biotinylated Liposome Stock Making

1. 1,2-Dioleoyl-sn-glycero-3-phosphocholine (DOPC, 850375P, Avanti Polar Lipids).
2. PEG-conjugated 1,2-dioleoyl-sn-glycero-3-phosphoethanolamine (mPEG 2000-DOPE, 880130P, Avanti Polar Lipids).
3. Biotinylated 1,2-Dipalmitoyl-sn-glycero-3-phosphoethanolamine (biotin-DPPE, 870273, Avanti Polar Lipids).

2.4 Single-Tethered DNA Curtain Assay for R-Loop

1. Nanoports (IDEX).
2. Double-sided tape.
3. Coverslips.
4. Lipid buffer: 20 mM Tris–HCl pH 8.0 and 100 mM NaCl.
5. BSA buffer: 40 mM Tris–HCl pH 8.0, 0.2 mg/ml bovine serum albumin (BSA), 50 mM NaCl, and 1 mM $MgCl_2$.
6. Streptavidin.

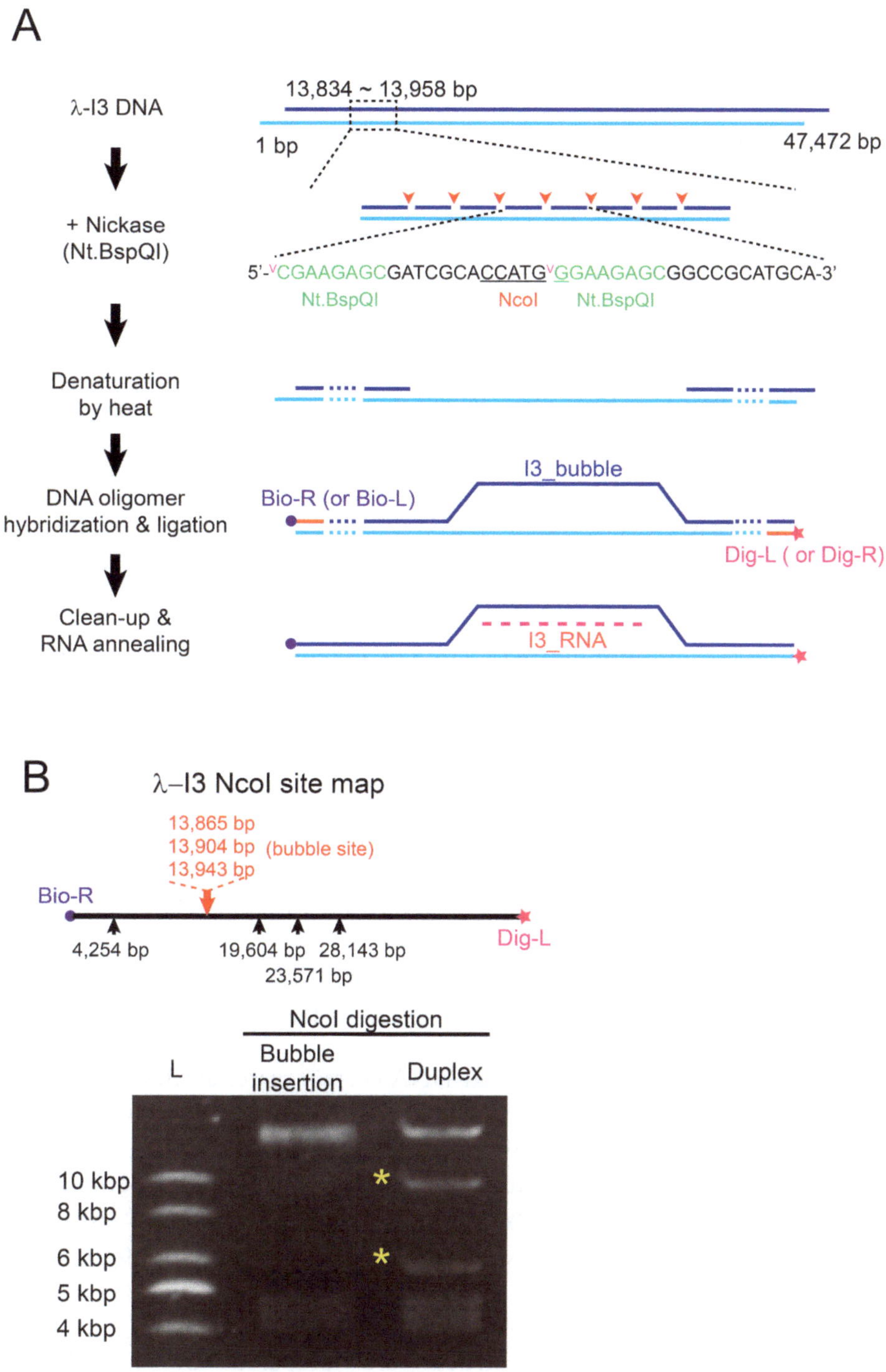

Fig. 2 Construction of λ-DNA containing R-loop. (**a**) A schematic of R-loop insertion into λ-DNA. Engineered λ-DNA (λ-I3), which contains three NcoI sites and seven nickase sites between 13,834 bp and 13,958 bp, is treated with Nt.BspQI nickase. The nicked fragments are denatured out of λ-I3 by heating 70 °C. An excessive amount of oligomers ("Bio-L/Dig-R or Bio-R/Dig-L" and I3_bubble) are hybridized and ligated to the λ-I3. After unligated oligomers are filtered out, Cy5-labeled RNA insert is hybridized. (**b**) Since the bubble structure inhibits the cleavage by restriction enzyme, the bubble insertion is ensured by the resistance to NcoI digestion. In the NcoI map of λ-I3, the three NcoI sites within the nickase sites (red) are not cleaved by NcoI when the bubble is properly inserted. The gel image shows that when the bubble is inserted, the NcoI sites within the nickase sites are not cleaved (yellow asterisks)

Table 1
List of oligomers for DNA curtain assays

Oligomer name	Sequences	Required steps
I3_Bubble	5′-[Phos]-TGC ATG CGG CCG CTC TTC CCA TGG TGC GAT TTT GTG GTT CCC ATA CCG TAT AAC CAT TTG GCT GTC CAA GCT CCG GGT GTT TGT TTT CCG CTC TTC CCA TGG TGC GAT CGC TCT TCG-3′	Subheading 3.1
I3_RNA	5′-[Cy5]-ACC CGG AGC UUG GAC AGC CAA AUG GUU AUA CGG UAU GGG- 3′	Subheading 3.1
Bio-R	5′-[Phos]-AGG TCG CCG CCC-[Biotin]-3′	Subheading 3.1
Bio-L	5′-[Phos]-GGG CGG CGA CCT-[Biotin]-3′	Subheading 3.1
Dig-R	5′-[Phos]-AGG TCG CCG CCC-[Dig]-3′	Subheading 3.1
Dig-L	5′-[Phos]-GGG CGG CGA CCT-[Dig]-3′	Subheading 3.1

[Phos] indicates phosphate labeling at 5′ of oligomer for ligation
[Cy5] indicates the labeling of Cy5 fluorescent dye

7. Biotinylated λ-DNA (λ-I3) containing R-loop (*see* Subheadings 2.1 and 3.1).
8. 1 μM anti-FLAG conjugated quantum dots (QDs) (SiteClick™ Qdot™ 565 antibody labeling kit, S10454, ThermoFisher Scientific) (*see* **Note 2**).
9. Purified 3× FLAG-TonEBP. The purification procedure of 3× FLAG-TonEBP follows the previous publication [4].
10. TonEBP buffer: 10 mM HEPES pH 7.5, 1 mM DTT, and 5% glycerol.
11. Imaging buffer: TonEBP buffer supplemented with 50 mM NaCl, 1.6% glucose, and 0.1× Gloxy (*see* **Note 3**).

2.5 Double-Tethered R-Loop DNA Curtain Assay

1. Single-tethered DNA curtain materials described in Subheading 2.4 are commonly used in the double-tethered curtain experiments.
2. A slide with a Cr diffusion barrier and pedestals, which is prepared by following the previous literature [4, 13].
3. Anti-digoxygenin Fab fragment (11333062910, Roche).

2.6 EMSA of TonEBP for R-Loop and Other DNA Constructs

1. Synthesized oligomers for EMSA (Table 2).
2. DNA constructs are prepared by hybridization (*see* Subheading 3.6).
3. Anti-DNA–RNA hybrid [S9.6] antibody (ENH002, Kerafast).
4. Reaction buffer: TonEBP buffer.
5. 5% non-denaturing PAGE gel.

Table 2
List of oligomers for EMSA

Oligomer name	Sequences	Required steps
DNA 1	5′-GCC GTC GCA TGA CGC TGC CGA ATT CTA CCA CGC GAT TCA TAC CTG TCG TGC CAG CTG CTT TGC CCA CCT GCA GGT TCA CCT CGT CCC TGG C-3′	Subheading 3.6
DNA 2	5′-[Cy3]-GCC AGG GAC GAG GTG AAC CTG CAG GTG GGC AAA GCA GCT GGC ACG ACA GGT ATG AAT CGC GTG GTA GAA TTC GGC AGC GTC ATG CGA CGG C-3′	Subheading 3.6
DNA 3	5′-[Cy3]-GCC AGG GAC GAG GTG AAC CTG CAG GTG GGC GGC TAC TAC TTA GAT GTC ATC CGA GGC TTA TTG GTA GAA TTC GGC AGC GTC ATG C GA CGG C-3′	Subheading 3.6
DNA 4	5′-GCA GCT GGC ACG ACA GGT ATG AAT C-3′	Subheading 3.6
RNA 1	5′-[Cy5]-GCA GCU GGC ACG ACA GGU AUG AAU C-3′	Subheading 3.6
RNA–DNA 1	5′-CCC ATA CCG TAT AAC CAT TTG GCT GTC CAA GCT CCG GGT-3′	Subheading 3.6
RNA–DNA 2	5′-[Cy5]-ACC CGG AGC UUG GAC AGC CAA AUG GUU AUA CGG UAU GGG-3′	Subheading 3.6
Fork 1	5′-[Cy3]-GCC AGG GAC GAG GTG AAC CTG CAG GTG GGC GGC TAC TAC TTT TTT TTT TTT TTT TTT TTT TTT TTT TTT TTT TTT TTT T-3′	Subheading 3.6
Fork 2	5′-TTT TTT TTT TTT TTT TTT TTT TTT TTT TTT TTT TTT TTT AGT AGT AGC CGC CCA CCT GCA GGT TCA CCT CGT CCC TGG C-3′	Subheading 3.6

[Cy3] indicates the labeling of Cy3 fluorescent dye

3 Methods

3.1 Preparation of R-Loop Containing λ-DNA

The following protocol describes the full steps for preparing λ-DNA containing R-loop for the DNA curtain assay (Fig. 2). This is a two-day procedure. In the first day, we form a bubble structure in λ-DNA. In the next day, after we test whether the bubble is correctly formed, we insert RNA into the bubble to make R-loop in λ-DNA.

1. Treat 0.1 mg/ml λ-I3 with 60 units of Nt.BspQI in 1X NEBuffer 3.1 at 50 °C for 1 h (total volume 100 μl).
2. Stop the nickase reaction by adding 1 μl of 20 mg/ml proteinase K and incubating at 50 °˜C for 1 h.
3. Add 1 μl of 100 μM Bio-R (or L) and 100 μM Dig-L (or R) for each and 2 μl of 100 μM I3_Bubble (Table 1).

4. Heat the mixture at 70 °C (*see* **Note 4**) for 15 min and slowly cool it down to room temperature (RT, 23 °C).
5. Add T4 DNA ligase and ATP up to 200 units and 1 mM, respectively.
6. Incubate at RT overnight (*see* **Note 5**).
7. The bubble site of λ-I3 has three NcoI recognition sequences (Fig. 2a). To test whether the bubble is properly formed, treat the bubble-inserted λ-I3 with 100 units NcoI at 37 °C for 1 h.
8. Run the digested fragments using 1% agarose gel electrophoresis. If the bubble is properly formed in λ-I3, it is not cleaved by NcoI (Fig. 2b).
9. If the bubble is correctly inserted, filter out the unligated oligomers using S-400 HR spin column.
10. Add I3_ RNA up to 2 μM to the filtered λ-I3 (*see* **Note 6**).
11. Heat the mixture at 65 °C for 15 min and slowly cool it down to RT to form the DNA-RNA hybrid at the bubble in λ-I3. This step should be done in the dark to avoid photobleaching of Cy5 on RNA.

3.2 Fabrication and Cleaning of Slides Containing Nano-Trench Patterns

The slides with nano-trench patterns are fabricated by following the previous protocol [12].

1. Two holes are drilled on each fused-silica slide.
2. Clean the fused-silica slides with piranha solution for 30 min (*see* **Note 7**).
3. Rinse the slides with D.I. and dry it with nitrogen gas.
4. Deposit 250 nm thick aluminum (Al) layer on each cleaned slide using DC sputter (SRM-120, Sorona) with 10 mTorr of argon gas.
5. Spin-coat 310 nm thick layer of 4% 950 K PMMA at 4000 rpm for 1 min.
6. Bake the PMMA layer at 180 °C on a hot plate for 3 min.
7. Draw the nano-trench patterns on the PMMA layer between two holes by electron beam (ebeam) lithography (NB3, Nanobeam Ltd) with 0.724 nA current at 80 kV.
8. Remove the ebeam-exposed PMMA areas by soaking the slides in developing solution for 2 min.
9. Rinse the slides with isopropanol.
10. Etch Al layers by inductively coupled plasma-reactive ion etching (ICP-RIE, FabStar, Top Technology Ltd.) using Cl_2 and BCl_3 gases, which remove Al of ebeam-exposed areas.
11. Further etch the slides using SF_4, CF_4, and O_2 gases to carve nano-trench patterns on slides.

12. Remove the remaining Al layer by soaking the slides in Al etchant (AZ-MIF300) for 10 min.
13. After rinsed with D.I., the slides are sonicated in acetone for 30 min, followed by cleaning in 2% Hellmanex III solution for at least 2 days with magnetic stirring.
14. The slides are sonicated in ethanol for 1 h and 1 M NaOH for 20 min successively. Finally rinse the slides with D.I.

3.3 Biotinylated Liposome Stock Making

Biotinylated liposome is prepared by following the previous protocol [15].

1. Rinse two 5 ml glass bottles with D.I., ethanol, and chloroform serially and completely dry the bottles using nitrogen gas.
2. Make 10× liposome stock (100 mg/ml DOPC, 8 mg/ml mPEG 2000-DOPE, and 0.5 mg/ml biotin-DPPE) in chloroform in one 5 ml glass bottle and store the 10× liposome stock at −20 °C.
3. Transfer 200 μl of 10× liposome stock into the other 5 ml glass bottle using a Hamilton syringe.
4. While tilting and rotating the glass bottle, mildly blow nitrogen gas into the bottle to evaporate chloroform and to form a thin layer of dried lipid on the bottle wall.
5. Place the bottle under vacuum for at least 2 days to completely remove the residual chloroform (*see* **Note 8**).
6. Add 2 ml of lipid buffer to the dried lipid and leave it at RT for at least 2 h. During this step, lipid molecules become hydrated.
7. Vortex the lipid solution for 3 min.
8. Sonicate the liposomes on ice using a tip sonicator (Q700, Qsonica): 15% output with 90 s ON pulse and 10 min OFF. This sonication needs to be repeated three times.
9. Filter the lipid solution using 0.22 μm syringe filter and store it at 4 °C until use.

3.4 Single-Tethered DNA Curtain Assay for R-Loop

All the following steps are performed at RT. The flowcell assembly protocol is modified from the previous literature [15].

1. Dry the nano-patterned slide prepared in Subheading 3.2 with nitrogen gas.
2. Put the slide on a clean table with the nano-patterns facing up.
3. On double-sided tape, put a clean and square paper that covers two holes and nano-patterns on the slide.
4. Excise out the paper-attached part from double-sided tape. The excised area will be a micro-channel.
5. Attach the double-sided tape to the slide while the excised part covers two holes and nano-patterns. The double-sided tape should not touch the nano-patterns.

6. Put a coverslip on top of the double-sided tape.
7. Rub the coverslip with a 1 ml pipette tip in order to tightly stick the slide and coverslip.
8. Sandwich the coverslip-attached slide with two cheap slide glasses and clip them.
9. Bake the clipped coverslip-attached slide assembly in 120 °C vacuum oven for 45 min. Under clipping, the double-sided tape slightly melts to more tightly stick the slide and coverslip.
10. Glue Nanoport on each hole of the slide using a glue-gun and then the assembled flowcell is stored in a vacuum chamber until use.
11. Two lines of luer-lock tubing are connected with a flowcell via Nanoports (Fig. 1c).
12. Wash the flowcell with 5 ml D.I. The solution injection into a flowcell is carried out using a luer-lock syringe. To avoid the injection of air bubbles, syringes should be linked with the flowcell by drop-by-drop connections.
13. Wash the flowcell with 3 ml of lipid buffer.
14. Add 40 μl of biotinylated lipid stock to 960 μl of lipid buffer.
15. Inject the diluted lipid solution into the flowcell in two shots. Each shot should be followed by 5 min of incubation. After this step, *no air bubbles are allowed in the flowcell.*
16. Wash the flowcell with 3 ml of lipid buffer and incubate for 20 min to promote lipid bilayer on the slide surface.
17. Inject 1 ml of BSA buffer. BSA will protect the slide surface that is not covered by the lipid bilayer.
18. Inject 0.02 mg/ml streptavidin in 1 ml of BSA buffer with two shots. Each shot should be followed by 5 min incubation.
19. Wash out unbound streptavidin with 3 ml of BSA buffer.
20. Dilute λ-I3 containing R-loop from Subheading 3.1 in 500 μl of BSA buffer up to 30 pM.
21. Inject the λ-DNA into the flowcell and incubate in the dark for 10 min.
22. While the λ-DNA is incubated in the flowcell, mix 200 nM 3× FLAG-TonEBP and 500 nM anti-FLAG conjugated QDs in 5 μl of TonEBP buffer and incubate on ice for 10 min.
23. Connect a syringe containing 10 ml of imaging buffer to a syringe pump in a microfluidic system (Fig. 1c).
24. Install the flowcell to the microfluidic system by drop-to-drop connections (Fig. 1c).

25. Place the flowcell on a prism-type TIRFM (Fig. 1c). The details of optical setup for a prism-type TIRFM are described in the previous literature [15].
26. Apply the flow and check if DNA curtain is well formed (*see* **Note 9**).
27. Inject 150 μl of 0.1–5 nM QD-labeled TonEBP into DNA curtain through the sample loop (100 μl) on the injection valve of the microfluidic system (Fig. 1c) (*see* **Note 10**).
28. After the QDs reach the DNA curtain, stop the flow and incubate for 10 min.
29. Resume the flow. When unbound QDs are washed out, take images via NIS-element software (Nikon). 488 nm and 637 nm lasers are used to image QD on TonEBP and Cy5 on RNA, respectively.
30. To make sure that Cy5-labeled R-loop or QD-labeled TonEBP is not stuck to the slide surface but binds to DNA, switch the flow on and off using the shut-off valve. When the flow is turned off, DNA molecules recoil out of the evanescent field and hence bound QDs (or proteins) disappear along with DNA (Fig. 3a, b).
31. Analyze the collected images using ImageJ (NIH). Since the R-loop is placed at a specific location on λ-DNA, Cy5-labeled R-loops are well aligned in the DNA curtain (Fig. 3a). The colocalization of TonEBP and R-loop in the DNA curtain demonstrates the preferential binding of TonEBP to R-loop. The bound positions of Cy5-labeled R-loops and QD-labeled TonEBP on λ-DNA are obtained by fitting them with a two-dimensional Gaussian function in ImageJ [15]. The position distribution manifests the colocalization of the R-loop and TonEBP (Fig. 3c).

3.5 Double-Tethered R-Loop DNA Curtain Assay

1. Assemble the flowcell with a double-tether patterned slide, which has a Cr nano-barrier and pedestal, as described from **steps 1** to **10**, Subheading 3.4.
2. Follow the same procedure as that from **steps 11** to **16**, Subheading 3.4.
3. Add 400 μl of 0.05 mg/ml anti-digoxygenin Fab fragments in lipid buffer to the flowcell and incubate for 40 min. Anti-digoxygenin Fab fragments will be adsorbed on the Cr pedestals.
4. Wash out unbound anti-digoxigenin with 1 ml of BSA buffer.
5. Inject 0.02 mg/ml streptavidin in 1 ml of BSA buffer with two shots. Each shot should be followed by 5 min of incubation.
6. Wash out the unbound streptavidin with 3 ml of BSA buffer.

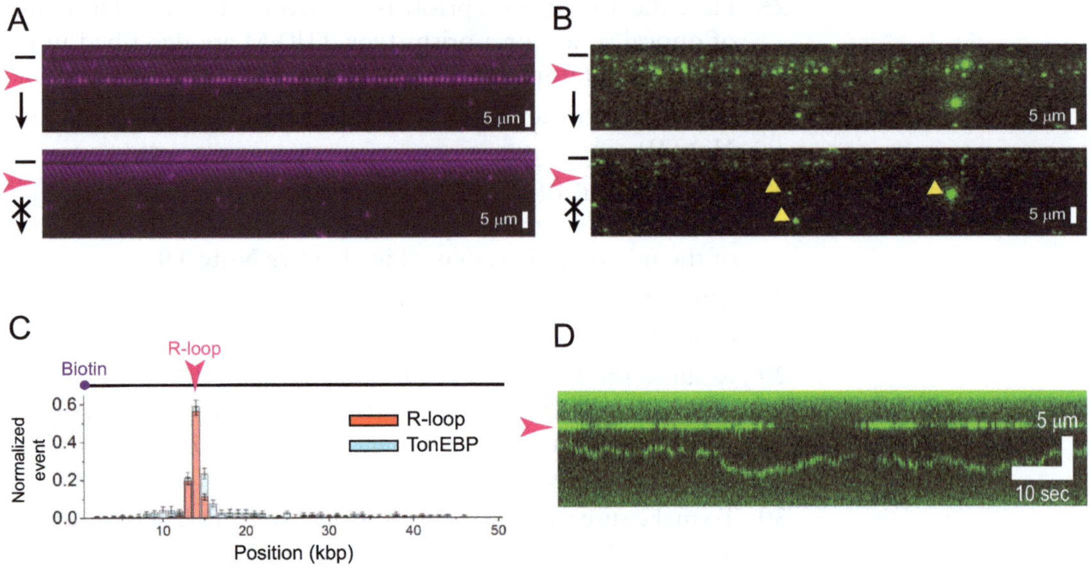

Fig. 3 DNA curtain images and analyses. (**a**) Cy5 image in the single-tethered DNA curtain with R-loop-containing λ-DNA. The magenta arrowhead indicates the location of the R-loop. The black line and arrow next to the image represent the nano-trench and the direction of buffer flow, respectively. Since RNA is labeled with Cy5, the line of Cy5 represents the R-loop in the presence of flow (top left). In the absence of flow, Cy5 signals disappear out of the evanescent field (bottom). (**b**) QD image in the single-tethered DNA curtain with R-loop containing DNA and QD-labeled TonEBP. The magenta arrowhead indicates the location of the R-loop. The black line and arrow next to the image represent the nano-trench and the direction of buffer flow, respectively. A line of QDs is shown in the presence of flow (top left). In the absence of flow, QDs disappear out of the evanescent field (bottom). The yellow triangles show nonspecific bindings of QDs to the surface. (**c**) Distribution histogram of QD and Cy5. The peaks of TonEBP and R-loop overlap well at the R-loop location, indicating that TonEBP is colocalized with the R-loop. The distribution is normalized by the total number (500 for each). The error bar is obtained from bootstrapping with a 70% confidence interval. (**d**) Representative kymograph of double-tethered DNA curtain. The magenta arrowhead indicates the location of the R-loop. One TonEBP binds to the R-loop and the other one diffuses along the DNA. The fluorescence signals intermittently turn off because of QD blinking, which is shown in single fluorophore and confirms that single QD-TonEBP is observed

7. Dilute λ-I3 containing the R-loop from Subheading 3.1 in 500 μl of BSA buffer up to 30 pM.
8. Inject the λ-DNA into the flowcell and incubate in the dark for 10 min.
9. While the λ-DNA is incubated in the flowcell, mix 200 nM 3× FLAG-TonEBP and 500 nM anti-FLAG conjugated QDs in 5 μl of TonEBP buffer on ice for 10 min.
10. Install the imaging buffer and flowcell to the microfluidic system and TIRFM as described from **steps 23** to **25** in Subheading 3.4 (Fig. 1c).

11. Continuously apply the buffer flow at 0.3 ml/min for 10–20 min to tether digoxigenin in λ-DNA to anti-digoxigenin coated on the pedestals. During this process, double-tethered lambda DNA is formed.
12. Inject 150 μl of 0.1–5 nM QD-labeled TonEBP into the DNA curtain through the sample loop (100 μl) connected with the injection valve of the microfluidic system (Fig. 1c).
13. Immediately after QDs reach the DNA curtain, stop the flow and collect images via NIS-element software (Nikon) without incubation. 488 nm and 637 nm lasers are used to image QD on TonEBP and Cy5 on RNA, respectively.
14. Analyze the data using ImageJ. Firstly, make kymographs for TonEBP movement on the individual DNA molecules (Fig. 3d). Then perform the analysis for diffusion by single particle tracking (MOSAICsuite ImageJ plug-in) [4, 8].

3.6 EMSA for TonEBP with R-Loop and DNA Constructs

1. For R-loop and other DNA constructs, appropriate oligomers are mixed in lipid buffer (total 20 μl) as shown in Table 3.
2. Place the mixture in the 95 °C heating block and incubate for 5 min.
3. Slowly cool it down to RT in the dark and store at −20 °C until use.
4. To test if the R-loop is properly formed, we observe the super-shift of the R-loop construct using S9.6, which is an R-loop-specific antibody. To this end, mix 30 nM R-loop (or 30 nM duplex DNA as a negative control) and 600 nM S9.6 antibody in TonEBP buffer.
5. Incubate the mixture at RT in the dark for 20 min.
6. Run 5% non-denaturing polyacrylamide gel electrophoresis (PAGE) in 0.5× TBE at 140 V and 4 °C in the dark for 30 min. Mix 4 μl of 6× orange G dye with 20 μl of reactant and then load 15 μl out of 24 μl.
7. Image fluorescence signal of the PAGE gel using Typhoon RGB (GE Healthcare) (Fig. 4a).
8. If R-loop is properly formed, mix 30 nM R-loop (or other construct) with purified TonEBP at different concentrations (0, 60, 120, 240, and 300 nM) in TonEBP buffer (total volume of each reaction is 20 μl) (Fig. 4b).
9. Incubate each mixture at RT in the dark for 1 h.
10. Run 5% non-denaturing PAGE in 0.5× TBE at 140 V and 4 °C in the dark for 30 min. Mix 4 μl of 6× orange G dye with 20 μl of reactant and then load 15 μl out of 24 μl.

Table 3
Hybridization of oligomers for EMSA

Substrate name	Mixture recipe (total 20 μl in lipid buffer)
Homoduplex	DNA 1: 6 μM, DNA 2: 5 μM
Bubble DNA	DNA 1: 6 μM, DNA 3: 5 μM
R-Loop	DNA 1: 6 μM, DNA 3: 5 μM, RNA 1: 5 μM
D-Loop	DNA 1: 6 μM, DNA 3: 5 μM, DNA 4: 5 μM
Fork	Fork 1: 5 μM, Fork 2: 6 μM
DNA–RNA hybrid	RNA-DNA 1: 6 μM, RNA-DNA 2: 5 μM

11. Image fluorescence signal using Typhoon RGB (GE Healthcare) (Fig. 4c).
12. Quantify the bound fraction based on band intensity using ImageJ (Fig. 4d). The bound fraction is calculated by dividing the bound band intensity by the sum of the bound and unbound band intensities.

4 Notes

1. Since RNA is more fragile than DNA and easily degraded by RNases, handling RNA must be done carefully. To avoid the degradation of RNA, DEPC-treated D.I. and agents for eliminating RNase contamination are strongly recommended.
2. To conjugate the antibody (anti-FLAG) to QD, we follow the protocol of the SiteClick™ Qdot™ antibody labeling kit. QD with 565 nm emission is selected to be distinguished from the Cy5 emission wavelength. A different QD emission wavelength may be chosen if needed.
3. 100× Gloxy is prepared as described in Cold Spring Harbor protocols.
4. Heating at 70 °C is strongly recommended for the heat inactivation of proteins and annealing of oligomers because 95 °C increases the nonspecific basepairing of oligomers with λ-I3.
5. Ligation with T4 DNA ligase is optimized at 16 °C. But the ligase is still active at RT, which may reduce nonspecific ligation.
6. It is strongly recommended that fresh λ-DNA containing R-loop is prepared for each DNA curtain experiment. Anneal the RNA with bubble-containing λ-I3 one day before the DNA curtain experiment.

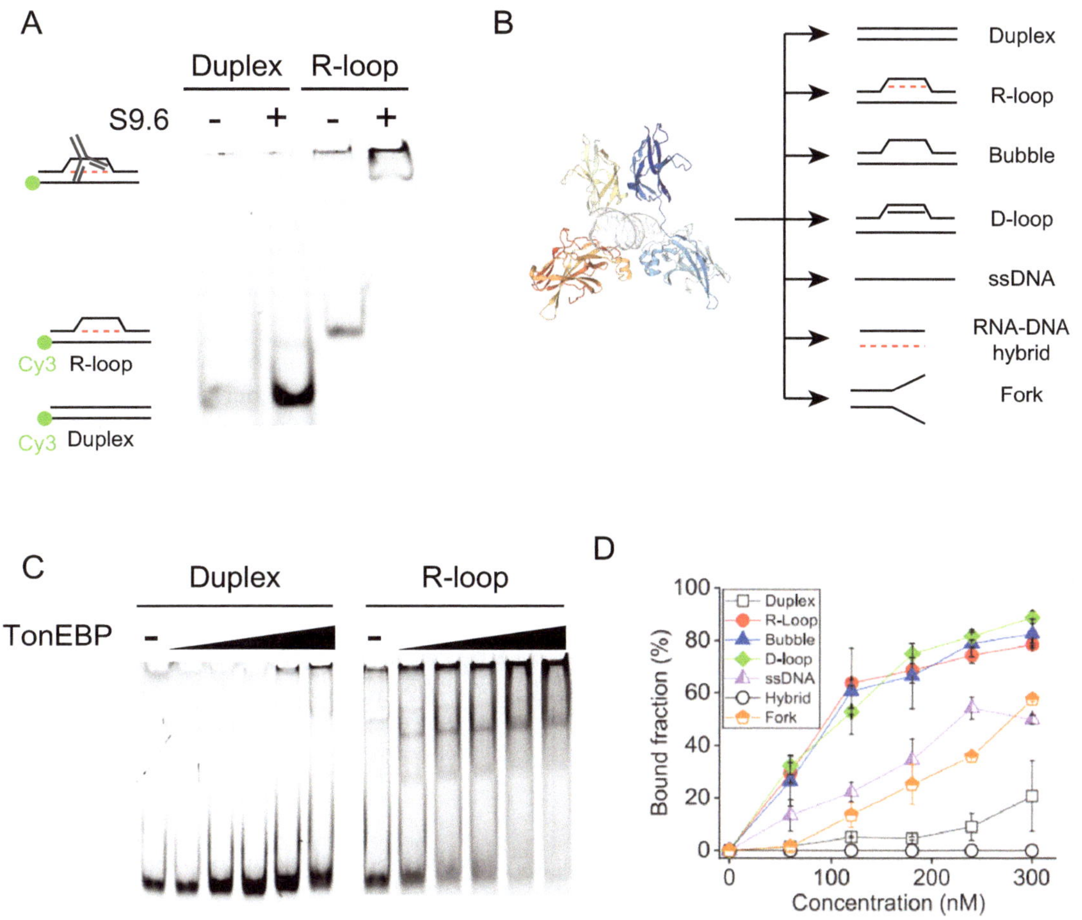

Fig. 4 Electrophoretic mobility shift assay (EMSA) with R-loop and other DNA constructs. (**a**) EMSA to confirm the R-loop formation. The R-loop band is delayed from duplex because it has three strands (first and third lanes). The R-loop is supershifted in the presence of the S9.6 antibody because it preferentially binds to the DNA–RNA hybrid whereas it does not bind to the duplex (second and fourth lane). The gel image is adopted from ref. 8. (**b**) Various DNA constructs tested for TonEBP binding affinity. The structure of TonEBP is adopted from PDB (1IMH) [9]. (**c**) EMSA gel images for duplex and R-loop depending on the concentration of TonEBP. Band shift for the R-loop occurs at lower TonEBP concentrations, indicating higher binding affinity to the R-loop than to the duplex. The gel image is adopted from ref. 8. (**d**) Quantified data for EMSA. TonEBP preferentially binds to the R-loop, bubble, and D-loop, whereas it does not bind to the DNA-RNA hybrid, suggesting that TonEBP identifies the displaced ssDNA in the R-loop. These data are adopted from ref. 8

7. Piranha solution is highly dangerous because it is a mixture of strong acid and oxidizer. This cleaning step should be done in a chemical or acid hood with proper protection. In addition, the solution must be handled very carefully.
8. Complete removal of residual chloroform is very important for lipid fluidity in DNA curtain.
9. DNA molecules in the DNA curtain assay can be seen by YOYO-1 staining. To check if the DNA curtain is well formed,

flow can be momentarily switched off using the shut-off valve. When the flow is turned off, DNA molecules recoil out of the evanescent field. Since YOYO-1 is an intercalation dye, it may disturb the protein-DNA interactions. So, a low concentration of YOYO-1 (50 pM or less) is used in the imaging buffer. There are two ways to exclude the YOYO-1 effect. The first way is post-staining with YOYO-1 after protein-DNA interaction experiments are finished. The second one is washing out YOYO-1 with a high salt buffer (20 mM Tris–HCl pH 8.0, 1 M $MgCl_2$ and 300 mM NaCl).

10. The flow rate should be properly set according to experiments. We usually use 0.1–0.8 ml/min. A higher flow rate stretches the DNA more and reduces thermal fluctuation of DNA molecules, whereas it exerts higher hydrodynamic force on QD-labeled proteins.

Acknowledgments

This work is supported by National Research Foundation Grant NRF-2020R1A2B5B01001792 and the Institute for Basic Science (IBS-R022-D1).

References

1. Aguilera A, Garcia-Muse T (2012) R loops: from transcription byproducts to threats to genome stability. Mol Cell 46(2):115–124. https://doi.org/10.1016/j.molcel.2012.04.009
2. Allison DF, Wang GG (2019) R-loops: formation, function, and relevance to cell stress. Cell Stress 3(2):38–46. https://doi.org/10.15698/cst2019.02.175
3. Ginno PA, Lott PL, Christensen HC, Korf I, Chedin F (2012) R-loop formation is a distinctive characteristic of unmethylated human CpG island promoters. Mol Cell 45(6):814–825. https://doi.org/10.1016/j.molcel.2012.01.017
4. Cheon NY, Kim HS, Yeo JE, Scharer OD, Lee JY (2019) Single-molecule visualization reveals the damage search mechanism for the human NER protein XPC-RAD23B. Nucleic Acids Res 47(16):8337–8347. https://doi.org/10.1093/nar/gkz629
5. Sollier J, Cimprich KA (2015) Breaking bad: R-loops and genome integrity. Trends Cell Biol 25(9):514–522. https://doi.org/10.1016/j.tcb.2015.05.003
6. Groh M, Gromak N (2014) Out of balance: R-loops in human disease. PLoS Genet 10(9): e1004630. https://doi.org/10.1371/journal.pgen.1004630
7. Choi SY, Lee-Kwon W, Kwon HM (2020) The evolving role of TonEBP as an immunometabolic stress protein. Nat Rev Nephrol 16(6): 352–364. https://doi.org/10.1038/s41581-020-0261-1
8. Kang HJ, Cheon NY, Park H, Jeong GW, Ye BJ, Yoo EJ, Lee JH, Hur JH, Lee EA, Kim H, Lee KY, Choi SY, Lee-Kwon W, Myung K, Lee JY, Kwon HM (2020) TonEBP recognizes R-loops and initiates m6A RNA methylation for R-loop resolution. Nucleic Acids Res 49(1):269–284. https://doi.org/10.1093/nar/gkaa1162
9. Stroud JC, Lopez-Rodriguez C, Rao A, Chen L (2002) Structure of a TonEBP-DNA complex reveals DNA encircled by a transcription factor. Nat Struct Biol 9(2):90–94. https://doi.org/10.1038/nsb749
10. Cha J, Lee JY (2021) A novel high-throughput single-molecule technique: DNA curtain. J Korean Phys Soc 78:442–448. https://doi.org/10.1007/s40042-020-00031-9
11. Collins BE, Ye LF, Duzdevich D, Greene EC (2014) DNA curtains: novel tools for imaging protein-nucleic acid interactions at the

single-molecule level. Methods Cell Biol 123: 217–234. https://doi.org/10.1016/B978-0-12-420138-5.00012-4

12. Kang Y, Cheon NY, Cha J, Kim A, Kim HI, Lee L, Kim KO, Jo K, Lee JY (2020) High-throughput single-molecule imaging system using nanofabricated trenches and fluorescent DNA-binding proteins. Biotechnol Bioeng 117(6):1640–1648. https://doi.org/10.1002/bit.27331
13. Lee JY, Finkelstein IJ, Crozat E, Sherratt DJ, Greene EC (2012) Single-molecule imaging of DNA curtains reveals mechanisms of KOPS sequence targeting by the DNA translocase FtsK. Proc Natl Acad Sci U S A 109(17): 6531–6536. https://doi.org/10.1073/pnas.1201613109
14. Kim Y, de la Torre A, Leal AA, Finkelstein IJ (2017) Efficient modification of lambda-DNA substrates for single-molecule studies. Sci Rep 7(1):2071. https://doi.org/10.1038/s41598-017-01984-x
15. Lee JY, Greene EC (2011) Assembly of recombinant nucleosomes on nanofabricated DNA curtains for single-molecule imaging. Methods Mol Biol 778:243–258. https://doi.org/10.1007/978-1-61779-261-8_16

Chapter 17

Immunostaining and Protein Pull-Down of Methyl-5-Cytosine-Marked RNA:DNA Hybrids

Haibo Yang and Li Lan

Abstract

RNA:DNA hybrids not only function in various physiological cellular processes but also represent a threat to genome integrity. The methyl-5-cytosine (m^5C) marked RNA:DNA hybrids containing the m^5C modified RNA strand add another layer of regulation to the R-loop structure. Here we describe the detection of m^5C modifications in the context of RNA:DNA hybrids using immunostaining, and identification of specific binding partners of m^5C marked RNA:DNA hybrids using pull-down method.

Key words m^5C, RNA:DNA hybrid, R-loop, Immunostaining, Biotin-pull down

1 Introduction

R-loops are three-stranded structures consisting of RNA:DNA hybrids and the displaced single-strand DNA. Physiological R-loops are required for cell proliferation, meanwhile, R-loops also play pathological roles under certain circumstances. Unscheduled R-loops interfere with transcription and DNA replication representing a threat to genome stability [1, 2]. The role of R-loops in DNA damage repair has been recently reported. Our lab recently found that the ROS damage at the actively transcribed regions induces R-loop formation [3]. With the recruitment of TRDMT1, the RNA m^5C methyltransferase, to the damage sites, RNA m^5C is induced by TRDMT1 in the context of RNA:DNA hybrids [4].

To study m^5C marked RNA:DNA hybrids, we adapted the previously established DNA damage at RNA transcribed sites (DART) system. In the DART system, KillerRed (KR), a light excitable chromophore, is used to release ROS upon light activation in a dose-dependent manner. Firstly, a tetracycline responsive elements—cassette was integrated at the X-chromosome of U2OS cells to generate the U2OS TRE cells [5]. Then the KR was fused to

Andrés Aguilera and Alexey Ruzov (eds.), *R-Loops: Methods and Protocols*, Methods in Molecular Biology, vol. 2528,
https://doi.org/10.1007/978-1-0716-2477-7_17,

a tet-repressor (tetR-KR, transcription off) or a transcription activator (TA-KR, transcription on) [6]. TA-KR or TetR-KR binds to the TRE cassette and induces ROS damage upon light exposure activation of KR, with transcription on or off. The ROS damage at the TA-KR damage sites will induce the m^5C marked RNA:DNA hybrids. RAD52 working as the RNA m^5C-reader is recruited to the damaged sites facilitating the RAD51 loading and promoting the RNA-dependent recombination repair [4].

2 Materials

2.1 Immunostaining of m^5C Marked RNA: DNA Hybrids

1. U2OS TRE cells.
2. pBROAD3/tetR-KR and pBROAD3/TA-KR plasmids.
3. Lipofectamine 2000.
4. 1× PBS.
5. 4% PFA.
6. 0.2% Triton in 1× PBS.
7. TE buffer:10 mM Tris–HCl, 2 mM EDTA, pH 9.0.
8. 0.1% PBST: 0.1% tween in 1× PBS.
9. 5% BSA: 5% BSA in 0.1% PBST.
10. Primary antibodies: S9.6, Cat# ENH001, Kerafast. m5C, Cat# ab10805, Abcam. KR, Cat# Ab961, Evrogen.
11. Secondary antibodies: Alexa Fluor 488 goat anti-mouse IgG; Alexa Fluor 594 goat anti-rabbit IgG.
12. 35 mm glass-bottom culture dishes (MatTek, catalog number: P35GC-1.5-14-C).
13. A stage UVP (Upland, CA) with 15 watts cool white fluorescent bulb (OSRAM SYLVANIA).
14. Heat block.
15. Olympus FV1000 confocal microscopy system (OLYMPUS, model: FV1000- FILTER DETECT; SYS.PACKAGE: IX81-1 405/M_AR/559/635N) with 488, 559 nm lasers.

2.2 Biotin Pull-Down of m^5C Marked RNA: DNA Hybrids

1. 30 nt oligos.
2. Annealing buffer: 10 mM Tris pH 7.5, 50 mM NaCl, 1 mM EDTA.
3. Flp-in 293 cells.
4. 1× PBS.
5. Lysis buffer: 10 mM Hepes pH 7.6, 50 mM Tris–HCl pH 7.5, 150 mM NaCl, 1% NP-40, 3 mM EDTA. Add protease inhibitors and 1 mM PMSF before use.

6. Pierce Nucleic-Acid Compatible Streptavidin Magnetic Beads.
7. 1× RNA Capture Buffer, 20 mM Tris pH 7.5, 1 M NaCl, 1 mM EDTA.
8. 20 mM Tris pH 7.5.
9. Protein-RNA Binding Buffer (10×): 0.2 M Tris (pH 7.5), 0.5 M NaCl, 20 mM $MgCl_2$, 1% Tween™-20 Detergent.
10. Wash Buffer (1×): 20 mM Tris pH 7.5, 10 mM NaCl, 0.1% Tween-20 Detergent.
11. 4× Western SDS loading buffer: 200 mM Tris–HCl pH 6.5, 400 mM DTT, 277 mM SDS, 6 mM Bromophenol blue, 4.3 M Glycerol.
12. Heat block.
13. Cell collector.
14. Magnetic stand.

3 Methods

3.1 Immunostaining of m^{5}C Marked RNA: DNA Hybrids

1. Seed 2.5 × 10^5 U2OS TRE cells per one 35 mm glass-bottom dish in 2 mL DMEM with 10% FBS, and culture for 24 h in the incubator at 37 °C with 5% CO_2 (*see* **Note 1**).
2. Transfect 1 μg of pBROAD3/tetR-KR or pBROAD3/TA-KR following the transfection protocol of Lipofectamine 2000. Culture for 36 h in the incubator at 37 °C with 5% CO_2.
3. Activation of KR is performed by exposing cells to a 15-watt SYLVANIA cool white fluorescent bulb for 25 min in a stage UVP (*see* **Note 2**).
4. After recovery for 1 h in the incubator, take out the dishes from the incubator and then remove the medium from the dishes. After washing once with 1× PBS, fix the cells by adding 1 mL of 4% PFA and incubate for 15 min at room temperature (*see* **Note 3**).
5. After washing three times with 1× PBS, permeabilize the cells by adding 1 mL of 0.2% Triton and incubate for 8 min at room temperature.
6. After washing three times with 1× PBS, add 1.5 mL of TE buffer (pH 9.0) and incubate the cells on a heat block at 95 °C for 20 min (*see* **Notes 4** and **5**).
7. After the heat treatment, allow the dishes to cool down for 10 min and then wash three times with 1× PBS.
8. Block the cells for 1 h at room temperature with adding 200 μL 5% BSA in 0.1% PBST.

9. After blocking, incubate the cells with 100 μL of primary antibodies [S9.6 or m^5C antibody (at 1:200 dilution) co-stained with KR antibody (at 1:4000 dilution)] overnight at 4 °C (*see* **Note 6**).
10. Wash the cells with 1.5 mL of 0.5% PBST for 5 min with agitation at room temperature. Repeat the wash three times.
11. Incubated the cells with 100 μL secondary antibodies (1:500 dilution for s9.6 or m^5C corresponding secondary antibody, 1: 4000 for KR secondary antibody) 1 h at room temperature.
12. Wash the cells with 1.5 mL 0.5% PBST for 5 min with agitation at room temperature. Repeat the wash three times.
13. Acquire the images using the Olympus FV1000 confocal microscopy system and FV1000 software.

3.2 Biotin Pull-Down of m^5C Marked RNA: DNA Hybrids

1. Biotin-labeled 30 nt RNA oligos with or without m^5C, and complementary DNA oligo were synthesized by IDT (*see* **Note 7**).
2. The oligos are suspended in the annealing buffer to a final concentration of 200 μM.
3. Mix 5 μL complementary DNA oligo with 5 μL RNA oligo or 5 μL m^5C modified RNA oligo in the microtube, and add 40 μL of annealing buffer to make a reaction system with a final volume of 50 μL.
4. Put the microtubes in the heat block at 95 °C for 5 min.
5. Allow the heat blocks with microtubes inside to cool down slowly to room temperature. Then put the annealed hybrids at 4 °C.
6. Split 1/4 of the 100% confluency Flp-in 293 cells into one 10 cm dish and culture for 2 days (*see* **Note 8**).
7. Wash the cells in the 10 cm dish with cold PBS once and drain the PBS (*see* **Note 9**).
8. Add 1 mL lysis buffer to the dish and use a cell collector to scrape the cells. Put the cells in a pre-cold microcentrifuge tube.
9. Incubate the microcentrifuge tube on ice for 30 min and then centrifuge for 15 min at 15,000 × *g*. at 4 °C. Aliquot the supernatant 100 μL per microtube and put it at −70 °C.
10. Add 50 μL of streptavidin magnetic beads to a microcentrifuge tube (*see* **Notes 10** and **11**).
11. Put the tube into a magnetic stand and remove the supernatant. After washing the beads three times with 50 μL of 20 mM Tris pH 7.5, resuspend the beads in 50 μL of 1× RNA capture buffer.

12. Add 50 pmol of the hybrids to the beads and incubate for 20 min at room temperature with agitation.
13. Put the tube into a magnetic stand and remove the supernatant. After washing the beads three times with 50 μL 20 mM Tris pH 7.5, resuspend the beads in 100 μL of 1× protein–RNA binding buffer

    Prepare a Master Mix of RNA-Protein Binding Reaction: 10 μL 10× Protein–RNA Binding Buffer, 5 μL 50% glycerol, 10–30 μL of cell lysate (protein conc. >2 mg/mL), nuclease-free water to 100 μL.
14. Put the tube into a magnetic stand and remove the supernatant. Add 100 μL of Master Mix to the RNA-bound beads and incubate 60 min at 4 °C with agitation.
15. Put the tube into a magnetic stand and transfer the supernatant to a microcentrifuge tube for later analysis. After washing beads three times with 1× Wash buffer, add 1× Western SDS loading buffer to the beads and put the tube in the heat block at 95 °C for 5 min.
16. Perform the Western blot analysis for the supernatant and pull-down fractions (*see* **Note 12**).

4 Notes

1. If specific protein needs to be checked on how it affects the RNA m^5C or RNA:DNA hybrids, the protein KO U2OS TRE cells or cells with knocking down or overexpression of a particular protein could be used.
2. Activation of KR is done by exposing cells to a 15 watt cool white fluorescent bulb. Cells are placed above a water bottle to prevent an increase in temperature. Ensure that the distance between the cells and the bulb is 15 cm, to achieve the desired amount of DNA damage.
3. Different recovery times can be applied to check the kinetics of damage-induced RNA m^5C or RNA:DNA hybrids.
4. When the cells are heated in the TE buffer on a heat block, the glass-bottom dish should be covered by the lid to prevent the evaporation of the TE buffer.
5. The heat treatment of the cells on a heat block is adapted from the heat-induced antigen retrieval method for PFA-fixed tissues using Tris-EDTA buffer.
6. After heat treatment, the fluorescent intensity of KR decreases a lot. Therefore, immunofluorescence staining with KR antibodies would make the KR foci brighter.

7. RNA oligos with different sequences or lengths could be applied.
8. If the protein level is low or no proper antibody is available, the protein could be expressed by transfection of plasmids.
9. Washing the Flp-in 293 cells with cold PBS must be gentle, as the cells are very easy to be floated.
10. Before taking the magnetic beads from the original bottle to the microcentrifuge tube, the beads should be mixed thoroughly.
11. The Methods from 2.10 to 2.16 are adapted from the protocol Pierce™ Magnetic RNA-Protein Pull-Down Kit.
12. The pull-down fraction could also be analyzed by mass spectrometry.

References

1. Crossley MP, Bocek M, Cimprich KA (2019) R-loops as cellular regulators and genomic threats. Mol Cell 73(3):398–411. https://doi.org/10.1016/j.molcel.2019.01.024
2. Garcia-Muse T, Aguilera A (2019) R loops: from physiological to pathological roles. Cell 179(3): 604–618. https://doi.org/10.1016/j.cell.2019.08.055
3. Teng Y, Yadav T, Duan M, Tan J, Xiang Y, Gao B et al (2018) ROS-induced R loops trigger a transcription-coupled but BRCA1/2-independent homologous recombination pathway through CSB. Nat Commun 9(1):4115. https://doi.org/10.1038/s41467-018-06586-3
4. Chen H, Yang H, Zhu X, Yadav T, Ouyang J, Truesdell SS et al (2020) M(5)C modification of mRNA serves a DNA damage code to promote homologous recombination. Nat Commun 11(1):2834. https://doi.org/10.1038/s41467-020-16722-7
5. Lan L, Ui A, Nakajima S, Hatakeyama K, Hoshi M, Watanabe R et al (2010) The ACF1 complex is required for DNA double-strand break repair in human cells. Mol Cell 40(6): 976–987. https://doi.org/10.1016/j.molcel.2010.12.003
6. Lan L, Nakajima S, Wei L, Sun L, Hsieh CL, Sobol RW et al (2014) Novel method for site-specific induction of oxidative DNA damage reveals differences in recruitment of repair proteins to heterochromatin and euchromatin. Nucleic Acids Res 42(4):2330–2345. https://doi.org/10.1093/nar/gkt1233

Chapter 18

In Vitro Binding of GADD45A to RNA:DNA Hybrids

Khelifa Arab and Christof Niehrs

Abstract

R-loops are three-stranded RNA:DNA hybrid structures that frequently form during transcription. While R-loop misregulation is associated with genome instability, cells also harness RNA–DNA hybrids in scheduled, "regulatory," R-loops to control gene expression. One regulatory role involves epigenetic gene regulation by the R-loop reader Growth Arrest and DNA Damage 45A (GADD45A). This small stress related protein promotes DNA demethylation by recruiting TET dioxygenase and Thymine DNA glycosylase to specific genomic loci. GADD45A requires adapters for its genomic localization. One such class of adapters are R-loops formed at certain CpG island promoters to which GADD45A binds directly, targets the demethylation machinery, and confers an open chromatin state. Here, we describe protocols for carrying out in vitro binding assays with GADD45A to RNA:DNA hybrids to biochemically study its direct binding to R-loops, specifically GADD45A pulldown and EMSA (electrophoretic mobility shift) assays.

Key words R-loops, RNA:DNA hybrids, GADD45A, EMSA

1 Introduction

R-loops are widespread in the genome and if not properly controlled, they can lead to genomic instability [1–3]. R-loops at promoters can interfere with transcription initiation, elongation, or termination [4–6]. However, cells exploit these very same features in a controlled manner to regulate gene expression in "regulatory R-loops" [7]. What structurally distinguishes both classes of R-loops, unscheduled and scheduled, is unclear. However, notably R-loops that form by natural antisense RNAs can epigenetically regulate transcription either positively [8, 9] or negatively [10–12]. If R-loops function as epigenetic marks, what are their "readers"? One such epigenetic R-loop reader is Growth Arrest and DNA Damage 45A (GADD45A), member of a small gene family that is conserved between flies and mammals and in humans contains three paralogs (alpha, beta, and gamma) [13]. *GADD45A* is a stress-response gene and the most studied among the family. It is

Andrés Aguilera and Alexey Ruzov (eds.), *R-Loops: Methods and Protocols*, Methods in Molecular Biology, vol. 2528,
https://doi.org/10.1007/978-1-0716-2477-7_18,

involved in a plethora of cellular pathways, including gene regulation, DNA damage response, cell cycle arrest, and apoptosis [14].

Studying *TCF21* gene activation and chromatin regulation, we showed that GADD45A binds R-loops in vitro and in vivo. The binding of GADD45A to R-loops represents the initial step for the recruitment of a regulatory complex, which involves TET1 (Ten eleven dioxygenase) and TDG (Thymidine DNA glycosylase) to demethylate CpG-rich promoters and promote a permissive chromatin state [13]. Here we detail two techniques, GADD45A pull-down and EMSA, which are both suitable to monitor the interaction between GADD45A and R-loops and relative affinities towards different types of nucleic acids.

In both methods, the principle relies on labeled nucleic acids and immunopurified FLAG-tagged GADD45A from mammalian cells.

2 Materials

2.1 General Reagents

1. Deionized water for buffers.
2. Autoclaved water for reactions (*see* **Note 1**).
3. HEK293 cells.
4. Soak the glass plates, apparatus, combs, and any reusable material in 3% H_2O_2 before usage for overnight.

2.2 Cell Culture and Related Equipment

1. Cell line HEK293.
2. Cell culture incubator at 5% CO_2.
3. Centrifuges for 1.5 mL and 15 mL tubes.
4. Lipofectamine 2000 (Thermo-scientific).
5. Purified plasmid DNAs (pCS2 + FLAG-tagged GADD45A, pcDNA3-FLAG-PTB, and pEGFP-N1-FLAG).
6. Diagenode Biorupter.
7. UVP UV-crosslinker.

2.3 Agarose Gel Electrophoresis

1. Agarose gel system with gel-sealing tape.
2. Agarose.
3. Glycerol.
4. 0.5× Tris-Borate buffer: 45 mM Tris base, 45 mM boric acid, pH 8.2–8.4.
5. 10× Tris-Borate EDTA (TBE): 900 mM Tris base, 900 mM boric acid, 20 mM EDTA, pH 8.0.
6. Precast gels Criterion TGX Stain-Free 12% (Biorad).
7. SDS-PAGE running buffer: 0.025 M Tris–HCl, pH 8.3, 0.192 M glycine, 0.1% SDS.

8. Tris-buffered saline (TBS; 10×): 1.5 M NaCl, 0.1 M Tris–HCl, pH 7.4.
9. Mini PROTEAN® 3 System glass plates (Biorad).
10. Western blot loading buffer (5×): 0.3 M Tris–HCl (pH 6.8), 10% SDS, 25% β-mercaptoethanol, 0.1% bromophenol blue, 50% glycerol.
11. Ammonium persulfate 10% solution in water.
12. Acrylamide 30% and Bis Acrylamid 2%.
13. *N*, *N*, *N*, *N'*-Tetramethyl-ethylenediamine (TEMED).
14. Saline-Sodium Citrate (SSC) 20× buffer 3 M NaCl, 0.3 M sodium citrate pH 7.0.

2.4 Reagents for the Binding Reactions

1. RNA and DNA Oligonucleotides.
2. [γ-^{32}P]-ATP (PerkinElmer).
3. Sephadex G-25 columns (Merck).
4. T4 Polynucleotide Kinase.
5. Benzonase.
6. DNAse I.
7. Purified protein of interest (GADD45A, PTB or EGFP).
8. Oligonucleotide dilution buffer: 10 mM Tris–HCl, pH 7.4.
9. Oligonucleotides annealing buffer [10 mM Tris–HCl (pH 7.4), 0.1 mM EDTA].
10. Cell lysis buffer [50 mM Tris–HCl (pH 7.9), 500 mM NaCl, 5 mM $MgCl_2$, 5 mM EDTA, 0.5% Triton X-100, 0.5 mM DTT].
11. Bovine serum albumin (BSA) 10% in water.
12. Washing buffer [20 mM Tris–HCl (pH 7.9), 500 mM KCl, 5 mM $MgCl_2$, 0.5 mM DTT, 5 mM EDTA].
13. FLAG-Protein elution buffer [20 mM Tris–HCl (pH 7.9), 300 mM KCl, 5 mM $MgCl_2$, 0.5 mM DTT, 0.1 mM EDTA, 20% glycerol containing 0.05% NP-40 and 20 μg of 3× FLAG® peptide. Anti-FLAG-Agarose beads (Sigma Aldrich, ANTI-FLAG® M2, #A2220).
14. Coomasie staining kit (Thermo-Fisher Scientific #LC6025).
15. Pulldown binding buffer [50 mM Tris–HCl (pH 7.9), 10% glycerol, 100 mM KCl, 5 mM $MgCl_2$, 10 mM β-mercaptoethanol, 0.1% NP-40].
16. EMSA binding buffer [20 mM Tris–HCl (pH 7.6), 100 mM NaCl, 1 mM EDTA, 1 mM DTT, 10% glycerol, 100 μg/mL BSA].

2.5 Quantification

1. Phosphorimaging system.
2. Gel Doc XR+ Gel System.
3. Imaging software capable of quantitation (ImageJ, or similar).
4. Software capable of curve fitting (GraphPad Prism, or similar).

3 Methods

Radio-labeled RNA is used to generate R-loops and RNA:DNA hybrids. In EMSAs, radiolabeled single stranded RNA or DNA and radiolabeled double stranded RNA or DNA were used as controls. In the following examples, we used an R-loop sequence derived from the *TCF21* promoter as well as a generic R-loop whose sequence corresponds to plasmid linker DNA.

3.1 Labeling of RNA Oligonucleotides

1. RNA and DNA oligonucleotides were ordered from Sigma or IDT as HPLC-purified grade per 100 nmole oligos for each. The oligo sequences were:
   (a) For *TCF21* region: DNAsense_ TCCTAGTGTCCAC CAAATTCCTCAGCGCTCGCTCACCCTCCTC TACGGCCACGACTCTGGGAGTG ; DNAantisense_ CACTCCCAGAGTCGTGGCCGTAGAGGAGGGT GAGCGAGCGCTGAGGAATTTGGTGGACAC TAGGA;
   RNA-antisense_GAGGGUGAGCGAGCG.
   (b) For a generic region:
   DNAsense_GTTGTAAAACGACGGCCAGTG CCAAGTTTCGGATCGACGCTCGACTCTAGAGG ATCCCCGGGTAC
   DNAantisense_ GTACCCGGGGATCCTCTAGAG TCGAGCGTCGATCCGAAACTTGGCACTGGCCG TCGTTTTACAAC
   RNAantisense_GUUUCGGAUCGACGC.
2. Dissolve the oligonucleotides in oligonucleotide dilution buffer to a final concentration of 100 μM. Store at −20 °C until use.
3. Prepare the post-labeling reaction of RNA or DNA oligos with gamma-phospho32-ATP (*see* **Note 2**).
4. Prepare the following reaction mixture: 5 pmoles RNA oligonucleotide, 10 μL 10× reaction buffer, 7.5 μL [γ-^{32}P]-ATP (3000 Ci/mmol), 25 U of T4 Polynucleotide Kinase, adjust the volume to 50 μL with water nuclease-free.
5. Incubate at 37 °C for 1 h, followed by 75 °C for 10 min.

6. Purify labeled RNA by gel filtration using Sephadex G-25 column following manufacturer instructions (*see* **Note 3**).
7. Evaluate the quality of purified labeled RNA by liquid scintillation. Typically expect around one million cpm/μL. Labeled RNA can be stored at 4 °C for 2–3 days or at −20 °C for long-term usage.

3.2 Generation of RNA:DNA Hybrid or R-Loops by Annealing Oligonucleotides

Labeled R-loops or RNA:DNA hybrid oligo-substrates were generated using ^{32}P-end-labeled antisense RNA and unlabeled DNA oligos. Unlabeled probes were generated similarly.

1. In the annealing reaction, (1.25 pmoles) 2.5 μL of labeled antisense RNA (*see* Subheading 3.1, **step** 7) and two or single complementary DNA oligonucleotide were mixed at 1:2:2 for R-loop formation or 1:2 at molar ratio for RNA:DNA hybrid formation, respectively.
2. Add the oligonucleotide mixture to the oligonucleotides annealing buffer to the final volume of 50 μL.
3. Heat up the annealing reaction to 95 °C for 5 min in the PCR machine and cool down slowly for 2 h after turning off the PCR machine. Store the probes at 4 °C for 1 week maximum. The labeled R-loops and RNA:DNA hybrids are stable for 1 week at 4 °C. Avoid freezing and thawing. The quality of annealing is shown in Fig. 1.

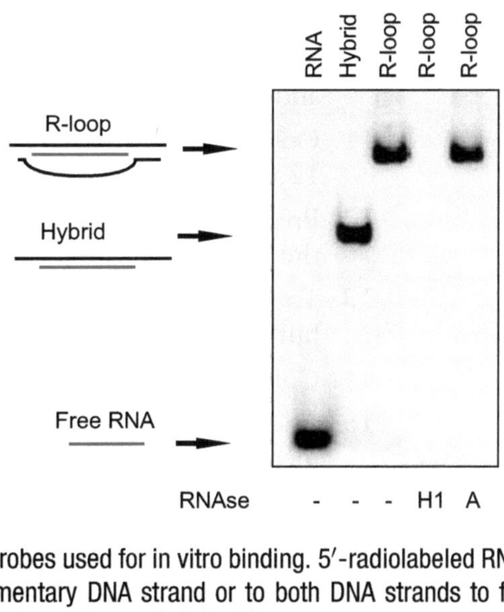

Fig. 1 R-loop probes used for in vitro binding. 5′-radiolabeled RNA was annealed to the complementary DNA strand or to both DNA strands to form a DNA:RNA hybrid or an R-loop. Where indicated, the products were treated with RNase H1 or RNase A. Products were resolved by PAGE and visualized by phosphorImaging

3.3 Purification of GADD45A, PTB, and GFP

Production of FLAG-tagged GADD45A, PTB (Polypyrimidine Tract-binding Protein), and EGFP in HEK293 cells and purification using affinity purification. Of note, recombinant human GADD45A expressed from E. coli appears to be biologically inactive (our unpublished data). PTB is an RNA binding protein used as positive control and EGFP as negative control.

1. Seed two millions of HEK293 cells per plate in five 15 cm dishes, 16 h before plasmid transfection for each purified protein.
2. After 12 h, transfect 5 μg of each plasmid DNA encoding GADD45A, PTB or EGFP per plate using Lipofectamine 2000 following manufacturer instructions.
3. 48 h post-transfection, harvest the cells by scrapping into 5 mL ice cold PBS in 15 mL falcon tubes.
4. Centrifuge the cells at 200 × *g*/3 min and collect all the pellets. Lyse the cells in a total volume of 5 mL of cell lysis buffer on ice.
5. Use a syringe with a 0.4 mm diameter needle to disrupt the cells by passing the lysate for 8–10 times.
6. Sonicate briefly using a Diagenode Biorupter for three cycles (30 s ON—30 s OFF).
7. Collect the cell lysate, and add 2500 U of Benzonase, and 100 U of DNAse I.
8. Incubate for 20 min at room temperature in a rotating device.
9. Meanwhile, take 150 μL of slurry of anti-FLAG-Agarose beads, wash 3 times with 500 μL of lysis buffer.
10. Resuspend anti-FLAG-Agarose beads in 200 μL of lysis buffer containing 10 μg/μL BSA and 10 μg/μL of yeast transfer RNA and rotate at 4 °C for at least 4 h.
11. Collect the supernatant of the cell lysate by centrifugation at 12,000 × *g* for 10 min.
12. Prepare 150 μL protein A/G mix 1:1 ratio and wash 3 times the beads with 500 μL of lysis buffer.
13. To clear the cell lysate, add protein A/G beads to the lysis buffer and incubate for 2 h at 4 °C on a rotating device.
14. Collect the cell lysate by centrifugation at 200 × *g* for 3 min.
15. Add the blocked anti-FLAG beads to 5 mL cell lysate and incubate overnight at 4 °C.
16. Next day, wash the beads extensively with 5 mL cell lysis buffer, 3 times. Rotate the beads 10 min for each washing and collect by centrifugation. All the washing steps follows 5 mL washing volume, 10 min rotation, and centrifugation at 200 × *g*/3 min.
17. Wash the beads with washing buffer for three times.

18. Wash the beads one time with washing buffer containing 1 M Urea.
19. Wash the beads 2 times with washing buffer containing 300 mM KCl.
20. For Pulldown assays, the protein bound beads were stored in 50 μL containing 40% glycerol at −20 °C.
21. For EMSAs the proteins were eluted in FLAG-protein elution buffer with 50 μL containing 20 μg of 3× FLAG peptide overnight at 4 °C in a rotating device. Collect the supernatant (eluted proteins) by centrifugation, add glycerol to final 40% and store at −20 °C.

3.4 Quantification of GADD45A, PTB and GFP Amounts Using Coomassie Staining

1. Prepare 12% precast gel and mount the Mini PROTEAN apparatus.
2. Prepare 5 μL of each purified protein; add 2 μL of the Western blot loading buffer (5×) and fill up to 10 μL each with lysis buffer.
3. Prepare dilutions of bovine serum albumin (BSA) from 10% stock solution to have standards for protein quantification ranging from 0.12, 0.25, 0.5, 1, and 2 μg.
4. Heat all protein samples, including BSA standards at 85 °C for 10 min.
5. Load directly the samples in 12% Precast gels SDS-PAGE and run the samples at 20 mA until the bromophenol bleu dye reached the bottom of the gel.
6. Open the apparatus and remove one plate and leave the gel on the other plate, rinse with pure water and poor the gel into a small container with pure water. Rinse three times with 100 mL H_2O after each 5 min time interval.
7. Cover the gel with 20 mL of SimplyBlue SafeStaining (Biorad) for 1 h at room temperature.
8. Remove the staining solution, wash with 100 mL pure water, three time, add an additional 100 mL of H_2O, and put the container in a shaker up to 12 h, until the gel is sufficiently destained.
9. Image the stained gel on the Gel Doc XR+ Gel Documentation System and quantify the bands.

3.5 R-Loop Pulldown Assay of GADD45A, PTB and GFP

1. Ten μL of M2 beads containing 250–500 ng immobilized FLAG- GADD45A, -PTB, or -EGFP were incubated for 1 h at RT with 2.5 pmoles of ^{32}P-labeled R-loop probes (2000 cpm/ng) in 100 μL pulldown binding buffer (*see* Subheading 2.4, **item 14**).
2. Both TCF21 R-loops and generic R-loops were tested.

3. Wash the beads five times 10 min each with 1 mL binding buffer on a rotating device.
4. Digest proteins by adding 20 μL binding buffer containing 20 μg of proteinase K, and 0.1% SDS to the beads and incubate at 37 °C for 20 min.
5. Released R-loops or RNA:DNA hybrids were analyzed by PAGE 10% and PhosphorImaging.
6. To prepare PAGE, soak the glass plates and combs in 3% H_2O_2 before usage for overnight.
7. Prepare all the solution with DEPC treated water.
8. Prepare 10% PAGE gel, mix 9.6 mL H_2O, 9.5 mL Acrylamide 30%, 7.5 mL Bis Acrylamid 2%, 3 mL 10× TBE, 400 μL TEMED.
9. Insert a 10-well gel comb and avoid introducing air bubbles. Polymerize the gel for 2 h.
10. Pre-run the gel for 1 h at 15 mA.
11. Load side by side the binding reaction and 10% as input of radiolabeled R-loop and labeled ssRNA as a gel marker.
12. Load labeled ssRNA and R-loops individually as markers next to proteinase K treated samples and run for 1.5 h at 15 mA.
13. After the run, carefully remove the radioactive buffer in a waste container.
14. Remove the gel from the casting system, and leave the gel attached to one of the plates.
15. Place the gel on two sheets of Whatmann 3MM paper and cover the other side with saran wrap; dry on a gel dryer at 85 °C for 1 h.
16. Expose the gel on a phosphorimager screen overnight. A typical pulldown gel autoradiogram is shown in Fig. 2.

3.6 EMSA for GADD45A Binding to R-Loop and RNA:DNA Hybrid

1. EMSA was performed with immunopurified GADD45A (*see* Subheadings 3.3 and 3.4).
2. Binding assays of GADD45A with P^{32}-labeled oligonucleotides are carried out in 20 μL reactions.
3. Immunopurified GADD45A (5 pmoles/10 μL, *see* Subheading 3.4, **step 9**) was incubated for 1 h with 0.2 pmoles of labeled R-loop or hybrid probe (2000 cpm/μL) in EMSA binding buffer (*see* Subheading 2.4, **item 15**).
4. For competition assays, add desired non-labeled competitor (ssRNA, ssDNA, dsRNA, dsDNA, RNA:DNA hybrid, R-loops) at increasing molar ratios with labeled R-loops.

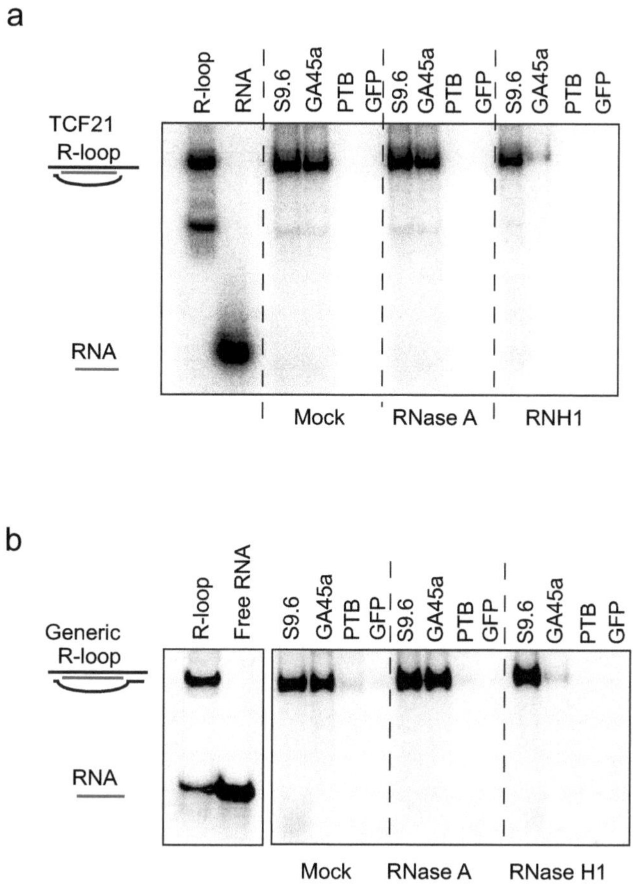

Fig. 2 (**a**) Pulldown assay showing GADD45A binding to *R-lo*ops in vitro. Bead-bound S9.6 antibody or GADD45A (GA45a), PTB or GFP were incubated with radiolabeled *TCF21* R-loops followed by RNase A or RNH1 treatment. Bound probes were visualized by PhosphorImaging. (**b**) The same assay as in (a) carried out using generic R-loops

5. After electrophoresis on 2.5% agarose gels in 0.5× TB buffer at 20 V/cm, remove the radioactive buffer and prepare the gel for the transfer.
6. Soak the gel in 100 mL 20× SSC for 15 min.
7. Cut Hybond-N$^+$ membranes to fit the size of the gel and soak the membranes in 2× SSC buffer.
8. Set up capillary transfer of the samples overnight to the Hybond-N$^+$ membranes (*see* **Note 4**).
9. Crosslink the membranes using UV-crosslinker with 2000 kJ for 30 s and expose the gel on a phosphorimager screen overnight. A typical result is shown in Fig. 3.

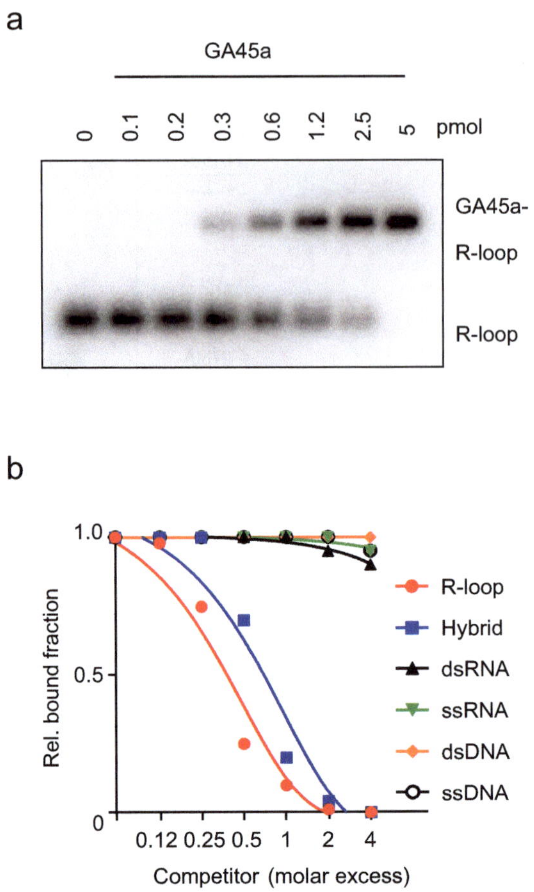

Fig. 3 (**a**) EMSA showing binding of GADD45A to R-loops. A radiolabeled *TCF21* R-loop probe was incubated with increasing amounts of GADD45A, followed by PAGE and PhosphorImaging. (**b**) Quantitation of competitive EMSAs using a GADD45A/ *TCF21* R-loop complex and the indicated unlabeled competitors. GADD45A binding to R-loops was set to 1

4 Notes

1. All the solutions and buffers were prepared using deionized water with at 0.035 S/cm resistance made from Millipore-system purified water and treated with Diethyl pyrocarbonate (DEPC). All electrophoresis combs and glass plates were cleaned with 3% H_2O_2 and DEPC treated water to inactivate RNAses and DNAses.
2. All the work was performed in specific dedicated radioactive area. Wear necessary protections, shield yourself from radiation with required instrumentations, and monitor for possible contamination after and before entering to the radioactive area.

3. Free [γ-^{32}P] ATP was removed by Sephadex G-25 Quick Spin Columns and the sample is collected in 50 μL volume.
4. Prepare a large container with 20× SSC filled to 1/5 vol. Put a tray corresponding at least to the size of the gel in the middle of the reservoir. Prepare a long piece of Whatman paper in 20× SSC for use as a strip. Lay Whatmann paper strip on tray with the ends dipping into the buffer in the container. Soak three pieces of Whatman paper in 20× SSC and put the gel upside down, the top-side faces the Whatmann paper. Put the Hybond membrane on top of the gel. Remove bubbles between the gel and the membrane and soak three pieces of Whatman paper in 2× SSC and lay them on top of the membrane. Put a stack towels papers with the size fitting Whatmann papers on the top. Place a glass plate on top and put heavy weigh. Leave the assembled transfer for overnight.

References

1. Skourti-Stathaki K, Proudfoot NJ (2014) A double-edged sword R loops as threats to genome integrity and powerful regulators of gene expression. Genes Dev 28:1384–1396
2. García-Muse T, Aguilera A (2019) R loops: from physiological to pathological roles. Cell 179:604–618
3. Aguilera A, Gómez-González B (2017) DNA-RNA hybrids: the risks of DNA breakage during transcription. Nat Struct Mol Biol 24: 439–443
4. Huertas P, Aguilera A (2003) Cotranscriptionally formed DNA:RNA hybrids mediate transcription elongation impairment and transcription-associated recombination. Mol Cell 12:711–721
5. Skourti-Stathaki K, Proudfoot NJ, Gromak N (2011) Human senataxin resolves RNA/DNA hybrids formed at transcriptional pause sites to promote Xrn2-dependent termination. Mol Cell 2011(42):794–805
6. El Hage A, French SL, Beyer AL, Tollervey D (2010) Loss of topoisomerase I leads to R-loop-mediated transcriptional blocks during ribosomal RNA synthesis. Genes Dev 24: 1546–1558
7. Niehrs C, Luke B (2020) Regulatory R-loops as facilitators of gene expression and genome stability. Nat Rev Mol Cell Biol 21:167–178
8. Gibbons HR, Shaginurova G, Kim LC, Chapman N, Spurlock CF 3rd, Aune TM (2018) Divergent lncRNA GATA3-AS1 regulates GATA3 transcription in T-helper 2 cells. Front Immunol 9:2512
9. Powell WT, Coulson RL, Gonzales ML, Crary FK, Wong SS, Adams S, Ach RA, Tsang P, Yamada NA, Yasui DH, Chédin F, LaSalle JM (2013) R-loop formation at Snord116 mediates topotecan inhibition of Ube3a-antisense and allele-specific chromatin decondensation. Proc Natl Acad Sci U S A 110:13938–13943
10. Sun Q, Csorba T, Skourti-Stathaki K, Proudfoot NJ, Dean C (2013) R-loop stabilization represses antisense transcription at the Arabidopsis FLC locus. Science 340:619–621
11. Beckedorff FC, Ayupe AC, Crocci-Souza R, Amaral MS, Nakaya HI, Soltys DT, Menck CF, Reis EM, Verjovski-Almeida S (2013) The intronic long noncoding RNA ANRASSF1 recruits PRC2 to the RASSF1A promoter, reducing the expression of RASSF1A and increasing cell proliferation. PLoS Genet 9:e1003705
12. Boque-Sastre R, Soler M, Oliveira-Mateos C, Portela A, Moutinho C, Sayols S, Villanueva A, Esteller M, Guil S (2015) Head-to-head antisense transcription and R-loop formation promotes transcriptional activation. Proc Natl Acad Sci U S A 112:5785–5790
13. Arab K, Karaulanov E, Musheev M et al (2019) GADD45A binds R-loops and recruits TET1 to CpG island promoters. Nat Genet 51:217–223
14. Niehrs C, Schäfer A (2012) Active DNA demethylation by Gadd45 and DNA repair. Trends Cell Biol 22:220–227

Chapter 19

Proximity Ligation Assay for Detection of R-Loop Complexes upon DNA Damage

Adele Alagia, Ruth F. Ketley, and Monika Gullerova

Abstract

In situ Proximity Ligation Assay (PLA) can be used to detect the close proximity (less than ~40 nm) of two biological molecules of interest in cells. Here we report the application of this method for the specific detection of R-loop interacting proteins and RNA modifications in close proximity to R-loops in non-damage and ionizing radiation (IR) induced DNA damage conditions.

Key words Proximity Ligation Assay, R-loops, RNA modification, DNA Damage, RNA:DNA hybrids, Double Strand breaks

1 Introduction

Understanding molecular interactions is vital to elucidate how biological pathways occur. Methods to detect protein–RNA interactions have become increasingly important as our understanding of RNA functionality has expanded. R-loops are RNA structures with an RNA:DNA hybrid and a single strand DNA, which can have opposing roles in the cell, as both a genomic threat and as important functional components in processes such as transcription [1] and in the regulation of gene expression [1, 2]. It is also becoming apparent that R-loops are important in the repair of the lethal DNA lesions double strand breaks (DSB) [3]. R-loops can form at DSBs and act as functional platforms in repair through the recruitment of repair proteins, to regulate both homologous recombination (HR) [4] and non-homologous end joining (NHEJ) [5]. R-loops are also now known to be modified, with the RNA strand carrying N6-methyladenosine (m^6A) [6] and 5-methylcytosine (m^5C) [7]. Although the presence of R-loops at DSBs has been shown by many, their exact role in DNA repair is not yet well understood. Hence, there is a great deal of interest in this

Andrés Aguilera and Alexey Ruzov (eds.), *R-Loops: Methods and Protocols*, Methods in Molecular Biology, vol. 2528, https://doi.org/10.1007/978-1-0716-2477-7_19,

field and consequently in the development of novel methodological tools that can further aid studies on R-loops in DNA repair.

In situ Proximity Ligation Assay (PLA) is a molecular biology technique first developed to detect the proximity of two proteins within ~40 nm of each other and visualize their location within the cell by observation of discrete fluorescent foci [8, 9]. The technique uses specific antibodies from two different species recognizing the proteins of interest. Secondary antibodies with oligonucleotide conjugates, known as PLA probes, are then added, and if these are in close proximity, i.e., the two proteins of interest are within ~40 nm of one another, the connector oligonucleotides will be ligated together after the addition of a ligase. A DNA polymerase is added, and the ligated oligonucleotides can act as primers for amplification, which can then be bound by complementary fluorescently labeled oligonucleotides. The result is a distinct PLA signal which can be visualized by microscopy and quantified using various software such as cell profiler (Fig. 1) [8–10].

Since its conception over a decade ago [11, 12], PLA has been optimized and adapted for use in a wide variety of applications, including but not limited to; the detection of protein–protein

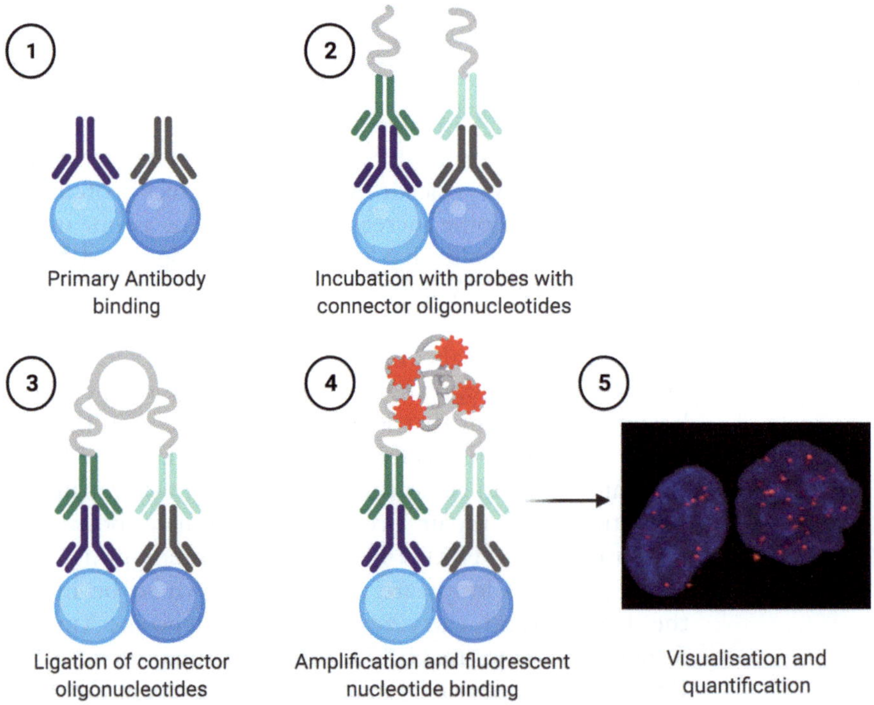

Fig. 1 Principles of in situ Proximity Ligation Assay. Cells seeded on coverslips are incubated with primary antibody, followed by incubation with secondary antibodies (PLA probes) with connector oligonucleotides. The oligonucleotides will ligate together upon addition of a ligase enzyme if in close proximity. This acts as a template for a DNA polymerase and rolling circle amplification (RCA) will take place, with binding of complementary fluorescent nucleotides resulting in a PLA foci visualised by confocal microscope

interactions [13]; monitoring post-translational modification of a protein of interest [14]; RNA–protein interactions [15]; RNA–DNA association [16]; and detection of proteins at DSBs [17].

We present here a method using the PLA Duolink® In Situ Red Fluorescent kit (https://www.sigmaaldrich.com/GB/en/products/protein-biology/duolink-proximity-ligation-assay) to detect both proteins and RNA modifications in close proximity to RNA:DNA hybrids using the S9.6 antibody. Additionally, we perform this in biologically relevant IR-induced DNA damage conditions. We show the specificity of this method by expressing a V5-tagged RNaseH1 enzyme, which binds the RNA:DNA hybrid and degrades the RNA strand, and catalytically dead (D210N) and binding (WKKD) RNaseH1 mutants [18]. As expected, performing PLA with V5 and S9.6 antibodies results in visible PLA signals in the presence of the D210N RNaseH1 mutant, which binds but does not cleave the RNA:DNA hybrids. These PLA signals are reduced in the presence of WT RNaseH1 which will bind and degrade the RNA:DNA hybrids. The PLA foci observed with the WKKD mutant are on the level of background, as this mutant will not bind the RNA:DNA hybrids (Fig. 2). This method can be applied to a wide variety of RNA:DNA hybrid interacting proteins (Fig. 3) and RNA modifications (Fig. 4) in different cellular contexts. As an example, for proof of concept, we chose to apply PLA to detect the interaction between DHX9, a known R-loop interactor [19], and R-loops. We observed the generation of PLA foci using antibodies for DHX9 and S9.6 (Fig. 3), and we validate the specificity of these PLA signals by expressing the D210N RNaseH1 and WKKD RNaseH1 mutants, where PLA signals are observed, and the WT RNaseH1, where the PLA signals are reduced due to degradation of the RNA:DNA hybrid. Furthermore, we performed this experiment in the context of DNA damage, where R-loops are enriched, and detected significantly increased numbers of PLA foci upon IR treatment (Fig. 3). Next, we wished to test the application of the same experimental principle for the detection of R-loop modifications. In order to do this, we employed PLA using an antibody for the RNA modification m^5C in combination with the S9.6 antibody in normal and IR treated samples, as R-loops have been shown to be methylated upon DNA damage. Indeed, we observed a significant increase in PLA signals in the presence of D210N and WKKD RNaseH1 expression in irradiated samples. These foci were significantly reduced by expression of RNaseH1 WT, further confirming the specificity of the PLA assay in the context of DNA damage (Fig. 4).

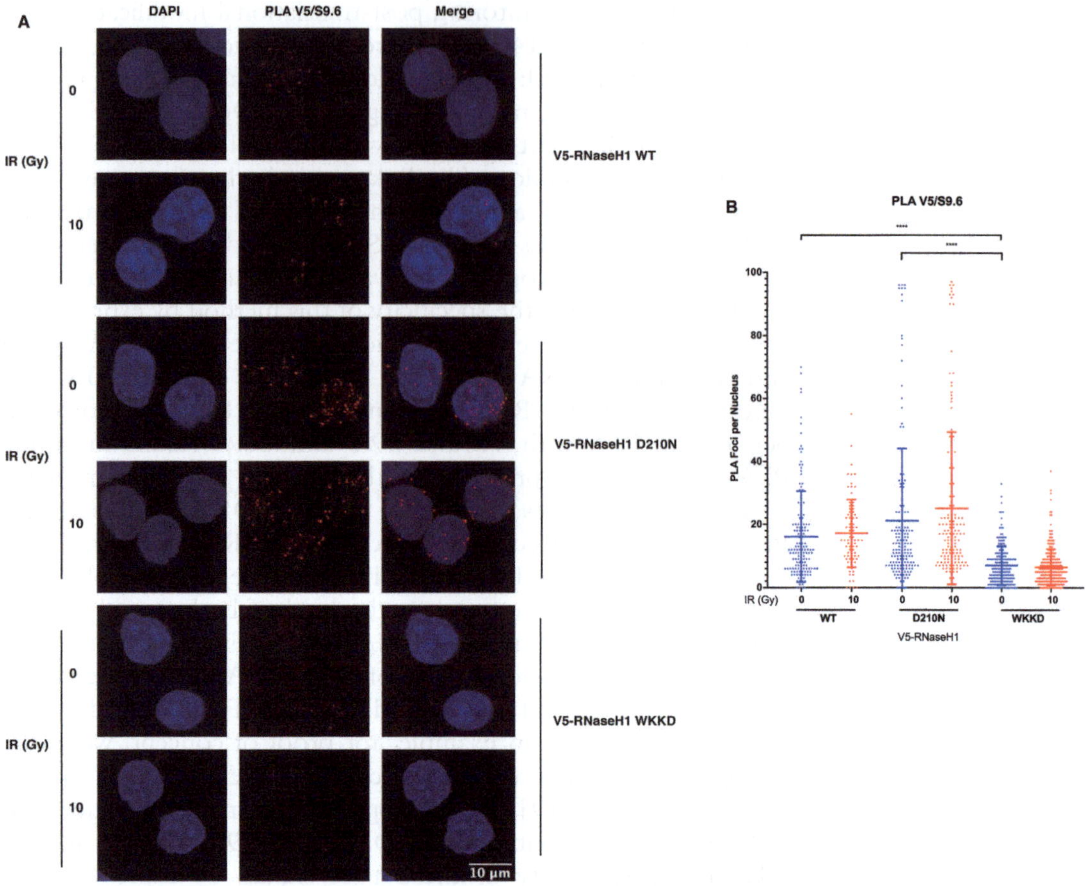

Fig. 2 Specific detection of RNA:DNA hybrids in close proximity to the RNA:DNA hybrid binding enzyme RNaseH1. (**a**) V5-tagged RNaseH1 wild type (WT) (top panel), RNaseH1 catalytically dead D210N mutant (middle panel), and RNaseH1 non-binding catalytically dead WKKD mutant (bottom panel) were expressed in cells. Antibodies for V5 and S9.6 were used for PLA to detect the RNaseH1 in close proximity to the RNA:DNA hybrids. Cells were exposed to IR (10 Gy) or not (0 Gy) before carrying out the PLA protocol. (**b**) Quantification of the results shown in panel **A**. Error bars show mean and standard deviation. The Mann–Whitney Statistical test was used as a test for significance

2 Materials

2.1 Equipment

1. Microscope slides.
2. Round coverslips (22 mm).
3. Humidity chamber (*see* **Note 3**).
4. Tissue culture plates (6-well).
5. Tissue culture dishes (35 mm).
6. Rocker.
7. Fume hood.
8. Tissue culture hood.

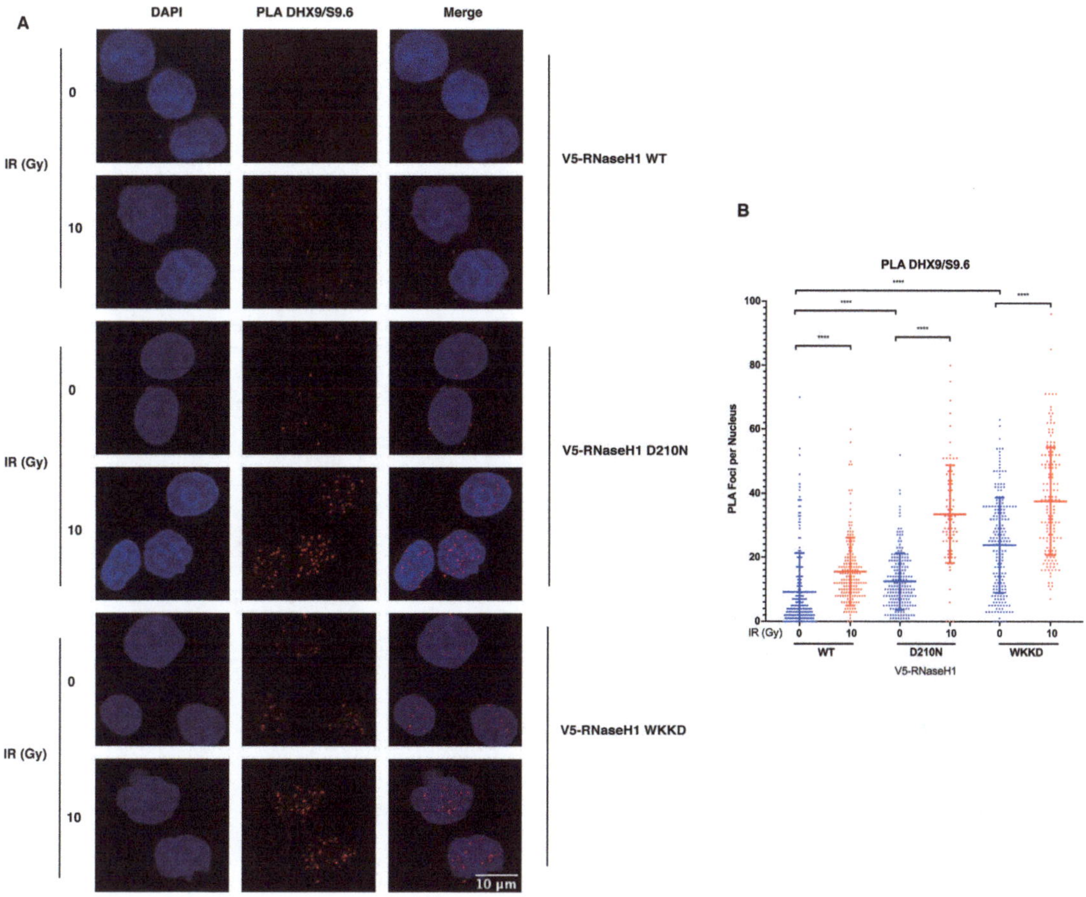

Fig. 3 Detection of DHX9/S9.6 PLA foci. (**a**) An antibody for endogenous DHX9, a known R-loop interactor, was used in combination with the S9.6 antibody for PLA. Cells were transfected with either WT (top), D210N mutant (middle), or WKKD mutant (bottom) RNaseH1. Cells were exposed to IR (10 Gy) or not (0 Gy) before carrying out the PLA protocol. (**b**) Quantification of the results shown in panel **A**. Error bars show mean and standard deviation. The Mann–Whitney Statistical test was used as a test for significance

9. Freeze block for enzymes.
10. 37 °C incubator.
11. Water bath.
12. Fluorescence microscope.
13. Analysis software (ImageJ and Cell profiler).

2.2 Reagents

1. Protease inhibitor cocktail tablets (Roche).
2. RNAse T1 (Thermo).
3. Shortcut RNAse III (New England Biolabs).
4. Duolink® In Situ Red Starter Kit Mouse/Rabbit (Sigma) containing:

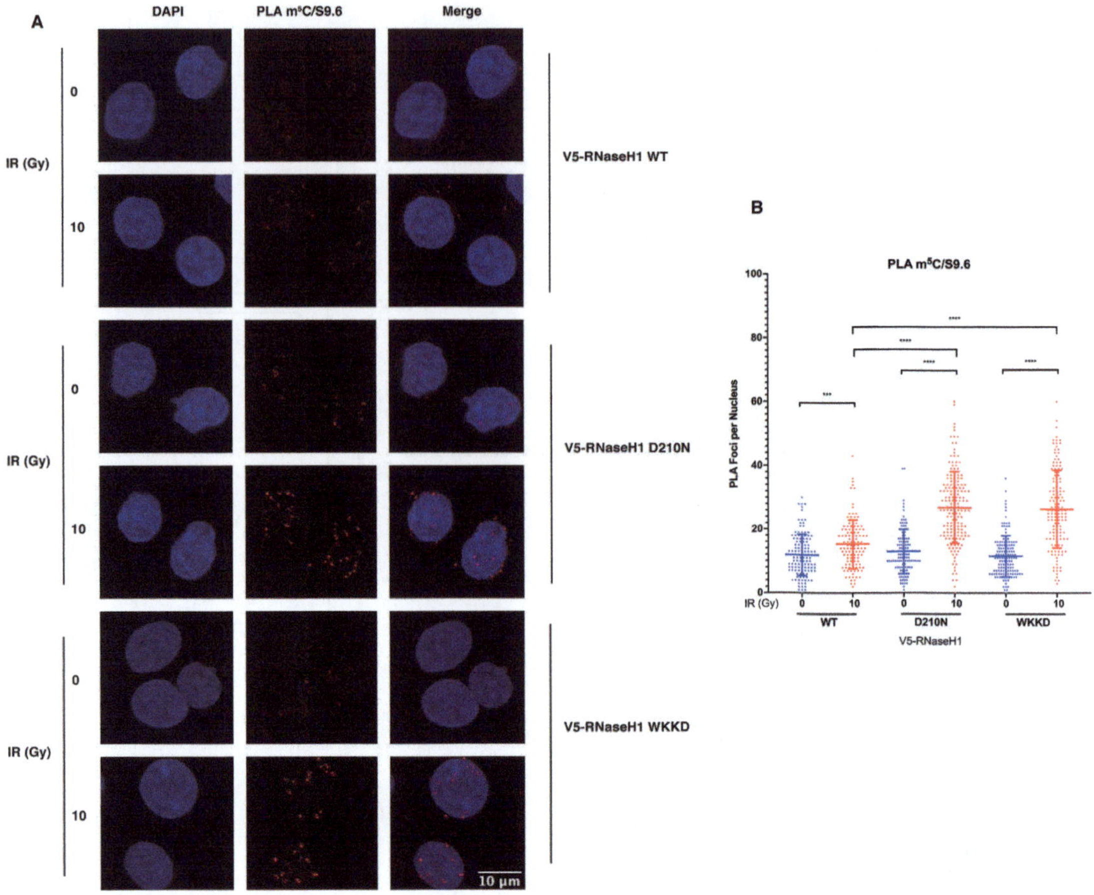

Fig. 4 Detection of m^{5}C/S9.6 PLA foci. (**a**) m^{5}C was detected in close proximity to RNA:DNA hybrids using antibodies for m^{5}C and S9.6. Cells were transfected with either WT (top), D210N mutant (middle), or WKKD mutant RNaseH1 (bottom). Cells were exposed to IR (10 Gy) or not (0 Gy) before carrying out the PLA protocol. (**b**) Quantification of the results shown in panel **A**. Error bars show mean and standard deviation. The Mann–Whitney Statistical test was used as a test for significance

(a) Duolink® In Situ Mounting Medium with DAPI.

(b) Duolink® In Situ Wash Buffers (Buffer A and B).

(c) Duolink® In Situ PLA® Probe Anti-Rabbit PLUS.

(d) Duolink® In Situ PLA® Probe Anti-Mouse MINUS.

(e) Duolink® In Situ Detection Reagents Red.

5. Primary antibodies: Anti-DNA–RNA hybrid antibody (clone S9.6) mouse monoclonal (MABE1095, Sigma-Aldrich), Anti-DNA–RNA hybrid antibody (clone S9.6) rabbit monoclonal (Ab01137-23.0, Absolute Antibody), Anti-V5 tag antibody rabbit monoclonal (ab206566, Abcam), Anti-RNA Helicase A antibody rabbit monoclonal (ab183731, Abcam), Anti-5-methylcytosine (5-mC) antibody (ab214727, Abcam).

6. Hela cells (CCL-2 ATCC).
7. Fetal bovine serum (FBS), heat inactivated.
8. Cell dissociation enzyme (e.g. TrypLE).
9. Dulbecco's Modified Eagle's Medium high glucose (DMEM).
10. Penicillin/Streptomycin solution (P/S).
11. Opti-MEM™ I Reduced Serum Medium.
12. Lipofectamine 3000 (Thermo).
13. Plasmids: ppYCAG_RNAseH1_wt (#111906, Addgene), ppY-CAG_RNAseH1_D210N (#111904, Addgene), and ppY-CAG_RNASEH1_WKKD (#111905, Addgene).
14. Clear nail varnish.

2.3 Solutions

Prepare all solution using ultra-pure water and DEPC-treated and analytical grade reagents. Autoclave or 0.22 μm filter to sterility. Store all reagents at room temperature (unless indicated otherwise).

1. Diethylpyrocarbonate (DEPC)-treated and sterile filtered H2O.
2. Phosphate-buffered saline without $CaCl_2$ and $MgCl_2$ (PBS).
3. 0.1% Poly-D-lysine solution (w/v).
4. 4% paraformaldehyde (PFA) solution in PBS (4% PFA).
5. 70% Ethanol solution in DEPC-H_2O.
6. 10% Triton X-100 in DEPC-H_2O.
7. 1 M Sucrose in DEPC-H_2O.
8. 1 M Pipes in DEPC-H_2O, store at 4 °C.
9. 5 M NaCl in DEPC-H_2O.
10. 1 M $MgCl_2$ in DEPC-H_2O.
11. 10% BSA in DEPC-H_2O store in dark at 4 °C.
12. 1 M ethylene glycol-bis(β-aminoethyl ether)-*N,N,N′,N′*-tetra-acetic acid (EGTA) in DEPC-H_2O.
13. 1 M Glycine in PBS.
14. Coating solution: 0.01% Poly-D-lysine in PBS, (store at 4 °C).
15. CSK buffer: 10 mM PIPES, 100 mM NaCl, 300 mM Sucrose, 3 mM $MgCl_2$, 1 mM EGTA, 0.7% Triton X-100, 1× protease inhibitor cocktail, (make fresh and store at 4 °C).
16. Quenching solution: 200 mM Glycine in PBS.
17. RNAse treatment solution: 0.1% BSA, 3 mM $MgCl_2$, 1:200 (v/v) RNAse T1, 1:200 (v/v) ShortCut RNAse III in PBS, (make fresh and store at 4 °C).

3 Methods

Day 1

3.1 Reverse Transfection with RNaseH1 Plasmids

1. Remove media from a 70% confluent flask of Hela cells and wash once with PBS.
2. Add tryPLE to the flask until detachment of cells has occurred and resuspend in DMEM to obtain a single cell suspension.
3. Count cells and resuspend in DMEM supplemented with 10% FBS without P/S.
4. Prepare RNaseH1 plasmids for transfection using Lipofectamine 3000 transfection reagents according to manufactures instructions. For each well of a 6-well plate the following volumes are used in Opti-MEM™ Reduced Serum Medium:
   (a) 3 μg of plasmid.
   (b) 4.5 μL of Lipofectamine 3000.
   (c) 6 μL of reagent 3000.
5. Vortex mixture for 10 s and incubate at room temperature (RT) for 20 min.
6. Add the plasmid-lipofectamine complexes at the bottom of each well (total 250 μL of Opti-MEM™).
7. Seed Hela cells to reach a confluency of 90% 16–18 h later.
8. Add DMEM supplemented with FBS 10% without P/S up to 2 mL.
9. Incubate cells at 37 °C for 16–18 h.

Day 2

3.2 Cover Glass Preparation

1. Place one microscope coverslip into a 35 mm dish.
2. Add 2 mL 70% ethanol to each dish and incubate at RT for 10 min.
3. Remove the 70% ethanol and wash twice with 2 mL PBS.
4. Add 2 mL of coating solution to each dish and incubate 20 min at 37 °C.
5. Remove the coating solution and wash twice with PBS.
6. Seed RNaseH1-tranfected HeLa cells to reach ~60% confluency after 24 h.

Day 3

3.3 Irradiation and Proximity Ligation Assay

1. Irradiate Hela Cells with a 10 Gy dose of IR (*see* **Note 1**).
2. Incubate cells for 15 min at 37 °C (*see* **Note 1**).

3. Remove media from the cells and wash twice with 2 mL ice cold PBS.
4. Add 2 mL of CSK buffer to each 35 mm dish and place on a rocker with gentle rocking for 15 min at 4 °C.
5. Remove the CSK buffer and wash three times with 2 mL ice cold PBS.
6. In a fume hood, add 2 mL of 4% PFA in PBS to each 35 mm dish and incubate for 10 min at 37 °C.
7. Remove the PFA and add 2 mL quenching solution, and incubate for 5 min at RT on the rocker with gentle rocking.
8. Prepare fresh RNase treatment solution (*see* **Note 2**).
9. Add 1 mL of RNase treatment solution to each 35 mm dish and incubate for 1 h at RT on a rocker with gentle rocking.
10. Wash three times with 2 mL PBS and incubate for an additional 5 min with rocking.
11. In prewarmed humidity chamber (*see* **Note 3**), add 100 μL of Duolink® blocking solution from the PLA kit on top of the parafilm for each coverslip sample.
12. Take out the coverslip from the 35 mm dish, tap off excess solution and place it cell side down on the blocking solution avoiding bubbles (*see* **Notes 4** and **5**).
13. Incubate for 1 h at 37 °C.
14. Prepare the primary antibody solution by diluting mouse and rabbit antibodies in 60 μL per sample of Duolink® Antibody Dilutant from the PLA kit (*see* **Notes 6–8**).
15. Add 60 μL of the diluted antibody solution to the parafilm surface of the humidity chamber.
16. Tap off excess buffer from the coverslip and place cell side down onto the antibody solution.
17. Incubate overnight at 4 °C in the humidity chamber.

Day 4

3.4 Proximity Ligation Assay

1. Prewarm PLA washing buffer A and B to RT in a water bath (*see* **Note 9**).
2. Place coverslips cell side up in 6-well plate with 2 mL of PLA washing buffer A and wash twice for 5 min with gently rocking at RT (*see* **Note 10**).
3. Pre-warm the humidity chamber to 37 °C.
4. Dilute the PLUS and MINUS PLA probes 1:5 in Duolink® Antibody Diluent.
5. Add 60 μL of probe solution on to the parafilm of the humidity chamber.

6. Tap off the excess of buffer from the coverslips and place coverslips cell side down onto the probe solution in the humidity chamber.
7. Incubate at 37 °C for 1 h.
8. Tap off the excess of probe solution and wash twice with 2 mL of washing buffer A for 5 min at RT.
9. Thaw Duolink ligase buffer and dilute ligase buffer at 1:5 ratio in DEPC-H_2O (*see* **Note 11**).
10. Add 12 μL of ligase enzyme (*see* **Note 12**) provided by the Duolink® Red Starter Kit to 48 μL diluted ligase buffer solution for a total of 60 μL for each coverslip.
11. Add 60 μL of Ligation reaction solution to the parafilm in the humidity chamber.
12. Incubate at 37 °C for 30 min.
13. Tap off the excess of ligation solution and wash twice with 2 mL of washing buffer A for 5 min at RT.
14. Thaw Duolink amplification buffer, dilute 1:5 in DEPC-H_2O and protect from light (*see* **Note 13**).
15. Add 6 μL Polymerase enzyme provided by the Duolink® Red Starter Kit to 54 μL diluted amplification buffer for a total of 60 μL for each coverslip.
16. Add 60 μL of Amplification reaction solution to the parafilm in the humidity chamber.
17. Place coverslips cell side down onto the amplification reaction solution.
18. Incubate for 1 h and 40 min at 37 °C.
19. Tap off the excess of amplification solution from the coverslip.
20. Wash twice with 2 mL of washing buffer B for 10 min at RT, protecting from light.
21. Perform the last wash with 0.01× washing buffer B for 1 min at RT, protecting from light.
22. Add 20 μL of Duolink in situ mounting medium with DAPI on a microscope slide.
23. Tap off the excess of washing solution and mount the coverslip on the microscope slide.
24. Seal the edges of coverslip with clear nail varnish (*see* **Note 14**).
25. Wait at least 15 min before acquiring images at fluorescent microscope using a 60× objective lens.
26. Slides can be stored in the dark at 4 °C (*see* **Note 15**).
27. Images can be processed using ImageJ, and PLA foci quantified using software such as Cell Profiler (*see* **Note 16**).

4 Notes

1. Ionizing radiation is delivered at a dose of 10 Gy and cells are fixed 15 min after exposure. The most appropriate time post-IR and IR dosage may differ depending on your experimental question and multiple different doses and time points can be tested.
2. It has been demonstrated that the S9.6 antibody can bind other RNA species distinct from RNA:DNA hybrids in immunofluorescence experiments [20]. Therefore, RNase treatment is crucial for the specific identification of the RNA:DNA hybrids (*see* Fig. 3 (with treatment) and Fig. 5 (without treatment) for comparison).
3. The humidity chamber can be set up according to Fig. 6.
4. Excess solution can be gently tapped off the slide on paper towel between wash and incubation steps.
5. Slides should not be allowed to dry out, always keep the coverslips moist. Drying out of the coverslips can cause excess background signals.
6. Some antibodies are not appropriate for use in PLA. Generally, if the antibody is suitable for immunofluorescence (IF) microscopy, it may be suitable for use in PLA. It is a good idea to have a single antibody control (where addition of a second antibody is omitted) (Fig. 7), in order to determine the background level of PLA signals (and therefore non-specific signals) generated by the antibody alone. Some antibodies will generate a very high background level of PLA signals and are, therefore, unsuitable.
7. A titration of each antibody can also be performed to determine the optimum concentration using at least three different concentrations, such as 1:250, 1:1000, and 1:2000.
8. If possible, it is good to test different antibodies raised for the same protein to determine the optimum (specific signal, minimal background) in your experiment. A successful PLA experiment can depend on the primary antibody used.
9. Wash buffer A and B can be warmed in a 37 °C water bath to RT prior to beginning the PLA protocol.
10. For washing steps, slides can be placed back in a 6-well plate cell side facing upwards and approximately 2 mL of wash solution used per well.
11. Ligase buffer should be thawed fully and mixed well before use. It may take some vortexing and pipetting to get the ligase buffer fully in solution. It can help to take the ligase buffer out of the −20 °C approximately half an hour in advance of

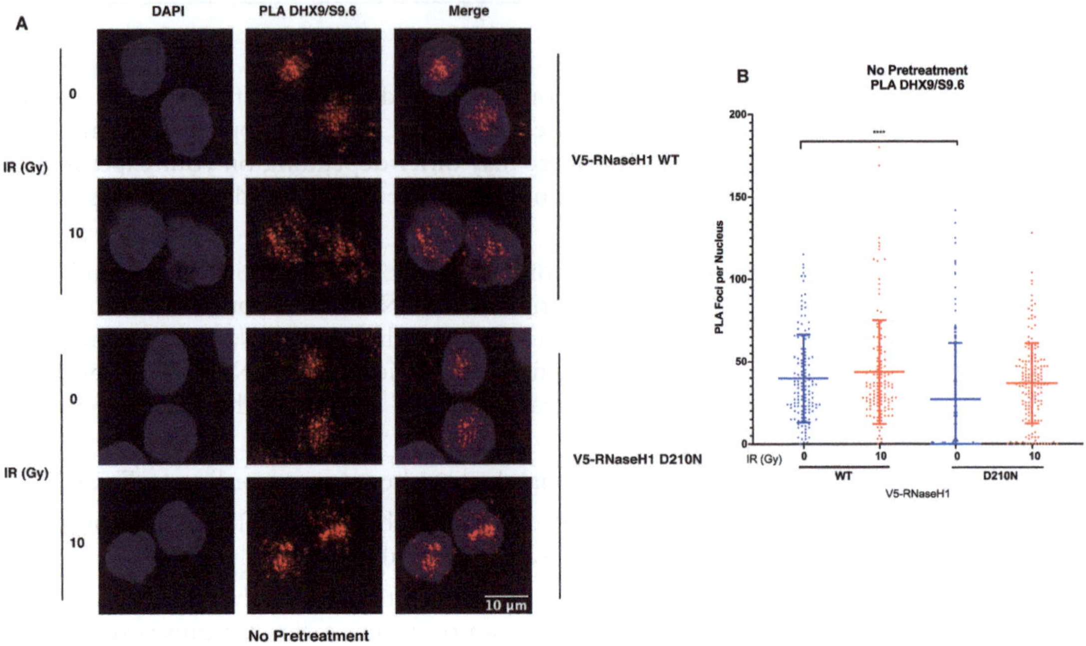

Fig. 5 The importance of the RNase pre-treatment step (*see* **Note 2**). The S9.6 antibody has been suggested to give non-specific signals which are insensitive to RNaseH1 treatment in immunofluorescence (IF) [20]. To overcome this, a pre-treatment step using both RNase III and RNase T1 was developed to ensure the S9.6 signals are specific for RNA:DNA hybrids in IF [20]. We employ this methodology in our set up for Proximity Ligation Assay with the S9.6 antibody in combination with the DHX9 antibody and show here the result without this pre-treatment step. The signals generated with DHX9 and S9.6 were insensitive to RNaseH1 WT expression and therefore likely unspecific. This is not the case when the RNase pre-treatment step is used (Fig. 3). (**b**) Quantification of the results shown in panel A. Error bars show mean and standard deviation. The Mann–Whitney Statistical test was used as a test for significance

use. Incomplete mixing of the ligase buffer can result in the generation of excess background signals.

12. All enzymes (ligase and polymerase) should be kept at −20 °C until just before use and kept in a freezer block or on ice when out of the −20 °C. Do not prepare the ligase or polymerase solutions in advance of use.
13. Protect the amplification buffer from light. Slides and reagents should be kept in the dark during and after the polymerase amplification step.
14. Coverslips can be mounted to slides and secured using clear nail varnish around the perimeter of the coverslip, taking care to seal the edges but covering as little of the coverslip surface as possible.
15. It is better to image the slides the same day or following day after mounting with DAPI. However, we have used slides which are 3–4 days old with similar quality to slides imaged the same day after mounting when stored at 4 °C.

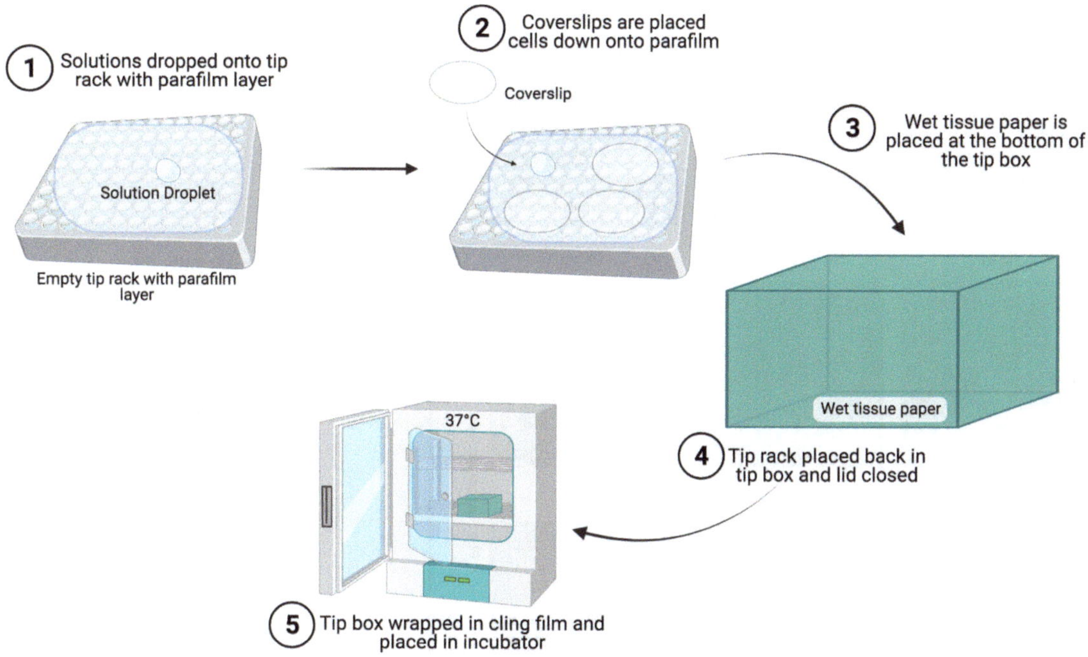

Fig. 6 PLA humidity chamber set up. Schematic of the preparation of the humidity chamber for Proximity Ligation Assay. An empty tip box was used to set up a DIY humidity chamber for coverslip incubation steps. Parafilm is placed on the tip box rack (at least 2 layers is best to give an even surface) and the edges taped down to secure this in place. Between reactions the parafilm surface can be wiped with 70% ethanol and dried to clean it. Reaction mixtures are pipetted onto the parafilm surface, and coverslips are placed cells down on the drops of solution. The rack is then placed in the tip box with wet paper towel placed at the bottom. The tip box lid is closed and the whole box wrapped in cling film to ensure the humid environment is maintained. During the polymerase reaction incubation step, which is light sensitive, the cling film-wrapped box can then be additionally wrapped in tin foil. The box can then be placed in a 37 °C incubator

16. The cell profiler speckle counter pipeline is used to quantify PLA foci. This can be downloaded from: https://cellprofiler.org/examples

Acknowledgments

The work is supported by funding from the Senior Research Fellowship by Cancer Research U.K. [grant number BVR01170 (to M.G.)]; the Lee Placito Trust [grant number BVR01070 (to M.G.)] and Edward Penley Abraham Research Fund [grant number BVR01670 (to M.G.)].

Figures 1 and 6 were created using BioRender.com.

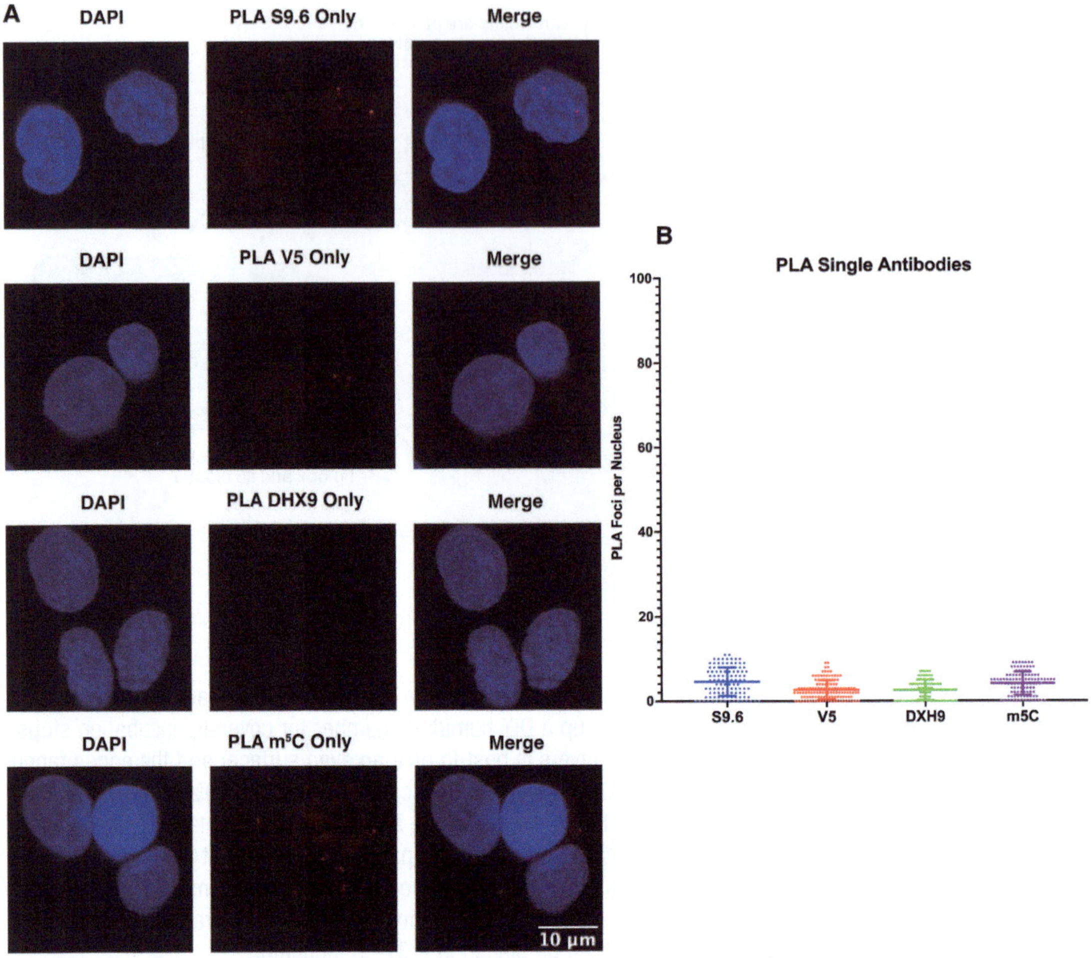

Fig. 7 Absence of PLA signals when individual antibodies are used. PLA was carried out with only a single antibody as indicated to verify the absence of background signal generated by the antibody alone. Error bars show mean and standard deviation

References

1. Skourti-Stathaki K, Proudfoot NJ (2014) A double-edged sword: R loops as threats to genome integrity and powerful regulators of gene expression. Genes Dev 28:1384–1396. https://doi.org/10.1101/GAD.242990.114
2. Hegazy YA, Fernando CM, Tran EJ (2020) The balancing act of R-loop biology: the good, the bad, and the ugly. J Biol Chem 295:905–913. https://doi.org/10.1016/S0021-9258(17)49903-0
3. Ketley RF, Gullerova M (2020) Jack of all trades? The versatility of RNA in DNA double-strand break repair. Essays Biochem 64:721–735. https://doi.org/10.1042/EBC20200008
4. D'Alessandro G, Whelan DR, Howard SM et al (2018) BRCA2 controls DNA:RNA hybrid level at DSBs by mediating RNase H2 recruitment. Nat Commun 9(1):5376. https://doi.org/10.1038/s41467-018-07799-2
5. Lu W-T, Hawley BR, Skalka GL et al (2018) Drosha drives the formation of DNA:RNA hybrids around DNA break sites to facilitate DNA repair. Nat Commun 9(1):532. https://doi.org/10.1038/s41467-018-02893-x
6. Zhang C, Chen L, Peng D et al (2020) METTL3 and N6-methyladenosine promote homologous recombination-mediated repair of DSBs by modulating DNA-RNA hybrid accumulation. Mol Cell 79:425–442.e7.

https://doi.org/10.1016/J.MOLCEL.2020.06.017

7. Chen H, Yang H, Zhu X et al (2020) m 5 C modification of mRNA serves a DNA damage code to promote homologous recombination. Nat Commun 11(1):2834. https://doi.org/10.1038/s41467-020-16722-7
8. Hegazy M, Cohen-Barak E, Koetsier JL et al (2020) Proximity ligation assay for detecting protein-protein interactions and protein modifications in cells and tissues in situ. Curr Protoc Cell Biol 89:e115. https://doi.org/10.1002/CPCB.115
9. How Proximity Ligation Assays (PLA) Work. https://www.sigmaaldrich.com/GB/en/technical-documents/technical-article/protein-biology/protein-and-nucleic-acid-interactions/how-pla-works. Accessed 24 July 2021
10. Carpenter AE, Jones TR, Lamprecht MR et al (2006) CellProfiler: image analysis software for identifying and quantifying cell phenotypes. Genome Biol 7:R100. https://doi.org/10.1186/GB-2006-7-10-R100
11. Söderberg O, Gullberg M, Jarvius M et al (2006) Direct observation of individual endogenous protein complexes in situ by proximity ligation. Nat Methods 3(12):995–1000. https://doi.org/10.1038/nmeth947
12. Fredriksson S, Gullberg M, Jarvius J et al (2002) Protein detection using proximity-dependent DNA ligation assays. Nat Biotechnol 20(5):473–477. https://doi.org/10.1038/nbt0502-473
13. Avolio R, Järvelin AI, Mohammed S et al (2018) Protein syndesmos is a novel RNA-binding protein that regulates primary cilia formation. Nucleic Acids Res 46:12067. https://doi.org/10.1093/NAR/GKY873
14. Jarvius M, Paulsson J, Weibrecht I et al (2007) In situ detection of phosphorylated platelet-derived growth factor receptor β using a generalized proximity ligation method *. Mol Cell Proteomics 6:1500–1509. https://doi.org/10.1074/MCP.M700166-MCP200
15. Zhang W, Xie M, Shu M-D et al (2016) A proximity-dependent assay for specific RNA--protein interactions in intact cells. RNA 22:1785. https://doi.org/10.1261/RNA.058248.116
16. Petruk S, Fenstermaker TK, Black KL et al (2016) Detection of RNA-DNA association by a proximity ligation-based method. Sci Rep 6:27313. https://doi.org/10.1038/srep27313
17. Galbiati A, d'Adda di Fagagna F (2019) DNA damage in situ ligation followed by proximity ligation assay (DI-PLA). Methods Mol Biol 1896:11–20. https://doi.org/10.1007/978-1-4939-8931-7_2
18. Chen L, Chen J-Y, Zhang X et al (2017) R-ChIP using inactive RNase H reveals dynamic coupling of R-loops with transcriptional pausing at gene promoters. Mol Cell 68:745. https://doi.org/10.1016/J.MOLCEL.2017.10.008
19. Cristini A, Groh M, Kristiansen MS, Gromak N (2018) RNA/DNA hybrid Interactome identifies DXH9 as a molecular player in transcriptional termination and R-loop-associated DNA damage. Cell Rep 23:1891. https://doi.org/10.1016/J.CELREP.2018.04.025
20. Smolka JA, Sanz LA, Hartono SR, Chédin F (2021) Recognition of RNA by the S9.6 antibody creates pervasive artifacts when imaging RNA:DNA hybrids. J Cell Biol 220(6):e202004079. https://doi.org/10.1083/JCB.202004079

Chapter 20

Biochemical Analysis of RNA–DNA Hybrid and R-Loop Unwinding Via Motor Proteins

Arijit Dutta, Youngho Kwon, and Patrick Sung

Abstract

R-loops, three-stranded RNA–DNA hybrids that arise mostly during transcription, could cause genomic instability via distinct routes. Detection of genomic RNA–DNA hybrids via immunofluorescence and RNA–DNA hybrid immunoprecipitation techniques have facilitated the discovery of many cellular factors that maintain R-loop homeostasis. One of multiple R-loop avoidance mechanisms is mediated by several nucleic acid motor proteins that utilize the energy from ATP hydrolysis to dissociate the R-loop structure. The biochemical activity of such motor proteins can be interrogated using synthetic R-loop substrates. Here, we describe methods to generate R-loop and RNA–DNA substrates for studying the activity of R-loop processing motor proteins such as human DHX9 and *S. cerevisiae* Pif1. Such studies provide valuable information regarding the directionality, nucleic acid strand preference, and processivity of these enzymes. Moreover, these substrates and companion biochemical assays provide the requisite tool for understanding the action of physiologically relevant regulators of these motor proteins.

Key words RNA–DNA hybrid, R-loop, DEXH-box helicase, DHX9, Pif1

1 Introduction

R-loops are three-stranded RNA–DNA hybrids that occur frequently during transcription. These structures harbor a base paired RNA–DNA duplex with an appended displaced single-stranded DNA (ssDNA) [1]. R-loops are found more frequently at genomic landscapes with high-GC content, repeat sequences, or regions prone to forming G-quadruplex structures [2]. They serve physiological functions in regulating gene expression [3] and class-switch recombination in immune cells [4]. However, elevated levels of R-loops, especially those arising from perturbations of transcription or transcription-coupled mRNA processing, lead to genomic instability via head on collision with replicative DNA polymerases [5] or generation of DNA strand breaks by endonucleases [6]. Moreover, R-loops could arise from the action of DNA binding proteins such as RAD51 [7, 8], RAD52 [9], and RPA [10]; whether and how

Andrés Aguilera and Alexey Ruzov (eds.), *R-Loops: Methods and Protocols*, Methods in Molecular Biology, vol. 2528,
https://doi.org/10.1007/978-1-0716-2477-7_20,

these R-loop structures serve physiological roles remain to be established. Abnormally high R-loop levels are etiologically linked to cognitive and neurodegenerative diseases such as amyotrophic lateral sclerosis (ALS) and fragile X-associated tremor/ataxia syndrome (FXTAS), as well as to multiple cancer types [11]. Cell-based studies have revealed distinct factors and mechanisms that affect R-loop homeostasis, namely: (a) proper functioning of the RNA biogenesis and export machinery (THO/ TREX, UAP56), splicing factors (SRSF1), and RNA exosome machinery as being crucial for preventing the accumulation of potentially pathogenic R-loops; (b) topoisomerase 1 prevents formation of R-loops via relaxing the DNA negative supercoils generated by the elongating RNA Pol II; (c) RNaseH1/2 recognizes RNA–DNA hybrids and digest the RNA strand [12]; **(iv)** ATP-dependent nucleic acid motor proteins, such as SETX, FANCM, DHX9, and Pif1 help resolve R-loops via unwinding the RNA–DNA hybrids. This chapter describes methods to generate RNA–DNA hybrids and R-loop substrates (Fig. 1) and to analyze the activity of DHX9 and Pif1.

DHX9, also known as RNA helicase A, is a DExD/H-Box superfamily II helicase. DHX9 associates with RNA Polymerase II and plays regulatory roles in transcription, translation, RNA processing [13], and in the maintenance of genomic stability via R-loop resolution [14, 15]. DHX9 is overexpressed in different cancers including breast, colorectal, and lung cancers, making it a potential therapeutic target for cancer treatment [13]. DHX9 has a conserved Asp-Glu-Ile-His (DEIH spanning amino acid residues 511–514) sequence in its helicase core domain that interacts with the NTP via a coordinated Mg^{2+} ion, that is required for NTP hydrolysis-coupled nucleic acid unwinding [13]. Mutation of Asp511 and Glu512 to alanine abolishes NTP binding and hence, the unwinding activity of DHX9. Herein, we test DHX9 for its activity on various RNA–DNA hybrid and R-loop substrates, confirming that it preferentially unwinds RNA–DNA hybrid and R-loop with 3′-RNA overhangs (Fig. 3) and possesses a 3′ to 5′ directionality when translocating on RNA, as previously reported [14].

Pif1 is a superfamily (SF) 1B helicase with seven conserved amino acid motifs (I, Ia, II, III, IV, V, and VI). Motifs I (Walker-A), II (Walker B), and IV are involved in binding NTP/Mg^{2+} and hydrolyzing ATP, while the other conserved motifs help couple ATP hydrolysis to nucleic acid unwinding [17–19]. Studies with *S. cerevisiae* Pif1 (scPif1) demonstrated that it can unwind both DNA duplex and RNA–DNA hybrid with a 5′ to 3′ directionality and a higher processivity for the latter [18, 19]. Pif1 also binds and unwinds G4 DNA structures [20]. Pif1 promotes Okazaki fragment maturation, DNA replication across G4 DNA structures, telomere maturation, and DNA double-strand break repair via break-induced replication and suppresses the accumulation of R-loops across tRNA genes [16, 21–23]. Herein, we confirmed

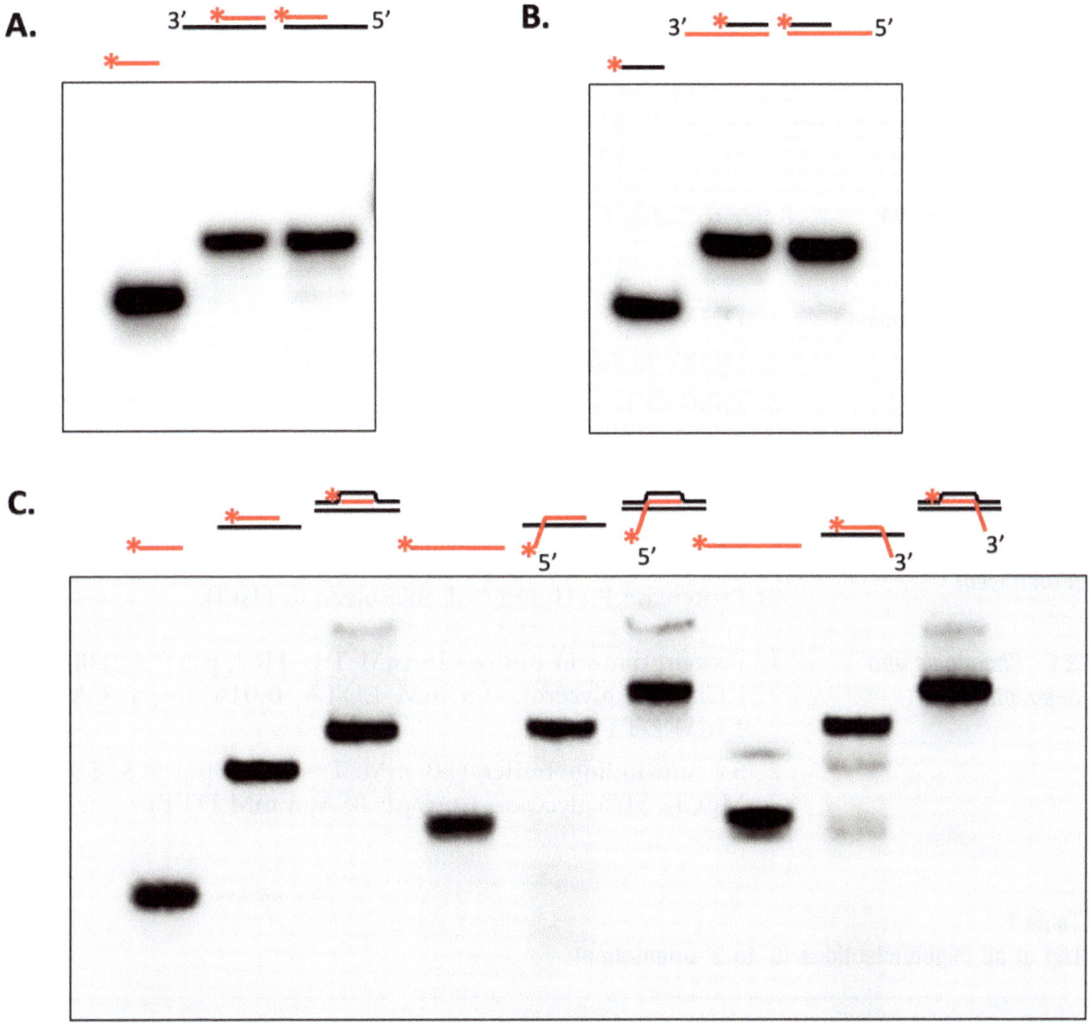

Fig. 1 Making of RNA–DNA hybrid and R-loop substrates: (**a**) 30 bases RNA (R3*), RNA–DNA hybrid with 5′-DNA overhang (R3* + D4), and RNA–DNA hybrid with 3′-DNA overhang (R3* + D5), separated in 15% polyacrylamide gel at 100 V for 60 min (**b**) 30 bases DNA (D3*), RNA–DNA hybrid with 5′-RNA overhang (D3* + R1), and RNA–DNA hybrid with 3′-RNA overhang (D3* + R2), separated in a 15% polyacrylamide gel at 100 V for 60 min (**c**) 30 bases RNA (R3*), RNA–DNA hybrid (R3* + D2), R-loop with no RNA overhang (R3* + D1 + D2), 60 bases RNA (R1*), RNA–DNA hybrid (R1* + D2), R-loop with 5′-RNA overhang (R1* + D1 + D2), 60 bases RNA (R2*), RNA–DNA hybrid (R2* + D2), R-loop with 3′-RNA overhang (R2* + D1 + D2), separated in a 10% polyacrylamide gel at 100 V for 60 min. "*" denotes ^{32}P-labeled oligonucleotide. RNA is represented in red color

that scPif1 unwinds RNA–DNA hybrid via translocation only on ssDNA in 5′ to 3′ direction (Fig. 4).

2 Materials

2.1 DNA and RNA Oligonucleotides

Refer to Table 1.

2.2 Purified Proteins (See Note 1)

1. DHX9-WT.
2. DHX9-helicase dead mutant (D511A, E512A).
3. scPif1-WT.
4. scPif1-helicase dead mutant (K264A).

2.3 Enzymes/ Reaction Buffers (Purchased)

1. T4 Polynucleotide Kinase (T4 PNK)/T4 PNK buffer.
2. T7 RNA polymerase/T7 RNA polymerase buffer.
3. Proteinase K (10 mg/mL, dissolved in H_2O).

2.4 Chemicals and Reagents

1. Protein dilution buffer (25 mM Tris–HCl, pH 7.5, 100 mM KCl, 10% glycerol, 0.5 mM EDTA, 0.01% Igepal CA-630, 1 mM DTT).
2. 5× unwinding buffer (50 mM Tris–HCl, pH 7.5, 50 mM $MgCl_2$, 50% glycerol, 1 μg/μL BSA, 5 mM DTT).

Table 1
List of all oligonucleotides (5′ to 3′ orientation)

Name	Size	Sequence
D1	90 nt	CATTGCATATTTAAAACATGTTGGATCCCACGTTGCATGCTGATAGCCTAC TAGAGCTGTATGAATTCAAATGACCTCTTATCAAGTGAC
D2	90 nt	GTCACTTGATAAGAGGTCATTTGAATTCATGGCTTAGAGCTTAATTGCTGAA TCTGGTGCTGGGATCCAACATGTTTTAAATATGCAATG
D3	30 nt	GGCTTAGAGCTTAATTGCTGAATCTGGTGC
D4	60 nt	GTCACTTGATAAGAGGTCATTTGAATTCATGGCTTAGAGCTTAATTGCTGAA TCTGGTGC
D5	60 nt	GGCTTAGAGCTTAATTGCTGAATCTGGTGCTGGGATCCAACATGTTTTAAATA TGCAATG
R1	60 nt	GUGCUACGAUGCUAGUCGUAGCUCGGGAGUGCACCAGAUUCAGCAA UUAAGCUCUAAGCC
R2	60 nt	GCACCAGAUUCAGCAAUUAAGCUCUAAGCCGCUGACGGCUCGAUGCUGA UCGUAGCAUCG
R3	30 nt	GCACCAGAUUCAGCAAUUAAGCUCUAAGCC

3. 10× annealing buffer (100 mM Tris–HCl, pH 7.5, 500 mM NaCl).
4. 10× loading buffer (100 mM Tris–HCl, pH 7.5, 10 mM EDTA, 50% Glycerol, 0.15% Orange G).
5. RNasin® Ribonuclease Inhibitor (Promega).
6. DEPC-treated water.
7. BSA: 20 μg/μL.
8. ATP: 100 mM, pH 7.5.
9. AMP-PNP: 100 mM.
10. $MgCl_2$: 50 mM.
11. [γ-^{32}P]ATP (6000 Ci/mmol, 10 μCi/μL) (Perkin Elmer).
12. Native gel loading buffer (15 mM Tris–HCl, pH 7.5, 25% glycerol, 0.05% Orange G).
13. SDS (10% dissolved in water).
14. Proteinase K (10 μg/μL, dissolved in H_2O).
15. Micro-BioSpin 6 columns (Bio-rad).
16. Whatman 3MM chromatography paper.
17. Hybond-N (GE Healthcare).
18. Acrylamide, N,N′-Methylene bis acrylamide (30% solution, 37.5: 1).
19. TEMED.
20. Ammonium persulfate, 10% solution in water.
21. Gel fixing solution (50% methanol, 20% polyethylene glycol 400).
22. 15% polyacrylamide gel with 1x TAE (Mini-PROTEAN 10.1 × 7.3 cm, 1.5 mm thick).

2.5 Apparatus

1. Cell Polyacrylamide gel electrophoresis apparatus: Mini-PROTEAN Tetra (Bio-Rad).
2. Mini-PROTEAN glass plates (short plate—10.1 × 7.3 cm, spacer plate—10.1 × 8.2 cm).
3. Gel dryer equipped with a vacuum pump.
4. Phosphorimager (e.g., Personal Molecular Imager system, Bio-Rad Laboratories or Amersham Typhoon Gel Imaging system, GE Healthcare) and imaging analysis software (Image lab, Bio-Rad Laboratories or ImageQuant, GE Healthcare).
5. Heating blocks or water-baths.
6. Microcentrifuge.
7. Storage Phosphor Screen.
8. Storage Cassette.

3 Methods

The following protocol describes a method to construct RNA–DNA hybrid and R-loop using ^{32}P-5′-end labeled DNA and RNA oligonucleotides. Alternatively, fluorescent-labeled oligonucleotides can be used to generate substrates. Unwinding of the substrates can be analyzed by polyacrylamide gel electrophoresis under native condition, followed by phosphor-imaging analysis. The unwound labeled DNA or RNA has a faster gel mobility compared to intact substrates. Heat treatment of the substrate releases the labeled strand to guide the identification of product resulting from helicase action (as shown in Figs. 3 and 4).

3.1 Labeling of 5′-OH of DNA and RNA Oligonucleotide with γ-^{32}P-ATP

Refer to Table 1 and Fig. 2 for labeling the appropriate RNA or DNA oligonucleotide for the specific RNA–DNA or R-loop substrate.

1. Set the labeling reaction mix as follows:

RNA or DNA oligonucleotide (100 μM)	1 μL
10× T4-PNK buffer	2 μL
T4-PNK	2 μL
γ-^{32}P-ATP (6000 Ci/mmol, 10 μCi/μL)	2 μL
H_2O	13 μL

2. Incubate the reaction mix at 37 °C for 1 h. Then incubate at 65 °C, 10 min to stop the reaction. Purify the radio-labeled oligonucleotide and remove excess γ-^{32}P-ATP with micro-BioSpin 6 column.

3.2 Generation of RNA–DNA and R-Loop Substrates

Refer to Fig. 2 for annealing combinations of labeled and unlabeled RNA and DNA oligonucleotides for generating specific RNA–DNA or R-loop substrate (*see* **Note 2**).

1. Set the labeling reaction mix in PCR tubes as follows:

^{32}P-labeled RNA or DNA oligonucleotide	1 pmole
Complementary RNA or DNA oligonucleotide	1 pmole
Third DNA oligonucleotide (for R-loops)	1 pmole
10× annealing buffer	2 μL
H_2O	To 20 μL

2. Incubate the tubes in a PCR machine with an annealing cycle (*see* **Note 3**). Check the annealed oligonucleotides by running 2–5 fmoles of the annealed product in a 15% polyacrylamide gel

R-loop with 5′-RNA overhang	D1+D2+R1*	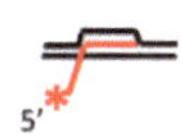
R-loop with 3′-RNA overhang	D1+D2+R2*	
R-loop with no-RNA overhang	D1+D2+R3*	
RNA-DNA hybrid with 5′-RNA overhang	D3*+R1	
RNA-DNA hybrid with 3′-RNA overhang	D3*+R2	
RNA-DNA hybrid with 5′-DNA overhang	R3*+D4	
RNA-DNA hybrid with 3′-DNA overhang	R3*+D5	

Fig. 2 Scheme for generating RNA–DNA hybrid and R-loop substrates

with 1× TAE at 100 V for 60 min. Correct annealing of the oligonucleotides can be confirmed from the difference in electrophoretic mobility of single, duplex, and triplex oligonucleotide as shown in Fig. 1 (*see* **Note 4**).

3. Dilute the annealed product to 50 nM with 1× annealing buffer, make 50 μL aliquots and store at −20 °C or 4 °C (*see* **Note 5**).

3.3 Nucleic Acid Unwinding Reactions

1. Make the following reaction mixture and a helicase (DHX9 or scPifl with final concentrations as indicated in Figs. 3 and 4).

5× unwinding buffer	2 μL
RNA–DNA hybrid or R-loop (50 nM)	1 μL
10 mM ATP or AMP-PNP (*see* **Note 6**)	1 μL
0.5 M KCl (*see* **Note 7**)	0.6 μL
RNasin® (*see* **Note 8**)	1 μL
Protein (*see* **Note 9**)	1 μL
H_2O	To make 10 μL

2. Incubate the tubes at 37 °C for 20 min. For certain helicases, the addition of 100 molar fold excess of unlabeled RNA/DNA oligonucleotide (same as what is labeled in the corresponding

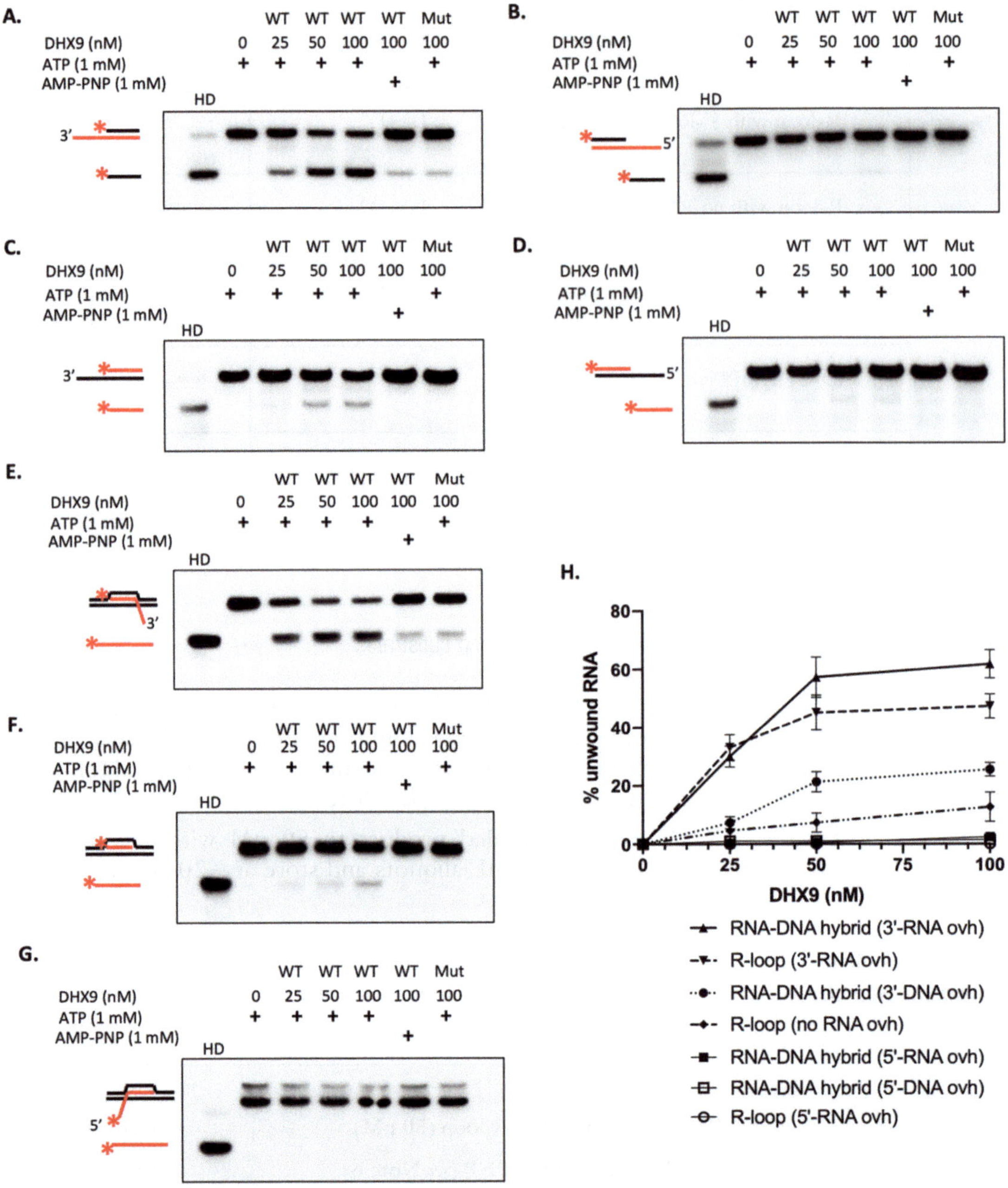

Fig. 3 DHX9 helicase activity on RNA–DNA hybrids with 3′-RNA overhang (**a**), 5′-RNA overhang (**b**), 3′-DNA overhang (**c**), 5′-DNA overhang (**d**); and R-loops with 3′-RNA overhang (**e**), no RNA overhang (**f**), and 5′-RNA overhang (**g**). DHX9 mut: DHX9 helicase dead mutant (D511A, E512A). "HD" denotes heat denatured substrate. Unwound RNA or DNA is represented as the "red" or "black" line, respectively. The quantification results are graphed in **h**

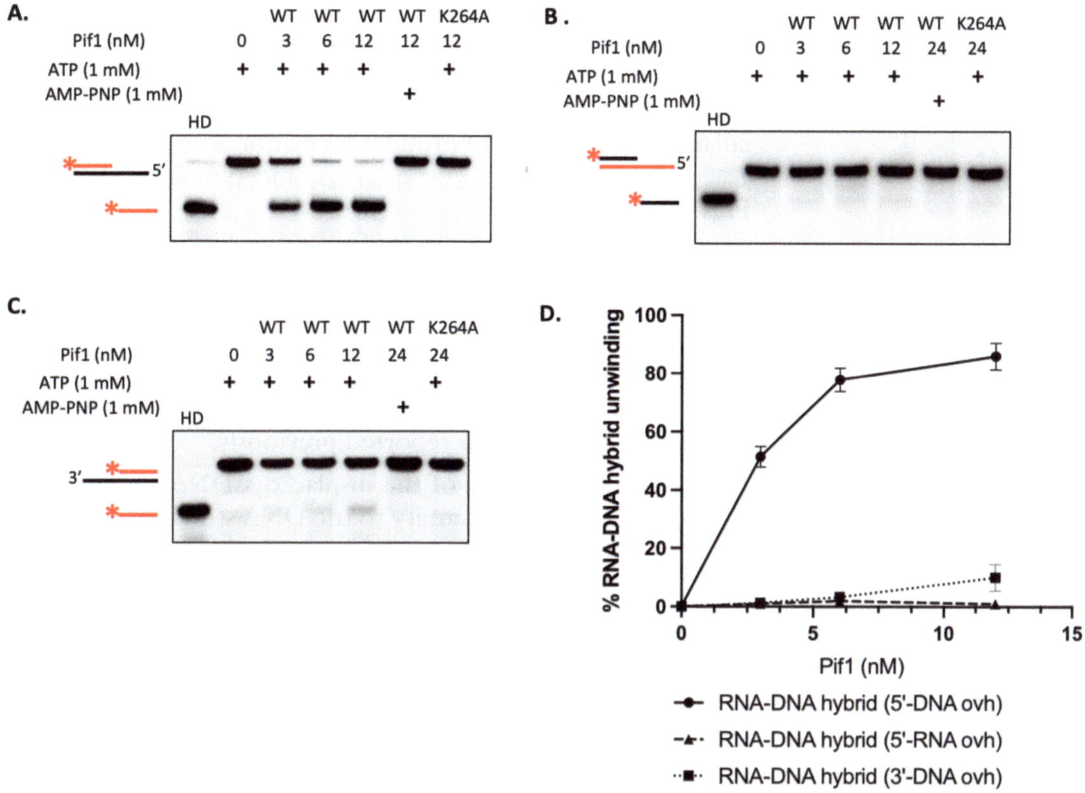

Fig. 4 scPif1 helicase activity on RNA–DNA hybrids with 5′-DNA overhang (**a**), 5′-RNA overhang (**b**), and 3′-DNA overhang (**c**). K264A: Pif1 helicase dead mutant. "HD" denotes heat denatured substrate. Unwound RNA or DNA is represented as the "red" or "black" line, respectively. The quantification results are graphed in **d**

R-loop/RNA–DNA substrate) in the reaction mix may be required to observe unwinding (*see* **Note 10**).

3. Stop the reactions by adding 1 μL 1% SDS and 1 μL 10 mg/mL Proteinase K and incubating at 37 °C for 5 min. Finally add 2 μL 10× loading buffer (*see* **Note 11**).

It is important to follow precautionary measures to avoid RNase contamination when carrying out these assays, which would result in degradation of the RNA (*see* **Note 8**).

3.4 Native Gel Electrophoresis

1. Immerse the gel electrophoresis apparatus in an ice bath to maintain low temperature during electrophoresis. Run 7 μL reaction mixes along with heat denatured substrate (heated at 95 °C, 5 min) in a 15% TAE native gel at 100 V for approximately 60 min, or until the dye front reaches the bottom of the gel. After electrophoresis, fix the polyacrylamide gel in fixing solution (50% methanol, 20% polyethylene glycol 400) for 15 min, and then place the gel on a sheet of Hybond-N on top of filter paper. Cover the gel with plastic wrap and dry the

gel under vacuum at 80 °C for 1 h. Expose the dried gel to a phosphor screen overnight, then scan the screen in the phosphorimager, and quantify band intensities using the image analysis software (ImageQuant TL 8.0, GE Healthcare). Calculate the percentage of product formation using the formula: $I_{product} \times 100/(I_{substrate} + I_{product})$, and plot it as shown in Figs. 3h and 4d.

4 Notes

1. scPif1-WT and the helicase mutant were purified from *E. coli* [23], and DHX9-WT and the helicase mutant were purified from insect cells [14], as reported previously.
2. Note that the sequence of the displaced ssDNA strand in the R-loops is not complementary to the DNA strand that is hybridized to RNA. Stable R-loop structures could not be generated with completely complementary DNA oligonucleotides because of spontaneous DNA branch migration that would result in the eviction of the RNA.
3. Set the annealing reaction cycle as 10 min at 95 °C for 10 min and gradually decreasing the temperature to 10 °C at 1 °C/min.
4. If the gel electrophoresis of the final product shows more than one band corresponding to duplex DNA or RNA–DNA hybrid, it indicates that the oligonucleotides were not completely annealed. This could be due to an unequal amount of oligonucleotides in the annealing mixture. Measure the oligonucleotide concentrations correctly and use the equimolar amount for annealing.
5. The RNA–DNA/R-loop substrates are stable at 4 °C for 4 weeks approximately. For long-term storage, keep as aliquots in a −20 °C freezer.
6. AMP-PNP is a non-hydrolyzable analog of ATP and would not support the nucleic acid unwinding activity of motor proteins.
7. Adjust the final KCl concentration to 50 mM considering the amount of KCl derived from protein stocks.
8. Addition of RNasin® Ribonuclease inhibitor prevents degradation of the unwound RNA by non-specific RNases that may be present in the purified proteins.
9. Proteins were diluted in a buffer containing 25 mM Tris–HCl, pH 7.5, 100 mM KCl, 10% glycerol, 0.5 mM EDTA, 0.01% Igepal CA-630, 1 mM DTT to make final concentrations as indicated in Figs. 3 and 4. Perform biochemical titration experiments to determine the optimum range of the protein

concentration used for the unwinding reactions, starting with a broad range of concentration and then narrowing the range. Reaction temperature and incubation time are also important experimental variables.

10. To prevent spontaneously annealing of the unwound ^{32}P-labeled nucleic acid strand to the complementary strand in substrates, add an excess of the unlabeled version of the nucleic acid to reactions before analysis. This is particularly important when testing scPif1 (data not shown).
11. Excess SDS in the stop buffer could result in smudgy bands. If this happens, decrease the percentage of SDS in the final mixture.

Acknowledgments

Studies in the laboratory of the authors have been supported by NIH grants R35 CA241801, RO1 CA168635, and RO1 ES007061. We thank Dr. Helmut Pospiech and Late Dr. Frank Große from Leibniz Institute for Age Research - Fritz Lipmann Institute (FLI) at Jena, Germany for providing P. Sung with the pFastBac-His-DHX9 construct.

References

1. Crossley MP, Bocek M, Cimprich KA (2019) R-loops as cellular regulators and genomic threats. Mol Cell 73:398–411
2. Skourti-Stathaki K, Proudfoot NJ (2014) A double-edged sword: R loops as threats to genome integrity and powerful regulators of gene expression. Genes Dev 28:1384–1396
3. Niehrs C, Luke B (2020) Regulatory R-loops as facilitators of gene expression and genome stability. Nat Rev Mol Cell Biol 21:167–178
4. Yu K, Chedin F, Hsieh CL, Wilson TE, Lieber MR (2003) R-loops at immunoglobulin class switch regions in the chromosomes of stimulated B cells. Nat Immunol 4:442–451
5. Hamperl S, Bocek MJ, Saldivar JC, Swigut T, Cimprich KA (2017) Transcription-replication conflict orientation modulates R-loop levels and activates distinct DNA damage responses. Cell 170:774–786.e19
6. Sollier J, Stork CT, Garcia-Rubio ML, Paulsen RD, Aguilera A, Cimprich KA (2014) Transcription-coupled nucleotide excision repair factors promote R-loop-induced genome instability. Mol Cell 56:777–785
7. Feretzaki M, Pospisilova M, Valador Fernandes R, Lunardi T, Krejci L, Lingner J (2020) RAD51-dependent recruitment of TERRA lncRNA to telomeres through R-loops. Nature 587:303–308
8. Wahba L, Gore SK, Koshland D (2013) The homologous recombination machinery modulates the formation of RNA-DNA hybrids and associated chromosome instability. elife 2: e00505
9. Mazina OM, Keskin H, Hanamshet K, Storici F, Mazin AV (2017) Rad52 inverse strand exchange drives RNA-templated DNA double-strand break repair. Mol Cell 67:19–29.e3
10. Mazina OM, Somarowthu S, Kadyrova LY, Baranovskiy AG, Tahirov TH, Kadyrov FA, Mazin AV (2020) Replication protein a binds RNA and promotes R-loop formation. J Biol Chem 295:14203–14213
11. Richard P, Manley JL (2017) R loops and links to human disease. J Mol Biol 429:3168–3180
12. Garcia-Muse T, Aguilera A (2019) R loops: from physiological to pathological roles. Cell 179:604–618
13. Lee T, Pelletier J (2016) The biology of DHX9 and its potential as a therapeutic target. Oncotarget 7:42716–42739

14. Chakraborty P, Grosse F (2011) Human DHX9 helicase preferentially unwinds RNA-containing displacement loops (R-loops) and G-quadruplexes. DNA Repair (Amst) 10:654–665
15. Cristini A, Groh M, Kristiansen MS, Gromak N (2018) RNA/DNA hybrid interactome identifies DXH9 as a molecular player in transcriptional termination and R-loop-associated DNA damage. Cell Rep 23:1891–1905
16. Tran PLT, Pohl TJ, Chen CF, Chan A, Pott S, Zakian VA (2017) PIF1 family DNA helicases suppress R-loop mediated genome instability at tRNA genes. Nat Commun 8:15025
17. Dehghani-Tafti S, Levdikov V, Antson AA, Bax B, Sanders CM (2019) Structural and functional analysis of the nucleotide and DNA binding activities of the human PIF1 helicase. Nucleic Acids Res 47:3208–3222
18. Boule JB, Zakian VA (2007) The yeast Pif1p DNA helicase preferentially unwinds RNA DNA substrates. Nucleic Acids Res 35:5809–5818
19. Chib S, Byrd AK, Raney KD (2016) Yeast helicase Pif1 unwinds RNA:DNA hybrids with higher processivity than DNA:DNA duplexes. J Biol Chem 291:5889–5901
20. Paeschke K, Bochman ML, Garcia PD, Cejka P, Friedman KL, Kowalczykowski SC, Zakian VA (2013) Pif1 family helicases suppress genome instability at G-quadruplex motifs. Nature 497: 458–462
21. Buzovetsky O, Kwon Y, Pham NT, Kim C, Ira G, Sung P, Xiong Y (2017) Role of the Pif1-PCNA complex in pol delta-dependent strand displacement DNA synthesis and break-induced replication. Cell Rep 21:1707–1714
22. Rossi ML, Pike JE, Wang W, Burgers PMJ, Campbell JL, Bambara RA (2008) Pif1 helicase directs eukaryotic Okazaki fragments toward the two-nuclease cleavage pathway for primer removal. J Biol Chem 283:27483–27493
23. Wilson MA, Kwon Y, Xu Y, Chung WH, Chi P, Niu H, Mayle R, Chen X, Malkova A, Sung P et al (2013) Pif1 helicase and Poldelta promote recombination-coupled DNA synthesis via bubble migration. Nature 502:393–396

Chapter 21

Topology of RNA:DNA Hybrids and R-Loops in Yeast

Yathish Jagadheesh Achar and Marco Foiani

Abstract

RNA:DNA hybrids are generated naturally behind the elongating RNA polymerase as a transcriptional intermediate. However, prolonged persistence of these structures challenges the integrity of the genome by creating R-loops and by interfering with DNA replication and other chromatin related processes. Precise mapping and characterization of their distribution across the genome has been a major challenge to understand the genesis of RNA:DNA hybrids and their conversion into genotoxic intermediates. Here we provide the detailed protocol for mapping RNA:DNA hybrid across the *Saccharomyces cerevisiae* genome.

Key words R-loop, RNA:DNA hybrid, S9.6 antibody, DRIP-chip

1 Introduction

Formation of RNA:DNA hybrids out of their natural context can lead to the formation of secondary DNA structure, known as R-loops. These are triplex nucleic acid structures, where nascent RNA is bound to one of the DNA strands by displacing the other strand. R-loops are detected in most living organisms and their formation and accumulation in vivo depends on several factors including DNA supercoil, GC content, and propensity to form alternative DNA structures in the displaced ssDNA such as G-quartets [1–3]. More often R-loops are considered to have adverse effects on biological processes except in situations such as in immunoglobulin (Ig) class-switch recombination (CSR). It has been shown both in yeast and human cells that deficiency in RNA and DNA metabolic factors may lead to the accumulation of RNA: DNA hybrids and effect several cellular process including replication, transcription, and recombination.

In recent years, R-loops have emerged as key regulators of cellular processes and also a major threat for genome integrity [4]. It has been particularly challenging to characterize those genetic, pharmacological, and environmental factors leading to

Andrés Aguilera and Alexey Ruzov (eds.), *R-Loops: Methods and Protocols*, Methods in Molecular Biology, vol. 2528,
https://doi.org/10.1007/978-1-0716-2477-7_21,

R-loop formation. In lower eukaryotes such as budding yeast the majority of RNA:DNA hybrids are non-harmful and form as a consequence of co-transcriptional events [5]. To better understand RNA:DNA hybrids role in genome instability several laboratories have developed independent protocols which allowed RNA:DNA hybrid mapping procedures at the genome scale. Here, we describe a DRIP-chip protocol, which will provide a population average view of RNA:DNA hybrid distribution in *Saccharomyces cerevisiae.* The following DRIP-chip relies first on the immunoprecipitation step using S9.6 monoclonal antibody, which displays a strong affinity for the RNA:DNA hybrid moiety, followed by hybridization to high-resolution tilling arrays.

2 Materials

Prepare all the buffers and solutions using ultrapure water and analytical grade reagents.

2.1 Preparation of S9.6 Magnetic Beads

1. Protein-A magnetic beads (Invitrogen, cat. no. 10001D).
2. Anti-DNA:RNA hybrid monoclonal antibody S9.6 serum [6].
3. 1.7 mL prelubricated Costar tubes (Corning, NY).
4. Magnet racks (DynaMag™ Invitrogen).
5. 2× Bead buffer: 2× Phosphate buffered saline (PBS), 2 mg/mL Bovine serum albumin, 0.2% Tween-20.

2.2 Crosslinking and Cells Collection

1. *Saccharomyces cerevisiae* W303 derivatives.
2. YPD medium.
3. 37% formaldehyde.
4. 2.5 M Glycine stock solution.
5. 1× TBS: 20 mM Tris–HCl, pH 7.5, 150 mM NaCl.

2.3 Cell Lysis and Chromatin Extract

1. Lysis buffer: 50 mM Hepes–KOH, pH 7.5, 140 mM NaCl, 1 mM EDTA, 1% Triton-X100, 0.1% Na-deoxycholate.
2. PMSF stock solution: 200 mM phenylmethanesulfonyl fluoride in isopropanol.
3. Antiproteolytic cocktail: Complete protease inhibitor tablets (Roche, Indianapolis, IN).
4. 2-mL O-ring screw cap tubes.
5. Zirconia beads.
6. Bead beater: FastPrep-24™ Classic bead beating grinder and lysis system (MP Biomedicals).
7. Branson Digital Sonifier SFX 250.

2.4 Bead Washing and Crosslink Reversal

1. Wash buffer: 10 mM Tris–HCl, pH 8.0, 250 mM LiCl, 0.5% NP-40, 0.5% Na-deoxycholate, 1 mM EDTA.
2. Elution buffer: 50 mM Tris–HCl, pH 8, 10 mM EDTA, 1% SDS.
3. TE: 10 mM Tris–HCl, pH 8, 1 mM EDTA.
4. 5 M NaCl stock solution.

2.5 DNA Purification and Library Amplification

1. Proteinase K (Roche): Stock 50 mg/mL.
2. RNase A (Sigma): Stock 10 mg/mL.
3. DNA clean & concentrator kit (Zymo research).
4. WGA2 GenomePlex complete Genome amplification (WGA) kit (Sigma).

2.6 DNA Digestion and End Labeling

1. 10× One-Phor-All-Buffer: 100 mM Tris-Acetate pH 7.5, 100 mM Mg-Acetate, 500 mM Potassium Acetate.
2. 25 mM $CoCl_2$.
3. DNAseI: Amplification Grade (Invitrogen)..
4. Terminal transferase (Roche).
5. Biotin-11-ddATP (NEN).

2.7 Microarray Hybridization

1. Hybridization buffer: 100 mM MES 1 M [Na+] 20 mM EDTA 0.01% Tween-20.
2. GeneChip® S. CEREVISIAE Tiling 1.0R cartridges (Affymetrix).
3. GeneChip hybrid oven (Affymetrix).
4. Affymetrix Fluidics Station 450 (Affymetrix).
5. Wash Buffer A (Nonstringent): 6× SSPE, 0.01% Tween-20.
6. Wash Buffer B (Stringent): 100 mM MES, 0.1 M [Na+], 0.01% Tween-20.
7. Stain Buffer: 100 mM MES, 1 M [Na+], 0.05% Tween-20.
8. SAPE Solution: 100 mM MES, 1 M [Na+], 0.05% Tween-20, 2 mg/mL acetylated BSA, 10 μg/mL Streptavidin Phycoerythrin.
9. Antibody Solution: 100 mM MES, 1 M [Na+], 0.05% Tween-20, 2 mg/mL acetylated BSA, 0.1 mg/mL Normal Goat IgG, 3 μg/mL biotinylated Anti-Streptavidin.
10. Array holding buffer: 100 mM MES, 1 M [Na+], 0.05% Tween-20.
11. Affymetrix GeneChip® Scanner 3000 7G.

3 Methods

Following DRIP-chip protocol was adopted and modified from published protocols [7, 8]. For this protocol we used *Saccharomyces cerevisiae* W303 derivatives. For each sample we parallelly process, the "Input DNA" and two immunoprecipitated replicates, IP-A and IP-B.

3.1 Preparation of S9.6 Antibody Beads

1. Mix thoroughly magnetic bead and transfer 100 μL of bead suspension to 1.7 mL of prelubricated microtubes.
2. Place the tube in a magnetic grid and wait till the beads attach to the magnet. Aspirate the clear solution without disturbing the bead and discard.
3. Wash beads once with 1 mL of cold 1× Bead buffer by gently rotating beads until all the beads are in suspension. Place the tube in a magnetic grid and remove the supernatant completely.
4. Resuspend the beads in ul of 2× Bead buffer and add μL of S9.6 antibody serum. Incubate with rotation at 4 °C overnight (*see* **Note 1**).
5. Before use, remove the antibody containing supernatant solution and was twice with ice cold 1× Bead buffer and finally resuspend in 100 μL of 1× Bead buffer.

3.2 Crosslinking and Cells Collection

1. Use 4×10^9 cells for each IP at a concentration of ~1×10^9 cells/mL (*see* **Note 2**).
2. Add 1% Formaldehyde and swirl gently at room temperature on a benchtop orbital shaker for 20 min.
3. Add 0.125 M glycine to quench the formaldehyde. Incubate at room temperature for 5 min with gentle and occasional swirl.
4. Transfer culture mix to prechilled 500 mL centrifuge tubes and incubate on ice for 10 min.
5. Centrifuge at 3200 × *g* for 5 min at 4 °C. Discard the supernatant and suspend cells in 30 mL of ice cold 1× TBS and transfer entire volume to 50 mL tube. Centrifuge at 3200 × *g* for 5 min at 4 °C and discard the supernatant and repeat the wash step. After the final wash, remove traces of liquid using vacuum and cell pellet is flash frozen in liquid nitrogen and stored at −80 °C.

3.3 Cell Lysis and Chromatin Extract Preparation for Immunoprecipitation

1. Thaw and dissolve the cell pellet in 4 mL of ice cold lysis buffer using pipette. Distribute equally into four 2 mL O-ring screw cap tubes and centrifuge at 3200 × *g* for 5 min at 4 °C. Carefully remove the supernatant using vacuum.

2. Resuspend cell pellet in 0.4 mL of lysis buffer (supplemented with 1 mM PMSF, 0.1% SDS and 1× antiproteolytic cocktail) and add zirconia beads up to 1 mm below the cell solution meniscus.
3. Lyse the cells using FastPrep-24 in cold room with the following set up: speed 6.5 m/s for 20 s. Repeat lysis step for 6 to 8 cycles. Between each round place the tubes on ice for 5 min (*see* **Note 3**).
4. Puncture the bottom of the 2 mL tube using syringe needle and insert tube into a chilled 15 mL collection tube. Centrifuge 1000 × *g* for 3 min at 4 °C.
5. Transfer complete lysate into a sterile 1.5 mL tube and centrifuge at 13,400 × *g* for 5 min at 4 °C.
6. Carefully remove the supernatant without disturbing the pellet and discard.
7. Resuspend the pellet in 1 mL of lysis buffer (supplemented with 1 mM PMSF, 0.2% SDS and 1× antiprotolytic cocktail).
8. Place the tubes on a rotating wheel at 4 °C for 90 min.
9. Centrifuge the tubes at 13400 × *g* for 5 min at 4 °C. Remove the supernatant without disturbing the pellet and discard.
10. Add 450 μL of lysis buffer (supplemented with 1 mM PMSF, 0.2% SDS, and 1× antiprotolytic cocktail) to each tube.
11. Shear the chromatin by applying six sonication cycles of 15 s at 15% output. In between each sonication cycles centrifuge at 2400 × *g* for 1 min at 4 °C (*see* **Note 4**).
12. Centrifuge the tubes at 16,000 × *g* for 5 min at 4 °C and transfer the supernatants to prechilled 1.7 mL of prelubricated tubes.
13. Collect 20 ul of supernatant as 'Input' cell extract and store at −20 °C (*see* **Note 5**).
14. Add 25 μL of previously washed S9.6 antibody coated magnetic beads from Subheading 3.1 to each four tubes and incubate on a rotating wheel at 4 °C overnight.

3.4 Bead Washing and Crosslink Reversal

1. Place the tubes in a magnetic grid and allow the beads attach to the magnet. Remove the supernatant using vacuum without disturbing the beads (*see* **Note 6**).
2. Wash the beads in the following order.
   (a) 2× with 1 mL of lysis buffer (without any supplements).
   (b) 2× with 1 mL of lysis buffer supplemented with 360 mM NaCl.
   (c) 2× with 1 mL of wash buffer.
   (d) 1× with 1 mL of 1× TE.

3. Remove the TE solution using a pipette and centrifuge beads at 1000 × *g* for 1 min at 4 °C. Place the tubes in a magnetic grid and allow the beads attach to the magnet and remove any liquid droplets remained using a vacuum.
4. Add 40 μL of freshly prepared elution buffer and incubate at 65 °C for 15 min with constant mixing.
5. Centrifuge at 11,400 × *g* for 1 min at room temperature.
6. Place the tubes back in the magnetic grid and collect supernatant to a fresh tube. At this stage all the four IP tubes can be pooled into one.
7. Add 140 μL of elution buffer to the 20 μL "Input" samples stored at −20 °C.
8. Incubate all the three tubes (Input, IP-A and IP-B) overnight at 65 °C for the crosslink reversal.

3.5 DNA Purification and Library Amplification

1. Consolidate the samples by brief spin at room temperature and add 2 μL of RNase A (10 mg/mL stock) and 10 μL of Proteinase K (50 mg/mL stock). Mix gently and incubate at 50 °C for 2 h.
2. DNA is purified using DNA clean & concentrator kit from Zymo research. Add 600 μL of DNA binding buffer to each tube, mix briefly by vertexing (*see* **Note** 7).
3. Transfer mixture to a Zymo-spin column and centrifuge at 15,000 × *g* for 30 s. Discard the flow-through.
4. Add 200 μL of DNA wash buffer from the kit to each column and centrifuge 15,000 × *g* for 30 s. Repeat the wash step one more time.
5. Spin the empty column at 15,000 × *g* for 3 min to remove any residual ethanol remaining in the column.
6. Transfer the column to a fresh 1.5 mL tube and elute by adding 12 μL elution buffer provided in the kit to column matrix and incubate at room for 2 min and centrifuge 15,000 × *g* for 1 min. Discard the column and store DNA on ice.
7. Transfer 10 μL of purified DNA into sterile PCR tubes and cool the samples on ice.
8. DNA is amplified using WGA2 GenomePlex complete Genome amplification kit (WGA) using manufacturer's instructions.
9. For each sample add.

   2 μL of 1× Library preparation buffer.

   1 μL of Library stabilization solution.
10. Vortex thoroughly and centrifuge very briefly. Place the tubes in a thermal cycler for 95 °C for 2 min.

11. Immediately place the tubes on ice, let the samples cool down completely and centrifuge briefly to consolidate.
12. Add 1 μL of Library preparation of Enzyme and mix thoroughly and centrifuge briefly and keep the tubes on ice.
13. Place the tube in a thermal cycle and incubate as follows.

    16 °C for 20 min.

    24 °C for 20 min.

    37 °C for 20 min.

    75 °C for 5 min

    4 °C On Hold.
14. Centrifuge briefly and place the tubes on ice.
15. For amplification add the following reagents from the kit to each tube.

    48.5 μL of Nuclease free water

    7.5 μL of 10× Amplification master mix

    5.0 μL of WGA DNA polymerase.
16. Vortex thoroughly and consolidate briefly and run the following cycles in a thermocycler.

    95 °C for 3 min.

    2 temperatures for 14 cycles

    94 °C for 15 s

    65 °C for 5 min

    4 °C On Hold.
17. Consolidate the samples and purify the amplified DNA using DNA clean & concentrator kit from Zymo research as described in Subheading 3.5.
18. Elute the DNA from column using 47 μL of MQ water and measure the DNA using a Nanodrop. Expected yield after the amplification will be >120 ng/μL (*see* **Note 8**).

3.6 DNA Digestion and End Labeling

To increase hybridization efficiency to the array, short DNAse treatment is performed on amplified DNA to reduce the fragment sizes. Shorter DNA fragments are labeled with Biotin-N11-ddATP.

1. Approximately 5000 ng of DNA is from amplification step is taken for DNAse reaction (*see* **Note 9**).
2. Prepare a DNAse reaction mixture by adding following in a prechilled microtube.

   14.8 μL of MQ water

   2.0 μL of 10× One-Phor-All-Buffer plus

   1.2 μL of 25 mM $CoCl_2$

   2.0 μL of DNAse I (1 U/μL Amplification grade).

3. Place the microtube containing 5000 ng of amplified DNA on ice and add the following.

   4.85 μL of 10× One-Phor-All-Buffer plus

   2.90 μL of 25 mM $CoCl_2$

   1.50 μL of DNAse reaction mixture.
4. Make up the volume to 50 μL by adding MQ water and vortex briefly.
5. Incubate at 37 °C for exactly 30 s and quickly transfer to 95 °C for 15 min to inactivate the enzyme.
6. Consolidate the samples and transfer the complete mixture to 1.5 mL tube and add:

   5.0 μL of TdT reaction buffer

   1.0 μL of Biotin-N11-ddATP

   1.0 μL of Terminal transferase (400 U/μL).
7. Vortex and centrifuge briefly. Incubate at 37 °C for 1 h.

3.7 Microarray Hybridization

Biotin-labeled DNA from the previous section is hybridized to Affymetrix GeneChip *S. cerevisiae* Tiling 1.0R Array (Sc03b_MR). Hybridization, washing, and scanning are performed as per manufacturer's instructions mentioned below:

1. Dilute the fragmented and labeled DNA in hybridization buffer at a concentration of 25–50 ng/μL and denatured at 99 °C for 10 min.
2. Centrifuge at maximum speed for 2 min and apply into the GeneChip® cartridge.
3. Use single GeneChip® *S. cerevisiae* Tiling 1.0R to hybridize with each biotin-labeled target.
4. Perform hybridization for 16 h at 45 °C in a rotisserie oven at 60 RPM rotational speed.
5. Wash and stain the GeneChip® cartridges in the Affymetrix Fluidics Station 450 following the FS450_0001 standard protocol, as follows.
   - (a) Wash ten cycles of 2 mixes/cycle with Wash Buffer A at 30 °C.
   - (b) Wash six cycles of 15 mixes/cycle with Wash Buffer B at 50 °C.
   - (c) Stain the probe array for 5 min in SAPE solution at 35 °C.
   - (d) Wash ten cycles of 4 mixes/cycle with Wash Buffer A at 30 °C.
   - (e) Stain the probe array for 5 min in antibody solution at 35 °C.
   - (f) Stain the probe array for 5 min in SAPE solution at 35 °C.

(g) Final wash 15 cycles of 4 mixes/cycle with Wash Buffer A at 35 °C.

(h) Fill the probe array with Array Holding buffer.

6. Scan GeneChip arrays using an Affymetrix GeneChip® Scanner 3000 7G with default parameters. Use Affymetrix GeneChip® Command Console software (AGCC) to acquire GeneChip® images and generate .DAT and .CEL files.

3.8 Data Processing and Visualization

1. Process the CEL files generated from the array scanner to identify enriched regions across the genome using rMAT [9] R package. Use the respective BPMAP (Sc03b_MR) file to annotate the probe id to the chromosome loci across the whole genome. Normalize the samples to negate the systematic biases such as probe effect, which are generated during the procedure. Smooth the probe intensities and calculate the values by using IP and Input. Apply the following parameters to detect the enriched regions based on the probe score:

   dMax = 300 (sliding window side),

   nProbesMin = 8 (Minimum number of probes to average),

   method = Score (calling enriched regions based on sliding widow scores),

   threshold = 1.5 (equal and greater than 1.5 are labeled as enriched region).

2. Use the enriched peak (.bed file) and logarithm base 2 IP/Input ratio (.bedgraph file) to visualize the DNA-RNA hybrid in the UCSC genome browser. The ChIPseeker R package (Yu, Wang, and He 2015) is used to annotate the peaks to other functional elements, average profile and heatmap of peaks binding to TSS and TTS regions.

3. Preform Meta-gene analysis to study the averaged enriched peak profile across all protein-coding genes (6706 genes from SacCer 2011 annotation). Along with ORF also consider with upstream (−500 bases from TSS) and downstream (+500 bases from TTS) regions of the genes. Use bedtools [10] to map the peak scores for every base of the gene including upstream (−500 b) and downstream (+500b) (*see* **Note 10**).

4. To perform Meta-gene analysis, we usually scale the length of each gene to 1000 bases. Scaling is performed using the following equation and is iterated for every base across all the genes;

$$(Z - x_i)/\left(y_i - x_i)\right) \times 1000.$$

   where x_i is the start position of the *i*th gene, $i = (1, 2, 3 \ldots$ total genes), y_i is the end position of the *i*th gene, $i = (1, 2, 3 \ldots$ total genes), and Z is base position $(1, 2, 3 \ldots . y_i - x_i)$.

5. To generate the average intensity plot, the IP/input values of the normalized position (1000 ORF, −500 upstream and +500 downstream) of each gene are aggregated using median.
6. For average gene density, the IP/input score are converted into a categorical value of either 1 or 0 based on a threshold of 1.5 ($\geq$1.5 is 1 and <1.5 is 0) of all the normalized positions (1000 ORF, −500 upstream and +500 downstream) of each gene and aggregated using the sum function.
7. For visualization, average intensity and average gene density are plotted with respect to normalized ORF position. The points are smoothened using the generalized additive model (GAM) to obtain a curve using ggplot2 R package.

4 Notes

1. Incubation time can be reduced to 2 h; however, it is advisable to carry out overnight.
2. Number of cells used for immunoprecipitation depends on strains. For most of the W303 mutants, we have recovered enough DNA for library amplification with 4×10^9 cells.
3. Number of cycles depends on the type of bead beater used. We generally get >90% lysis between 6 and 8 cycles. It is important to cool the samples on ice between each lysis cycle.
4. Number of sonication rounds depends on type of sonicator used and its output. Efficiency of sonication can be verified by running cleaned DNA extract on a 1.2% agarose gel. Fragments are observed from as a smear from 200 to 1000 bp with a maximum peak at 400–600 bp.
5. "Input" fraction needs to be processed parallelly from the DNA purification step.
6. Bead washing can be carried out either in cold room or outside using ice. It is essential to keep both the tubes and the washing solutions chilled on ice during all the washing procedures.
7. Any DNA purification kit which will allow low volume elution (10–15 μL) can used at this step.
8. If the DNA yield is lower, the purified DNA can be used for further amplification for two cycles using GenomePlex WGA reamplification Kit from Sigma. Increasing amplification cycle more than two rounds might cause non-specific amplification, hence not advisable.
9. We prefer to use 5000 ng; however, we had found no difference if the DNA amount varies slightly at this stage.
10. It is advisable to consider both average intensity plot and average gene density plot for visualization.

Acknowledgments

We thank Mohamood Adhil for bioinformatic analysis, Daniele Piccini for S9.6 antibody and all our lab members for discussions. We thank Cogentech and Claudia Valli, Mirko Riboni and Simone Minardi for microarray. Research in MF lab was supported by grants from the Associazione Italiana per la Ricerca sul Cancro (AIRC), the European Union, MIUR, Worldwide Cancer Research, and Telethon-Italy.

Data and code availability: The raw and processed data for DRIP-chip microarray shown in Fig. 1 are available at the Gene Expression Omnibus (GEO) with the accession numbers GSE114410.

Custom-made scripts used for DRIP-chip microarray analysis is made available in the GitHub repositoryhttps://github.com/adhilmd/TopologyCustomAnalysis

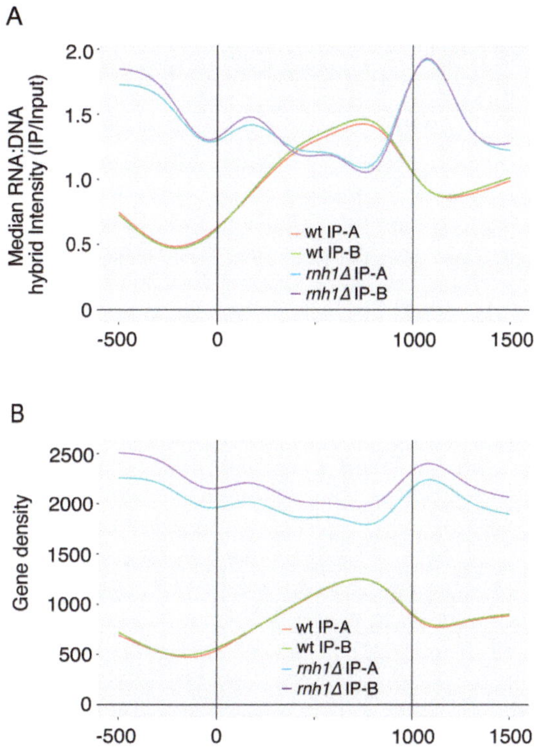

Fig. 1 RNA:DNA hybrid accumulation comparison in wildtype and *rnh1Δ* cells. (**a**) Meta-gene plot showing average intensity comparison in wildtype and *rnh1Δ* cells with two replicates. RNA Pol II coding regions (meta-gene analysis $n = 6706$ genes) were scaled to 1 kbp and flanking 0.5 kbp from TSS and TTS were plotted against median intensity on the Y-axis. (**b**) Meta-gene profiles showing average gene density comparison in wild type and *rnh1Δ* cells with two replicates. RNA Pol II genes (meta-gene analysis $n = 6706$ genes) were plotted against gene density on the Y-axis

References

1. Duquette ML et al (2004) Intracellular transcription of G-rich DNAs induces formation of G-loops, novel structures containing G4 DNA. Genes Dev 18(13):1618–1629
2. Roy D, Lieber MR (2009) G clustering is important for the initiation of transcription-induced R-loops in vitro, whereas high G density without clustering is sufficient thereafter. Mol Cell Biol 29(11):3124–3133
3. Aguilera A, Garcia-Muse T (2012) R loops: from transcription byproducts to threats to genome stability. Mol Cell 46(2):115–124
4. Crossley MP, Bocek M, Cimprich KA (2019) R-loops as cellular regulators and genomic threats. Mol Cell 73(3):398–411
5. Achar YJ et al (2020) Negative supercoil at gene boundaries modulates gene topology. Nature 577(7792):701–705
6. Boguslawski SJ et al (1986) Characterization of monoclonal antibody to DNA.RNA and its application to immunodetection of hybrids. J Immunol Methods 89(1):123–130
7. Chan YA et al (2014) Genome-wide profiling of yeast DNA:RNA hybrid prone sites with DRIP-chip. PLoS Genet 10(4):e1004288
8. Bermejo R et al (2009) ChIP-on-chip analysis of DNA topoisomerases. Methods Mol Biol 582:103–118
9. Droit A, Cheung C, Gottardo R (2010) rMAT--an R/Bioconductor package for analyzing ChIP-chip experiments. Bioinformatics 26(5):678–679
10. Quinlan AR, Hall IM (2010) BEDTools: a flexible suite of utilities for comparing genomic features. Bioinformatics 26(6):841–842

Chapter 22

Detecting and Mapping N6-Methyladenosine on RNA/DNA Hybrids

Abdulkadir Abakir, Fahad Alenezi, and Alexey Ruzov

Abstract

N6-methyladenosine (m^6A) is an RNA modification essential for posttranscriptional regulation of gene expression in eukaryotes. We recently demonstrated that m^6A decorates the RNA components of R-loops, specific nucleic acid structures consisting of an RNA/DNA hybrid and a single strand of non-template DNA, that represent a major source of genetic instability and, at the same time, contribute to regulation of gene expression in mammalian cells. According to growing body of experimental evidence, adenosine methylation affects stability of these structures and potentially influences various aspects of their metabolism. Here, we present two methods for detection and analysis of m^6A-containing RNA/DNA hybrids: an immunostaining protocol allowing investigation of their spatial distribution in eukaryotic cells and m^6A-DNA immunoprecipitation (DIP), an antibody-based technique that permits their genome mapping and locus-specific analysis. In addition to the m^6A-focused studies, these methodologies can also contribute to elucidating the functional roles of other RNA modifications in R-loop biology.

Key words R-loops, RNA/DNA hybrids, RNA modifications, N6-methyladenosine, Immunostaining, RNase H, Genome mapping techniques

1 Introduction

N6-methyladenosine (m^6A) is an RNA modification that regulates stability and translation of mRNAs in eukaryotes [1, 2]. In mammals, this modification is introduced to its target RNAs by METTL3/METTL14 m^6A methyltransferase complex, removed from them by specific demethylases such as fat mass- and obesity-associated protein (FTO) or AlkB homolog H5 (ALKBH5), and interpreted by a number of nuclear and cytoplasmic “reader” proteins that specifically recognize methylated adenosine in RNA [1]. A large volume of recent experimental data suggests that such dynamic and reversible RNA methylation represents one of the key layers of posttranscriptional regulation of gene expression being of crucial importance for animal development, cellular differentiation, and pathogenesis of several diseases [3, 4]. In addition to

Andrés Aguilera and Alexey Ruzov (eds.), *R-Loops: Methods and Protocols*, Methods in Molecular Biology, vol. 2528,
https://doi.org/10.1007/978-1-0716-2477-7_22,

mRNAs, m^6A has recently been found on other RNA species including transfer, ribosomal, and long noncoding RNAs [5–7]. Moreover, we and others have recently demonstrated [8–11] that m^6A decorates the RNA components of R-loops, specific nucleic acid structures consisting of an RNA/DNA hybrid and a single strand of non-template DNA [12, 13] (Fig. 1), that are abundant in eukaryotic genomes representing a major source of genetic instability [12, 14] and, at the same time, contributing to the regulation of gene expression in mammalian cells [15]. Importantly, m^6A is likely to influence several aspects of R-loop biology being involved in regulation of stability of these structures [8] as well as in processes of homologous recombination and DNA repair associated with R-loop metabolism [16].

Here, we present two experimental techniques for detection and analysis of m^6A on R-loops. The first method is a modification of the immunostaining protocol that we previously employed for detection of low-abundance DNA modifications in mammalian, fish, and amphibian cells and tissues [17–21] supplemented by pretreatment of the control samples with RNase H, an enzyme that specifically degrades RNA molecules in the context of RNA/DNA hybrids. This additional step allows to discriminate the m^6A signal produced by this modification within the RNA/DNA hybrids from those generated by m^6A-decorated non-hybrid RNAs, or by N6-methyldeoxyadenosine (6mA), a DNA modification that is recognized by the same antibody and may be present in mammalian DNA according to several studies [22]. This technique can be instrumental for assessing the spatial distribution and relative levels of m^6A-containing RNA/DNA hybrids in different cells and tissues.

The second technique is m^6A-DNA immunoprecipitation (m^6A-DIP) that we used for genome-wide mapping and locus-specific quantitative PCR (qPCR) analysis of m^6A-marked RNA/DNA hybrids. This method is also based on previously published DNA immunoprecipitation (DIP) protocols [23] modified by additional pretreatment of the control samples with RNase H. Similar to our immunostaining technique, subsequent comparison of the experimental samples untreated by RNase H with RNase H-treated control samples allows distinguishing the DIP signal produced by the m^6A-marked RNA/DNA hybrids from the DIP signal that may potentially be generated by DNA 6mA.

Both techniques described here may be instrumental in examining the dynamics and spatial distribution of m^6A present on RNA/DNA hybrids in different physiological and pathological contexts. Moreover, comparison of the m^6A-DIP results with previously published "6mA-DIP" datasets performed without RNase H pretreated controls [22, 24, 25] should contribute to the current debates on the existence of DNA 6mA in the genomes of higher eukaryotes [22, 26–28].

Fig. 1 Artistic representation of an R-loop stylized as Arabic letters written in classic Diwani script, courtesy of Fahad Alenezi. Red fonts represent RNA and black fonts DNA

Furthermore, given that several recent studies suggest that other RNA modifications such as RNA 5-methylcytosine (m^5C) [29] and, possibly, 3-methylcytidine (m^3C) [16] may also be present on the RNA moieties of RNA/DNA hybrids, methodologies described in this chapter would be useful for elucidating the distribution of these modifications and their potential roles in R-loop biology.

2 Materials

The protocols described here are based on the use of human pluripotent stem cells (hPSCs). However, the immunostaining technique has been validated on several different cell lines [8], and both methods can potentially be employed for the analysis of m^6A-containing RNA/DNA hybrids in various animal models.

2.1 Cell Culture

1. A vial of hPSCs frozen in liquid N_2.
2. Essential 8TM medium (Life Technologies).
3. Corning® Matrigel® (Corning).
4. ROCK inhibitor (Tocris Bioscience).
5. Nonessential Amino Acids.
6. TrypLE™ Select Enzyme (Thermo Fisher).
7. Cell culture grade phosphate-buffered saline (PBS).
8. 100 μM β-mercaptoethanol.
9. GlutaMAX (Gibco™).
10. DMEM-F12.
11. Knockout Serum Replacement.
12. BGK solution: DMEM-F12, 100 μM β-mercaptoethanol, 1% GlutaMAX, 1% Nonessential Amino Acids, 1% GlutaMAX, 15% Knockout Serum Replacement.

2.2 Immunostaining of m^{6}A-Marked RNA/DNA Hybrids

1. 8-well glass chamber slides (*see* **Note 1**).
2. Phosphate-buffered saline (PBS): 37 mM NaCl, 2.7 mM KCl, 8 mM Na_2HPO_4, and 2 mM KH_2PO_4 pH 7.5.
3. Fixative: Freshly made ice-cold 4% paraformaldehyde in PBS (*see* **Note 2**).
4. Glass Coplin jar (*see* **Note 3**).
5. PBX: 0.5% Triton X-100 solution in PBS (*see* **Note 4**).
6. RNase A (QIAGEN, cat No. 79254).
7. RNase H (NEB, cat No. M0297S).
8. DNase I (QIAGEN, cat No. 79254).
9. 37 °C incubator.
10. 4 N HCl (*see* **Note 5**).
11. Moisture chamber.
12. 10% bovine serum albumin (BSA) solution in PBS (*see* **Note 6**).
13. PBT solution: 0.01% Tween 20 in PBS.
14. Glass coverslips 22 × 32 mm.
15. Primary antibodies: Anti-m^{6}A rabbit polyclonal antibody (Synaptic Systems, cat No. 202003) (*see* **Note 7**), anti-DNA-RNA Hybrid S9.6 antibody (Merck Millipore, cat. No. MABE1095), anti-5hmC mouse monoclonal antibody (Active Motif) (*see* **Note 8**).
16. Secondary antibodies: Peroxidase-conjugated anti-rabbit secondary antibody, Alexa Fluor 555-conjugated goat anti-mouse secondary antibody (*see* **Note 9**).
17. Tyramide Signal Amplification (TSA) kit (e.g., Thermo Fisher) (*see* **Note 10**).
18. Nail polish.

2.3 m^{6}A-DIP

1. Lysis buffer: 20 mM Tris pH 8.0, 4 mM EDTA, 20 mM NaCl, 1% SDS.
2. RNase A (QIAGEN, cat. No. 19101).
3. Phenol/Chloroform/Isoamyl Alcohol, 25:24:1.
4. Proteinase K (PK) Solution.
5. Phosphate-buffered saline (PBS).
6. Ethanol, absolute.
7. RNase free 5 M NaCl.
8. TE Buffer: 10 mM Tris pH 7.5, 1 mM EDTA.
9. Nuclease free water.
10. Covaris S2 ultrasonicator with the corresponding tubes (Covaris) or equivalent.

11. Qubit™ Flex Fluorometer (Invitrogen) or equivalent.
12. The Agilent 2200 TapeStation system (Agilent Technologies) or equivalent.
13. RNase H (NEB).
14. Anti-m^{6}A rabbit polyclonal antibody (Synaptic systems, cat. No. 202003).
15. 10× IP buffer: 100 mM Sodium Phosphate pH 7.0, 1.4 M NaCl, 0.5% Triton X-100 (*see* **Note 11**).
16. Tube rotator.
17. Eppendorf micro-centrifuge.
18. Dynabeads™ M-280 Sheep Anti-Rabbit IgG (Invitrogen).
19. Dynabeads™ M-280 Sheep Anti-Mouse IgG (Invitrogen).
20. Mouse IgG1 kappa Isotype control antibody or equivalent.
21. Magnetic rack.
22. QIAquick PCR purification kit (QIAGEN) or equivalent.
23. Parafilm tape.
24. Kapa Library Quantification Kit (Kapa Biosystems).
25. NEBNext® Multiplex Oligonucleotides for Illumina ® (New England Biolabs) or equivalent.
26. NEBNext DNA Ultra Library Preparation Kit (New England Biolabs) or equivalent.
27. Real-time qPCR machine.

3 Methods

3.1 Immunochemical Detection of m^{6}A-Marked RNA/DNA Hybrids

3.1.1 hPSC Culture

Carry out all cell culture procedures in a mycoplasma-free Class II safety cabinet.

1. Thaw a stock of Matrigel™ solution by incubating it at 4 °C overnight.
2. Aseptically, transfer the thawed Matrigel™ solution to a 50 mL conical tube containing sterile DMEM/F-12 (*see* **Note** 12).
3. Add 6 mL of Matrigel™ solution to a sterile T75 cell culture flask.
4. Incubate the Matrigel™ coated flask for a minimum of 30 min at room temperature. Alternatively, the coated flask can also be incubated for 30 min at 37 °C in 5% CO_2 to accelerate polymerization (*see* **Note** 13).
5. Immediately before use, carefully aspirate the excess Matrigel™ solution, without dislodging the Matrigel™ coating from the flask.

6. Add 10 mL of freshly prepared E8 medium supplemented with 10 μM rock inhibitor.
7. Plate the cell culture flask with 25,000 hPSCs per cm^2.
8. Rock the flask gently to ensure even distribution of cells across the flask and incubate at 37 °C with 5% CO_2.
9. Replace the E8 medium every 24 h without the rock inhibitor supplementation until the cells are ready to passage (*see* **Note** 14).

3.1.2 Preparation of the Slides for Immunostaining: Fixation and Permeabilization of the Cells

Prepare at least four 8-well chamber slides for each experimental condition. Each experimental condition should include the samples preprocessed with (a) RNase A, (b) RNases A and H, and (c) untreated control samples (*see* Fig. 2).

1. Add 200 μL of Matrigel™ solution to each well of an 8-well chamber slide.
2. Incubate the Matrigel™ coated 8-well chamber slide for a minimum of 30 min at room temperature.
3. Carefully aspirate the excess of Matrigel™ solution without dislodging the Matrigel™ coating from the wells.
4. Add 350 μL of E8 medium supplemented with 10 μM rock inhibitor to each well.
5. Seed approximately 13 × 10^3 cells per well, gently rock the slide to ensure even distribution of the cells, and incubate the 8-well chamber slide at 37 °C in 5% CO_2 cell culture incubator.
6. Replace E8 medium every 24 h without the rock inhibitor supplementation.
7. Upon reaching 70% confluency (usually on the second day of incubation), aspirate E8 medium and replace it with equal volume of PBS to wash the cells.
8. Replace the PBS with 200 μL of ice-cold 4% PFA and incubate the cells at room temperature for 10 min to fix the cells.
9. Remove the fixative, and wash the cells with 400 μL ice-cold PBS.
10. Aspirate PBS, add 400 μL PBX solution to each well, and incubate for 30 min at room temperature to permeabilize the cells.
11. Without touching the cells, carefully aspirate excess PBX solution and wash the cells twice with PBS for 5 min at room temperature.

3.1.3 Pretreatment of the Samples with RNases A and H or DNase I, and Depurination of DNA

1. After permeabilization, label the slides into 3 or 4 separate groups: *Group 1*: No RNases ("untreated"), *Group 2*: RNase A only, *Group 3*: RNases A and H (RNases A + H), and *Group 4*: DNase I (optional).

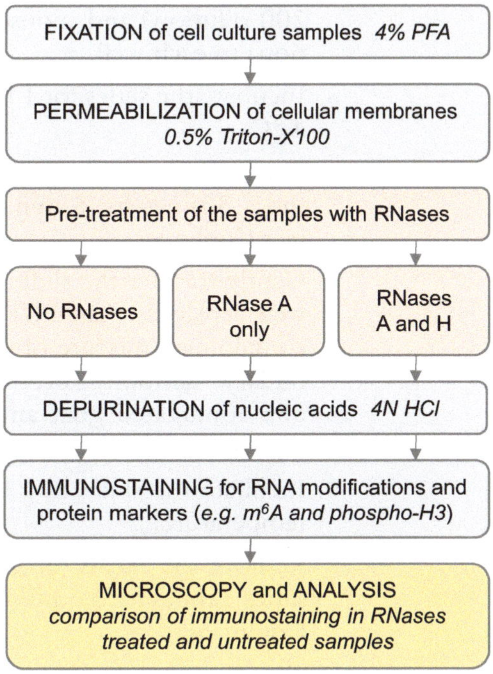

Fig. 2 Flowchart of the immunostaining procedure for detection of RNA modifications decorating RNA/DNA hybrids

2. Process the "untreated" slides in the same solutions in parallel with other groups of slides without addition of any RNases or DNases.
3. For RNase A treatment, incubate the slides in 25 mg/mL RNase A in PBS for 1 h at 37 °C.
4. For combined RNases A and H treatment, incubate the RNase A-treated slides in 1× RNase H buffer containing 10 U of RNase H per well overnight at 37 °C. Process "RNase A only" slides in parallel in the same buffer without addition of RNase H (*see* **Note 15**).
5. Incubate the "DNase I only" slide in 20 U of DNase I solution in TE buffer per well for 4 h at room temperature.
6. Wash all the slides 3 times for 5 min each with PBS.
7. Incubate the slides in 4 M HCL solution in a Coplin jar in safety cabinet for 1 h at room temperature (*see* **Note 5**).
8. Discard the 4 M HCl solution, wash the slides 5 times for 5 min each, with PBS at room temperature in a Coplin jar.

3.1.4 Co-immunostaining Cells for m^{6}A and 5hmC

1. Gently transfer the slides into a humidified chamber. Incubate the slides in blocking buffer for 1 h at room temperature.
2. Remove the blocking buffer and add 200 μL of a blocking buffer containing a mixture of rabbit anti-m^{6}A antibody (1:

200 dilution) and mouse anti-5hmC antibody (1:1000 dilution) to each well.

3. Incubate the slides for 1 h at room temperature or overnight at 4 °C.
4. Remove the excess primary antibody mixture and wash the slides 3 times for 5 min each with PBT at room temperature in a Coplin jar.
5. Gently transfer the slides into a humidified chamber. For each well, replace the last PBT wash with 200 μL of blocking buffer containing a mixture of goat anti-rabbit HRP-conjugated secondary antibody (1:400 dilution) and an Alexa Fluor 555-conjugated goat anti-mouse secondary antibody (1:400 dilution).
6. Incubate the slides in a humidified chamber for 1 h at room temperature.
7. Remove the excess secondary antibody mixture and wash the slides 3 times for 5 min each with PBT at room temperature in a Coplin jar.
8. Transfer the slides into humidified chamber. Incubate the slides for 2 min at room temperature in TSA amplification buffer containing Tyramide at 1:200 dilution (*see* **Note 10**).
9. Remove the TSA solution and immediately wash the slides 5 times in PBS for 5 min each at room temperature in a Coplin jar.
10. Remove the excess liquid and rubber grommets from the slides without letting the slide to dry. Apply a drop of mounting medium to the slide.
11. Place a coverslip on the slide and gently apply even pressure, without creating air bubbles.
12. Seal the coverslip with nail polish.
13. Store the slides in the dark for ~2 h before examination under the microscope. Compare the staining patterns of RNAse A only, RNAses A and H pretreated and control (untreated and DNase I treated) slides. The m^{6}A-RNA/DNA hybrid signal should be present in RNase A-treated and both control slides but disappear upon pretreatment of the samples with RNase H (in RNases A + H treated slides) (Fig. 3).

3.2 m^{6}A-DIP for Mapping the m^{6}A-Marked RNA/DNA Hybrids

3.2.1 Extraction of Genomic DNA from hPSCs

1. Once the hPSCs reach 70–80% confluency (approximately on the third day of plating), aspirate the spent E8 medium.
2. Wash the cells with 10 mL of PBS.
3. Aspirate excess PBS, add 3 mL of TrypLE Select Enzyme, and incubate the cells for 4 min at 37 °C in a humidified incubator with 5% CO_2 to dissociate the cells.

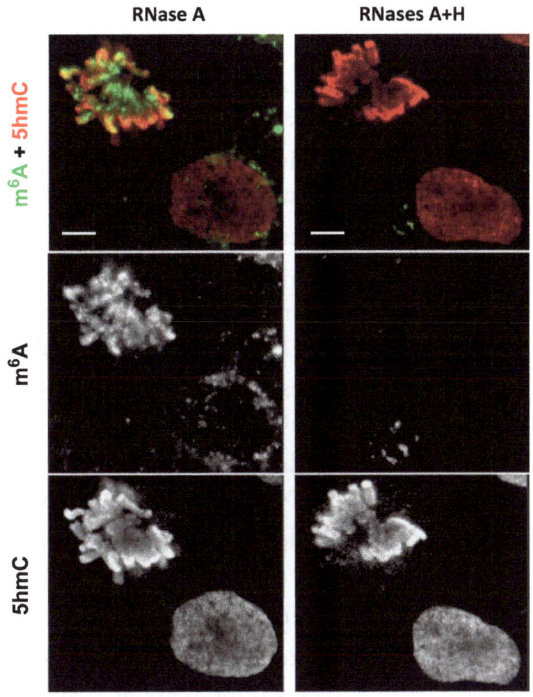

Fig. 3 The results of immunostaining of hPSCs for m6A and 5hmC upon pretreatment of the slides with RNase A only and RNases A and H. The views with dividing (top) and interphase (bottom) cells are shown for both conditions. The m6A-RNA/DNA staining in mitotic chromatin is evident upon RNase A pretreatment and disappears in RNases A and H pretreated cells. 5hmC staining is unaffected by treatment with RNases as this modification is present on DNA and does not seem to decorate RNA/DNA hybrids (at least in high quantities) in hPSCs according to this analysis. Scale bars are 10 μM

4. Inactivate TryPLE by adding 7 mL of BGK solution.
5. Pellet the dissociated cells in 15 mL conical tube by centrifugation for 4 min at 160 × *g*.
6. Without removing the pellet, aspirate the supernatant containing excess BGK solution and add 10 mL of PBS. Resuspend the cells in PBS.
7. Pellet the cells in 15 mL conical tube by centrifugation for 4 min at 160 × *g*.
8. Resuspend the cell in 300 μL of the lysis buffer containing 20 μL of proteinase K (10 mg/mL stock) solution.
9. Incubate the cells at 55 °C for at least 5 h, or O/N.
10. Add 1 volume of phenol/chloroform, vortex to mix well.
11. Centrifuge for 5 min at 15871 × *g*, take the aqueous phase, and place it in a fresh tube.
12. Precipitate the DNA with 1 volume of isopropanol supplemented with 1/10 (v/v) of 5 M NaCl.

13. Incubate the tube at −20 °C for 1 h, centrifuge at 18407 × *g* for 10 min.
14. Remove the isopropanol, make sure DNA pellet remains in the tube.
15. Add 1 volume of 70% ethanol to the DNA pellet, gently invert the tube ~10 times (do not vortex).
16. Centrifuge the tube at 18407 × *g* for 5 min to pellet DNA.
17. Without touching pelleted DNA, quick spin to collect any remaining ethanol. Remove all ethanol and leave the tube for 5–10 min at room temperature to dry.
18. Resuspend the DNA pellet in 200 μL of RNase/DNase free water.
19. Add 15 μL of RNase A (100 μg/mL stock) to the resuspended DNA—mix well and incubate at 37 °C for 1 h.
20. Add 1 volume of phenol/chloroform, vortex and centrifuge for 5 min at 18407 × *g*.
21. Remove the aqueous phase and place it into fresh tube.
22. Add 1× volume of isopropanol containing 1/10 (v/v) of 5 M NaCl, incubate at −20 °C for 1 h.
23. Centrifuge the tube for 10 min at 18407 × *g*, remove the supernatant so that the DNA pellet remains.
24. Add 1 volume of 70% ethanol, centrifuge for 5 min at 18407 × *g*.
25. Remove the ethanol, quickly recentrifuge the tube to remove any remaining ethanol. Leave to air dry for 3–5 min, but do not let pellet dry out completely.
26. Resuspend the DNA pellet in 100 μL of nuclease free water, place at 37 °C for 15 min.
27. Determine the DNA concentration using Qubit with BR reagents.

3.2.2 Sonication of Genomic DNA

1. For each sample, dilute ~5–7 μg of genomic DNA in 130 μL of TE buffer.
2. Sonicate genomic DNA using Covaris S2 sonicator with fragmentation parameters: Duty cycle 5%, Intensity 1, cycles/burst 200, time 85 s, bath temperature 4 °C (*see* **Note 16**). Sonication using these parameters should lead to a distribution of the length of DNA fragments peaking at 300–450 bp (Fig. 4).
3. Keep 1/10th of the sample as input.

3.2.3 RNase H Treatment

Following the sonication, an RNase H-treated sample must be included as a control for RNA/DNA hybrid detection for each experimental setup.

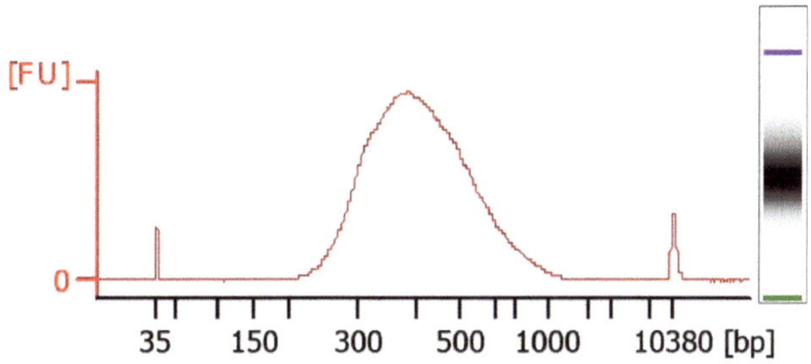

Fig. 4 Representative bioanalyzer profile of an m6A-DIP-seq library generated from the DNA sonicated with parameters suggested in the protocol

1. Dilute both control and experimental samples to 400 μL of TE buffer.
2. Add 10 U of RNase H and 45 μL of 10× RNase H buffer to the control sample and incubate overnight at 37 °C with gentle shaking.
3. Add 45 μL of 10× RNase H buffer to the untreated DNA sample and incubate it in parallel with the control samples, overnight at 37 °C with gentle shaking, without addition of RNase H.

3.2.4 Immunoprecipitation

1. In case m6A-DIP is to be analyzed by deep sequencing, adaptors should be ligated to genomic DNA at this stage as instructed by a particular library preparation protocol (*see* **Note 17**). Proceed straight to the next step in case the samples are to be analyzed by quantitative PCR (qPCR) only.
2. Incubate the samples for 10 min at 95 °C for mild DNA denaturation (*see* **Note 18**).
3. Place the samples immediately on ice. Incubate on ice for at least 10 min prior to the addition of antibody.
4. Add 51 μL of 10× IP Buffer to each sample.
5. Add 6 μL of anti-m6A antibody to the untreated and RNase H-treated samples (use an additional sample for no antibody or IGG only control).
6. Incubate the samples overnight at 4 °C while shaking/spinning at 20 rpm.
7. On the next day, for each sample, prepare 20 μL of sheep anti-rabbit magnetic beads. Wash the beads in PBS twice, for 5 min each.
8. Resuspend the beads, volume to volume in 1× IP Buffer.
9. Add 20 μL of the beads to each sample.
10. Incubate the beads/DNA/antibody mixture at 4 °C for 2 h while shaking or spinning on a rotating wheel.

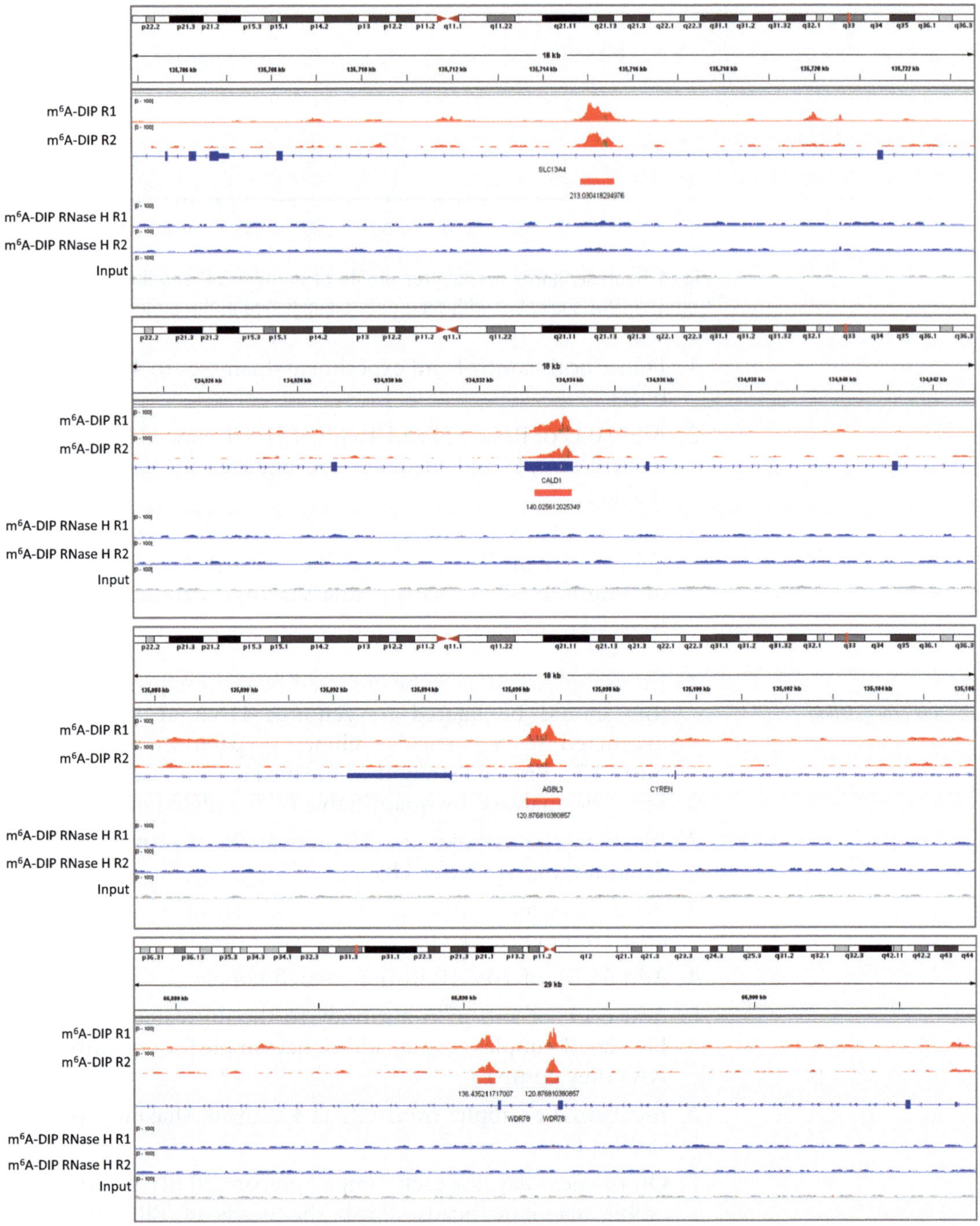

Fig. 5 Four representative genome browser views of the read coverages for m^{6}A-DIP performed on hPSCs genomic DNA. m^{6}A-DIP tracks are shown in red, the profiles of m^{6}A-DIP performed on RNase H pretreated samples are depicted in blue, and input coverage is shown in gray. Tracks for two individual replicates (R1 and R2) are demonstrated for m^{6}A-DIP and m^{6}A-DIP RNase H datasets. Red rectangles designate the m^{6}A-DIP peaks

11. Place the sample on a magnetic rack for 5 min or until the solution clears. Remove and discard the supernatant.
12. Remove the samples from the magnet, add 700 μL of 1× IP Buffer to each sample, and incubate for 10 min.
13. Place the sample on a magnetic rack for 5 min or until the solution clears. Remove and discard the supernatant.
14. Repeat the wash.
15. After the wash, add 250 μL of Proteinase K buffer containing 5 μL of 20 mg/mL Proteinase K to the beads.
16. Incubate the beads overnight (or at least for 5 h) at 55 °C in a shaking incubator. Make sure the Eppendorf tubes are well sealed to prevent evaporation of the solution.
17. Next day, purify the DNA from the samples using the QIAquick PCR purification kit or equivalent.
18. Determine DNA concentration using Qubit with HS reagents.
19. At this stage, the samples can be analyzed by qPCR to evaluate the m^{6}A enrichment over input comparing it to "no antibody" or IGG only, and RNase H-treated controls. Alternatively, m^{6}A-DIP reactions performed using adaptor-ligated DNA (*see* Subheading 3.2.4, **step 1**) can be used for library preparation and analyzed by high-throughput sequencing (Fig. 5) (*see* **Note 19**).

4 Notes

1. Although we routinely use 8-well chamber slides to culture hPSCs for immunostaining, depending on the cell type, the cells can also be grown on appropriately coated glass coverslips placed in 6-well cell culture plates.
2. 4% PFA can also be replaced by 4% formaldehyde. Both formaldehyde and PFA are toxic. The solutions should be made in the fume chamber with gloves, safety glasses, and a face mask.
3. Other glass equipment designed to hold glass slides can also be used instead.
4. Make sure Triton X-100 is evenly dissolved in 1× PBS solution before use.
5. HCL is corrosive, wear appropriate protective equipment. Alternatively, the slides can be incubated in 4 M HCl for 30 min at 37 °C.
6. We recommend to use freshly made ice-cold 10% BSA solution in 1× PBS.
7. Although we routinely source the anti-m^{6}A antibody from Synaptic Systems, other anti-m^{6}A antibodies can also be used,

provided their specificity and affinity for m^6A are experimentally validated.

8. Other combinations of primary antibodies can also be used for co-immunostaining m^6A with other DNA modifications or its co-localization with various protein markers (e.g., phosphorylated histone H3 [8]).
9. Similarly, where appropriate, other combinations of secondary antibodies can be used.
10. Although our protocol is optimized for this TSA system, other TSA Kits can also be used. We recommend to optimize the time of incubation with tyramide independently for each TSA kit. Long times of incubation result in high background/signal ratio [19].
11. 10× and 1× IP buffer can be kept at RT for several weeks.
12. Gently pipette up and down (without generating bubbles) to obtain a homogenous solution.
13. Unless used immediately, the coated flasks can be stored at 4 °C for a maximum of 1 week. Ensure stored vessels are equilibrated to room temperature 30 min prior to use.
14. For consistent and accurate results, daily monitoring of the cells in culture under the light microscope (e.g., using a phase contrast microscopy) to check for abnormal changes in morphology, growth rate, presence of differentiated cells, etc. is critical. Immediately discard cultures in case of any abnormal changes and restart with a fresh vial of hPSCs with a relatively low passage, in aseptic cell culture conditions.
15. In our experience, the overnight incubation can be replaced by 1 h incubation at 37 °C.
16. At this point, the samples can be frozen at −20 °C.
17. We use NEBNext® Multiplex Oligonucleotides for Illumina ® (New England Biolabs) for adaptor ligation.
18. Shorter times of incubation decrease the DIP efficiency, whereas substantially longer times of denaturation (e.g., 30 min) seem to lead to complete dissociation of the RNA and DNA in the hybrids and, therefore, loss of m^6A-DIP signal [8].
19. We generate our libraries using NEBNext DNA Ultra Library Preparation Kit (11 cycles of amplification) and quantify them using Kapa Library Quantification Kit. The high-throughput sequencing is performed on Illumina NextSeq500 platform to generate 2 × 150-bp reads.

Acknowledgements

This work was supported by a Medical Research Council IMPACT DTP PhD Studentship (grant no. MR/N013913/1) to A.A and by the Biotechnology and Biological Sciences Research Council (grant no. BB/N005759/1) to A.R. A.A. thanks Wolf Reik (Babraham Institute) for support.

References

1. Shi H, Wei J, He C (2019) Where, when, and how: context-dependent functions of RNA methylation writers, readers, and erasers. Mol Cell 74:640–650. https://doi.org/10.1016/j.molcel.2019.04.025
2. Yue Y, Liu J, He C (2015) RNA N6-methyladenosine methylation in post-transcriptional gene expression regulation. Genes Dev 29:1343–1355. https://doi.org/10.1101/gad.262766.115
3. Roundtree IA, Evans ME, Pan T, He C (2017) Dynamic RNA modifications in gene expression regulation. Cell 169:1187–1200. https://doi.org/10.1016/j.cell.2017.05.045
4. He PC, He C (2021) m^6 a RNA methylation: from mechanisms to therapeutic potential. EMBO J 40:e105977. https://doi.org/10.15252/embj.2020105977
5. Meyer KD, Saletore Y, Zumbo P, Elemento O, Mason CE, Jaffrey SR (2012) Comprehensive analysis of mRNA methylation reveals enrichment in 3′ UTRs and near stop codons. Cell 149:1635–1646. https://doi.org/10.1016/j.cell.2012.05.003
6. Dominissini D, Moshitch-Moshkovitz S, Schwartz S, Salmon-Divon M, Ungar L, Osenberg S, Cesarkas K, Jacob-Hirsch J, Amariglio N, Kupiec M, Sorek R, Rechavi G (2012) Topology of the human and mouse m6A RNA methylomes revealed by m6A-seq. Nature 485:201–206. https://doi.org/10.1038/nature11112
7. He R-Z, Jiang J, Luo D-X (2020) The functions of N6-methyladenosine modification in lncRNAs. Genes Dis 7:598–605. https://doi.org/10.1016/j.gendis.2020.03.005
8. Abakir A, Giles TC, Cristini A, Foster JM, Dai N, Starczak M, Rubio-Roldan A, Li M, Eleftheriou M, Crutchley J, Flatt L, Young L, Gaffney DJ, Denning C, Dalhus B, Emes RD, Gackowski D, Corrêa IR, Garcia-Perez JL, Klungland A, Gromak N, Ruzov A (2020) N6-methyladenosine regulates the stability of RNA:DNA hybrids in human cells. Nat Genet 52:48–55. https://doi.org/10.1038/s41588-019-0549-x
9. Yang X, Liu Q-L, Xu W, Zhang Y-C, Yang Y, Ju L-F, Chen J, Chen Y-S, Li K, Ren J, Sun Q, Yang Y-G (2019) m6A promotes R-loop formation to facilitate transcription termination. Cell Res 29:1035–1038. https://doi.org/10.1038/s41422-019-0235-7
10. Marnef A, Legube G (2020) m6A RNA modification as a new player in R-loop regulation. Nat Genet 52:27–28. https://doi.org/10.1038/s41588-019-0563-z
11. Kang HJ, Cheon NY, Park H, Jeong GW, Ye BJ, Yoo EJ, Lee JH, Hur J-H, Lee E-A, Kim H, Lee K-Y, Choi SY, Lee-Kwon W, Myung K, Lee JY, Kwon HM (2021) TonEBP recognizes R-loops and initiates m6A RNA methylation for R-loop resolution. Nucleic Acids Res 49: 269–284. https://doi.org/10.1093/nar/gkaa1162
12. García-Muse T, Aguilera A (2019) R loops: from physiological to pathological roles. Cell 179:604–618. https://doi.org/10.1016/j.cell.2019.08.055
13. Chedin F, Benham CJ (2020) Emerging roles for R-loop structures in the management of topological stress. J Biol Chem 295: 4684–4695. https://doi.org/10.1074/jbc.REV119.006364
14. Gómez-González B, Aguilera A (2019) Transcription-mediated replication hindrance: a major driver of genome instability. Genes Dev 33:1008–1026. https://doi.org/10.1101/gad.324517.119
15. Niehrs C, Luke B (2020) Regulatory R-loops as facilitators of gene expression and genome stability. Nat Rev Mol Cell Biol 21:167–178. https://doi.org/10.1038/s41580-019-0206-3
16. Zhang L-H, Zhang X-Y, Hu T, Chen X-Y, Li J-J, Raida M, Sun N, Luo Y, Gao X (2020) The SUMOylated METTL8 induces R-loop and tumorigenesis via m3C. iScience 23:100968. https://doi.org/10.1016/j.isci.2020.100968

17. Wheldon LM, Abakir A, Ferjentsik Z, Dudnakova T, Strohbuecker S, Christie D, Dai N, Guan S, Foster JM, Corrêa IR, Loose M, Dixon JE, Sottile V, Johnson AD, Ruzov A (2014) Transient accumulation of 5-carboxylcytosine indicates involvement of active demethylation in lineage specification of neural stem cells. Cell Rep 7:1353–1361. https://doi.org/10.1016/j.celrep.2014.05.003
18. Alioui A, Wheldon LM, Abakir A, Ferjentsik Z, Johnson AD, Ruzov A (2012) 5-carboxylcytosine is localized to euchromatic regions in the nuclei of follicular cells in axolotl ovary. Nucleus 3:565–569. https://doi.org/10.4161/nucl.22799
19. Abakir A, Wheldon L, Johnson AD, Laurent P, Ruzov A (2016) Detection of modified forms of cytosine using sensitive immunohistochemistry. J Vis Exp (114):54416. https://doi.org/10.3791/54416
20. Almeida RD, Loose M, Sottile V, Matsa E, Denning C, Young L, Johnson AD, Gering M, Ruzov A (2012) 5-hydroxymethylcytosine enrichment of non-committed cells is not a universal feature of vertebrate development. Epigenetics 7:383–389. https://doi.org/10.4161/epi.19375
21. Blythe MJ, Kocer A, Rubio-Roldan A, Giles T, Abakir A, Ialy-Radio C, Wheldon LM, Bereshchenko O, Bruscoli S, Kondrashov A, Drevet JR, Emes RD, Johnson AD, McCarrey JR, Gackowski D, Olinski R, Cocquet J, Garcia-Perez JL, Ruzov A (2021) LINE-1 transcription in round spermatids is associated with accretion of 5-carboxylcytosine in their open reading frames. Commun Biol 4:691. https://doi.org/10.1038/s42003-021-02217-8
22. Bochtler M, Fernandes H (2021) DNA adenine methylation in eukaryotes: enzymatic mark or a form of DNA damage. BioEssays 43:e2000243. https://doi.org/10.1002/bies.202000243
23. Weber M, Davies JJ, Wittig D, Oakeley EJ, Haase M, Lam WL, Schübeler D (2005) Chromosome-wide and promoter-specific analyses identify sites of differential DNA methylation in normal and transformed human cells. Nat Genet 37:853–862. https://doi.org/10.1038/ng1598
24. Yao B, Cheng Y, Wang Z, Li Y, Chen L, Huang L, Zhang W, Chen D, Wu H, Tang B, Jin P (2017) DNA N6-methyladenine is dynamically regulated in the mouse brain following environmental stress. Nat Commun 8:1122. https://doi.org/10.1038/s41467-017-01195-y
25. Kweon S-M, Chen Y, Moon E, Kvederaviciutė K, Klimasauskas S, Feldman DE (2019) An adversarial DNA N^6-methyladenine-sensor network preserves polycomb silencing. Mol Cell 74:1138–1147.e6. https://doi.org/10.1016/j.molcel.2019.03.018
26. Schiffers S, Ebert C, Rahimoff R, Kosmatchev O, Steinbacher J, Bohne A-V, Spada F, Michalakis S, Nickelsen J, Müller M, Carell T (2017) Quantitative LC-MS provides no evidence for m^6 dA or m^4 dC in the genome of mouse embryonic stem cells and tissues. Angew Chem Int Ed Engl 56:11268–11271. https://doi.org/10.1002/anie.201700424
27. O'Brown ZK, Boulias K, Wang J, Wang SY, O'Brown NM, Hao Z, Shibuya H, Fady P-E, Shi Y, He C, Megason SG, Liu T, Greer EL (2019) Sources of artifact in measurements of 6mA and 4mC abundance in eukaryotic genomic DNA. BMC Genomics 20:445. https://doi.org/10.1186/s12864-019-5754-6
28. Lowe P, Olinski R, Ruzov A (2021) Evidence for noncytosine epigenetic DNA modifications in multicellular eukaryotes: an overview. Methods Mol Biol 2198:15–25. https://doi.org/10.1007/978-1-0716-0876-0_2
29. Chen H, Yang H, Zhu X, Yadav T, Ouyang J, Truesdell SS, Tan J, Wang Y, Duan M, Wei L, Zou L, Levine AS, Vasudevan S, Lan L (2020) m5C modification of mRNA serves a DNA damage code to promote homologous recombination. Nat Commun 11:2834. https://doi.org/10.1038/s41467-020-16722-7

Chapter 23

Genome-Wide Native R-Loop Profiling by R-Loop Cleavage Under Targets and Tagmentation (R-Loop CUT&Tag)

Honghong Wang, Conghui Li, and Kaiwei Liang

Abstract

R-loops are three-stranded nucleic acid structures that consist of a DNA–RNA hybrid and a displaced single-stranded DNA. R-loops occur during transcription and participate in multiple physiological processes such as DNA repair, modulating DNA topology, and regulation of gene transcription. Dysfunctional R-loops associate with several human diseases such as neurological disorders and cancer. Therefore, accurately and comprehensively profiling native R-loops is crucial to understand their functions under both physiological and pathological conditions. Here, we describe a convenient native R-loop profiling method, R-loop CUT&Tag, which combines a DNA–RNA hybrid sensor (GST-His_6–2 × HBD or S9.6 antibody) with a pA-Tn5-based cleavage under targets and tagmentation approach. R-loop CUT&Tag starts with 0.5 million cells and can sensitively detect native and specific R-loops at the promoter, gene body, and enhancer regions.

Key words R-loop, Hybrid-binding domain (HBD), CUT&Tag, DNA–RNA hybrid, pA-Tn5

1 Introduction

The R-loop is a unique three-stranded nucleic acid structure formed by newly transcribed RNA invading double-stranded DNA during gene transcription, resulting in a displaced single-strand DNA and a DNA–RNA hybrid [1]. R-loops were overlooked for a while and considered as mere "by-products" of transcription [2]. Later studies have revealed that R-loops actively participate in various critical biological processes, such as DNA damage repair and genome instability, transcriptional regulation, and regulation of chromatin landscape [3, 4]. Additionally, dysfunction in the regulation of R-loops has been linked to multiple human diseases, including neurological diseases, cancers, and immune diseases [5–7].

R-loops are evolutionarily conserved and are widely present from bacteria to mammals. In mammals, R-loops are prevalent

Andrés Aguilera and Alexey Ruzov (eds.), *R-Loops: Methods and Protocols*, Methods in Molecular Biology, vol. 2528, https://doi.org/10.1007/978-1-0716-2477-7_23,

and collectively occupy up to 5% of mammalian genomes [8]. R-loops are distributed in specific sequences and areas of the genome such as gene promoters, gene bodies, and enhancers [4, 8–10]. Investigating the R-loop biology in physiological and pathological processes relies on accurate and comprehensive profiling of R-loops in the human genome. During the last decade, the detection of R-loops was heavily dependent on the S9.6 monoclonal antibody for specific DNA–RNA hybrid (a defining feature of R-loops) binding and capturing [6]. Recently, we have reported an artificial DNA–RNA hybrid recognition sensor protein GST-His$_6$–2 × HBD that interacts with DNA–RNA hybrids specifically. Furthermore, by combining the sensor protein GST-His$_6$–2 × HBD or S9.6 antibody with the recently developed CUT&Tag technology [11, 12], we established a native R-loop mapping method named R-loop CUT&Tag.

R-loop CUT&Tag is a straightforward native R-loop mapping method. This method takes less than a day from cell collection to library preparation, and it does not require fixation, sonication, or stable transgenic cell lines [4]. This method takes advantage of the unique features of pA-Tn5 transposase, which allows for rapid library preparation and low sample input [12–14]. Additionally, Tn5 transposase was reported to transpose DNA–RNA hybrids, which also could be used for library preparation after strand displacement and avoid many laborious steps [13, 15]. Furthermore, R-loop CUT&Tag is sensitive and reproducible for native R-loop mapping with high resolution. R-loop CUT&Tag provides highly specific native R-loop signals at gene promoters and could detect transient R-loops at the gene body and enhancer regions [4].

2 Materials

2.1 Recombinant pA-Tn5 and pA-Tn5 Transposome Assembly

1. His$_6$-pA-Tn5 plasmid (Addgene #171871).
2. BL21(DE3) competent *E. coli*.
3. 100 mg/mL Ampicillin.
4. IPTG (isopropyl-β-D-thiogalactopyranoside).
5. Imidazole (1, 3-diazacyclopenta-2, 4-diene).
6. Glycerol.
7. 50% (w/v) Poly(ethyleneimine) solution in H_2O (Sigma-Aldrich # P3143).
8. cOmplete, EDTA-free protease inhibitor tablets (Roche).
9. Amicon Ultra-15 centrifugal filter unit (Millipore, 30 kDa cutoff).
10. Pierce BCA protein assay kit (Thermo Fisher Scientific).

11. Ni-NTA beads 6FF (Smart-Lifesciences # SA005010) or Ni-NTA agarose (Qiagen # 30210).
12. LB (Luria-Bertani) liquid medium (5 g/L NaCl, 10 g/L tryptone, 5 g/L yeast extract).
13. PBS: 137 mM NaCl, 2.7 mM KCl, 8 mM Na_2HPO_4, and 2 mM KH_2PO_4.
14. HXG buffer: 20 mM HEPES-KOH pH 7.2, 0.8 M NaCl, 10% glycerol, 0.2% Triton X-100.
15. 2× dialysis buffer: 100 mM HEPES-KOH pH 7.2, 0.2 M NaCl, 0.2 mM EDTA, 2 mM DTT, 0.2% Triton X-100, 20% glycerol.
16. Storage buffer: 50 mM HEPES-KOH pH 7.2, 0.1 M NaCl, 0.1 mM EDTA, 1 mM DTT, 0.1% Triton X-100, 60% glycerol.
17. TE buffer: 10 mM Tris–HCl pH 8.0, 1 mM EDTA.
18. 5× TAPS-DMF buffer: 50 mM TAPS-KOH pH 8.3, 25 mM $MgCl_2$, 50% DMF.
19. Stopping buffer: 250 mM EDTA, 0.2% SDS.
20. Tn5ME-rev: 5′-[phos] CTG TCT CTT ATA CAC ATC T-3′.
21. Tn5ME-A: 5′-TCG TCG GCA GCG TCA GAT GTG TAT AAG AGA CAG-3′.
22. Tn5ME-B: 5′-GTC TCG TGG GCT CGG AGA TGT GTA TAA GAG ACA G-3′.

2.2 Purification of Recombinant GST-His_6–2 × HBD Protein

1. pGEX-GST-His_6–2 × HBD plasmid (Addgene #171870).
2. T7 Express *lysY/I*^{q} competent *E. coli* cells.
3. Amicon Ultra-15 centrifugal filter unit (Millipore, 10 kDa cutoff).
4. 2× YT liquid medium (5 g/L NaCl, 16 g/L tryptone, 10 g/L yeast extract).

2.3 Cell Preparation and Binding Cells to Concanavalin A-Coated Beads

1. Accutase solution.
2. Concanavalin A-coated magnetic beads (Smart-Lifesciences # SA028025).
3. 2 M spermidine.
4. 5% Digitonin in DMSO.
5. Binding buffer: 20 mM HEPES-KOH pH 7.5, 10 mM KCl, 1 mM $CaCl_2$, 1 mM $MnCl_2$.
6. Wash buffer: 20 mM HEPES-KOH at pH 7.5, 150 mM NaCl, 0.5 mM spermidine, 1× protease inhibitors.
7. Antibody buffer: 20 mM HEPES-KOH pH 7.5, 150 mM NaCl, 0.5 mM spermidine, 1× protease inhibitors, 0.05% digitonin, 0.01% NP-40, 2 mM EDTA.

2.4 Incubation with DNA–RNA Hybrid Sensors and Secondary Antibody

1. Dig-wash buffer: 20 mM HEPES-KOH pH 7.5, 150 mM NaCl, 0.5 mM spermidine, 1× protease inhibitors, 0.05% digitonin, 0.01% NP-40.
2. GST-His_6–2 × HBD protein.
3. Anti-DNA–RNA hybrid antibody, clone S9.6.
4. Mouse Anti-Histag antibody.
5. Rabbit anti-mouse IgG (1 mg/mL).
6. Mouse anti-rabbit IgG (1 mg/mL).
7. RNase A or RNase H.

2.5 Incubation with pA-Tn5 Adapter Complex and Tagmentation

1. 37.5 μM pA-Tn5 adapter complex.
2. Dig-300 buffer: 20 mM HEPES-KOH pH 7.5, 300 mM NaCl, 0.5 mM spermidine, 1× protease inhibitors, 0.05% digitonin, 0.01% NP-40.
3. Tagmentation buffer: 10 mM TAPS-KOH pH 8.3, 10 mM $MgCl_2$, 10% DMF.
4. 20 mg/mL proteinase K.
5. 10% SDS.
6. 0.5 M EDTA.
7. 0.1% Tween-20.
8. AMPure XP beads (Beckman Coulter # A63881) or home-made SPRI beads with Sera-Mag carboxylate-modified magnetic beads (Sigma-Aldrich #GE24152105050350).

2.6 R-Loop CUT&Tag Library Preparation

1. 10 mM dNTPs.
2. Q5 high-fidelity DNA polymerase (NEB).
3. 5× Q5 reaction buffer.
4. 5× Q5 high GC enhancer.
5. Bst 2.0 WarmStart polymerase (NEB).
6. SYBR Green I—10,000× concentrate in DMSO.
7. Ad1-S502: 5′-AAT GAT ACG GCG ACC ACC GAG ATC TAC ACC TCT CTA TTC GTC GGC AGC GTC AGA TGT G-3′.
8. Ad2.1: 5′-CAA GCA GAA GAC GGC ATA CGA GAT TCG CCT TAG TCT CGT GGG CTC GGA GAT GT-3′.
9. Ad2.2: 5′-CAA GCA GAA GAC GGC ATA CGA GAT CTA GTA CGG TCT CGT GGG CTC GGA GAT GT-3′.
10. Ad2.3: 5′-CAA GCA GAA GAC GGC ATA CGA GAT TTC TGC CTG TCT CGT GGG CTC GGA GAT GT-3′.

11. Ad2.4: 5′-CAA GCA GAA GAC GGC ATA CGA GAT GCT CAG GAG TCT CGT GGG CTC GGA GAT GT-3′.

2.7 DNA Sequencing and Data Analysis

1. Human genome (UCSC hg38).
2. NCBI *E. coli* genome (*Escherichia coli* str. K12 substr. MG1655 U00096.3).
3. Fastp.
4. Bowtie2 version 2.4.4.
5. Deeptools 3.3.0.
6. Integrative Genomics Viewer (IGV).
7. UCSC Genome Browser.

3 Methods

3.1 Recombinant pA-Tn5 and pA-Tn5 Transposome Assembly

1. Chemically transform the His_6-pA-Tn5 plasmid into BL21 (DE3) competent *E. coli* cells. Plate on ampicillin selection plates and incubate overnight at 37 °C. The next day, pick a single colony and inoculate it into 5 mL LB liquid medium containing 100 μg/mL ampicillin. After overnight culture at 37 °C and 200 rpm, transfer 5 mL bacterial culture to 1 L LB medium supplemented with 100 μg/mL ampicillin and culture at 37 °C, 200 rpm.
2. Incubate at 37 °C with shaking until OD_{600} reaches 0.6, add IPTG to a final concentration of 0.5 mM. The culture is further incubated at 37 °C, 200 rpm for 5 h to induce the pA-Tn5 expression.
3. Centrifuge the BL21 *E. coli* cells at 5000 × *g* for 10 min to collect cells. Wash the cells with PBS once. Resuspend cells in 20 mL HXG buffer supplemented with 1× cOmplete EDTA-free protease inhibitor cocktails. High-pressure homogenization at 600 psi for 5 min is used to break the cells and extract proteins (*see* **Note 1**).
4. Centrifuge at 13,523 × *g*, 4 °C for 30 min to harvest the supernatant, add 0.1 mL of 10% poly(ethyleneimine) solution to precipitate bacterial DNA, which is further removed by centrifugation at 13,523 × *g*, 4 °C for 10 min.
5. Equilibrate 2 mL Ni-NTA beads 6FF or Qiagen Ni-NTA agarose with HXG buffer.
6. Incubate the supernatant with 2 mL Ni-NTA beads 6FF or Qiagen Ni-NTA agarose at 4 °C, overnight with rotation. The beads are washed with 100 mL HXG buffer with 20 mM imidazole and 6 mL HXG buffer with 50 mM imidazole. The His-tagged pA-Tn5 is eluted with the HXG buffer containing 250 mM imidazole.

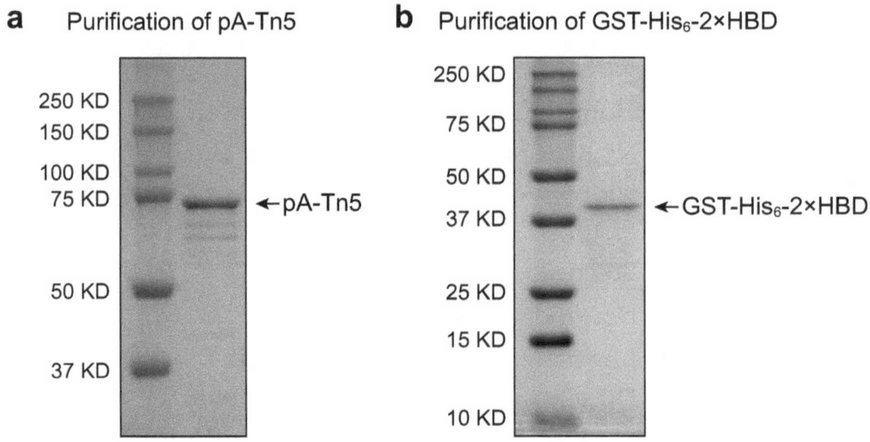

Fig. 1 SDS-PAGE analysis of purified pA-Tn5 (**a**) and GST-His6–2 × HBD (**b**)

7. The eluted pA-Tn5 protein is dialyzed against 1 L 2× dialysis buffer three times and concentrated with an Amicon Ultra-15 centrifugal filter unit with a 30 kDa cutoff. Dilute the concentrated pA-Tn5 at 1:1 ratio with 100% glycerol for storage at −20 °C.
8. Analyze the pA-Tn5 protein by SDS-PAGE and Coomassie blue staining (Fig. 1), and quantify the protein by BCA protein assays (*see* **Note 2**).
9. Dilute the oligonucleotides Tn5ME-rev, Tn5ME-A, and Tn5ME-B with TE buffer to 400 μM.
10. The mosaic end double-stranded oligonucleotides (Tn5MEDS-A/-B) are prepared by mixing equal volume of Tn5MErev with Tn5ME-A or Tn5ME-B. Boil 400 mL of water in a large glass beaker on a hotplate and incubate the tube of oligonucleotides in the boiling water for 5 min. Turn off the hotplate and leave the oligonucleotides in the beaker to slowly cool down to room temperature.
11. The 37.5 μM pA-Tn5 adapter transposome is assembled by adding 37.5 μM annealed oligonucleotides, 37.5 μM purified pA-Tn5, and storage buffer. The mixture is incubated on a 3D rotator at room temperature for 60 min. Store the assembled pA-Tn5 transposome at −20 °C (*see* **Note 3**).
12. The activity of the pA-Tn5 transposome can be checked by tagmentation of plasmid DNA. Mix 1 μg plasmid DNA and 1 μL assembled pA-Tn5 transposome in 20 μL 1 × TAPS-DMF buffer. The transposition reaction is set up by incubation at 55 °C for 15 min, and the reaction could be stopped by adding 2 μL stopping buffer. The DNA products can be resolved on 1.5% agarose gel (*see* **Note 4**).

3.2 Purification of Recombinant GST-His_6–2 × HBD Protein

1. The pGEX-GST-His_6–2 × HBD plasmid is chemically transformed into the T7 Express *lysY/I^q* competent *E. coli* cells. One single colony is picked and inoculated with 5 mL 2× YT liquid medium containing 100 μg/mL ampicillin. The culture is grown at 200 rpm in a 37 °C shaker overnight.
2. Next day, 5 mL bacterial culture is transferred to 1 L 2× YT medium supplemented with 100 μg/mL ampicillin and cultured at 37 °C, 200 rpm. Protein expression is induced by the addition of 0.5 mM IPTG when the OD_{600} reaches 0.6. The culture is further grown for 5 h at 200 rpm in a 37 °C shaker (*see* **Note 5**).
3. Bacterial cells are collected and lyzed in 40 mL HXG buffer supplemented with 1× cOmplete, EDTA-free protease inhibitor cocktails. Homogenization of lysates is performed with a high-pressure homogenizer at 800 psi for 5 min. The supernatant is collected by centrifuging at 13,523 × *g* at 4 °C for 30 min (*see* **Note 1**).
4. Equilibrate 2 mL Ni-NTA beads 6FF or Ni-NTA agarose with HXG buffer.
5. The supernatant is incubated with 2 mL Ni-NTA beads 6FF or Ni-NTA agarose at 4 °C overnight. Wash the beads with 50 mL HXG buffer containing 20 mM imidazole.

   Then wash the beads with 50 mL HXG buffer containing 50 mM imidazole and with 10 mL HXG buffer containing 100 mM imidazole. The GST-His_6–2 × HBD protein is eluted with HXG buffer containing 250 mM imidazole. The eluted protein is dialyzed with 1 L HXG buffer three times. The GST-His_6–2 × HBD protein is concentrated with an Amicon Ultra-15 centrifugal filter unit (10 kDa cutoff).
6. The purity of GST-His_6–2 × HBD protein is analyzed by SDS-PAGE and Coomassie blue staining (Fig. 1). Perform protein quantification assays with the Pierce BCA protein assay kit. The purified protein is diluted 1:1 with 100% glycerol (*see* **Note 6**).

3.3 Cell Preparation and Binding Cells to Concanavalin A-Coated Beads

1. Harvest fresh and healthy cells at room temperature and count cells. The cells can be digested with the accutase solution or blown down directly, such as HEK293T cells (*see* **Note 7**).
2. Centrifuge at 600 × *g* for 3 min at room temperature and drain the liquid. Wash the cells twice with wash buffer. Every 5 × 10^5 cells are resuspended in 0.1 mL wash buffer and then transferred into a 0.2 mL PCR tube (*see* **Note 8**).
3. Transfer 10 μL Concanavalin A-coated magnetic beads into 150 μL binding buffer in a 0.2 mL PCR tube and mix by pipetting. Place the 0.2 mL tube of Concanavalin A-coated beads on a magnet stand to clear (30 s to 2 min). Withdraw

the liquid completely, and remove the tube from the magnet stand (*see* **Note 9**).

4. Add 150 μL binding buffer to the 0.2 mL PCR tube and mix by pipetting. Place the 0.2 mL tube on a magnet stand and drain the liquid completely. Resuspend the beads in 10 μL binding buffer.
5. Retrieve the tube with cells and vortex gently, slowly add the Concanavalin A-coated magnetic bead slurry into the cells. Place the tube on a 3D rotator for 10 min.
6. Quick spin to remove liquid from the cap (<100 × *g*), place the tube on a magnet stand to clear, and discard the liquid. Resuspend the beads in 50 μL antibody buffer (*see* **Note 10**).

3.4 Incubation with DNA–RNA Hybrid Sensors and Secondary Antibody

1. Add the appropriate amount of DNA–RNA hybrid sensors (20 μg/mL GST-His$_6$–2 × HBD or 20 μg/mL S9.6). Rotate at room temperature for 2 h or overnight at 4 °C (*see* **Notes 11** and **12**).
2. Spin to remove liquid from the cap, place the tubes on a magnet stand to clear, and completely remove the liquid. Wash the beads with 150 μL Dig-wash buffer twice.
3. For GST-His$_6$–2 × HBD groups, the mouse Anti-Histag antibody at dilution of 1:100 is prepared with antibody buffer, and each sample is incubated with 100 μL Anti-Histag antibody for 1 h at room temperature. For S9.6 antibody groups, the rabbit anti-mouse IgG at a dilution of 1:100 (10 μg/mL) is prepared with antibody buffer, and each sample is incubated with 100 μL rabbit anti-mouse antibody for 1 h.
4. Quick spin to remove liquid from the cap, place the tubes on a magnet stand to clear, and completely withdraw the liquid. Wash the beads with 150 μL Dig-wash buffer once to remove excess antibody.
5. Secondary antibody could be used to amplify the R-loop signals. For GST-His$_6$–2 × HBD groups, the rabbit anti-mouse IgG at a dilution of 1:100 (10 μg/mL) is prepared with antibody buffer, and each sample is incubated with 100 μL rabbit anti-mouse antibody for 1 h at room temperature. For S9.6 antibody groups, the mouse anti-rabbit IgG at a dilution of 1: 100 (10 μg/mL) is prepared with antibody buffer, and each sample is incubated with 100 μL mouse anti-rabbit antibody for 1 h (*see* **Note 13**).
6. Spin to remove liquid from the cap, place the tubes on a magnet stand to clear, and drain the liquid. Wash with 150 μL Dig-wash buffer three times and rotate for 5 min each time.

3.5 Incubation with pA-Tn5 Adapter Complex and Tagmentation

1. Prepare a 1:250 dilution of pA-Tn5 adapter complex in Dig-300 buffer. Mix the beads with 100 μL diluted pA-Tn5 complex. Rotate at room temperature for 1 h.
2. Quick spin to remove liquid from the cap, place the tubes on a magnet stand, and withdraw the liquid. Wash with 100 μL Dig-300 buffer four times to remove unbound pA-Tn5 adapter complex and rotate 5 min for each wash.
3. Resuspend beads in 40 μL tagmentation buffer and incubated at 37 °C for 1 h. To stop the tagmentation reaction, add 1.5 μL 0.5 M EDTA, 0.5 μL 10% SDS, 1 μL 20 mg/mL proteinase K, and 5 μL H_2O, and incubate at 55 °C for 60 min and 75 °C for 20 min (*see* **Note 14**).
4. Purify the tagmentation products with 1.8× AMPure XP beads or homemade SPRI beads. Elute with 10 μL H_2O containing 0.1% Tween-20 (*see* **Note 15**).

3.6 R-Loop CUT&Tag Library Preparation

1. The eluent is mixed with 0.5 μL 10 mM dNTPs, 0.4 U Q5 high-fidelity DNA polymerase, and 10 U Bst 2.0 WarmStart DNA polymerase in 20 μL of 1× Q5 polymerase reaction buffer with 1× Q5 high GC enhancer. The mixture is incubated at 65 °C for 30 min to complete the strand displacement reaction (*see* **Note 16**).
2. Inactivate the Bst 2.0 WarmStart polymerase by heating up to 80 °C for 20 min.
3. Use 2 μL of strand-displaced products to perform qPCR to determine how many PCR cycles are needed. Set up the reaction with 4 μL 5× Q5 reaction buffer, 4 μL 5× Q5 high GC enhancer, 0.4 μL 10 mM dNTPs, 1 μL 10 μM Ad1-S502, 1 μL 10 μM Ad2.1, 2 μL template DNA, 0.2 μL Q5 polymerase, 0.4 μL 1:2000 diluted SYBR Green I, and 7 μL H_2O. Perform the PCR with the program: 72 °C 5 min, 98 °C 45 s, and then 30 cycles of 98 °C 15 s, 63 °C 15 s, and 72 °C 15 s with final extension at 72 °C for 2 min (*see* **Note 17**).
4. Calculate the number of PCR cycles needed for samples, by determining the number of cycles needed to reach 1/3 of the maximum reaction (*see* **Note 18**).
5. Use 18 μL of strand-displaced products to perform library preparation. The strand-displaced products are mixed with 500 nM of Ad1-S502 primer, 500 nM of uniquely barcoded i7 primer (Ad2.1), and 0.4 U of Q5 high-fidelity DNA polymerase. Perform the PCR with the calculated cycle numbers.
6. Size select the libraries with 0.56–0.85× AMPure XP beads. Determine the size distribution and concentration of libraries by capillary electrophoresis using an Agilent 2100 or equivalent (Fig. 2) (*see* **Note 15**).

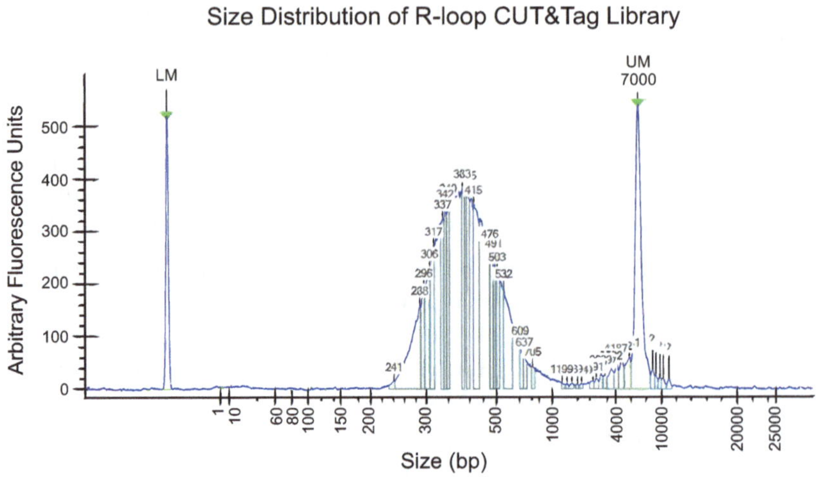

Fig. 2 Size distribution analysis of R-loop CUT&Tag library by LabChip. The library size ranges from 240 bp to 700 bp with an average size of 400 bp. *UM* upper marker, *LM* lower marker

7. Mix barcoded libraries to achieve equal representation as desired. Perform an AMPure XP bead cleanup if needed to get rid of any residual PCR primers.

3.7 DNA Sequencing and Data Analysis

1. Perform paired-end Illumina sequencing with the barcoded libraries following the manufacturer's instructions (e.g., Nova-Seq 6000) (*see* **Note 19**).
2. Preprocess FASTQ raw files using fastp with parameters: fastp -w 16.
3. Align paired-end reads to human genome (UCSC hg38) using Bowtie2 with parameters: --end-to-end --very-sensitive --no-mixed --no-discordant --phred33 -I 10 -X 700.
4. The track files are made with the bamCoverage command from deeptools 3.3.0. The aligned reads are normalized to total reads aligned (reads per million, r.p.m.).
5. The track files can be viewed in Integrative Genomics Viewer (IGV) or UCSC Genome Browser (Fig. 3).
6. For spike-in normalization, align paired-end reads to NCBI *E. coli* genome (*Escherichia coli* str. K12 substr. MG1655 U00096.3) using Bowtie2 with parameters: --end-to-end --very-sensitive --no-mixed --no-discordant --phred33 -I 10 -X 700. To calibrate samples in a series for samples performed in parallel, we use counts of *E. coli* fragments carried over with the pA-Tn5 for spike-in normalization.

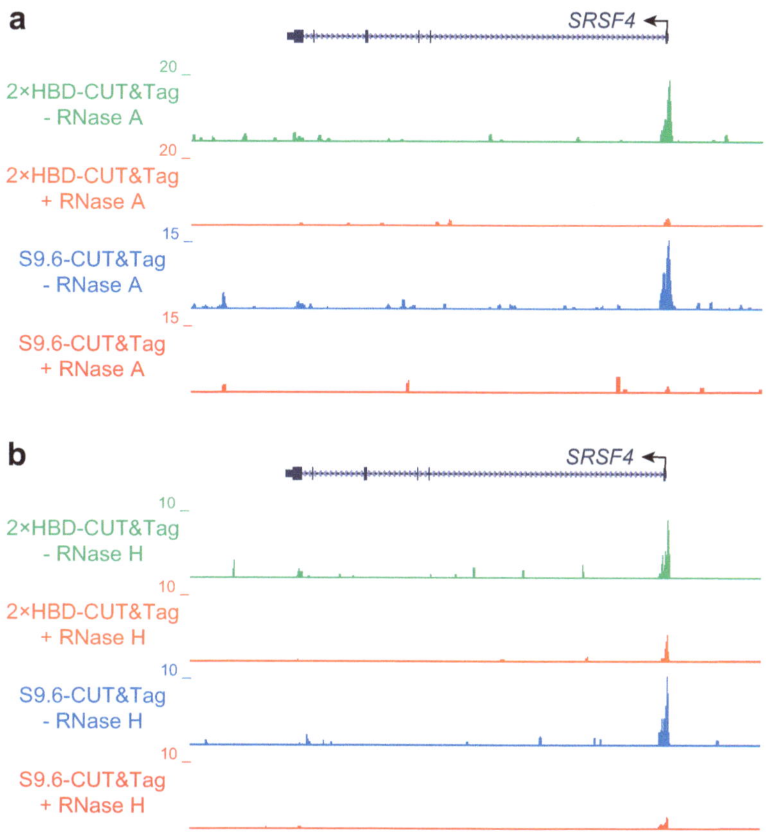

Fig. 3 UCSC Genome Browser tracks of R-loop CUT&Tag signals at the SRSF4 locus. The tracks were normalized by reads per million, and the RNase A-(**a**) or RNase H-(**b**) treated groups were further normalized with the *E. coli* spike-in control [4]

4 Notes

1. Ultrasonic sonication could also be used for protein extraction and genomic DNA breakage. However, efficient sonication depends on the instrument types and proper sonication parameters.
2. Generally, we could purify 3–8 mg pA-Tn5 protein from 1 L culture and store the pA-Tn5 at a concentration of around 5–6 mg/mL.
3. The pA-Tn5 adapter transposome is stable and can be stored at −20 °C for about half a year.
4. The assembled pA-Tn5 transposome complex yielding DNA fragments with an average size less than 500 bp pass quality control could be used in the following experiments.

5. Using 2× YT liquid medium other than LB medium could dramatically increase the expression of GST-His_6–2 × HBD protein.
6. Usually, we could purify 1–2 mg GST-His_6–2 × HBD protein from 1 L bacterial culture, and the purified protein could be stored at −20 °C with a final glycerol concentration of 55%.
7. It is not recommended to digest cells with trypsin, which can reduce the binding efficiency of cells to Concanavalin A-coated beads.
8. All the things that deal with cells should be returned to room temperature before use to minimize the pressure on the cells.
9. Take the Concanavalin A-coated beads from 4 °C, mix them evenly, and determine the appropriate volume of the beads according to the cell numbers and cell types. For 5×10^5 HEK293T cells, we use 10 μL Concanavalin A-coated beads for the binding reaction.
10. The antibody buffer needs to be precooled on ice in advance.
11. Either GST-His_6–2 × HBD or Anti-DNA–RNA hybrid antibody (clone S9.6) could be used for the recognition of DNA–RNA hybrids.
12. For negative controls, RNase A or RNase H treatment could be used in the binding step to digest the DNA–RNA hybrids. Add 5 mM $MgCl_2$ to the antibody buffer and incubate at 37 °C for 2 h when using RNase H (Fig. 3).
13. The secondary antibody incubation is required for R-loop CUT&Tag to increase the binding sites of protein A.
14. The tagmentation buffer is crucial for the tagmentation reaction. The activity of Tn5 may vary from batch to batch, and the activity of Tn5 could be modulated with increasing concentration of DMF (from 0% to 10%). Besides, tagmentation with Dig-300 buffer supplemented with 5 mM $MgCl_2$ and an optimal concentration of DMF could be tried to improve the signal/noise ratio.
15. Use the DNA cleanup beads according to the manufactory's instructions.
16. Tn5 transposase is reported to randomly bind DNA–RNA hybrids and transpose adapters onto both strands of the DNA–RNA hybrids. Additionally, the transposed products could have the strand displaced and be directly used for sequencing library preparation, which saves many time-consuming steps.
17. The goal here is to stop amplification well before saturation and avoid variation among samples due to PCR bias.

18. To minimize the contribution of large DNA fragments and excess primers, PCR should be performed for 13–17 cycles.
19. For maximum economy, we sequence each sample with 10–20 million paired-end reads on a NovaSeq 6000. Around 10–20 million paired-end reads should be sufficient to effectively profile native R-loops in the human genome.

Acknowledgments

This work was funded by the National Key Research and Development Program of China Stem Cell and Translational Research (NO. 2019YFA0111100), NSFC Excellent Young Scientists Fund (82122006), and the China National Natural Science Foundation (82172641) awarded to K.L.

References

1. Garcia-Muse T, Aguilera A (2019) R loops: from physiological to pathological roles. Cell 179(3):604–618
2. Aguilera A, Garcia-Muse T (2012) R loops: from transcription byproducts to threats to genome stability. Mol Cell 46(2):115–124
3. Marnef A, Legube G (2021) R-loops as Janus-faced modulators of DNA repair. Nat Cell Biol 23(4):305–313
4. Wang K, Wang H, Li C et al (2021) Genomic profiling of native R loops with a DNA-RNA hybrid recognition sensor. Sci Adv 7(8): eabe3516
5. Skourti-Stathaki K, Proudfoot NJ (2014) A double-edged sword: R loops as threats to genome integrity and powerful regulators of gene expression. Genes Dev 28(13): 1384–1396
6. Crossley MP, Bocek M, Cimprich KA (2019) R-loops as cellular regulators and genomic threats. Mol Cell 73(3):398–411
7. Santos-Pereira JM, Aguilera A (2015) R loops: new modulators of genome dynamics and function. Nat Rev Genet 16(10):583–597
8. Sanz LA, Hartono SR, Lim YW et al (2016) Prevalent, dynamic, and conserved R-loop structures associate with specific Epigenomic signatures in mammals. Mol Cell 63(1): 167–178
9. Yan Q, Shields EJ, Bonasio R et al (2019) Mapping native R-loops genome-wide using a targeted nuclease approach. Cell Rep 29(5): 1369–80 e5
10. Chen L, Chen JY, Zhang X et al (2017) R-ChIP using inactive RNase H reveals dynamic coupling of R-loops with transcriptional pausing at gene promoters. Mol Cell 68(4):745–57 e5
11. Kaya-Okur HS, Wu SJ, Codomo CA et al (2019) CUT&Tag for efficient epigenomic profiling of small samples and single cells. Nat Commun 10(1):1930
12. Wang Q, Xiong H, Ai S et al (2019) CoBATCH for high-throughput single-cell epigenomic profiling. Mol Cell 76(1): 206–16 e7
13. Lu B, Dong L, Yi D et al (2020) Transposase-assisted tagmentation of RNA/DNA hybrid duplexes. elife 9:e54919
14. Carter B, Ku WL, Kang JY et al (2019) Mapping histone modifications in low cell number and single cells using antibody-guided chromatin tagmentation (ACT-seq). Nat Commun 10(1):3747
15. Di L, Fu Y, Sun Y et al (2020) RNA sequencing by direct tagmentation of RNA/DNA hybrids. Proc Natl Acad Sci U S A 117(6):2886–2893

Chapter 24

Mapping R-Loops Using Catalytically Inactive RNaseH1 (R-ChIP)

Xuan Zhang, Yajing Hao, and Xiang-Dong Fu

Abstract

R-loops, three-stranded structures containing double-stranded DNA invaded by single-stranded RNA, have been linked to diverse biological processes. They play important roles in regulating gene regulation and DNA repair, contributing to a wide range of diseases. Understanding the formation and dynamic regulation of R-loops is thus a gateway to address many fundamental questions in regulatory biology, which requires the elucidation of the R-loop landscape at the genome scale. To aid in such efforts, this article provides an overview on R-loop mapping strategies along with a detailed protocol based on the use of catalytically inactive RNaseH1, an evolutionarily conserved protein responsible for R-loop recognition and resolution.

Key words R-loop, R-ChIP, RNaseH1, Immunoprecipitation, Transcription, DNA damage

1 Introduction

R-loops were first observed during experiments attempting to map DNA topology relative to expressed RNA. This technology was first applied to study the transcription unit for rRNA. By using purified rRNA to anneal to the corresponding DNA template, thereby looping out specific single-stranded DNA regions, it was possible to directly visualize the R-loop structure by electronic microscopy to deduce the gene structure [1, 2]. This strategy played a vital role in the discovery of introns in a protein-coding gene from an adenovirus genome [3, 4]. It was soon recognized that such structures are also formed on chromatin during important biological processes, such as immunoglobulin class switch and transcription [5, 6]. Further studies demonstrated that the formation of R-loops provides an important mechanism to regulate gene expression during cell fate determination [7].

Mapping R-loop locations along the genome is key to understand many critical biological processes. In the past decade,

Andrés Aguilera and Alexey Ruzov (eds.), *R-Loops: Methods and Protocols*, Methods in Molecular Biology, vol. 2528,
https://doi.org/10.1007/978-1-0716-2477-7_24,

multiple methods have been developed to map R-loops genome-wide. In general, R-loops could be recognized by using either a specific monoclonal antibody (S9.6) or catalytically inactive RNaseH1 [8]. The most original S9.6-based method to capture R-loop containing DNA for deep sequencing is DRIP-seq (DNA–RNA immunoprecipitation sequencing) [9]. This strategy has been further refined to improve the specificity and resolution, as represented by ssDRIP-seq, DRIPc-seq, and bisDRIP-seq [10–12]. A major concern on S9.6-based approaches is related to the binding specificity of the antibody, as it also has significant affinity for double-stranded RNAs, which may contribute to explanation of some major discrepancies in literature [13]. Alternatively, R-ChIP (R-loop mapping by chromatin immunoprecipitation) was developed by taking advantage of the high binding affinity of RNaseH1 for DNA/RNA hybrids [14]. In this technology, the expression of a catalytically inactivated version of this enzyme in cells enables efficient capture of R-loops for deep sequencing by using a standard ChIP protocol.

Like the DRIP-based method, the R-ChIP method has been evolved to multiple versions. For example, RR-ChIP sequences RNAs associated with RNaseH1-captured R-loops [15]. A major shortcoming for both R-ChIP and RR-ChIP is the requirement to express a mutant RNaseH1 in cells, which may not be feasible in certain applications and/or cause potential artifacts due to overexpressed RNaseH1 that may be dominant over the endogenous enzyme. These concerns have prompted the development of additional RNaseH1-based methods (Fig. 1). MapR was developed by fusing mutant RNaseH1 with a general DNA cutting enzyme MNase [16]. BisMapR is a derivative of MapR for constructing strand-specific libraries [17]. These technologies essentially use RNaseH1 to guide MNase to cleave DNA in the genome where R-loops are located. R-loop CUT & Tag combines inactive RNaseH1 with the CUT & Tag sequencing technique to map R-loops by using the transposase Tn5 for simultaneous DNA cutting and primer ligation [18]. However, in general, these techniques generate much fewer peaks compared to R-ChIP, which may result from compromised affinity and specificity of purified RNaseH1 in the absence of potential co-factors, such as the single strand DNA binding protein RPA, involved in enhancing RNaseH1 targeting to R-loops in mammalian cells [19].

Each R-loop mapping strategy may on the one hand improve the existing technologies and on the other hand introduce additional shortcomings, thus leading to significant discrepancies in literature. For example, R-ChIP predominately detects signals at gene promoters; however, RR-ChIP additionally generates abundant signals in gene bodies (Fig. 2). Such extra gene-body signals may result from chromatin associated RNAs clustered around gene promoters. While MapR and R-loop CUT & Tag overcome the

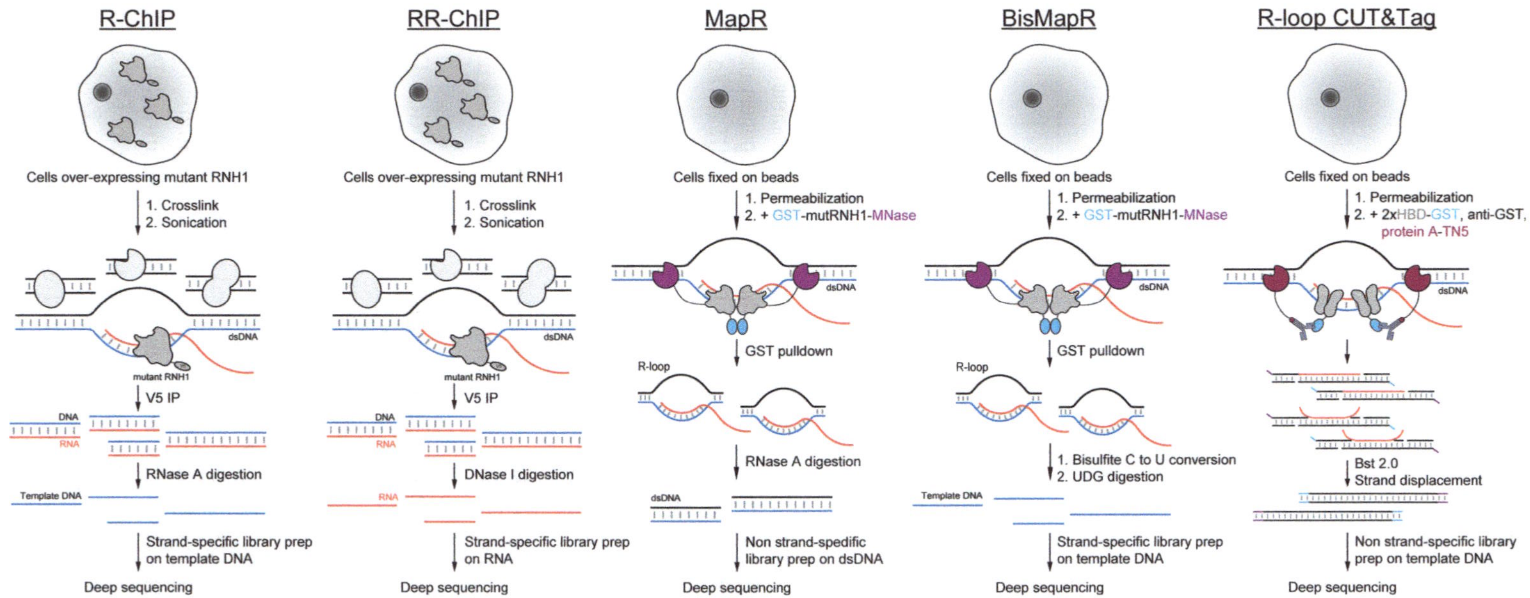

Fig. 1 Schematic of R-loop methods based on RNaseH1. In R-ChIP and RR-ChIP, a cell line is first constructed to express a catalytically inactive RNaseH1, and R-loops are captured using a ChIP-seq protocol. In R-ChIP the template DNA is used for strand-specific library construction; in RR-ChIP, RNA enriched in R-loops is isolated for library construction. In MapR, BisMapR, and R-loop CUT & Tag, cells are fixed on beads followed by permeabilization. In MapR, R-loops are cut and released by GST tagged inactivated RNaseH1 fused to MNase and released R-loops are enriched by GST pulldown. In BisMapR, captured R-loops are treated with bisulfite to convert C to U on the ssDNA. RNA within R-loops is removed by RNaseH1 and second strand synthesis is performed in the presence of dUTP. After UDG digestion, the remaining template DNA is used to construct a strand-specific library. In R-loop CUT & Tag, two hybrid binding domains of RNase H1 fused to GST (2xHBD-GST) are used to bind R-loops. Anti-GST linked to Tn5 is next used to cut and tag R-loops as in the CUT & Tag protocol

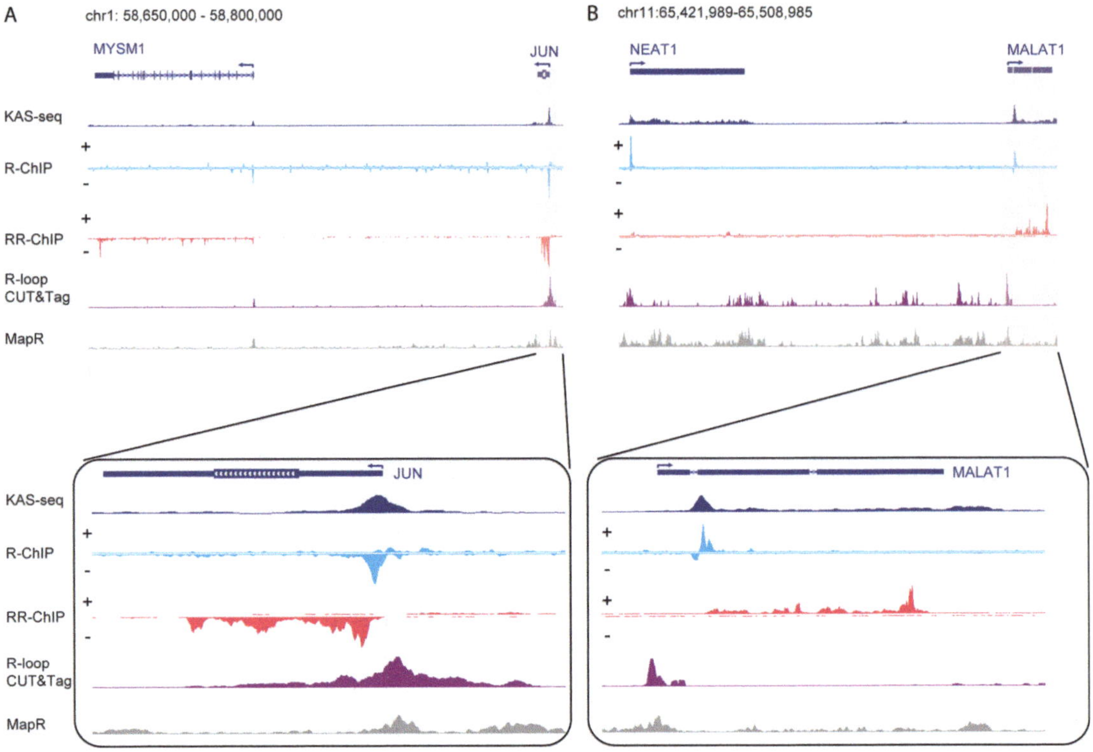

Fig. 2 Genome browser view of R-loop signals detected by different methods. (**a**) The genomic region containing JUN and MYSM1. (**b**) The genomic region containing NEAT1 and MALAT1. Both panels display R-loop signals (normalized to reads per million, RPM) detected by R-ChIP, RR-ChIP, R-loop CUT & Tag, and MapR. KAS-seq signals are provided for comparison on top of each panel

need to express an exogenous mutant RNaseH1, both show extensive signals in intergenic regions (Fig. 2), which may result from nonspecific targeting of purified RNaseH1. Judging which signals represent bona fide R-loops is a quite challenging problem because of the lack of a "gold standard" for mapping genuine R-loops in the genome.

Because a key feature of R-loops is the single-stranded DNA replaced from the double helix, the ability to detect such single-stranded DNA may be leveraged to compare between different R-loop mapping methods. However, it is important to keep in mind that detectable genomic regions associated with single-stranded DNA do not necessarily represent R-loop formation activities. According to this rationale, we utilized the data generated by KAS-seq (Kethoxal-Assisted Single-stranded DNA sequencing), a recently developed technique for mapping single-stranded DNA along the genome [20], to compare between RNaseH1-based methods. It appears that the signals detected with KAS-seq show the highest coincident with those detected with R-ChIP (Fig. 2).

Given the robust correlation between the data generated by R-ChIP and KAS-seq, we believe that R-ChIP still represents one of the most robust R-loop mapping strategies developed to date. Therefore, we herein describe a detailed protocol for using R-ChIP to aid in its application, as outlined in Fig. 1. As aforementioned, this technology requires stable expression of a tagged catalytically inactive RNaseH1.

2 Materials

Prepare all well-mixed buffers and solutions by filtering through a 0.22 μm syringe.

2.1 Crosslinking and Sonication

1. 1× PBS.
2. 37% Formaldehyde.
3. 1.375 M Glycine.
4. RNaseOut (Invitrogen).
5. Cell Lysis Buffer: 10 mM Tris–HCl pH 8.0, 10 mM NaCl, 0.5% NP-40, Protease Inhibitor Cocktail (Sigma, added immediately prior to use).
6. Nuclear Lysis Buffer: 50 mM Tris–HCl pH 8.0, 1% SDS, 10 mM EDTA, Protease Inhibitor Cocktail (added immediately prior to use).
7. TE Buffer: 10 mM Tris–HCl pH 8.0, 1 mM EDTA.

2.2 Chromatin Immunoprecipitation

1. Dilution Buffer: 20 mM Tris–HCl pH 8.0, 150 mM NaCl, 2 mM EDTA, 1% Triton X-100.
2. Blocking Buffer: 10 mg/mL Glycogen, 10 mg/mL BSA, 20 mg/mL tRNA.
3. Magnetic beads (Dynabeads Protein G, Invitrogen).
4. Antibody against the V5 tag (Invitrogen, cat. no. R960-25).
5. TSEI Buffer: 20 mM Tris–HCl pH 8.0, 150 mM NaCl, 2 mM EDTA, 1% Triton X-100, 0.1% SDS, Protease Inhibitor Cocktail (added immediately prior to use).
6. TSEII Buffer: 20 mM Tris–HCl pH 8.0, 500 mM NaCl, 2 mM EDTA, 1% Triton X-100, 0.1% SDS, Protease Inhibitor Cocktail (added immediately prior to use).
7. Buffer III: 10 mM Tris–HCl pH 8.0, 250 mM LiCl, 1 mM EDTA, 1% NP-40, 1% sodium deoxycholate, Protease Inhibitor Cocktail (added immediately prior to use).
8. Elution Buffer: 10 mM Tris–HCl pH 8.0, 1 mM EDTA, 1% SDS.
9. RNase A.

10. Proteinase K.
11. Phenol/chloroform/isoamyl alcohol (25:24:1).

2.3 Library Construction and Sequencing

1. dNTP mix.
2. Phi29 DNA polymerase (NEB).
3. dATP.
4. Klenow fragment (3′- > 5’exo-) (NEB).
5. T4 DNA Ligase and T4 DNA Ligase Buffer (NEB).
6. N9 random primer (Illumina HiSeq 2500 system):
   5′-/Invddt/ CAAGCAGAAGACGGCATAC GAGNNNNNNNNN-3′.
7. Adapter oligos (Illumina HiSeq 2500 system):
   Oligo A: 5′-/Phos/ GATCGGAAGAGCGTCGTGTAGG GAAAGAGTGT-3′.
   Oligo B: 5′-AGACGTGTGCTCTTCCGATCT-3′.
8. Annealing Buffer: 10 mM Tris–HCl pH 8.0, 50 mM NaCl, 1 mM EDTA.
9. PCR Primer Mix (Illumina HiSeq 2500 system):
   PCR primer: 5′-CAAGCAGAAGACGGCATACGAG-3′.
   Barcode primer: 5′- AATGATACGGCGACCACCGAGATC TACACNNNNNACACTCTTTCCCTACAC GACGCTCTTCCGATCT-3′.
10. Phusion DNA Polymerase (NEB).
11. DNA Cleanup Kit (Zymo).
12. 10% TBE Gel.
13. Gel loading buffer (NEB).
14. SYBR gold dye (Invitrogen).
15. GlycoBlue (Invitrogen).
16. Gel Elution Buffer: 10 mM Tris–HCl pH 8.0, 300 mM NaCl, 1 mM EDTA, 0.1% Tween 20.

2.4 General Supplies and Equipment

1. Cell scraper (if working with adherent cells).
2. 0.22 μm syringe filters.
3. Probe sonicator or equivalent (*see* **Note 1**).
4. Magnetic stand.
5. 1.5 mL microcentrifuge tubes, preferably low protein and DNA binding.
6. Qubit fluorometer or equivalent (Thermo Fisher).
7. Qubit dsDNA HS Assay Kit (Invitrogen).
8. Costar column (Corning).

3 Methods

R-ChIP requires the construction of stable cell line expressing inactive RNaseH1, which can be delivered via either lentiviral virus or transposon. We used the pPyCAG vector to construct stable cell lines on HEK293T and K562 cells. Western blot should be used to confirm the inactive RNaseH1 expression with V5 antibody before immunoprecipitation. Buffers should be freshly prepared and kept on ice before use.

3.1 Crosslinking and Sonication

Preparation of antibody-conjugated magnetic beads (Day 1)

1. Add 50 μL of well-resuspended magnetic beads to 1.5 mL tube.
2. Wash the beads twice in ChIP Dilution Buffer. For each wash, resuspend the beads in 1 mL ChIP Dilution Buffer, spin the tube briefly to collect cap and sidewall liquid, incubate on the magnetic stand for 5 min, remove the supernatant.
3. Resuspend the beads in 1 mL Blocking Buffer and incubate for 2 h on a tube rotator at room temperature (RT).
4. Place the beads on a magnetic stand for 5 min and remove the supernatant.
5. Wash the beads three times with Blocking Buffer as described above. Resuspend beads in 1 mL Blocking Buffer.
6. Add 5–10 μg of V5 antibody to the beads.
7. Incubate overnight with rotation at 4 °C.

   Crosslinking

   For each sample, 1–2 × 10^7 cells are enough for robust R-loop signal in K562 and HEK293T cells. Volumes below are described for one sample.
8. Wash cells 3 times with ice-cold PBS, 10 mL PBS for each wash.
9. Add 270 μL 37% formaldehyde in 10 mL PBS to achieve final concentration at 1% (v/v). Mix well with cells and incubate for 10 min at RT.
10. Add 1 mL 1.375 M glycine to achieve final concentration at 0.125 M to quench the reaction. Mix well with cells and incubate for 15 min at RT.
11. Wash cells 3 times with ice-cold PBS, 10 mL PBS for each wash.
12. For adhesive cells, scrape the cells off the plate.
13. Collect cells by centrifugation at 600 × *g* at 4 °C for 5 min in 1.5 mL microcentrifuge tube. Discard the supernatant.

14. The half-dry cell pellet can be immediately frozen in liquid nitrogen and stored at −80 °C. The fixed R-loop stored at −80 °C is stable within 1 month.

    Sonication.

15. Add 1 mL ice-cold Cell Lysis Buffer (with Protease Inhibitor Cocktail and RNaseOut added) to resuspend the frozen or fresh cell pellet in a 1.5 mL microcentrifuge tube. Incubate on ice for 10–20 min to release the nuclei.
16. Pellet the nuclei by centrifugation at 800 × *g* at 4 °C for 5 min. Discard the supernatant.
17. Wash the nuclei once with 1 mL Cell Lysis Buffer (with Protease Inhibitor Cocktail and RNaseOut added) and pellet again as in **step 16**.
18. Resuspend the nuclei in 500 μL Nuclear Lysis Buffer (with Protease Inhibitor Cocktail and RNaseOut added). Mix well and incubate on ice for 10 min.
19. Sonicate in ice bath, by immersing the probe sonicator tip in the tube, using 7 cycles of 10 s sonication, 30 s rest between cycles, and power level of 4.0. Adjust the depth of the sonicator tip to avoid foaming. Check the temperature after each cycle to make sure that the samples are not overheated. The desired chromatin fragment size is between 200 and 500 bp before immunoprecipitation, which can be confirmed on 1.5% agarose gel.
20. Bury the sample tube in ice for 30–60 min to precipitate out the extra SDS from fragmented chromatin.
21. Centrifuge at 16,000 × *g* at 4 °C for 15 min to pellet the cell debris and extra crystalized SDS.
22. Collect the supernatant and adjust the sample volume to 1 mL with TE buffer.

Stock Solution	Concentration	Volume/tube
10% triton X-100	1%	130 μL
10% sodium deoxycholate	0.1%	13 μL
100× protease inhibitor	1×	13 μL
1× TE	1×	144 μL

23. Add the following additional reagents to bring the volume to 1.3 mL for each sample.
24. Save 50 μL "input protein" for western blot validation of IP efficiency, and 50 μL "input DNA" to provide input control for sequencing (*see* **Note 2**).

3.2 Immunoprecipitation

1. Place the antibody-conjugated beads prepared in Subheading 3.1, **step** 7 on the magnetic stand for 5 min and remove the supernatant.
2. Add 1.2 mL diluted chromatin from Subheading 3.1, **step 24** to the beads. Incubate overnight with rotation at 4 °C.

   Beads washing and decrosslinking (Day 2)

   Beads Washing*:* For each wash, resuspend the beads in 1 mL buffer, pipette the beads solution 20 times gently to mix well, place the beads solution on magnetic stand for 5 min, remove the supernatant, specifically,
3. Place the beads on a magnetic stand for 5 min. Collect the supernatant as the depleted R-loop solution for IP efficiency validation by western blot.
4. Wash the beads 3 times with TSEI Buffer.
5. Wash the beads 3 times with TSEII Buffer.
6. Wash the beads once with Buffer III.
7. Wash the beads once with 1 mL TE Buffer.
8. Centrifuge the beads at 1500 × *g* for 1 min. Remove the residual TE Buffer and collect the semi-dry beads.
9. Resuspend the beads in 170 μL Elution Buffer.
10. Incubate the beads solution at 65 °C for 30 min on the thermomixer to dissociate antibodies from the beads, vortex at 1200 rpm for 15 s with 2 min interval.

    Decrosslinking.
11. Place the beads on the magnetic stand for 5 min to collect the supernatant as immunoprecipitated R-loop sample.
12. Centrifuge at 16,000 × *g* for 5 min at RT. Combine the residual supernatant to the IPed R-loop solution in the previous step. Aliquot 20 μL of the solution as the IPed protein for western blot analysis, together with the input protein collected at Subheading 3.1, **step 24**.
13. Add 120 μL Elution Buffer to the 50 μL "input DNA" prepared in Subheading 3.1, **step 24**.
14. Incubate both IPed R-loop solution and Input DNA at 65 °C overnight in a thermomixer, vortex at 1200 rpm for 15 s with 2 min interval.

    DNA cleanup (day 3).
15. Add 150 μL TE Buffer and 6 μL RNase A (10 mg/mL) to IPed R-loop and input samples.
16. Mix well and incubate for 2 h at 37 °C in a thermomixer, vortex at 1200 rpm for 15 s with 2 min interval.
17. Add 7 μL of 20 mg/mL Proteinase K to IPed R-loop and input samples.

18. Mix well and incubate for 2 h at 65 °C in a thermomixer, vortex at 1200 rpm for 15 s with 2 min interval.
19. Extract the DNA with phenol/chloroform/isoamyl alcohol twice in both IPed and input samples. For each extraction, add 300 μL phenol/chloroform/isoamyl alcohol, vortex for 20s and centrifuge for 5 min at 16,000 × *g* at RT, transfer the top (aqueous) phase to a fresh 1.5 mL DNA low-binding tube.
20. Add 750 μL ethanol, 30 μL of 3 M NaAc, and 1 μL GlycoBlue to IPed and input samples. Precipitate the DNA overnight at −20 °C or at −80 °C for 15 min.
21. Centrifuge at 16,000 × *g* for 15 min at 4 °C to pellet the DNA.
22. Wash the pellet with 1 mL ice-cold 70% ethanol twice.
23. Dissolve the DNA pellet in 20 μL water.
24. Measure DNA concentration using the Qubit dsDNA HS assay kit.

3.3 Sequencing Library Construction

dsDNA Extension (day 4).

dsDNA extension converts the ssDNA generated from RNase A-digested DNA/RNA hybrids to dsDNA by one cycle extension using N9 primer, specifically,

1. Prepare dsDNA extension reaction without enzyme in 200 μl PCR tube for each sample (annealing):

Reagent	Volume (19 μL)	Concentration
10 ng DNA (Subheading 3.2, **step 23**) diluted with water to 11 μL	11 μL	10 ng
10× phi29 DNA polymerase buffer	2 μL	1×
BSA (1 mg/mL)	4 μL	200 μg/mL
3 mM dNTP mix	1 μL	150 μM
20 μM N9 primer	1 μL	1 μM

2. Centrifuge briefly and incubate for 5 min at 95 °C, followed by 5 min at 25 °C in a thermal cycler.
3. Add 1 μL phi29 DNA polymerase and incubate for 20 min at 30 °C, followed by 10 min at 65 °C to inactive the enzyme.
4. Purify DNA using the Zymo DNA Cleanup Kit. Elute in 20 μL water.

dA Addition.

5. Prepare dA addition reaction in a 200 μL PCR tube for each sample:

Reagent	Volume (30 μL)
DNA sample from **step 4**	20 μL
10× Klenow buffer	3 μL
Klenow exo (3′ to 5′ exo minus)	1 μL
1 mM dATP	6 μL

6. Centrifuge briefly and incubate at 37 °C for 30 min in a thermal cycler.
7. Purify DNA using the Zymo DNA Cleanup Kit. Elute in 13 μL water.

   Adapter Ligation.

8. Adapter needs to be pre-annealed before use. Resuspend adapter oligo A and B at a concentration of 100 μM in Annealing Buffer, respectively.
9. Mix 20 μL of 100 μM oligo A with 20 μL of 100 μM oligo B.
10. Add 60 μL Annealing Buffer.
11. Anneal oligos on a thermal cycler with the following program: 2 min at 95 °C, ramping down at the rate of 0.1 °C/s from 95 to 25 °C. The annealed adapters stock solution (20 μM) can be stored in −80 °C freezer.
12. Prepare 2 μM adapter solution by diluting 20 μM adapter stock solution 10 times with water.
13. Prepare Adapter Ligation reaction in a 200 μL PCR tube for each sample:

Reagent	Volume (20 μL)
DNA sample from **step 7**	13 μL
10× ligation buffer	2 μL
Adapter A + B (2 μM)	1 μL
T4 DNA ligase	4 μL

14. Centrifuge briefly and incubate at RT for 2 h.

   PCR amplification.

15. Prepare PCR reaction in a 200 μL PCR tube for each sample:

Reagent	Volume (50 μL)
DNA ligation product (**step 14**)	8 μL
5× Phusion buffer	10 μL

(continued)

Reagent	Volume (50 μL)
dNTP mix (10 mM each dNTP)	1.5 μL
PCR primer (10 μM)	1 μL
Barcode primer (10 μM)	1 μL

16. Centrifuge briefly and incubate in a thermal cycler using the following program:

Steps	Temp/°C	Time
Initial denaturation	98	30s
14 to 18 cycles depend on the concentration of library	98	10s
	65	30s
	72	30s
Final extension	72	5 min

17. Purify DNA using Zymo DNA Cleanup Kit. Elute in 20 μL water.

    Size selection.

18. Load the sample with loading buffer on 10% TBE Gel.
19. Run the gel at 100 V constant voltage until the pink dye migrates to the bottom of the gel.
20. Dilute the SYBR gold in 1:10000 in TBE Buffer. Soak the gel in the staining buffer and incubate for 10 min at RT.
21. Cut out the smear as the size range of 150–400 bp on the gel.
22. Add 360 μL Gel Elution Buffer to the gel slice to elute the DNA. Incubate for 2 h with rotation at RT.
23. Collect the supernatant by filtering out gel debris using a Costar column.
24. Add 1000 μL ethanol, 36 μL 3 M NaAc, and 1 μL GlycoBlue to the eluted library. Precipitate the DNA overnight at −20 °C or − 80 °C for 15 min.
25. Centrifuge at 16,000 × *g* for 15 min at 4 °C to pellet the DNA.
26. Wash the pellet with 1 mL ice-cold 70% ethanol twice.
27. Dissolve the DNA pellet in 10 μL water.
28. Measure DNA concentration using the Qubit dsDNA HS assay kit.
29. The library is ready for sequencing (*see* **Note 3**).

4 Notes

1. Multiple methods for shearing chromatin are available. Except for the probe sonicator described in the protocol, Bioruptor and the Covaris instruments can also be used. The detailed sonication settings should be optimized for different cells, according to the desired final chromatin size and temperature controls. During sonication, high temperature should be avoided by adjusting the interval waiting time between each pulse.
2. This protocol includes using the sonicated input as control to correct for the background of enriched R-loops. Besides this type of control, using irrelevant antibody "Normal IgG" can also provide good control for any bias introduced during the R-ChIP process.
3. All primers are designed for Illumina HiSeq2500 system. Sequencing as single-end 50 bp reads is usually sufficient and cost-effective. The optimal sequencing depth depends on the number of R-loops of different targets. For HEK293T and K562 cells, 40 million reads per sample (at least 15 million unique mapped reads) is sufficient to generate a high confidence R-loop profile. Controls should be sequenced deeper than the R-ChIP samples (~ 50 million reads).

Acknowledgements

We thank Dongyang Zhang for critical reading of the manuscript. This project was supported by NIH grants (HG004659 and DK098808).

References

1. White RL, Hogness DS (1977) R loop mapping of the 18S and 28S sequences in the long and short repeating units of Drosophila melanogaster rDNA. Cell 10:177–192
2. Thomas M, White RL, Davis RW (1976) Hybridization of RNA to double-stranded-DNA: formation of R-loops. Proc Natl Acad Sci U S A 73:2294–2298
3. Chow LT, Gelinas RE, Broker TR et al (1977) An amazing sequence arrangement at the 5′ ends of adenovirus 2 messenger RNA. Cell 12:1–8
4. Berget SM, Moore C, Sharp PA (1977) Spliced segments at the 5′ terminus of adenovirus 2 late mRNA. Proc Natl Acad Sci U S A 74: 3171–3175
5. Belotserkovskii BP, Tornaletti S, D'souza AD et al (2018) R-loop generation during transcription: formation, processing and cellular outcomes. DNA Repair (Amst) 71:69–81
6. Roy D, Yu K, Lieber MR (2008) Mechanism of R-loop formation at immunoglobulin class switch sequences. Mol Cell Biol 28:50–60
7. Niehrs C, Luke B (2020) Regulatory R-loops as facilitators of gene expression and genome stability. Nat Rev Mol Cell Biol 21:167–178
8. Vanoosthuyse V (2018) Strengths and weaknesses of the current strategies to map and characterize R-loops. Noncoding RNA 4:9
9. Ginno PA, Lott PL, Christensen HC et al (2012) R-loop formation is a distinctive

characteristic of unmethylated human CpG island promoters. Mol Cell 45:814–825

10. Xu W, Xu H, Li K et al (2017) The R-loop is a common chromatin feature of the Arabidopsis genome. Nat Plants 3:704–714
11. Sanz LA, Hartono SR, Lim YW et al (2016) Prevalent, dynamic, and conserved R-loop structures associate with specific Epigenomic signatures in mammals. Mol Cell 63:167–178
12. Dumelie JG, Jaffrey SR (2017) Defining the location of promoter-associated R-loops at near-nucleotide resolution using bisDRIP-seq. Elife 6:e28306
13. Konig F, Schubert T, Langst G (2017) The monoclonal S9.6 antibody exhibits highly variable binding affinities towards different R-loop sequences. PLoS One 12:e0178875
14. Chen L, Chen JY, Zhang X et al (2017) R-ChIP using inactive RNase H reveals dynamic coupling of R-loops with transcriptional pausing at gene promoters. Mol Cell 68:745–757.e5
15. Tan-Wong SM, Dhir S, Proudfoot NJ (2019) R-loops promote antisense transcription across the mammalian genome. Mol Cell 76: 600–616.e6
16. Yan Q, Sarma K (2020) MapR: a method for identifying native R-loops genome wide. Curr Protoc Mol Biol 130:e113
17. Wulfridge P, Sarma K (2021) A nuclease- and bisulfite-based strategy captures strand-specific R-loops genome-wide. elife 10:e65146
18. Wang K, Wang H, Li C et al (2021) Genomic profiling of native R loops with a DNA-RNA hybrid recognition sensor. Sci Adv 7:eabe3516
19. Nguyen HD, Yadav T, Giri S et al (2017) Functions of replication protein a as a sensor of R loops and a regulator of RNaseH1. Mol Cell 65:832–847.e4
20. Wu T, Lyu R, You Q et al (2020) Kethoxal-assisted single-stranded DNA sequencing captures global transcription dynamics and enhancer activity in situ. Nat Methods 17: 515–523

Chapter 25

Targeted Nuclease Approaches for Mapping Native R-Loops

Phillip Wulfridge, Qingqing Yan, and Kavitha Sarma

Abstract

R-loops are three-stranded, DNA:RNA hybrid-containing structures that form naturally throughout the genome as a consequence of transcription. Accurately determining the genomic locations and strand of origin of R-loops is critical to understanding their roles in gene regulation and disease. Here, we describe a nuclease-based protocol for genome-wide and strand-specific R-loop detection, which we term MapR. This method targets native R-loops for cleavage and release using a modified RNase H enzyme, followed by deep sequencing. An extension of the protocol, BisMapR, can additionally introduce strand specificity via non-denaturing bisulfite conversion of the R-loop's single-stranded DNA component. MapR and BisMapR identify R-loops with high resolution and low background, can be performed with low cell input, and require short experimental time.

Key words R-loop, DNA:RNA hybrid, Strand-specific, Deep sequencing

1 Introduction

R-loop structures are most commonly formed co-transcriptionally when a nascent mRNA molecule reinvades upstream DNA, binding to the template strand to form a DNA:RNA hybrid and displacing the opposite strand as single-stranded DNA [1]. R-loops can participate in regulation of gene expression [2], but their accumulation can also result in DNA damage and contribution to genomic instability [3, 4]. Whether R-loops will form at specific loci is determined by many factors including G/C skew and presence of G-quadruplex forming sequences. Additionally, a multitude of cellular regulators coordinate to promote, inhibit, or resolve R-loops [5].

Methods for R-loop detection generally leverage the affinity of either the S9.6 antibody [6–9] or *Escherichia coli* nuclease RNase H [10–13] for DNA:RNA hybrids. In this chapter, we discuss MapR, an RNase H based strategy to identify native R-loops [11]. MapR bears many similarities in methodology and advantages to the CUT & RUN method for transcription factor profiling [14, 15]. In this

Andrés Aguilera and Alexey Ruzov (eds.), *R-Loops: Methods and Protocols*, Methods in Molecular Biology, vol. 2528, https://doi.org/10.1007/978-1-0716-2477-7_25,

method, a fusion protein GST-RHΔ-MNase, comprising a catalytically deficient RNase H (RHΔ) conjugated to MNase, recognizes and binds R-loops in situ within intact cells (Fig. 1). Activation of the MNase moiety with calcium chloride cleaves and releases R-loop-containing fragments for isolation and sequencing [11, 12]. As the vast remainder of the genomic DNA remains within the cells, no further enrichment for R-loop containing

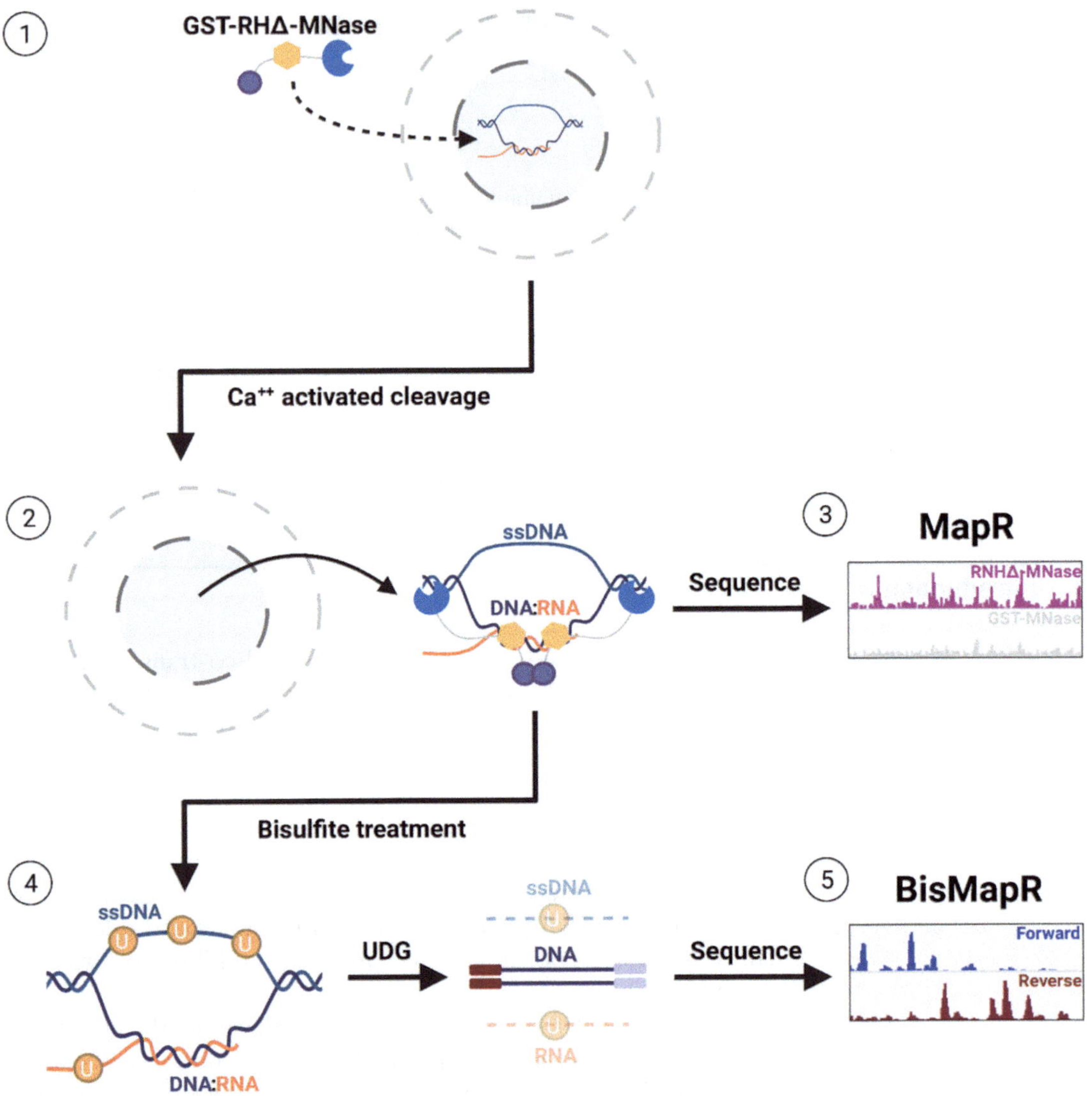

Fig. 1 Schematic overview of MapR and BisMapR. GST-RHΔ-MNase diffuses into the nucleus and binds to genomic R-loops (1). Addition of calcium chloride activates GST-RHΔ-MNase and cleaves R-loops for release (2). Isolated R-loops can be sequenced to produce MapR data (3) or further processed by sodium bisulfite treatment (4). Conversion of cytosines to uracils by bisulfite marks the ssDNA strand for degradation by uracil–DNA glycosylase, while the DNA:RNA hybrid-forming strand is sequenced to produce strand-specific BisMapR data (5). Created with BioRender.com

fragments is required, resulting in high resolution and signal-to-background ratio. Furthermore, MapR can be performed within 24 h and can be optimized for low cell numbers.

We also describe an extension to MapR, BisMapR, which in addition to detecting R-loop locations in the genome can identify the strand on which the DNA:RNA hybrid forms [13]. BisMapR separates the single-stranded and hybrid-containing components of R-loops released by MapR by leveraging the chemical properties of sodium bisulfite. Under non-denaturing conditions, sodium bisulfite converts cytosine bases to uracil on exposed single-stranded DNA, but not double-stranded DNA or DNA:RNA hybrids [16]. Next, the RNA component of the DNA:RNA hybrid is replaced by uracil-containing DNA via second-strand synthesis. Finally, treatment with uracil–DNA glycosylase during library construction degrades uracil-containing molecules and the remaining hybrid-forming DNA fragments can be sequenced, aligned, and computationally separated by strand of origin.

2 Materials

2.1 R-Loop Isolation (MapR)

1. Binding buffer: 9.5 ml H_2O, 200 μl 1 M HEPES-KOH pH 7.9, 100 μl 1 M KCl, 100 μl 100 mM $CaCl_2$, 100 μl 100 mM $MnCl_2$.
2. Wash buffer: 47.5 ml H_2O, 1 ml 1 M HEPES-Na^+ pH 7.5, 1.5 ml 5 M NaCl, 25 μl 1 M spermidine. Dissolve 1 × cOmplete Protease Inhibitor Cocktail Tablet (Roche) into buffer.
3. Digitonin-wash buffer: 40 ml wash buffer, 160 μl 5% digitonin.
4. Heterologous spike-in DNA, sonicated to 200 bp fragments. For example, genomic DNA extracted from *Drosophila melanogaster* or *Saccharomyces cerevisiae* can be used for human or murine cells.
5. 2× Stop buffer: 852 μl H_2O, 68 μl 5 M NaCl, 40 μl 0.5 M EDTA, 20 μl 0.2 M EGTA, 4 μl 5% digitonin, 10 μl 5 mg/ml linear acrylamide, 1 μl 1 ng/μl heterologous spike-in DNA. If performing MapR, also add 5 μl 10 mg/ml RNase A; for BisMapR, add 5 μl H_2O instead (*see also* **Note 1**).
6. BioMag Plus Concanavalin A beads, Polysciences.
7. Purified recombinant GST-RHΔ-MNase and GST-MNase enzymes.
8. 100 mM Calcium chloride ($CaCl_2$) solution.
9. 10% Sodium dodecyl sulfate (SDS).
10. 20 mg/ml Proteinase K.
11. Phenol–chloroform–isoamyl alcohol.
12. 7.5 M Ammonium acetate (NH_4OAc) solution.

13. 100% ethanol and 75% ethanol in nuclease-free water.
14. Tris-EDTA buffer, nuclease-free.
15. Phase-lock gel heavy tubes, QuantaBio.

2.2 Bisulfite Conversion and Second-Strand Synthesis

1. EZ DNA Methylation-Gold Kit, Zymo.
2. NEBNext Second-Strand Synthesis Enzyme Mix, NEB.
3. NEBNExt Second-Strand Synthesis Reaction Buffer with dUTP, NEB.
4. AMPure XP beads, Beckman Coulter.
5. 80% ethanol in nuclease-free water.
6. Tris-EDTA buffer, nuclease-free.

3 Methods

3.1 R-Loop Isolation (MapR)

*Steps prior to overnight incubation (**step 6**) are performed at room temperature to minimize cell stress and rupturing and/or fragmentation of DNA.*

1. Collect two to ten million cells into a 1.5 ml microcentrifuge tube or 15 ml polypropylene tube, depending on volume. Centrifuge at room temperature (25 °C) at 600 × *g* for 3 min. Carefully aspirate media to avoid disturbing the cell pellet.
2. Wash cells twice with 1 ml room temperature wash buffer. Centrifuge at room temperature at 600 × *g* for 3 min between each wash. Gently resuspend cells in 1 ml wash buffer after second wash.
3. Take 20 μl of BioMag Plus Concanavalin A beads and wash twice with 1 ml binding buffer. Separate beads on a magnetic tube rack between each wash. Resuspend beads in 20 μl wash buffer after second wash (*see* **Note 2**).
4. Add 20 μl of washed Concanavalin A beads to resuspended cells from **step 2** and place on rotator for 1 h at room temperature.
5. Divide sample evenly into two separate tubes, such that each tube contains an equal number of cells. Place tubes onto magnetic tube rack and remove buffer. Gently resuspend in 50 μl of digitonin-wash buffer. Avoid introducing bubbles (cavitation) into liquid during resuspension.
6. Mix 2.8 μg GST-RHΔ-MNase into one tube (final concentration 1 μM) for binding to native R-loops. Mix 2 μg GST-MNase into the second tube (final concentration 1 μM) as a control to evaluate nonspecific binding and cleavage. Place tubes on rotator in a cold room and rotate overnight at 4 °C (*see* **Note 3**).

7. Spin down tubes briefly in a microcentrifuge to collect liquid that may have accumulated in the cap and place onto magnetic tube rack. Wash cells three times with 1 ml digitonin-wash buffer. Separate beads on magnetic tube rack between each wash. Gently resuspend cells in 100 μl digitonin-wash buffer after third wash.
8. Chill cells to 0 °C in wet ice (approx. 2 min incubation). Mix 2 μl of 100 mM calcium chloride into each tube and immediately place back into wet ice. Incubate samples at 0 °C for 30 min, resuspending beads every 10 min with gentle flicking (*see* **Note 4**).
9. Add 100 μl of 2× stop buffer to each tube, mix thoroughly by flicking, and incubate at 37 °C on a heat block for 10 min to release R-loop chromatin fragments. Centrifuge at 4 °C at 16,000 × *g* for 5 min. Place tubes on magnetic tube rack and carefully transfer R-loop containing supernatant to a fresh microcentrifuge tube (*see* **Note 5**).
10. Add 2 μl 10% SDS and 2.5 μl 20 mg/ml proteinase K to each tube and incubate at 70 °C on a heat block for 10 min. Spin down samples briefly to collect any evaporated liquid and add 300 μl of phenol–chloroform–isoamyl alcohol. Mix thoroughly through pipetting or vortex, then transfer into phase-lock gel heavy tubes, and centrifuge at room temperature at 16,000 × *g* for 5 min. Transfer aqueous phase (top) to a fresh microcentrifuge tube.
11. Add 4 μl linear acrylamide (5 mg/ml), 100 μl 7.5 M ammonium acetate, and 750 μl cold 100% ethanol to each tube. Mix thoroughly by inversion and store at −80 °C for at least 60 min. Centrifuge at 4 °C at 16,000 × *g* for 30 min to pellet R-loops. Wash pellets once with 1 ml cold 75% ethanol and recentrifuge at 4 °C at 16,000 × *g* for 10 min. Air-dry pellets for 2 min, then resuspend gently in 20 μl Tris-EDTA buffer.

3.2 Bisulfite Conversion and Second-Strand Synthesis

1. Prepare CT Conversion Reagent from Zymo EZ DNA Methylation-Gold Kit according to manufacturer's instructions. Add 130 μl CT Conversion Reagent to 20 μl of sample in Tris-EDTA (from Subheading 3.1, **step 11**) in a fresh microcentrifuge tube and mix thoroughly by pipetting. Incubate at room temperature for 3 h (*see* also **Note 6**).
2. Add 600 μl M-Binding buffer to the sample, mix through inversion, load the sample into a Zymo-Spin IC column, and centrifuge. Wash once with 100 μl M-Wash buffer and centrifuge. Add 200 μl M-Desulfonation buffer to the column and incubate it at room temperature for 20 min, then centrifuge. Wash two more times with 200 μl M-Wash buffer, centrifuging and discarding flowthrough each time. Move the column to a

microcentrifuge tube, add 20 μl of M-Elution buffer, and incubate it at room temperature for 1 min, then centrifuge to collect final bisulfite-converted R-loops. All centrifugation procedures in this step are performed at room temperature at 10,000 × *g* for 30 s.

3. Transfer 20 μl of elution product from **step 2** to a PCR tube. Add 8 μl of NEBNext Second-Strand Synthesis Reaction Buffer, 4 μl NEBNext Second-Strand Synthesis Enzyme Mix, and 48 μl H_2O (total volume 80 μl). Incubate tube in a thermocycler at 16 °C for 1 h with the heated lid off or below 40 °C.
4. Clean up the second-strand product by adding 144 μl of AMPure XP beads (1.8× sample volume) to the tube and mixing thoroughly by flicking or vortexing. Incubate for 5 min and place onto a magnet tube rack. Wash the beads twice with 80% ethanol, air-dry for 5 min, and resuspend in 30 μl 0.1× Tris-EDTA buffer (dilute with nuclease-free water).
5. Bisulfite-converted R-loops can be sequenced with any library preparation method that contains a uracil–DNA glycosylase (UDG) treatment prior to, or concurrent with, PCR amplification. UDG will degrade uracil-containing molecules including ssDNA and second-strand DNA corresponding to the RNA component of DNA:RNA hybrids. BisMapR can be used for both single-end and paired-end sequencing applications (*see* also **Notes** 7 and **8**).

4 Notes

1. The stop buffer used in BisMapR differs from the buffer used in MapR, namely through omission of RNase A; this is critical for preserving the RNA strand in released R-loops to obtain strand specificity.
2. Resuspend the beads in wash buffer to avoid premature DNA cleavage by MNase that can be activated by the presence of Ca^{2+} in binding buffer. If processing multiple samples, up to 80 μl beads can be processed together without change in buffer volumes.
3. To prevent liquid from spreading onto cap and side during rotation, samples should be transferred into a fresh microcentrifuge tube; the low volume should allow the sample to stay at the bottom of the tube during rotation.
4. Calcium activates MNase activity, the rate of which is temperature dependent; keep samples as uniformly on ice as possible to minimize variation in cleavage.

5. If proceeding to BisMapR, ensure that 2× stop buffer does not contain RNase A.
6. Do not denature sample with a higher temperature incubation, as it is done in other bisulfite protocols, as this will separate the DNA and RNA strands of R-loop hybrids.
7. As bisulfite conversion in the BisMapR protocol functions to eliminate unwanted molecules from the sequencing library rather than alter the sequence of any reads, alignment can be performed to a standard reference genome.
8. In a single-end sequencing experiment, the aligned reads in a BisMapR library originate from the strand forming the DNA component of an R-loop's DNA:RNA hybrid. For example, reads will originate from the template strand of a co-transcriptionally formed R-loop. For paired-end sequencing experiments, which contain paired reads that each align to opposite strands, the strand of the first mate (first read in a pair) corresponds to the DNA component of the DNA:RNA hybrid; conversely, the reverse strand to the second mate also corresponds to the DNA component strand. Software such as samtools [17] can be used to separate reads in the alignment file by mate. For example, SAM flags 99 and 147 correspond to paired-end reads that originate from the forward strand, while reads with SAM flags 83 and 163 originate from the reverse strand.

Acknowledgments

This work was supported by NIH grants T32CA009171 (to P.W.) and DP2-NS105576 (to K.S.)

References

1. Thomas M, White RL, Davis RW (1976) Hybridization of RNA to double-stranded-DNA: formation of R-loops. Proc Natl Acad Sci 73(7):2294–2298. https://doi.org/10.1073/pnas.73.7.2294
2. Niehrs C, Luke B (2020) Regulatory R-loops as facilitators of gene expression and genome stability. Nat Rev Mol Cell Biol 21(3):167–178. https://doi.org/10.1038/s41580-019-0206-3
3. Crossley MP, Bocek M, Cimprich KA (2019) R-loops as cellular regulators and genomic threats. Mol Cell 73(3):398–411. https://doi.org/10.1016/j.molcel.2019.01.024
4. Skourti-Stathaki K, Proudfoot NJ (2014) A double-edged sword: R loops as threats to genome integrity and powerful regulators of gene expression. Genes Dev 28(13):1384–1396. https://doi.org/10.1101/gad.242990.114
5. García-Muse T, Aguilera A (2019) R loops: from physiological to pathological roles. Cell 179(3):604–618. https://doi.org/10.1016/j.cell.2019.08.055
6. Ginno Paul A, Lott Paul L, Christensen Holly C, Korf I, Chédin F (2012) R-loop formation is a distinctive characteristic of unmethylated human CpG Island promoters. Mol Cell 45(6):814–825. https://doi.org/10.1016/j.molcel.2012.01.017
7. Dumelie JG, Jaffrey SR (2017) Defining the location of promoter-associated R-loops at

near-nucleotide resolution using bisDRIP-seq. elife 6:e28306. https://doi.org/10.7554/eLife.28306

8. Nadel J, Athanasiadou R, Lemetre C, Wijetunga NA, Broin PÓ, Sato H, Zhang Z, Jeddeloh J, Montagna C, Golden A, Seoighe C, Greally JM (2015) RNA:DNA hybrids in the human genome have distinctive nucleotide characteristics, chromatin composition, and transcriptional relationships. Epigenetics Chromatin 8:46. https://doi.org/10.1186/s13072-015-0040-6
9. Wahba L, Costantino L, Tan FJ, Zimmer A, Koshland D (2016) S1-DRIP-seq identifies high expression and polyA tracts as major contributors to R-loop formation. Genes Dev 30(11):1327–1338. https://doi.org/10.1101/gad.280834.116
10. Chen L, Chen JY, Zhang X, Gu Y, Xiao R, Shao C, Tang P, Qian H, Luo D, Li H, Zhou Y, Zhang DE, Fu XD (2017) R-ChIP using inactive RNase H reveals dynamic coupling of R-loops with transcriptional pausing at gene promoters. Mol Cell 68(4):745–757.e745. https://doi.org/10.1016/j.molcel.2017.10.008
11. Yan Q, Shields EJ, Bonasio R, Sarma K (2019) Mapping native R-loops genome-wide using a targeted nuclease approach. Cell Rep 29(5):1369–1380.e1365. https://doi.org/10.1016/j.celrep.2019.09.052
12. Yan Q, Sarma K (2020) MapR: a method for identifying native R-loops genome wide. Curr Protoc Mol Biol 130(1):e113. https://doi.org/10.1002/cpmb.113
13. Wulfridge P, Sarma K (2021) A nuclease- and bisulfite-based strategy captures strand-specific R-loops genome-wide. eLife 10:e65146. https://doi.org/10.7554/eLife.65146
14. Skene PJ, Henikoff S (2017) An efficient targeted nuclease strategy for high-resolution mapping of DNA binding sites. eLife 6:e21856. https://doi.org/10.7554/eLife.21856
15. Skene PJ, Henikoff JG, Henikoff S (2018) Targeted in situ genome-wide profiling with high efficiency for low cell numbers. Nat Protoc 13(5):1006–1019. https://doi.org/10.1038/nprot.2018.015
16. Gough GW, Sullivan KM, Lilley DM (1986) The structure of cruciforms in supercoiled DNA: probing the single-stranded character of nucleotide bases with bisulphite. EMBO J 5(1):191–196
17. Li H, Handsaker B, Wysoker A, Fennell T, Ruan J, Homer N, Marth G, Abecasis G, Durbin R (2009) The Sequence Alignment/Map format and SAMtools. Bioinformatics 25(16):2078–2079. https://doi.org/10.1093/bioinformatics/btp352

Chapter 26

Quantitative DNA–RNA Immunoprecipitation Sequencing with Spike-Ins

Magdalena P. Crossley and Karlene A. Cimprich

Abstract

R-loops are three-stranded nucleic acid structures, comprising an RNA–DNA hybrid and a displaced strand of ssDNA. R-loops have important physiological roles in cells, but deregulation of R-loop dynamics can also have harmful cellular outcomes. The genome-wide mapping of R-loops offers an unbiased approach to study R-loop biology in a wide range of contexts. Here we present a protocol to sequence RNA–DNA hybrids genome-wide with strand-specificity and high resolution. We also include information on how to prepare and incorporate into the workflow appropriate internal spike-in standards which facilitate accurate normalization of the sequencing signal, thereby providing quantitative insights into R-loop formation between different experimental samples.

Key words R-Loops, qDRIP (quantitative DNA–RNA immunoprecipitation), S9.6 antibody, RNA–DNA hybrids, Spike-in, Next-generation sequencing

1 Introduction

1.1 RNA–DNA Hybrid Sequencing

R-loops are three-stranded nucleic acid structures, comprising an RNA–DNA hybrid and a displaced strand of ssDNA. R-loops have important physiological roles in cells, such as regulating the initiation and termination steps of gene transcription. However, deregulation of R-loop dynamics can also have harmful cellular outcomes by interfering with DNA replication and causing DNA damage [1]. Indeed, R-loop perturbations have been linked to several human pathologies, including cancer, autoimmune and neurodegenerative diseases [1]. As such, there is a need to quantify R-loop levels genome-wide in order to identify sites of R-loop formation, correlate these with other genomic features or sequencing datasets, and evaluate how R-loop formation and levels change upon various cellular perturbations. RNA–DNA immunoprecipitation (DRIP) is a widely used method to isolate chromatin-bound RNA–DNA hybrids [2]. DRIP can be combined with quantitative PCR

Andrés Aguilera and Alexey Ruzov (eds.), *R-Loops: Methods and Protocols*, Methods in Molecular Biology, vol. 2528, https://doi.org/10.1007/978-1-0716-2477-7_26,

(qPCR) in order to quantify hybrid levels for specific regions of interest, or with sequencing to map RNA–DNA hybrids genome-wide [2].

1.2 Potential Inaccuracies in the Normalization of Sequencing Data

Next-generation sequencing is a powerful approach to look at the genome-wide distribution of a feature of interest in an unbiased manner and without requiring previous knowledge about sites of interest. In many sequencing-based approaches, the signal across the genome is presented as a normalized value, using the total read count per sample for this normalization. While this adjusts the sequencing depth between samples to enable more accurate comparisons, this approach also makes certain assumptions about the signal that may not always hold true. One key assumption is that the total signal between samples—in this case total RNA–DNA hybrid content—remains constant. Therefore, biological perturbations which cause an overall increase or decrease in sequencing signal may lead to inaccurate normalization. In some contexts, and especially when increases or decreases in total signal between samples are large, read-count normalization can introduce biases and even lead to misleading biological interpretations due to inaccurate inter-sample comparisons [3–5]. A theoretical example is illustrated in Fig. 1a, in which read-count normalization may introduce artifactual signal increases or decreases between samples. The use of internal standards or spike-in controls—synthetic nucleic acid sequences or chromatin added in an equal amount to each experimental sample—can overcome these potential issues and make comparisons between samples more accurate [4–7]. Generally, spike-ins are introduced as early as possible into the experimental workflow and serve as internal standards for the downstream steps. Since the same amount of spike-in is added across all samples, a normalization factor can be calculated based on the spike-in signal ultimately obtained and then applied to the experimental signal. Such approaches are increasingly used in RNA sequencing and chromatin immunoprecipitation sequencing [4, 6, 8], but have not been commonly applied to R-loop and DRIP-based sequencing studies.

Our recent study showed how changes in RNA–DNA hybrid content between samples can lead to misleading interpretations when normalizing by total read count, in this case after a global decrease in hybrid content following transcription inhibition with 5,6-dichloro-1-β-D-ribofuranosylbenzimidazole (DRB) [7]. It was reasoned that DRB would reduce the total number of hybrid-containing sites on the genome since R-loop formation requires transcription. Therefore, since the total mapped reads will be similar in each sample regardless of overall hybrid content, read counts across remaining areas would be correspondingly increased. If normalized by total read counts, this could lead to an artifactually high signal at unaffected regions after DRB treatment relative to

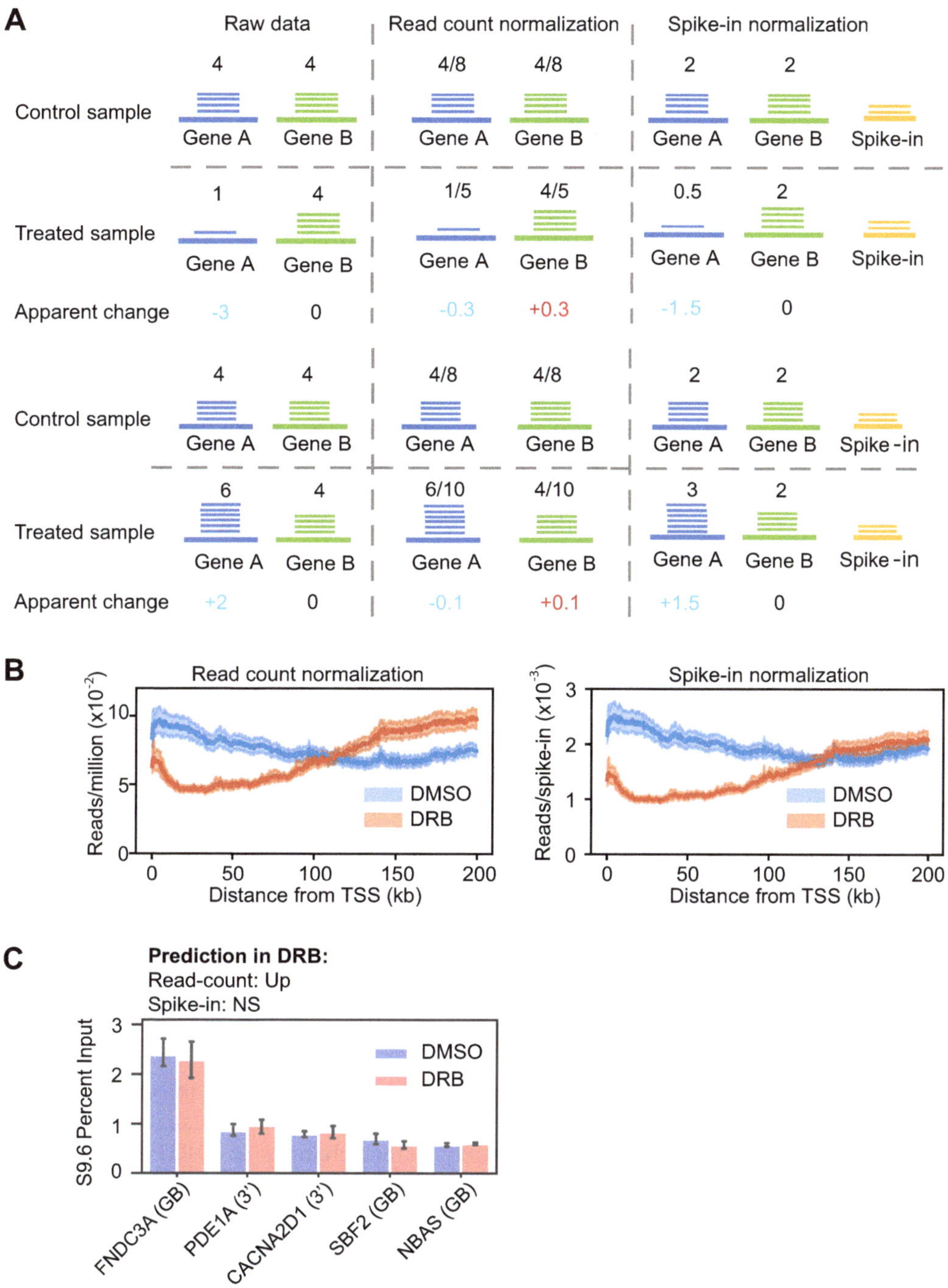

Fig. 1 Comparisons between total read count and spike-in normalization. (**a**) Adapted from [5]. Top. When signal decreases happen at a specific genomic region (Gene A), normalizing total sequencing reads between samples introduces artifactual increases in the number of reads from other regions of the genome (Gene B), which is falsely interpreted as being increased under the specific experimental condition. Such artificial changes can be avoided by using spike-in controls as a reference for normalization. Bottom: same but for a signal increase at Gene A, which by read-count normalization artifactually causes a small decrease in signal in Gene B. (**b**) From [7]. Metaplots of DMSO (blue) and DRB (red) S9.6 IP signal over the first 200 kb of all genes longer than 200 kb and shown to be expressed in HeLa cells by GRO-seq, normalized using total read counts

controls. DRB inhibits the transition to processive RNAPII elongation, preventing initiating polymerases close to the 5′ pause site from progressing through gene bodies. A treatment time of ~40 min with DRB would be expected to reduce RNA–DNA hybrid content at the 5′ ends of genes but the 3′ ends of long genes would not be affected, as polymerases would continue to traverse through these regions, forming R-loops co-transcriptionally. Consistent with this hypothesis, the authors inspected the RNA–DNA hybrid signal at the 3′ ends of long genes and found that by read-count normalization this signal appeared to significantly increase following DRB treatment. However, by normalizing with spike-ins, the signal was corrected to show no change at these sites (Fig. 1b). Importantly, DRIP-qPCR, which is not affected by normalization issues, confirmed that there was indeed no significant change in RNA–DNA hybrid signal after DRB treatment at the 3′ ends of long genes (Fig. 1c).

While other R-loop perturbations may not induce such dramatic and global changes in RNA–DNA hybrid content as transcription inhibition, accurate normalization may still be important. Many R-loop resolution factors are involved in co-transcriptional processes such as splicing and mRNA processing [1, 9]. While depletion of these factors may indeed reduce R-loop resolution, this may also reduce transcription [10], reducing the total hybrid content (analogous to DRB treatment) and leading to artifactual changes at some sites.

Normalization using read counts from unaffected genomic regions has been used effectively as a natural internal standard in some sequencing experiments; for example, some have used read counts from non-RNA-Pol-II-transcribed sites such as rDNA in GRO-seq [11]. However, it is not currently clear that this approach is suitable for RNA–DNA hybrid mapping since recent reports have found an interplay between RNAPI, RNAPII, and R-loops at rDNA loci [12].

1.3 The Use of Spike-In Controls in RNA–DNA Hybrid Sequencing Approaches

Two recent DRIP-seq studies involved spike-ins in the experimental workflow [7, 13], and using these for normalization, the authors were able to accurately detect changes in DRIP signal at certain genomic locations that would otherwise have been missed or inaccurately represented using read-count normalization. Additionally, the use of spike-ins has also been used for absolute quantification of RNA–DNA hybrids, revealing new biological insights [7].

Fig. 1 (continued) (left) or spike-in read counts (right). (**c**) From [7]. qPCR measurements over genes called as significantly up by read-count normalization, but nonsignificantly changed by spike-in normalization. Error bars represent 95% CI of mean value. In the primer name, GB indicates gene body, 3′ indicates TES proximal regions

Two different spike-in approaches have been implemented in DRIP-seq [7, 13]. One involves the use of cells from a heterologous species. Specifically, *Drosophila* cells were mixed with samples of human cells. Following immunoprecipitation with the S9.6 antibody to enrich for RNA–DNA hybrids, sequencing reads were aligned to both human and *Drosophila* genomes [7, 13]. A high-confidence peak set was derived from the *Drosophila* signal, and the reads mapping within these peaks were then summed and used to normalize the experimental samples. Such cellular spike-ins are advantageous because they contain a diverse set of RNA–DNA hybrids—the whole complement from the *Drosophila* genome—and are easy to integrate into the DRIP-seq workflow. However, a disadvantage is that the amount of *Drosophila* RNA–DNA hybrids used cannot be quantified since they are derived from intact cells. Thus, this approach does not allow for absolute quantification of hybrids in the experimental samples.

Alternatively, synthetic RNA–DNA hybrids derived from *E. coli* genome sequences with little homology to the human genome—were used after synthesis, purification, and addition to human cells [7]. Reads mapping to the synthetic spike-ins in each sample were then summed and used to normalize sequencing signal. Since the synthetic hybrids can be quantified, this approach allows for absolute quantification of hybrid signal in the experimental samples. Synthetic hybrids are harder to generate and characterize technically than cell-based hybrids, but collaborative or commercial efforts may alleviate these challenges, as demonstrated for RNA-seq [8].

In this chapter, we describe methods to prepare both types of spike-ins—synthetic RNA–DNA hybrid molecules (Subheading 3.1) and bulk cellular RNA–DNA hybrids from *Drosophila* (Subheading 3.2); how to integrate spike-ins into a high-resolution and stranded DRIP-seq workflow (termed qDRIP-seq) (Subheading 3.3); how to probe RNA–DNA hybrid enrichment by qPCR (Subheading 3.4); a method for sequencing library preparation (Subheading 3.5), and approaches for bioinformatic analysis of spike-in signal and subsequent normalization (Subheading 3.6).

2 Materials

2.1 Selecting Regions for Spike-In Controls from E. coli

1. *E. coli* genome sequence file (available at: https://www.ncbi.nlm.nih.gov/nuccore/U00096.3/).

2.2 Isolation of Genomic DNA from E. coli

1. LB medium: 1% tryptone, 0.5% yeast extract, 200 mM NaCl.
2. TE buffer: 10 mM Tris-Cl pH 8.0, 1 mM EDTA pH 8.0.

3. Lysis buffer: TE buffer, 0.6% SDS, Proteinase K (120 μg/mL final concentration), 200 proof ethanol, RNase A (1 mg/mL final concentration) (*see* **Note 1**).
4. Phenol:chloroform:isoamyl alcohol (25:24:1).
5. Tabletop centrifuge.
6. Incubator at 37 °C.
7. 2 mL phase-lock tubes (*see* **Note 2**).
8. 70% ethanol solution.

2.3 Amplification and Purification of PCR Fragments from E. coli Genomic DNA

1. Phusion High Fidelity DNA polymerase and reaction buffer.
2. Solution of mixed dNTPs, diluted to 25 mM each in water.
3. Forward and reverse primers for the spike-in, also containing the T7 RNAP promoter sequence (TAATACGACTCACTATAGGG) (Table 1).
4. *E. coli* genomic DNA from the previous step (*see* Subheading 2.2).
5. Gel extraction kit and reagents to make 2% agarose gel.
6. NanoDrop Microvolume Spectrophotometer.

2.4 In Vitro Transcription to Obtain Complementary RNA

1. DNA template from Subheading 2.3.
2. T7 RNA Polymerase and transcription-optimized buffer.
3. rNTPs (ATP, GTP, CTP, UTP): equimolar mix of all four each at 2.5 mM.
4. DTT.
5. Tween-20.
6. Incubator at 37 °C.
7. RNA purification kit.
8. DNase I.
9. NanoDrop.

2.5 Annealing and Purification of Synthetic RNA–DNA Hybrids

1. NEB buffer 2.1.
2. PCR cycler.
3. RNase H enzyme and RNase H reaction buffer (from NEB).
4. Agarose and agarose gel electrophoresis machinery.
5. UV transilluminator.
6. Single edge razor blade.
7. Syringe needle.
8. Cotton wool.
9. Benchtop centrifuge.

Table 1
Sequences and primers used to generate synthetic RNA–DNA hybrid spike-ins and negative internal controls

Name	Sequence	Description
T7 RNAP promoter	TAATACGACTCACTATAGGG	T7 RNAP promoter sequence
T7 RNAP promoter + upstream DNA	GGATCCTGCAGTAATACGACTCACTATAGGG	T7 RNAP promoter sequence + short region of upstream DNA to help transcription efficiency
L286_F	GGATCCTGCAGTAATACGACTCACTATAGGGCCG TGATGGTTTTCGCCTTT	Primer used for PCR amplification from E.coli to make IVT template - hybrid spike-in
L286_R	ACTGCGACAATAATCCCGGA	Primer used for PCR amplification from E.coli to make IVT template - hybrid spike-in
H281_F	GGATCCTGCAGTAATACGACTCACTATAGGGAATG TCGCTGTGTTGTTGCT	Primer used for PCR amplification from E.coli to make IVT template - hybrid spike-in
H281_R	CCAGAACACCATCAACACCC	Primer used for PCR amplification from E.coli to make IVT template - hybrid spike-in
LDNA_F	CATCCCAGAGCGTTGTTGAG	Primer used for PCR amplification from E.coli to make IVT template - dsDNA spike-in
LDNA_R	AGCCCCAAAACCTAATCCCA	Primer used for PCR amplification from E.coli to make IVT template - dsDNA spike-in
HDNA_F	ACTCCGAGCAGCCAGATATC	Primer used for PCR amplification from E.coli to make IVT template - dsDNA spike-in
HDNA_R	CGTGGTAATGGCCGAAATCA	Primer used for PCR amplification from E.coli to make IVT template - dsDNA spike-in

(continued)

Table 1
(continued)

Name	Sequence	Description
L286_T	ACTGCGACAATAATCCCGGAAATTGACAACCATGC TGGGGCGTTCGGCAGCATAGAAAGCGCCGA TAAAATGGTGACCCCTTCTTCTTTAGCAGC TTCAATATACGAAGGTGGAA TAGAAAGCAGGCAGCTAAAGACAAAGAACAG TACACTTATGCAGATGATGAGATAAGCGAC TTTCATAATTTTTTTGCATTTATCCATAGCGTG TTCGCCATATTTTTCACGTCTGTCTA TGGCAAACGTAGAAATAATGGGCGTATGGC TAAAGGCGAAAACCATCACGGCCCTATAGTGAG TCGTATTACTGCAGGATCC	Template for in vitro transcription (amplified by PCR)
H281_T	CCAGAACACCATCAACACCCAGCA TTCGCGCCGGATGCAGCGACGCCATATGGA TGGCTTCCGCAGGCGTTACGCCCGTCAAC TCGACCATATTGCGCACTGCCGCA TCAACAGACAGCGTAC TGCCCGCCAGCCCACCAGACGCGG TACGGACAACGCCACCGTGCATCTGCACTTC TTCACCACATAACGTATAGCGACCATCCGGCA TCCCAGCTGCCTGCATCGCGTCGGTGATCAG TACGATTCTCTC TTTCGCACAGCAACAACACAGCGACATTCCCTA TAGTGAGTCGTATTACTGCAGGATCC	Template for in vitro transcription (amplified by PCR)
L286_S	ACTGCGACAATAATCCCGGAAATTGACAACCATGC TGGGGCGTTCGGCAGCATAGAAAGCGCCGA TAAAATGGTGACCCCTTCTTCTTTAGCAGC TTCAATATACGAAGGTGGAA TAGAAAGCAGGCAGCTAAAGACAAAGAACAG TACACTTATGCAGATGATGAGATAAGCGAC TTTCATAATTTTTTTGCATTTATCCATAGCGTG TTCGCCATATTTTTCACGTCTGTCTA TGGCAAACGTAGAAATAATGGGCGTATGGC TAAAGGCGAAAACCATCACGG	Spike-in sequence and oligo purhased for RNA: DNA hybrid annealing
H281_S	CCAGAACACCATCAACACCCAGCA TTCGCGCCGGATGCAGCGACGCCATATGGA TGGCTTCCGCAGGCGTTACGCCCGTCAAC TCGACCATATTGCGCACTGCCGCA TCAACAGACAGCGTAC TGCCCGCCAGCCCACCAGACGCGG TACGGACAACGCCACCGTGCATCTGCACTTC TTCACCACATAACGTATAGCGACCATCCGGCA TCCCAGCTGCCTGCATCGCGTCGGTGATCAG TACGATTCTCTC TTTCGCACAGCAACAACACAGCGACATT	Spike-in sequence and oligo purhased for RNA: DNA hybrid annealing

(continued)

Table 1
(continued)

Name	Sequence	Description
ssDNA_S	GGCGGTAATGACAGCCAGTACGCCCGGCGC TTTTTGCGCGGCGTCCGTATCAAGGGCGG TGAGGCGTCCTTTGGCAA TGGCGGAACCGACGATATAGCCATAGGCGGCG TTGGGGGCTTCTTCATGCCATTCGTAGGC	Spike-in sequence (oligo purchased and used directly as ssDNA)
LDNA_S	CATCCCAGAGCGTTGTTGAGTGTAAATG TAGGAGGGAAGATAGTTACAGATCGTGTAA TGAGGTCTGTATGGAGTTTCTTTTTTCTTTA TACTCTCTTCACGGTGTTTTTTATACTGGTG TTAAATGGTATGGGATATGATTTTCTTACATCA TTTGCAACAGTGGCTGCATGTATTAATAATA TGGGATTAGGTTTTGGGGCT	Spike-in sequence (generated by PCR)
HDNA_S	ACTCCGAGCAGCCAGATA TCGGAACCGAGCGGCAGCTGTTGCATACGCA TATCCAGCAGGCGCAGATCCAGCGTGCCG TAACGCTGCCACAGCAGCCAGCAAGCAA TCGCCAGCAGCAGAG TGCCAAGACGCCCCAGCGCAAACCACAG TTTGCCCTCTTTGCTGTTGCTGG TGAGGAACACCGCGCACAGGGCCATGA TTTCGGCCATTACCACG	Spike-in sequence (generated by PCR)

10. Qubit Fluorometer and High Sensitivity RNA assay kit (or other sensitive method to accurately quantify RNA).
11. TE buffer: 10 mM Tris-Cl pH 8.0, 1 mM EDTA pH 8.0.
12. −20 °C or −80 °C freezer.

2.6 Drosophila *Schneider Cell Culture*

1. Schneider's Drosophila Medium.
2. Fetal Bovine Serum.
3. Water bath.
4. Cell culture flasks.

2.7 DNA–RNA Immunoprecipitation (DRIP)

1. 20% SDS.
2. Phosphate-buffered saline (PBS): 137 mM NaCl, 2.7 mM KCl, 10 mM Na_2HPO_4, 1.8 mM KH_2PO_4.
3. TE buffer.
4. Proteinase K 20 mg/mL.
5. 15 mL and 2 mL phase-lock tubes.
6. Phenol:chloroform:isoamyl alcohol (25:24:1), pH 6.7.
7. RNase H and RNase H reaction buffer.

8. S9.6 antibody (available from multiple sources such as Kerafast, Abcam, or Sigma-Aldrich).
9. LoBind Eppendorf tubes.
10. Dynabeads Protein G.
11. Magnetic strip for Eppendorf tubes.
12. Water bath.
13. Benchtop centrifuge.
14. Agarose for gel electrophoresis.
15. Gel electrophoresis running apparatus.
16. NanoDrop.
17. −20 °C or −80 °C freezer.
18. Heat block or incubator.
19. Rotating shaker.
20. Molecular biology grade ethanol (100% and 75% solutions).
21. 3 M sodium acetate pH 5.2.
22. GlycoBlue co-precipitant.
23. Covaris sonicator (or other sonicator) and associated microtubes.
24. Binding buffer 10×: 100 mM $NaPO_4$ pH 7, 1.4 M NaCl, 0.5% Triton X-100, store at room temperature (RT). 1 M $NaPO_4$ is made by mixing 39 mL 2 M monobasic sodium phosphate (NaH_2PO_4) with 61 mL 2 M dibasic sodium phosphate (Na_2HPO_4). Dilute 10× binding buffer with TE to make 1×.
25. Elution buffer: 50 mM Tris-HCl pH 8.0, 10 mM EDTA, 0.5% SDS, store at RT.

2.8 Quantitative PCR on Genomic and Spike-In RNA–DNA Hybrids (DRIP-qPCR)

1. TE buffer.
2. 10 mM Tris-HCl pH 8.0.
3. Forward and reverse primers (diluted to 10 μM) for desired R-loop loci and spike-in sequences.
4. qPCR machine/LightCycler.
5. qPCR reaction mix such as SYBR Green.
6. Ultrapure water.
7. qPCR plates and plate sealing film.

2.9 Library Preparation from ssDNA of RNA–DNA Hybrids

1. Accel-NGS 1S Plus DNA kit and index sequences.
2. Lobind 0.2 mL and 1.5 mL tubes.
3. AMPure XP beads (or other SPRI beads).
4. Magnetic strip for 1.5 mL tubes.
5. Thermocycler.

6. Agilent BioAnalyzer.
7. 80% ethanol solution (made fresh).

3 Methods

3.1 Preparation of Synthetic RNA–DNA Hybrid Spike-Ins from E. coli

3.1.1 Spike-In Design

Select RNA–DNA hybrid sequences from a species with little homology to the experimental samples. For example, if RNA–DNA hybrid sequencing from human cells is desired, RNA–DNA hybrids based on *E. coli* sequences can be synthesized, since these can be clearly distinguished from human sequences by qPCR and sequencing approaches. NCBI BLASTn (https://blast.ncbi.nlm.nih.gov/) should be used to ensure selected sequences are not represented in the genome of interest. Ideally, at least 2–10 synthetic hybrid sequences should be designed (*see* **Note 3**). Sequences of less than 150 bp may not be efficiently or reliably recovered during library preparation and should be avoided or tested carefully (*see* **Note 4** and Subheading 3.4). Genomic DNA sequences may be chosen to have different lengths and/or nucleotide content. These can be amplified from genomic *E. coli* DNA, and the complementary RNA molecules are then synthesized by in vitro transcription. To increase the purity of spike-ins and minimize spurious annealing products, however, it is recommended to purchase the ssDNA sequence and use this directly for annealing with the purified, complementary RNA strand.

3.1.2 Isolation of Genomic DNA from E. coli

This method is adapted from https://bio-protocol.org/bio101/e97.

1. Grow a culture of *E. coli* in 1.5 mL of LB medium overnight at 37 °C.
2. Transfer culture to Eppendorf tube and centrifuge at maximum speed (≥15,000 × *g*) at RT for 1 min to pellet cells.
3. Discard supernatant and resuspend pellet in 600 μL lysis buffer. Vortex to resuspend the pellet.
4. Incubate for 1 h at 37 °C. Transfer to a 2 mL phase-lock tube.
5. Add 600 μL of phenol:chloroform:isoamyl alcohol and mix by inversion (*see* **Note 5**).
6. Spin at maximum speed for 5 min at RT. Organic phase with phenol should move below the gel, and the aqueous phase should remain above the gel.
7. Transfer upper aqueous layer to new Eppendorf tube.
8. Add 2.5 volume of cold 200 proof ethanol, invert to mix, and incubate the tube at −20 °C for at least 30 min.
9. Spin at maximum speed for 15 min at 4 °C. A DNA pellet should be visible.

10. Remove supernatant and add 1 mL 70% ethanol to rinse the pellet.
11. Spin at maximum speed for 2 min, discard supernatant, and air-dry DNA pellet until no droplets of ethanol are visible.
12. Resuspend in TE buffer and use this genomic DNA for PCR. This can be stored at −20 °C for later use.

3.1.3 Amplification and Purification of PCR Fragments from E. coli Genomic DNA

1. Set up PCR reactions with Phusion polymerase following the manufacturer's guidelines using 1 μL of E. coli genomic DNA from Subheading 3.1.2 as the template DNA, dNTP mix (ATP, GTP, CTP, TTP, 200 μM final concentration of each), and Phusion polymerase buffer. Include forward and reverse primers (0.5 μM each) specific to the spike-in sequence and also designed to contain the T7 RNA Polymerase promoter sequence (*see* **Note 6**, Table 1 for primer and spike-in sequences).
2. Select the annealing temperature based on the specific primers used: (http://tmcalculator.neb.com).
3. Take 2–5 μL of PCR product to run on an agarose gel (typically 2% to separate spike-ins of 150–500 bp). Ensure that the main band is at the correct size (*see* **Note 7**).
4. Run the remainder of the PCR product on a 1% agarose gel and excise the gel band at the correct size (*see* **Note 8**).
5. Extract the product from the agarose gel using a gel extraction kit according to manufacturer's instructions (*see* **Note 9**). Use an elution volume of 30 μL to maximize the concentration. Measure DNA concentration on NanoDrop.

3.1.4 In Vitro Transcription to Obtain Complementary RNA

1. Set up in vitro transcription reactions with the dsDNA templates obtained in Subheading 3.1.3, using 30 ng of dsDNA template per reaction. Follow manufacturer's guidance for components required for T7 RNA polymerase. Typically 30 ng of dsDNA template is used per reaction, rNTP mix (ATP, GTP, CTP, UTP, final concentration of 50 μM each), transcription buffer, DTT (10 mM final concentration), Tween-20 (0.05% final concentration), 5 μL T7 RNA polymerase, up to a final reaction volume of 100 μL. Incubate the reactions at 37 °C for 2–16 h. Depending on the yield, it may be helpful to scale up to 2–4 reactions per spike-in.
2. Purify RNA with RNA purification kit according to manufacturer's instructions, e.g., QIAGEN RNeasy kit. If multiple reactions per spike-in are set up, these can all be pooled by applying them sequentially to the same binding column. Perform DNase treatment to remove DNA on the columns for 15 min. Elute in 30 μL of ultrapure water. Measure concentration of RNA on NanoDrop (*see* **Note 10**).

3.1.5 Annealing and Purification of Synthetic RNA–DNA Hybrids

1. Set up annealing reactions with RNA (from Subheading 3.1.4) and complementary ssDNA oligonucleotides (*see* **Note 11**). Per reaction, use 100–400 ng of DNA/RNA, with the ratio of DNA:DNA dependent on hybrid annealing efficiency (*see* **Note 12**), in 45 μL of 1× NEB buffer 2.1. Begin with a denaturation step for 10 min at 95 °C, followed by 70 cycles of 90 s (hold) at decreasing intervals of 1 °C until the temperature reaches 25 °C.
2. To visualize the annealed material and to determine hybrid formation, take ~200 ng of annealed hybrids and divide material in two. Set up RNase H reactions with 1 μL of RNase H (or water for mock treatment), 100 ng of annealed hybrids in 1× RNase H reaction buffer and incubate at 37 °C for 1 h. Run material on 2% agarose gel. An RNaseH-sensitive gel shift should be visible, indicating hybrid formation.
3. To purify RNA–DNA hybrids, load the annealing reactions on a 1% agarose gel and excise the upper (RNA–DNA hybrid) band with a clean blade. A gel crushing procedure is used to isolate the RNA–DNA hybrids [14] (*see* **Note 13**) (Fig. 2).
4. Use a syringe needle to make a hole in the bottom of a 0.5 mL Eppendorf tube (hole should be from the inside of the tube to the outside).
5. Add some cotton wool to the bottom of the tube to make a ~4 mm filter.
6. Place the excised gel on the cotton wool filter in the tube (*see* **Note 14**).
7. Cap the tube and place into a 1.5 mL Eppendorf tube (uncapped). Centrifuge for 10 min at 2000 × *g*. The DNA

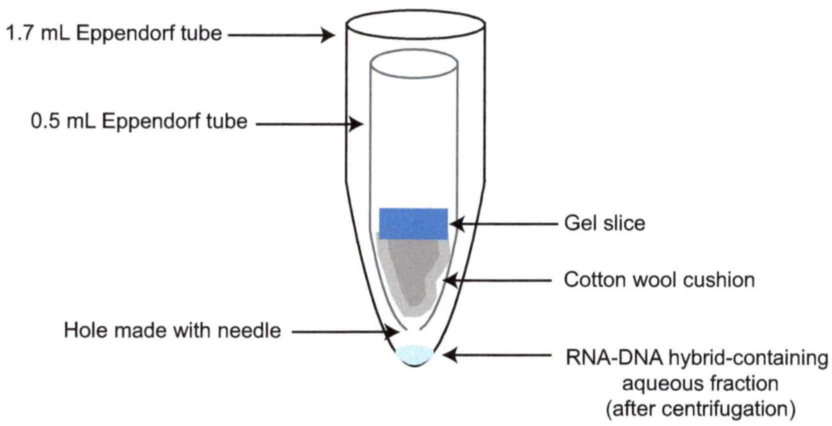

Fig. 2 Illustration of cushion-filter assembly for gentle purification of RNA–DNA hybrids from an agarose gel slice

and aqueous components spin down into the Eppendorf tube through the hole.

8. Measure RNA–DNA hybrid concentrations after cleanup by fluorimetry using the Qubit HS RNA assay kit to facilitate equimolar mixing (*see* **step 10** and **Note 15**).
9. As negative spike-in controls which should not be significantly recovered following immunoprecipitation with the S9.6 antibody, dsDNA or ssDNA sequences from *E. coli* can also be included. The same ssDNA sequences can be used as those used for RNA–DNA hybrid annealing. For dsDNA, amplified fragments from Subheading 3.1.3 can be purified in a similar way to the annealed hybrid molecules.
10. Once all synthetic spike-in sequences are purified, mix them together in a 1:1 molar ratio to make a stock in TE buffer. For HeLa cells a working stock mix of 2.5 pM of each spike-in can be made. For longer term storage, a 10× stock can be made and frozen, and a 1× working solution then diluted as required in TE buffer. Store at −80 °C as small aliquots and avoid freeze thaw cycles. Five microliter of working stock of spike-in mix (2.5 pM each spike-in) is added per cell lysis tube. This gives a final ratio of ~3.75:1 spike-in molecules:cell count (7.5 million:2 million). This means the spike-ins are roughly equal molarity to the HeLa genome (since HeLa cells are approximately hypertriploid). A similar approach can be taken for other cell types (*see* **Note 16**).

3.2 Drosophila S2 Cells as Cell-Based Spike-Ins

1. Culture *Drosophila* Schneider 2 (S2) cells at 26–28 °C (non-humidified, in an ambient air regulated incubator) in a cell culture flask or Schneider's *Drosophila* Medium containing 10% heat-inactivated fetal bovine serum (*see* **Note 17**) and 50 units penicillin G and 50 μg streptomycin sulfate/mL of medium.
2. S2 cells should be cultured to a density of 1 to 5 × 10^6 cells/mL. Passage cells every ~2–3 days and do not seed cells below a density of 0.5 × 10^6 cells/mL, as splitting cells too thin can cause them to become tetraploid (*see* **Note 18**).
3. Harvest S2 cells on the same day as experimental samples. Count the S2 cells and add to each experimental sample. Given the relative genome sizes of *Drosophila* and human, mixing cells with a ratio of 1:1.5 *Drosophila*:human cells will result in ~10% of reads mapping to the *Drosophila* genome.

3.3 DNA–RNA Immunoprecipitation (DRIP)

This method is an adapted version of DRIP-seq [15]. Instead of restriction enzyme digestion to fragment genomic DNA, sonication is used prior to immunoprecipitation. Since sonicated DNA fragments are significantly shorter than fragments following restriction digestion, this ultimately increases the resolution of hybrid mapping. However, it should be noted that sonication likely also disrupts some R-loop structures, and this can reduce the overall enrichment of signal over background compared to restriction digestion. As described above, spike-in controls are added at the beginning of the protocol, during cell lysis. Since the spike-ins are carried through the whole DRIP protocol, their overall recovery reflects the losses of genomic hybrids throughout the workflow. Below we outline a DRIP method that combines spike-ins together with high-resolution and strand-specific hybrid mapping (qDRIP-seq) [7]. Both cellular spike-ins (in this case genomic RNA–DNA hybrids from Drosophila, *see* Subheading 3.2) and synthetic RNA–DNA hybrids (*see* Subheading 3.1) can be added together to experimental samples.

1. Harvest cells, wash and resuspend in PBS. Count cells and transfer ~2 million to 15 mL falcon tube (*see* **Note 19**). Pellet cells by centrifugation and remove the supernatant.
2. Resuspend cell pellets in 1.6 mL of TE. Carefully add cell and/or synthetic hybrids to each sample tube. Add 50 μL of 20% SDS and 5 μL of Proteinase K 20 mg/mL to each tube and invert the tube until solution becomes viscous. Incubate for ~3 h or overnight at 37 °C in a water bath (*see* **Note 20**).
3. Add genomic DNA from the previous step by pouring it directly into a 15 mL phase-lock tube. Then add 1 volume (1.6 mL) of phenol:chloroform:isoamyl alcohol, invert gently, and spin down in a benchtop centrifuge at 15,000 × *g* for 5 min. Phases should separate with a viscous aqueous phase containing genomic material remaining above the gel.
4. Add 160 μL volume 3 M NaOAc pH 5.2 and 4 mL 100% EtOH to a new 15 mL tube and pour in the top aqueous phase from the phase-lock tube.
5. Invert until the DNA precipitates (may require ~20–30 inversions depending on number of cells used) and transfer spooled DNA by pipetting with a cut tip to an Eppendorf tube.
6. Wash DNA 3 times by adding 1 mL of 70% ethanol, then remove carefully with a pipette without disturbing the pellet. After the final wash, remove as much ethanol as possible with a small pipette tip. A small white spool of DNA should be visible at the bottom of the tube (*see* **Note 21**).
7. Allow to air-dry completely while inverting tubes (to give a clear pellet). This usually takes ~10 min. If more cells are lyzed this can take longer. Add 135 μL of TE and carefully dislodge the cell pellet from the side of the tube with a pipette tip. Do not pipette or vortex. Leave for ~20–30 min on ice and then

transfer to Covaris microtubes, avoiding air bubbles. DNA will be viscous and will not dissolve.

8. Sonicate the genomic DNA to achieve a peak fragment size of ~300 bp. The sonication power and time will need to be optimized depending on the instrument, cell line, and number of cells used. For ~1-2 million HeLa cells, the following settings are a good starting point for a Covaris S220 focused-ultrasonicator: peak incident power: 140; duty factor: 10%; cycles per burst: 200, sonication time: 60 s. Solution should now be clear and not viscous.
9. After sonication, transfer the sample to a fresh Eppendorf tube. Take 3 μL of material to run on an agarose gel to determine the fragment size.
10. Measure the DNA concentration using a Nanodrop. If required, set up reactions for RNase H digestion to show that RNA–DNA hybrids are isolated in the subsequent immunoprecipitation, otherwise proceed to **step 17**. Add 4–10 μg of sonicated DNA with 5–10 μL RNase H in 1× RNase H digestion buffer. The final reaction volume should be at least ten times the volume of RNase H added to dilute the enzyme storage buffer. Incubate overnight at 37 °C.
11. After incubation, transfer reaction mix to a 2 mL phase-lock tube. Add 1 volume phenol:chloroform:isoamyl alcohol and invert the tube to mix. Centrifuge at 15,000 × *g* at room temperature for 5 min. Remove aqueous phase and transfer into a fresh Eppendorf tube.
12. Add 1 μL glycoblue reagent, 1/10 volume 3 M NaOAc pH 5.2 and 2.5–3 volumes 100% EtOH. Invert to mix.
13. Incubate the tubes at −20 °C or −80 °C for 1–2 h (or overnight if preferred).
14. Centrifuge the tubes at 4 °C for 30 min at maximum speed—a pellet should be visible on the side of the tube. Wash the pellet with 1 mL 70% ethanol. Centrifuge at 4 °C for 5 min at maximum speed and remove the supernatant.
15. Air-dry the pellets for 10 min at RT until they become transparent.
16. Add 50 μL TE and leave the tubes on ice for 30 min to resuspend the pellets.
17. Measure the DNA concentration (OD_{260}) of each sample on the Nanodrop.
18. Dilute 4.4 μg DNA in 500 μL TE. For DRIP-qPCR 1–2 IPs can be set up per sample (*see* **Note 22**). For DRIP-seq 2–3 IPs should be set up and are then pooled before qPCR and library preparation (*see* Subheadings 3.4 and 3.5).

19. Remove 50 μL of each tube as input for qPCR or sequencing. The IP can then be calculated as % input.
20. Add 50 μL of 10× binding buffer and 20 μL of S9.6 antibody per IP.
21. Incubate tubes overnight at 4 °C inverting on a rotating shaker.
22. The next day, wash 50 μL of Dynabeads Protein G slurry 2 times with 700 μL of 1× binding buffer (*see* **Note 23**).
23. Apply to the magnet to remove supernatant and then add DNA/antibody complexes to prewashed beads, incubate 2 h at 4 °C on a rotating shaker.
24. Spin down beads briefly to remove liquid from lid and sides of the tube, apply to magnet, and discard supernatant.
25. Wash the beads:antibody complexes with 700 μL of 1× binding buffer, invert 5–10 min at room temperature on a rotating shaker. Spin down as in **step 24** and discard supernatant.
26. Repeat **step 25** twice (3 washes total).
27. Remove supernatant and add 250 μL elution buffer and 7 μL Proteinase K and incubate at 55 °C for 45 min with inversion.
28. Clean up elution with 250 μL phenol/chloroform extraction, followed by ethanol precipitation (as in **step 12**).
29. Centrifuge the tubes at 4 °C for 30 min at maximum speed—a pellet should be visible on the side of the tube. Wash the pellet with 1 mL 70% ethanol. Centrifuge at 4 °C for 5 min at maximum speed and remove the supernatant.
30. Allow tubes to air-dry (5–10 min).

3.4 Quantitative PCR on Genomic and Spike-In RNA–DNA Hybrids (DRIP-qPCR)

1. For DRIP-qPCR, resuspend each pellet in 50 μL of TE buffer. Dilute samples as required (*see* **Note 24**). For ease of calculation, dilute inputs in the same way.
2. For samples that will be used for sequencing, resuspend in 15 μL 10 mM Tris-HCl pH 8.0 and combine all tubes (up to 45 μL). Withdraw 5 μL to test enrichment by qPCR and use 2 μL per locus (for example, 3 positive and 2 negative loci can be tested, *see* **Note 25**). If spike-ins are used, also include primers to test for spike-in enrichment. Suitable primer sequences are listed in Table 2.
3. Perform qPCR using 2 μL DNA per locus (typically diluted 1:3 to 1:6). A typical qPCR reaction for a 384-well plate will contain: 0.5 μL of each primer (diluted to 10 μM), 6.25 μL of SYBR Green qPCR master mix, and 3.25 μL of ultrapure water. Amplifications are typically performed on a LightCycler (45 cycles alternating between 95 °C, 60 °C, and 72 °C, each step lasting 15 s).

Table 2
Primer sequences for DRIP-qPCR to detect human R-loop loci, Drosophila R-loop loci, and synthetic spike-ins

Primer name	Sequence 5′-3′	Description
THUMPD2_F	CAGTGGAGCACTTCTGACTGT	qPCR - R-loop positive (human)
THUMPD2_R	TGAGTTGGCCATTGGACCTC	qPCR - R-loop positive (human)
OPN3_F	GGAAGAAAGGGTCTGTCGCA	qPCR - R-loop positive (human)
OPN3_R	ACCGAACCTGGTTTGCAGAG	qPCR - R-loop positive (human)
EFNB2_F	TGAGCTTTGCTGTAGGGTGG	qPCR - R-loop positive (human)
EFNB2_R	AAAGACTTCCACGATGGCGT	qPCR - R-loop positive (human)
RPL13A_F	CTCCTCCTGCCTGCATTCTT	qPCR - R-loop positive (human)
RPL13A_R	CACACTTCCTGCCCACTTGT	qPCR - R-loop positive (human)
ACTB_F	ATTGCAGTGTCCCTCTGCTG	qPCR - R-loop positive (human)
ACTB_R	AGCATCGTGCCAGAAGCTAA	qPCR - R-loop positive (human)
MYC_F	CCCTGTCTTGGTGGATGGTC	qPCR - R-loop positive (human)
MYC_R	GCTGTATCCGTCCCTCATGG	qPCR - R-loop positive (human)
ZNF544_F	ATGACTGGTCTTCCCGCTTG	qPCR - R-loop negative (human)
ZNF544_R	CGCTTCTGAACCTGACACCA	qPCR - R-loop negative (human)
MLKL_F	TGTAGCTGGGAAGAGCTGGT	qPCR - R-loop negative (human)
MLKL_R	TGCACTTAAACCCCATTGGT	qPCR - R-loop negative (human)
Tub-F	ACCTGAACCGTCTGATTGGC	qPCR - R-loop positive (Drosophila)
Tub-R	GCAGAGAGGCGGTAATCGAG	qPCR - R-loop positive (Drosophila)
CG9662-F	CCCTGCATTGGTGTATTTCC	qPCR - R-loop positive (Drosophila)
CG9662-R	TGACATGGCTTACGATGCAG	qPCR - R-loop positive (Drosophila)

L286_F	GGATCCTGCAGTAATACGACTCACTATAGGGCCGTGATGGTTTTCGC CTTT	qPCR - synthetic spike-in hybrid
L286_R	ACTGCGACAATAATCCCGGA	qPCR - synthetic spike-in hybrid
H281_F	GGATCCTGCAGTAATACGACTCACTATAGGGAATGTCGCTGTGTTG TTGCT	qPCR - synthetic spike-in hybrid
H281_R	CCAGAACACCATCAACACCC	qPCR - synthetic spike-in hybrid
LDNA_F	CATCCCAGAGCGTTGTTGAG	qPCR - synthetic spike-in dsDNA
LDNA_R	AGCCCCAAAACCTAATCCCA	qPCR - synthetic spike-in dsDNA
HDNA_F	ACTCCGAGCAGCCAGATATC	qPCR - synthetic spike-in dsDNA
HDNA_R	CGTGGTAATGGCCGAAATCA	qPCR - synthetic spike-in dsDNA
ssDNA_F	TAATGACAGCCAGTACGCCC	qPCR - synthetic spike-in ssDNA
ssDNA_R	CTACGAATGGCATGAAGAAGC	qPCR - synthetic spike-in ssDNA

4. Calculate enrichments using the delta-delta Ct or Pfaffl method using Ct values from input and IP samples to obtain enrichment as % inputs (*see* **Note 26**). Relative enrichments between a negative locus can also be calculated. A positive locus is usually 1–5% of input (and 10- to 30-fold enriched over a negative locus) (*see* **Note 27**). Synthetic RNA–DNA hybrid spike-in sequences are typically enriched at 20–60% input.
5. If proceeding to library construction and sequencing, ensure that DRIP enrichment is sufficiently high, and that spike-in sequences are also highly enriched (reproducible yields of % input described should be obtained in multiple experiments).

3.5 Library Construction from ssDNA of RNA–DNA Hybrids

Different approaches have been taken to sequence RNA–DNA hybrids. Library preparation from dsDNA ends adjacent to the R-loop or hybrid has typically been used in DRIP-seq, but this does not provide stranded information of hybrid formation. To do so, the single-stranded RNA or DNA strands of the hybrid can be sequenced (*see* **Note 28**). Sequencing of the RNA strand of the hybrid has been used successfully [15]. However, this approach is technically challenging and, since the S9.6 antibody also binds cellular RNAs, can result in sequencing of contaminating RNAs [16]. Given these RNAs are much more abundant than RNA–DNA hybrids in cells, this may be problematic [16]. Alternatively, the DNA strand of the hybrid can be sequenced. This is appealing, since S9.6 does not bind appreciably to ssDNA in DRIP [7], thereby alleviating any issues of nonspecific binding. One such method for library preparation from ssDNA is the highly sensitive Accel-NGS 1S Plus DNA library preparation kit (Swift Biosciences).

1. Prior to a full sequencing run, it is advisable to ensure that spike-in hybrids are efficiently amplified and recovered during library preparation. Library preparation can be carried out on synthetic hybrid spike-ins alone and the resulting libraries analyzed on an Agilent Bioanalyzer—spike-ins should be detectable at the correct size (plus the length of adapters). Additionally, smaller scale sequencing using MiSeq can be performed to calibrate an appropriate quantity of spike-ins to be added to experimental samples, by calculating the proportion of total sequencing reads derived from spike-ins (*see* **Note 29**).
2. Take approximately half of each IP sample and ~1–10 ng of input for library construction. If required, the library preparation can be repeated on the remaining samples. Follow the manufacturer's guidelines for the tailing and ligation of adapters. Avoid overdrying magnetic beads—it is sufficient to remove ethanol with a pipette tip and directly proceed to the next step. For the indexing PCR (using the cell numbers and IP

conditions described above) inputs typically require ~7–9 cycles and IPs ~10 (*see* **Note 30**).

3. Following PCR, perform a double-sided cleanup with AMPure beads to select fragment sizes ~200–500 bp. To 50 μL PCR reactions, add 50 μL 10 mM Tris-HCl pH 8.0 and 60 μL AMPure beads (0.6× ratio with supernatant). Vortex to mix, spin down to gather liquid to the bottom of the tube, and leave for 5 min at RT.
4. Apply to the magnet and transfer the supernatant to a fresh tube. Add 40 μL AMPure beads (1× ratio with original supernatant). Vortex to mix, spin down to gather liquid to the bottom of the tube, and leave for 5 min at RT.
5. Apply to the magnet and remove the supernatant. Wash the beads twice with 500 μL 80% ethanol.
6. Spin down to gather any residual ethanol and remove with a tip. Add 12–20 μL TE. Vortex to mix, spin down to gather liquid to the bottom of the tube, and leave for 2 min at RT. Apply to the magnet and collect the eluate.
7. Confirm the library construction has worked by running 1 μL of each library on an Agilent BioAnalyzer. A successful library trace will have a symmetric curve without primer or adapter contamination (visible as peaks at ~70–80 bp or ~130 bp, respectively) (*see* Fig. 3a). If contaminants are present, the AMPure bead cleanup step can be repeated.
8. If pooling multiple libraries for sequencing, the concentration of each library should be determined, for example, using the Agilent BioAnalyzer or qPCR (for example, using NEBNext Library Quant Kit for Illumina). Pool libraries together to give a final concentration of 5–10 nM and submit for sequencing, aiming for ~30 million or more reads per IP sample. Paired end sequencing is recommended.

3.6 Bioinformatic Analysis and Normalization with Spike-In Counts

The workflow described in this chapter enables high-resolution and strand-specific DRIP-seq (Fig. 3b, c). Moreover, it is comparable with the use of synthetic RNA–DNA hybrids and *Drosophila* cell-based hybrids as internal standards, or spike-ins. These spike-ins offer alternative normalization strategies, independent of the conventional total mapped reads per sample. A normalization factor can be calculated based on the spike-in read counts and then applied to the experimental signal. In this section, we describe how sequencing reads mapping to either synthetic RNA–DNA hybrids or cell-based hybrids from a heterologous species can be extracted from sequencing files and used for normalization.

3.6.1 Read Trimming

During library construction with the 1S Plus DNA library kit, a low-complexity, G-rich tail of ~8–13 bases is added to the 3′ end of

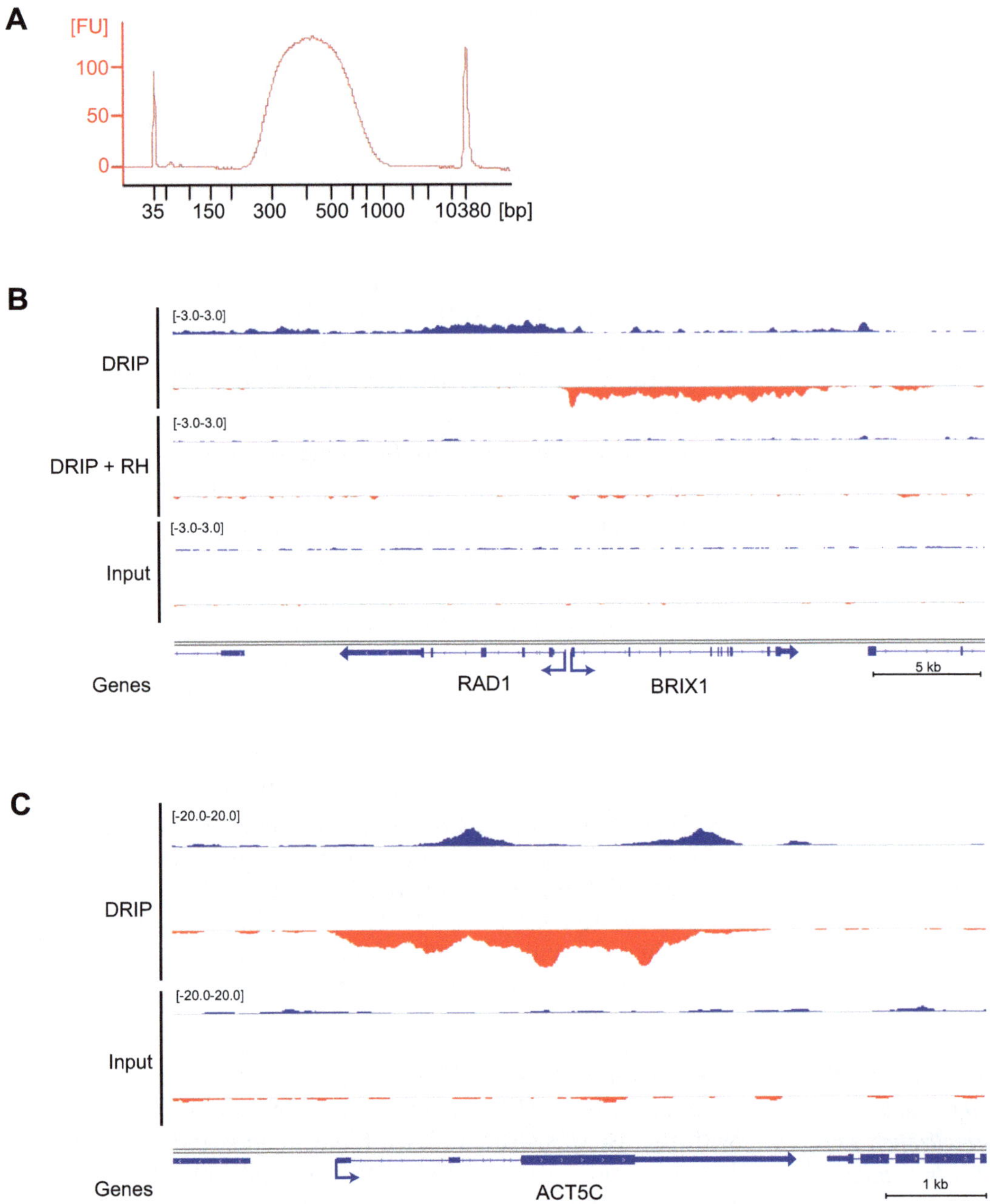

Fig. 3 qDRIP-seq. (**a**) BioAnalyzer trace of a PCR-amplified DRIP library from ssDNA. (**b**) Representative genome browser view of qDRIP-seq signal at human RAD1 and BRIX1 genes. From top to bottom: qDRIP-seq, RNase H digested sample pooled prior to IP, and input. All tracks normalized by reads per million mapped. Negative strand signal is in red, positive in blue. Bent arrows represent the transcription start site (TSS), while large triangular arrows represent the transcription end site (TES). (**c**) As in (**b**), but for qDRIP-seq signal at the *Drosophila* ACT5C gene. From top to bottom: qDRIP-seq and input

each DNA fragment. This tail is present at the beginning of Read 2 (R2) and may also be present at the end of Read 1 (R1) if R1 length is similar to library insert size (for paired end sequencing). Due to these low-complexity tails, running FastQC on fastq files raises "per base sequence content" and "per base GC content" flags at the beginning of R2. These tails should be trimmed to improve mapping efficiency.

1. First trim Illumina adapter sequences from R1 and R2, e.g., using "cutadapt -j $N_CORES -minimum-length 30 -a $ADAPTER_FWD -A $ADAPTER_REV."
2. On the adapter-trimmed files, then remove the low-complexity tails from the end of R2, e.g., using "cutadapt -j $N_CORES –minimum-length 30 -U 15."

3.6.2 Making Chimeric Genome Files for Mapping

1. A custom genome can be created, for example, combining hg38 with the *Drosophila* genome and/or synthetic spike-in sequences. This can be created using bowtie2, e.g.,

```
"$FAFILES = hg38.fa,dm6.fa,spike_ins.fa"
"bowtie2-build -f $FAFILES $CHIMERA_GENOME_BASE."
```

This approach can be modified to include *Drosophila* genomic (or other species' genome) and/or synthetic spike-in sequences, as required. Genome files can be downloaded from:http://hgdownload.soe.ucsc.edu/. A fasta file with synthetic hybrid sequences (reverse complement of the actual ssDNA sequence) should be generated, for example (*see* **Note 31**):

```
>L286
CCGTGATGGTTTTCGCCTTTAGCCATACGCCCATTATTTCTACGTTTGCCATAGACAGACG
TGAAAAATATGGCGAACACGCTATGGATAAATGCAAAAAAATTATGAAAGTCGCTTATCT
CATCATCTGCATAAGTGTACTGTTCTTTGTCTTTAGCTGCCTGCTTTCTATTCCACCTT
CGTATATTGAAGCTGCTAAAGAAGAAGGGGTCACCATTTTATCGGCGCTTTCTATGCT
GCCGAACGCCCCAGCATGGTTGTCAATTTCCGGGATTATTGTCGCAGT

>H281
AATGTCGCTGTGTTGTTGCTGTGCGAAAGAGAGAATCGTACTGATCACCGACGCGATG
CAGGCAGCTGGGATGCCGGATGGTCGCTATACGTTATGTGGTGAAGAAGTGCAGATG
CACGGTGGCGTTGTCCGTACCGCGTCTGGTGGGCTGGCGGGCAGTACGCTGTCTGTTGATG
CGGCAGTGCGCAATATGGTCGAGTTGACGGGCGTAACGCCTGCGGAAGCCATCCATATGG
CGTCGCTGCATCCGGCGCGAATGCTGGGTGTTGATGGTGTTCTGG
```

2. Align read files to the chimeric genome. Both local alignment with bwa-mem or end-to-end alignment with bowtie2, e.g.,

```
"bowtie2 -p $N_CORES --very-sensitive --no-unal --no-mixed --
no-discordant."
```

Typical alignment rates are approximately 90%. An example browser view of stranded RNA–DNA hybrid signal on the human genome (hg38) is shown in Fig. 3b.

3.6.3 Normalization Using Synthetic Spike-Ins

1. Count the number of reads mapping to each synthetic spike-in sequence, e.g.,

```
"samtools example.bam | awk '($2 == 83 || $2 == 163){print}' |
cut -f 3 | grep -v 'chr' | sort | uniq -c."
```

   This ensures only reads mapping to the reverse strand (i.e., the actual ssDNA of the spike-in hybrid) are counted. For paired end reads, divide this value by two to get true read counts.

2. Sum the reads for all spike-ins and calculate a mean count for all the spike-ins. This is the synthetic spike-in normalization factor (*see* **Note 32**). Repeat this process for all experimental samples.

3.6.4 Normalization Using Cell-Based Spike-Ins

Rather than simply counting all reads mapping to the *Drosophila* genome in each sample, for increased accuracy and reproducibility, a high-confidence consensus peak set is derived from the *Drosophila* DRIP signal, and the reads mapping within these peaks are then summed to give a normalization factor for each sample.

1. Extract reads mapping to either the human or Drosophila genomes, e.g.,

```
"SPLIT_HG='(/chr/ || / ^@[^(SQ)]){print}'
SPLIT_DM='(/dm/ || / ^@[^(SQ)]){print}'
samtools view -h $FILENAME | awk '${SPLIT_HG}' | samtools view
-bSh > ${OUTDIR}/hg_$FILENAME
samtools view -h $FILENAME | awk '${SPLIT_DM}' | samtools view
-bSh > ${OUTDIR}/dm_$FILENAME"
```

   Drosophila genome files are available here: http://hgdownload.soe.ucsc.edu/goldenPath/dm6/bigZips/

2. After removing duplicate alignments, these reads can be used to make genome browser tracks of Drosophila genomic RNA–DNA hybrid signal (Fig. 3c) (*see* **Note 33**).
3. Using bam files containing only reads mapping to the Drosophila genome, merge all input files together and all IP samples together, e.g., using samtools merge.
4. Call peaks, e.g., using MACS2 (broadpeak settings). The peak intervals should be viewed on the genome browser and compared to areas of enrichment on the genome browser tracks of hybrid signal. The peak set may need to be refined to only

include the peaks that are highly statistically enriched above input—previously an FDR cutoff of 1e-14 was used (9):

```
"awk 'BEGIN{OFS="\t"}($9 >= 14){print}' dm_merged_IP_vs_IN.
broadPeak".
```

5. The filtered Drosophila peak set can then be assigned strands according to the annotated direction of transcription, e.g., by intersecting them with Drosophila genes (available from the UCSC Table Browser). In the case that a peak intersects more than one strand, the gene with the most base overlap is selected. Then use bedtools coverage to count reads over the peaks for each sample:

```
"bedtools coverage -sorted -counts -a peak_file.bed -b reads.
bed."
```

6. Reads from the sense-strand within each peak are then summed and converted to reads per million to get the normalization factor for each sample. This factor is then used to normalize the RNA–DNA hybrid signal mapping to the human genome.

3.6.5 Applications of Normalization Factors from Spike-Ins

Read counts from synthetic or cellular spike-ins can then be used instead of reads per million mapped to normalize, for example, genome browser tracks, aggregate plots over the beginning and ends of genes, and differential peak calling between samples. In addition, for synthetic hybrids, since these were added at levels similar to genome copy number (or can be scaled down to this), the spike-in read counts are similar to a genomic site that forms R-loops 100% of the time. The synthetic spike-in counts can therefore be used to derive absolute hybrid frequencies at each R-loop (peak) site [7].

4 Notes

1. Make fresh and add to the cell pellet immediately.
2. Phase-lock tubes are not essential, but make phenol/chloroform extraction much easier and more efficient.
3. In other contexts where synthetic spike-ins are used, for example, when spike-in libraries were developed and made commercially available for RNA-seq, a blend of up to a hundred individual spike-ins have been used, of varying lengths and sequences to mimic the variation in genomic RNA in the experimental samples. However, the time and cost to develop such an extensive panel of hybrid sequences are likely prohibitive for many laboratories. Previously, two synthetic spike-ins were sufficient to correctly normalize hybrid levels in samples

upon transcription inhibition, but ideally more spike-ins would be used to improve statistical certainty.

4. SPRI bead cleanup steps are standard in library preparation protocols, and the conventional size cutoffs used can lead to a greater loss of sequences <150 bp in length, compared to longer sequences. While hybrid sequences of 150 bp (or shorter) can still be recovered, their recovery may be variable between samples and experiments. Since the synthetic hybrids are added at relatively small amounts, this can lead to high variability in spike-in recovery and therefore inaccuracies in normalization.

5. Care should be taken as phenol/chloroform is toxic and can cause burns. This step should be performed in a fume hood and tubes capped tightly. Phase-lock tubes should be used when possible, to minimize contamination of phenol/chloroform in samples.

6. The T7 RNAP promoter sequence is as follows: TAATAC GACTCACTATAGGG. The sequence upstream of the promoter may impact the efficiency of T7 RNAP transcription, so it may be desirable to add a short DNA sequence to the in vitro DNA template (*see* Table 1).

7. If the PCR product is the incorrect size, or a smear is visible, adjust the PCR conditions (such as the annealing temperature) or design different primer sequences and repeat the PCR step.

8. DNA fragments can be excised by being visualized on a UV illuminator. A face shield should be worn to protect eyes against UV. This step should be performed quickly to avoid UV damage of the DNA in the gel. Using lower energy wavelengths and settings may also help reduce DNA damage during this step.

9. Performing the elution step of the gel extraction protocol with the elution buffer heated to 50 °C may increase the gel extraction yield.

10. Purified RNA can be stored frozen at −20 °C or −80 °C short term but is not stable in water solution for long term storage and should be used for annealing reactions as quickly as possible.

11. The complementary DNA sequence can be ordered directly as single-stranded oligonucleotides. Spike-ins can also be annealed using the same dsDNA template as that used for in vitro transcription reactions plus the purified RNA. However, this can produce a heterogeneous mix of spike-ins, with some spurious annealing products, which are detectable by sequencing. To increase the purity of the spike-ins (which makes read counts and hence normalization more accurate) it

is therefore recommended to perform annealing reactions with purified ssDNA and RNA.

12. Hybrid annealing may require optimization depending on the size and sequence of the hybrids. Ratios of RNA–DNA can be modified—in some cases 2:1 or 4:1 may increase the annealing efficiency.

13. Ethanol precipitation can cause some DNA–DNA hybrid dissociation. The described gel crushing procedure [14] effectively extracts the annealed hybrids, avoiding ethanol precipitation and increasing their recovery yield.

14. Excised gel pieces should be as small as possible to avoid contaminants and dilution of the DNA.

15. Since the RNA–DNA hybrid cannot be directly quantified by fluorimetry, we found the RNA quantification kit could be used with hybrids as an adequate substitute.

16. The exact amount of spike-in hybrids that should be added to each sample will need to be optimized. The number of spike-in molecules recovered will depend on many factors, including efficiency of pull down, losses during the workflow, and library preparation. At least several hundred molecules should be recovered in inputs to reduce sampling error. A small-scale sequencing run, for example, with MiSeq, can be used to confirm the spike-in quantity is sufficient.

17. Heat-inactivate FBS by incubating for 30 min in a water bath at 56 °C.

18. Ideally, flow cytometry should be used to confirm that the ploidy of the S2 cells is kept constant between experiments. If hybrid content changes across the cell cycle, then altered ploidy levels of S2 cells may change the hybrid content and hence could affect normalization between experiments. While mostly diploid, S2 can become tetraploid over time if split too thin or if cells become stressed.

19. This cell count works well for human cells such as HeLa, as ~30–50 μg of genomic DNA can be extracted and used for multiple IPs. If fewer IPs are needed, the cell number can be reduced. Cells should be sub-confluent at time of harvesting, ensuring that transcription (and hence co-transcriptional R-loop formation) is not suppressed. Do not use an excessive number of cells. This makes handling and drying of genomic material more challenging, as it increases drying times, increases viscosity, and can lead to reduced IP efficiency.

20. Cell lysis and chromatin digestion can be performed for 2–3 h. For convenience, lysis can also be performed overnight. While the procedure should be shortened as much as possible to

increase recovery of RNA–DNA hybrids, we have not detected a significant loss by incubating lysis reactions for longer times.

21. Remove ethanol with a pipette—do not spin down tubes in between wash steps as this may increase RNA contamination.
22. For sequencing, perform 2–3 IPs in parallel to obtain sufficient material for library preparation and to minimize technical variability. For DRIP-qPCR alone, 1–2 IPs per sample may be performed. Having multiple IPs can help reduce the chance of technical "drop outs" of samples and are a good way to assess reproducibility. If enrichment between technical replicates varies significantly, the DRIP procedure should be repeated and optimized.
23. Use safe-lock Eppendorf tubes for this step (using LoBind tubes during the IP step is recommended to prevent nonspecific binding to the tube wall). This prevents leakage during the elution step when tubes are heated. Lids on normal tubes may loosen causing sample loss.
24. Depending on the R-loop enrichment at a given site, samples can usually be diluted 1:3–1:8 for qPCR amplification.
25. Standard primer design tools can be used, such as NCBI primer blast (https://www.ncbi.nlm.nih.gov/tools/primer-blast/). Amplicons should ideally be 70–150 bp in length. Examples of R-loop positive and negative loci and corresponding primers are provided in Table 2. These are suitable for a range of human cell lines, including HeLa cells. Also included are primer sequences for *Drosophila* R-loop positive sites, and to amplify synthetic hybrid and dsDNA/ssDNA spike-ins (ssDNA/dsDNA serve as negative internal controls).
26. Sonication disrupts some R-loop structures, and thus enrichment is generally lower than that observed with fragmentation by DRIP digestion. However, smaller fragment size also enables RNA–DNA hybrids to be probed at a higher resolution on the genome.
27. It is recommended to present DRIP-qPCR data as % input whenever possible, rather than only as normalized values (such as fold change to a negative control locus or to a control-treated sample). This ensures that DRIP yields are robust and reproducible between experiments and also that data can be more easily compared between different conditions and laboratories.
28. Sonication disrupts the non-template DNA strand of R-loops. Therefore, the ssDNA from the template DNA strand is enriched and captured from RNA–DNA hybrids during library constructions, resulting in strand-specific sequencing libraries.
29. To minimize sampling error of spike-ins (especially in input samples in which hybrids are not enriched), ideally at least

several hundred reads should map to each spike-in. For cellular spike-ins (such as from *Drosophila* genomic hybrids) several million reads, or approximately 10% of total sequencing reads, should map to the heterologous genome. This ensures a balance of enough spike-in reads for accurate counting, but not too many to unnecessarily reduce the reads mapping to the genome of interest.

30. If multiple libraries are to be pooled and sequenced, each library requires a unique indexed adapter.
31. These spike-ins sequences are those used previously [7].
32. To give spike-in normalization constants of similar magnitudes to conventional total read-count normalization, the spike-in read counts can be adjusted by a scale factor, for example, 1000 was used previously to give an adjusted normalization factor [7].
33. The *Drosophila* genome tends to have weaker hybrid signal within peaks, compared to the human genome [7, 13]. Due to this overall lower enrichment of RNA–DNA hybrids, careful selection of high-confidence peak regions within the *Drosophila* genome is required to avoid counting background signal in the normalization.

Acknowledgments

We thank Stephan Hamperl and Joshua Brickner for valuable feedback. This work was supported by the Leukemia and Lymphoma Society [5455-17 to M.P.C.]; National Institutes of Health [GM119334 to K.A.C., S10OD018220 to the Stanford Functional Genomics Facility]. K.A.C. is an ACS Research Professor.

References

1. Crossley MP, Bocek M, Cimprich KA (2019) R-loops as cellular regulators and genomic threats. Mol Cell 73:398–411
2. Ginno PA, Lott PL, Christensen HC, Korf I, Chedin F (2012) R-loop formation is a distinctive characteristic of unmethylated human CpG island promoters. Mol Cell 45:814–825
3. Loven J, Orlando DA, Sigova AA, Lin CY, Rahl PB, Burge CB, Levens DL, Lee TI, Young RA (2012) Revisiting global gene expression analysis. Cell 151:476–482
4. Hu Z, Chen K, Xia Z, Chavez M, Pal S, Seol JH, Chen CC, Li W, Tyler JK (2014) Nucleosome loss leads to global transcriptional up-regulation and genomic instability during yeast aging. Genes Dev 28:396–408
5. Chen K, Hu Z, Xia Z, Zhao D, Li W, Tyler JK (2015) The overlooked fact: fundamental need for spike-in control for virtually all genome-wide analyses. Mol Cell Biol 36:662–667
6. Orlando DA, Chen MW, Brown VE, Solanki S, Choi YJ, Olson ER, Fritz CC, Bradner JE, Guenther MG (2014) Quantitative ChIP-Seq normalization reveals global modulation of the epigenome. Cell Rep 9:1163–1170
7. Crossley MP, Bocek MJ, Hamperl S, Swigut T, Cimprich KA (2020) qDRIP: a method to quantitatively assess RNA–DNA hybrid formation genome-wide. Nucleic Acids Res 48:e84
8. Jiang L, Schlesinger F, Davis CA, Zhang Y, Li R, Salit M, Gingeras TR, Oliver B (2011)

Synthetic spike-in standards for RNA-seq experiments. Genome Res 21:1543–1551
9. Santos-Pereira JM, Aguilera A (2015) R loops: new modulators of genome dynamics and function. Nat Rev Genet 16:583–597
10. Castillo-Guzman D, Hartono SR, Sanz LA, Chédin F SF3B1-targeted splicing inhibition triggers global alterations in transcriptional dynamics and R-loop metabolism. https://doi.org/10.1101/2020.06.08.130583
11. Laitem C, Zaborowska J, Isa NF, Kufs J, Dienstbier M, Murphy S (2015) CDK9 inhibitors define elongation checkpoints at both ends of RNA polymerase II-transcribed genes. Nat Struct Mol Biol 22:396–403
12. Abraham KJ et al (2020) Nucleolar RNA polymerase II drives ribosome biogenesis. Nature 585:298–302
13. Svikovic S, Crisp A, Tan-Wong SM, Guilliam TA, Doherty AJ, Proudfoot NJ, Guilbaud G, Sale JE (2019) R-loop formation during S phase is restricted by PrimPol-mediated repriming. EMBO J 38:e99793
14. Sun Y, Sriramajayam K, Luo D, Liao DJ (2012) A quick, cost-free method of purification of DNA fragments from agarose gel. J Cancer 3: 93–95
15. Sanz L, Chedin F (2019) High-resolution, strand-specific R-loop mapping via S9.6-based DNA–RNA immunoprecipitation and high-throughput sequencing. Nat Protoc 14: 1734–1755
16. Chedin F, Hartono SR, Sanz LA, Vanoosthuyse V (2021) Best practices for the visualization, mapping and manipulation of R-loops. EMBO J 40:e106394

Chapter 27

DNA:RNA Immunoprecipitation from *S. pombe* Cells for qPCR and Genome-Wide Sequencing

Laetitia Vachez, Camille Teste, and Vincent Vanoosthuyse

Abstract

By temporarily distorting the DNA double helix, the moving RNA polymerases can lead to the formation of non-B DNA structures. One of the most abundant and largest non-B DNA structures in the genome is the R-loop, a three-stranded structure forming when the nascent RNA hybridizes with its DNA template, thereby extruding the non-template DNA strand. Growing evidence suggests that at least a subset of R-loops could induce transcription stress and genome instability, although the direct, primary consequences of R-loop formation on the surrounding chromatin are still unclear.

To understand the direct impact of R-loops on transcription and genome stability, accurate and quantitative mapping of R-loops is essential. R-loop mapping is commonly achieved using the antibody-based-DNA:RNA Immunoprecipitation (DRIP) strategy. While it is reasonably straightforward to obtain robust DRIP enrichments from human cells, this has proved harder in yeast, where DRIP signals are often relatively weak, with a poor signal-to-noise ratio. Although it is unclear whether such weak signals stem from a technical or a biological reality, they make the accurate quantification of DRIP signals all the more important, especially when deep sequencing is used to monitor and quantify the distribution of R-loops genome-wide. Here we propose a DRIP protocol that has been optimized for the mapping and the quantification of R-loops in *Schizosaccharomyces pombe* but that can also be used in *Saccharomyces cerevisiae*. As a result, this protocol can be used to generate calibrated DRIP-seq data, where genomic DNA extracted from *S. cerevisiae* serves as spike-in reference.

Key words *Schizosaccharomyces pombe*, *Saccharomyces cerevisiae*, DRIP, ssDRIP-seq, S9.6, RNase H1

1 Introduction

Most R-loops form when the nascent RNA hybridizes with its DNA template through Watson–Crick interactions, thereby preventing the re-annealing of the non-template DNA strand [1]. Recent data estimated that ~1000 R-loops form every hour in the human genome, each with an average half-life of ~10 min [2]. Single-molecule footprinting has shown that the median size of R-loops is ~300 bp but that individual R-loops can extend up to several kilobases [3]. R-loops are therefore an abundant class of large

Andrés Aguilera and Alexey Ruzov (eds.), *R-Loops: Methods and Protocols*, Methods in Molecular Biology, vol. 2528, https://doi.org/10.1007/978-1-0716-2477-7_27,

non-B DNA structures in the genome. However, the primary consequences of R-loop formation on the surrounding chromatin are still largely unclear and no unifying model has been proposed to date to explain the direct impact of R-loops on key DNA transactions such as transcription, DNA replication, or DNA repair.

Two main types of methods currently exist to map R-loops (please refer to recent reviews for a more in-depth discussion of current strategies, refs. 1, 4, 5). The first type maps R-loops ex vivo on purified genomic DNA; the second maps R-loops in situ, after formaldehyde crosslinking or on permeabilized live cells. Both families of methods use either of two main types of baits to enrich for DNA:RNA hybrids: either the S9.6 antibody that has a strong affinity for DNA:RNA hybrids in a largely sequence-independent manner or a catalytically inactive form of RNase H1 [or more specifically its Hybrid-Binding Domain (HBD)], which is an enzyme that degrades the RNA moiety of DNA:RNA hybrids. The most commonly used strategy so far is the DNA:RNA Immunoprecipitation (DRIP) method popularized by the Chédin lab, which uses the S9.6 antibody to pull down RNA:DNA hybrids ex vivo from sheared genomic DNA [6]. An essential specificity control of the DRIP method is to treat half of the purified genomic DNA with active RNase H before immunoprecipitation with the S9.6 antibody, in order to thoroughly eliminate the RNA:DNA hybrids present in the sample.

The main difficulty of DRIP is to extract and to shear the genomic DNA in order to preserve a maximum amount of RNA: DNA hybrids and to have it free of contaminants, thereby allowing the complete removal of RNA:DNA hybrids by exogenous RNase H treatments. Although this is reasonably straightforward in human cells, this has proved harder in yeast, possibly because of the presence of thick cell walls and/or of contaminants. As a result, DRIP signals in yeast tend to be relatively low with a poor signal-to-noise ratio and a variable sensitivity to exogenous RNase H treatments.

We first used DRIP-qPCR in the fission yeast *Schizosaccharomyces pombe* to demonstrate the formation of R-loops at RNA Polymerase III (RNAP3)-transcribed genes [7]. Our first DRIP protocol relied on the mechanical disruption of frozen fission yeast cells to extract genomic DNA and, although this protocol yielded clean RNase H-sensitive signals by qPCR, the amount of material immunoprecipitated was small and unpractical for preparing sequencing libraries. We then established an alternative protocol based on the enzymatic digest of cell walls to extract genomic DNA [8]. Although this protocol allowed the genome-wide mapping of the RNA moiety of R-loops in cycling, unchallenged cells [8], it proved less satisfactory when using qPCR to quantify local enrichments as the DRIP signals had variable to poor RNase H sensitivity. Here we propose an alternative way of extracting genomic DNA from fission yeast that gave satisfactory results both in DRIP-qPCR and in strand-specific DRIP-seq. The same strategy can be applied to budding yeast, which means that genomic DNA extracted from budding yeast can easily serve as spike-in calibrator for fission yeast DRIP-seq maps. The genomic DNA can be sheared using either sonication or restriction enzymes. The protocol can be used on fresh or frozen cells.

Briefly, ~10^9 cells are collected and their cell wall is digested enzymatically. The resulting spheroplasts are resuspended in a CTAB-based buffer containing Proteinase K and incubated at 55 °C. Phenol/chloroform extraction followed by Ethanol precipitation is then used to purify nucleic acids, which can then be sheared by either sonication or restriction enzymes depending on the objectives of the experiment. The purified genomic DNA (gDNA) is then accurately quantified by qPCR and incubated with a known amount of S9.6 antibody, always using the same ratio gDNA/antibody. After extensive washes, the immunoprecipitated DNA is purified and quantified by qPCR or processed for sequencing library preparation. This protocol works for both fission and budding yeasts (*see* Fig. 2).

2 Materials

2.1 Yeast Culture

1. YES+A (for 1 L): 5 g Yeast extract, 30 g Glucose, 100 mg Leucine, 100 mg Uracil, 100 mg Histidine, 100 mg Lysine, 100 mg Arginine, 150 mg Adenine.
2. YPD (for 1 L): 10 g Bacto Peptone, 10 g Yeast Extract, 20 g Glucose.
3. Liquid Nitrogen.
4. 50 mL tubes for centrifugation.
5. Round-bottomed 14 mL tubes for centrifugation.
6. Culture of *Schizosaccharomyces pombe*: For each experiment, aim for 10^7 cells/mL in 100 mL (i.e., 10^9 cells total). We recommend to systematically grow both rnh1 + rnh201+ and *rnh1Δrnh201Δ* cells as positive control. rnh1 + rnh201+ has a generation time of ~2.4 h in YES+A, while *rnh1Δrnh201Δ* has a generation time of ~3 h.
7. Culture of *Saccharomyces cerevisiae* for calibration purposes: For calibration purposes, we also grow *rnh1Δrnh201Δ* cells from *S. cerevisiae* in YPD.

2.2 gDNA Extraction and Purification

1. QIAGEN MaXtract High Density tubes (15 mL and 2 mL) or equivalent.
2. Spheroplasting buffer (for 100 mL): 2.94 g Sodium citrate, 0.69 g NaH_2PO_4, 21.9 g Sorbitol pH 6.0.

3. Zymolyase 100 T and Zymolyase 20 T.
4. ß-mercaptoethanol.
5. CTAB buffer (for 100 mL): 2 g CTAB (hexadecyltrimethylammonium bromide), 10 mL 1 M Tris–HCl pH 8.0, 4 mL 0.5 M EDTA pH 8.0, 28 mL 5 M NaCl.
6. Proteinase K (20 mg/mL).
7. Triton X-100.
8. Tween® 20.
9. Phenol/Chloroform/IAA (25:24:1).
10. Glycogen.
11. 3 M Sodium acetate pH 5.2.
12. Ethanol.
13. Isopropanol.
14. 10 mM Tris–HCl pH 8.0
15. AMPure XP beads or equivalent.

2.3 gDNA Quantification by qPCR

1. Suggested primer pairs to quantify gDNA from *S. pombe* and *S. cerevisiae* against dilutions of known gDNA standards:

Primers to quantify gDNA from *S. pombe*	
sum3 qL1	AGGTCGCCGTGAATACAATC
sum3 qR1	GCCAAATCCACGGTTGTTAC
ecm33 qL4	TTGGCAAAGATGAGACATAAGG
ecm33 qR4	AATAAACTCGGTAGTCTTGCAATC
Primers to quantify gDNA from *S. cerevisiae*	
Pma1_F	GTCCACTTCTGAAGCCTTTGAC
Pma1_R	TGCATAGCAGCCATGAAGTC
Yef3_F	GTTGTCTTCTGCGGAATTGAG
Yef3_R	CGTCAGAAGAAACGTAAGCATC

2.4 Shearing the gDNA

1. *(optional):* microTUBE AFA Fiber Pre-Slit Snap-Cap 6 × 16 mm tubes.
2. 10× CutSmart Buffer (NEB) or equivalent.
3. 100× BSA (20 mg/mL).
4. 0.1 M Spermidine.
5. 50% Glycerol.
6. Restriction enzymes: HpaII (10 U/μL), AluI (10 U/μL), DdeI (10 U/μL), MboI (5 U/μL).

7. 10 mg/mL RNase A.
8. 5 U/μL RNase H.
9. 2 U/μL RNase III.

2.5 Immunoprecipitation

1. Dynabeads™ Protein A for Immunoprecipitation or equivalent.
2. Corning® Costar® low binding microcentrifuge tubes or equivalent.
3. Eppendorf® LoBind microcentrifuge tubes or equivalent.
4. Chelex 100 Molecular Biology Grade Resin.
5. S9.6 antibody.
6. IP buffer: 10 mM Tris, 140 mM NaCl, 0.5% Triton.
7. IP High Salt buffer:10 mM Tris, 500 mM NaCl, 0.5% Triton.
8. IP Wash buffer: 10 mM Tris, 1 mM EDTA, 1% Igepal, 0.5% Sodium deoxycholate, 0.25 M LiCl.
9. PBS-T: 140 mM NaCl, 2.7 mM KCl, 10 mM $Na_2HPO_4{\cdot}2H_2O$, 1.7 mM KH_2PO_4, 0.02% Tween® 20.

2.6 Quantification of DRIP Signals by qPCR

1. Suggested primer pairs to quantify DRIP signals by qPCR:

Primers for positive sites in *S. pombe*	
sum3 qL1	AGGTCGCCGTGAATACAATC
sum3 qR1	GCCAAATCCACGGTTGTTAC
srp7 qL2	AAGAGTAGTCTTCGTGGCAACTG
srp7 qR2	GATGTGCATTGTTTCCAACC
tyr04 FW	TGGTGTAGTTGGTTATCACATCC
tyr04 RV	AATCTCCTGAGCCAGAATCG
ala07 qL1	ATGTTACCGCTACACCATCG
ala07 qR1	AAACAGTTTGGATTGTTTTTGG
fba1 qL1	TCAAGACCACCAACGACAAG
fba1 qR1	AGGCGAATTGGGTATCAGTG
Primers for a negative site in *S. pombe*	
cen3 qL1	CGTGTCTGATTCACGCAAAG
cen3 qR1	CGCCGCCTATATTAGCAAAC
Primers for positive sites in *S. cerevisiae*	
Pma1_F	GTCCACTTCTGAAGCCTTTGAC
Pma1_R	TGCATAGCAGCCATGAAGTC

(continued)

Yef3_F	GTTGTCTTCTGCGGAATTGAG
Yef3_R	CGTCAGAAGAAACGTAAGCATC
Pyk1_F	CAGAGGTGTCTTCCCATTCG
Pyk1_R	ACGGGCTTCAACATCATCAG
Pdc1_F	GTTGATTCACGGTCCAAAGG
Pdc1_R	CACCGAAAGTTGGCAACAAG

2.7 Equipment

1. Apparatus to run agarose gels.
2. Magnet.
3. Water baths at 37 °C and 50 °C.
4. Thermomixer.
5. Rotating wheel for 1.5 mL tubes.
6. Apparatus to run quantitative PCR (qPCR).
7. *(optional):* Covaris S220 sonicator or equivalent.
8. Fume hood.

3 Methods

3.1 Cell Culture

1. Once the cultures have reached the desired concentration, pellet the cells (2000 × *g*, 2 min, 4 °C).
2. Wash the pellets twice with 90 mL of cold distilled water.
3. After the second wash, eliminate carefully all the water. A quick, additional centrifugation might be necessary to collect all the remaining water and fully dry the pellet.
4. Resuspend the cells at 10^9 cells/mL in Spheroplasting buffer.
5. Make 1 mL aliquots in round-bottomed 14 mL tubes. Pellet the cells again and remove all the excess liquid.
6. Proceed immediately or freeze the pellets in liquid Nitrogen (*see* **Note 1**).
7. Culture of *Saccharomyces cerevisiae* for calibration purposes: When the cultures reach the right concentration, proceed as above to make cell pellets containing 5×10^8 cells.

3.2 Spheroplasting

1. Prepare a water bath at 37 °C. Prepare another water bath at 30 °C if you plan to use *S. cerevisiae* cells for calibration. Refrigerate a centrifuge for 14 mL tubes at 4 °C.
2. If you start from frozen *Schizosaccharomyces pombe* cells, thaw the aliquots of 10^9 cells on ice. Add 5 mL of spheroplasting buffer containing 2 mg/mL of zymolyase 100T to the thawed cell pellets and resuspend well by pipetting. Under the fume hood, add 5 μL of ß-mercaptoethanol and mix thoroughly.

3. Incubate for 2 h 30 min at 37 °C in a water bath. Agitate regularly. Keep the spheroplasting buffer on ice for later (*see* **Notes 2** and **3**).
4. If you plan to use *Saccharomyces cerevisiae* cells for calibration: 2 h after you started incubating the *S. pombe* cells at 37 °C, resuspend 5×10^8 *S. cerevisiae* cells into 750 μL of spheroplasting buffer containing 1 mg/mL of zymolyase 20T. Add 0.75 μL of ß-mercaptoethanol and mix thoroughly. Incubate for 30 min at 30 °C in a water bath.
5. At the end of the incubations, spin down all tubes at 5000 × *g* at 4 °C for 5 min.
6. Discard carefully the supernatant under the fume hood and wash with 5 mL of cold spheroplasting buffer. Spin down again for 5 min at 5000 × *g*.
7. Repeat **steps 5** and **6**. Make sure you discard the supernatant completely.

3.3 Cell Lysis

1. Prepare a water bath at 55 °C.
2. Resuspend thoroughly but gently the spheroplasts pellets in CTAB buffer. Use 100 μL of CTAB buffer for 10^8 cells. Avoid foaming.
3. Add Triton (1% final) and 40 μg of proteinase K per 100 μL of CTAB buffer. Mix gently. Avoid foaming.
4. Incubate for 45 min at 55 °C. Agitate the tubes regularly (*see* **Note 4**).
5. During the incubation at 55 °C, label enough MaXtract High Density 15 mL tubes. Spin them down for 3 min at 1500 × *g* to pellet the gel.
6. At the end of the incubation at 55 °C, transfer the samples to the MaXtract High Density tubes without pipetting (simply tip the content of the tube into the MaXtract tubes). Under the fume hood, add 1 volume of phenol/chloroform/IAA (25: 24:1). Mix thoroughly by inverting the tubes until the obtention of a homogenous white micelle. Let the tubes rest for 2 minutes and mix another fifteen times or so (*see* **Note 5**).
7. Spin down for 6–10 min at 1500 × *g* at room temperature (the supernatant should be clear).
8. (*Optional*) You can add a chloroform extraction step to get rid of the remaining traces of phenol. In that case, repeat **steps 6** and **7** by replacing the volume of phenol/chloroform/IAA (25:24:1) with a volume of chloroform.

3.4 Nucleic Acids Precipitation

1. During the centrifugation, transfer 1/10 volume of 3 M sodium acetate pH 5.2 and 2.5 volumes of absolute ethanol to a clean 15 mL tube.
2. At the end of the centrifugation, transfer the supernatant without pipetting into the 15 mL containing the sodium acetate and the ethanol. Mix slowly by inverting gently the tubes to precipitate the nucleic acids. Continue until white precipitate forms (no translucent part should remain).
3. Keep on ice for at least 15 min to make sure the precipitation is complete.
4. During that time, prepare three 2 mL tubes containing 1 mL of fresh 80% ethanol.
5. If the precipitate forms a pellet that seems easy to catch, transfer the precipitate with a p1000 tip into a clean 2 mL tube containing 1 mL of fresh 80% ethanol. In case the precipitate is not easy to catch with a p1000 tip, go to **step 10**.
6. Invert the tube a few times. Discard the supernatant without disturbing the pellet.
7. Repeat **steps 5** and **6** at least twice (three washes in total).
8. Remove all remaining ethanol and let the pellet dry for ~20 min at room temperature.
9. Resuspend the pellet in 50 μL of DNase and RNase-free 10 mM Tris–HCl pH 8.0.
10. In case the precipitate is not easy to catch with a p1000 tip, gently pellet the precipitate for <10 s at 1500 × *g*. As above, wash the pellet at least three times with 80% ethanol without pipetting. Remove all the remaining ethanol, let the pellet dry for ~20 min at room temperature and resuspend the pellet in 50 μL of DNase and RNase-free 10 mM Tris–HCl pH 8.0.
11. Let the pellet resuspend fully for 30 min on ice (*see* **Note 6**).

3.5 Confirm the Integrity of the gDNA

1. Take 3 μL of gDNA to a new tube.
2. Add 1 μL of RNase A (10 mg/mL) to digest the RNAs. Incubate at 37 °C for 10–15 min.
3. Run the reaction on an agarose gel (0.8%, 70 V, 40 min) (*see* Fig. 1a). If your genomic DNA is intact, you can proceed with shearing the gDNA (Subheading 3.6).

3.6 Shearing the gDNA

Several options are available to shear the gDNA. For regular DRIP-qPCR experiments, we recommend using sonication because it is quicker and most amplicons can be quantified (*see* Subheading 3.6.1). If you need greater resolution, use restriction digests with frequent cutters instead (*see* Fig. 1b, c and Subheading 3.6.2). Note, however, that DRIP-qPCR after shearing the gDNA with frequent cutters leads to enrichments (% IP) that are roughly half those obtained after sonication (*see* Fig. 1d, e).

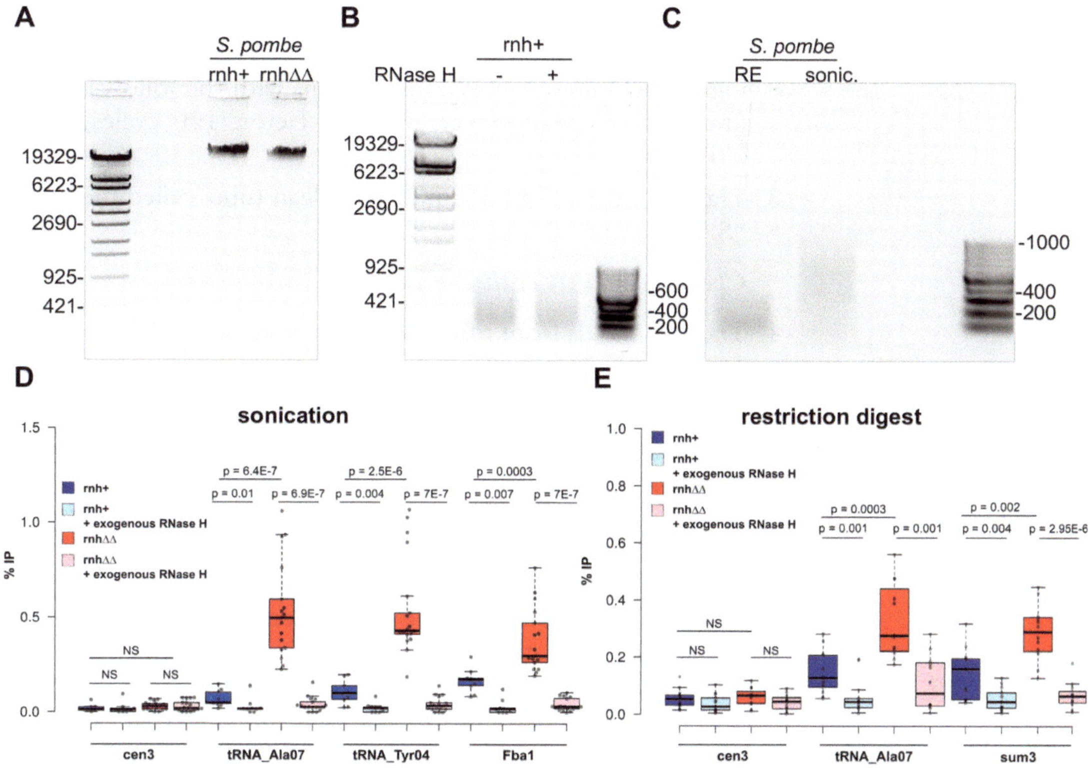

Fig. 1 DRIP-qPCR in fission yeast. (**a**) gDNA extracted from rnh1 + rnh201+ cells (rnh+) or *rnh1Δrnh201Δ* cells (rnhΔΔ) was run on an agarose gel. (**b**) gDNA extracted from rnh1 + rnh201+ cells (rnh+) was sheared using restriction enzymes and run on an agarose gel. Adding RNase H did not alter the restriction pattern. (**c**) Comparison of the shearing pattern after restriction enzyme digest (RE) or sonication (sonic.). (**d**) DRIP enrichments (% IP) after sonication of gDNA extracted from the indicated strains. A negative control site (centromeric cen3) and three positive control sites (tRNA_Ala07, tRNA_Tyr04 and Fba1) are shown. 9 biological replicates were used for rnh + cells and 17 for rnhΔΔ cells. P-values (*p*) were obtained after performing a Wilcoxon–Mann Whitney test. (**e**) DRIP enrichments (% IP) after restriction digest of gDNA extracted from the indicated strains. A negative control site (centromeric cen3) and two positive control sites (tRNA_Ala07 and Sum3) are shown. 13 biological replicates were used for rnh + cells and 12 for rnhΔΔ cells. P-values (*p*) were obtained after performing a Wilcoxon–Mann Whitney test. For (**d**) and (**e**), thick black lines show the medians; box limits indicate the 25th and 75th percentiles as determined by R software; whiskers extend 1.5 times the interquartile range from the 25th and 75th percentiles, outliers are represented by dots

3.6.1 Option #1: Shearing the gDNA Using Sonication

This protocol has been optimized for the use of a Covaris S220 sonicator with the microTUBE AFA Fiber Pre-Slit Snap-Cap 6 × 16 mm tubes (*see* **Notes 7** and **8**).

1. Add 83 μL of DNase and RNase-free Tris–HCl pH 8.0 10 mM to your samples to reach a final volume of 130 μL. Homogenize gently each sample by flicking the tube. (*Optional*) You might want to incubate your DNA at 37 °C for 30 min to improve its solubility.

2. Transfer the whole sample to a fresh microTUBE AFA Fiber Pre-Slit Snap-Cap 6 × 16 mm tube. Use the sonicator according to the manufacturer's instructions with the following program: 75 W peak power, 5% duty factor, 120 cycles/burst, 2′05″.
3. Transfer the sonicated sample to a clean tube, called thereafter "S."

Check the quality of the sonication.

4. Transfer 10 μL from tube "S" to a new tube.
5. Add 1 μL of RNase A (10 mg/mL) and incubate at 37 °C for 10–15 min.
6. Run the reaction on an agarose gel (1.2%, 70 V, 40 min) (*see* Fig. 1c).

Digest the sonicated gDNA with RNases.

7. Split the remaining sample in tube "S" into two fresh tubes with 60 μL each. One will be the test sample and the other one the RNase H-treated control sample.
8. To the test tube, add 10 μL of 10× CutSmart buffer (NEB), 3 μL RNase III, 5 μL of 50% glycerol, and 22 μL of DNase & RNase-free water.
9. To the control tube, add 10 μL 10× CutSmart buffer (NEB), 3 μL RNase III, 5 μL RNase H, 22 μL DNase and RNase-free water.
10. Incubate both tubes for 2 h at 37 °C.

(Optional) Remove soluble RNAs.

11. To both tubes, add 15 μL 5 M NaCl, 10 μL 0.1 mg/mL RNase A, 25 μL DNase and RNase-free water.
12. Incubate 30′ at 37 °C.

Purify the sonicated gDNA.

13. Add 50 μL of DNase and RNase-free water to both tubes. Mix gently by flicking.
14. Transfer the content of each tube to a fresh MaXtract High Density tube (2 mL).
15. Under the fume hood, add 200 μL of phenol/chloroform/IAA (25:24:1). Mix thoroughly by inverting the tubes until the obtention of a homogenous white micelle. Let the tubes rest for 2 min and mix another fifteen times or so.
16. Centrifuge under the fume hood at 13,000 × *g* for 10 min.
17. Collect the supernatant by tipping it to a fresh DNA Lobind tube. Proceed to Subheading 3.7.

3.6.2 Option #2: Shearing the gDNA Using Restriction Enzymes

1. Prepare two DNA Lobind tubes. One will contain the test sample and the other one the control, RNase H-treated sample.
2. To the test tube, add 15 μL 10× CutSmart buffer, 1.5 μL 100× BSA, 1.5 μL 0.1 M spermidine, 1.5 μL HpaII (10 U/μL), 1.5 μL AluI (10 U/μL), 1.5 μL DdeI (10 U/μL), 3 μL MboI (5 U/μL), 10 μL 50% glycerol, 94.5 μL DNase and RNase-free water. Mix gently by flicking the tube.
3. To the control tube, add 15 μL 10× CutSmart buffer, 1.5 μL 100× BSA, 1.5 μL 0.1 M spermidine, 1.5 μL HpaII (10 U/μL), 1.5 μL AluI (10 U/μL), 1.5 μL DdeI (10 U/μL), 3 μL MboI (5 U/μL), 10 μL RNase H (5 U/μL), 94.5 μL of DNase and RNase-free water. Mix gently by flicking the tube.
4. Add 20 μL of gDNA to each tube. Resuspend gently by flicking the tube.
5. Incubate overnight at 37 °C with gentle agitation (Thermomixer 700 rpm).

RNase III digest to remove dsRNAs.

6. To each sample, add 1 μL 10× CutSmart buffer, 3 μL RNase III (2 U/μL), 6 μL DNase and RNase-free water. Mix gently.
7. Incubate at 37 °C for 1 h.
8. During the incubation, prepare enough MaXtract High Density 2 mL tubes by spinning them down to pellet the gel according to the manufacturer's recommendations.
9. Add 40 μL of DNase and RNase-free water. Mix gently.
10. Transfer the reactions to MaXtract High Density 2 mL tubes.
11. Under the fume hood, add 200 μL of phenol/chloroform/IAA (25:24:1). Mix thoroughly by inverting the tubes until the obtention of a homogenous white micelle. Let the tubes rest for 2 min and mix another 15 times or so.
12. Centrifuge under the fume hood at 13,000 × *g* for 10 min.
13. Collect the supernatant to a fresh DNA Lobind tube.

3.7 Precipitation of the Sheared gDNA

1. To each sample, add 20 μL 3 M Sodium acetate pH 5.2, 2 μL glycogen, 200 μL isopropanol.
2. Invert the tubes until their contents become homogeneous.
3. Store for 1 h at −20 °C.
4. Spin down for 30 min at ≥20,000 × *g* at 4 °C.
5. Discard the supernatant.
6. Add 500 μL of fresh 80% ethanol. Invert gently two or three times. Keep at RT for 5 min. Carefully discard the ethanol.
7. Repeat twice.

8. Carefully remove the excess ethanol and let dry ~20 min at RT.
9. Gently resuspend the pellets in 20 μL of 10 mM Tris by flicking the tube. Leave on ice for 1 h.

3.8 Check the Quality of the Shearing

1. Take 5 μL to a new tube.
2. Add 1 μL of RNase A (10 mg/mL) and incubate at 37 °C for 10–15 min.
3. Run the reaction on an agarose gel (1.2%, 70 V, 40 min) (*see* Fig. 1b, c).

3.9 gDNA Quantification by Quantitative PCR (qPCR)

1. Dilute the gDNA 1/500 (take 2 μL of gDNA into 998 μL of 10 mM Tris).
2. We perform qPCR reactions with 4 μL of diluted gDNA in 10 μL final volume. We quantify our qPCR signals against serial dilutions of highly purified genomic DNA whose concentration has been determined using a Nanodrop. We use two different sets of primers for corroboration, each amplifying two different unique loci (*see* Table in Subheading 2.3, **item 1**). This is to make sure that we always use the same ratio [μg of gDNA/μg of antibody] during the immunoprecipitation (*see* **Note 9**).

3.10 DNA:RNA Immunoprecipitation

3.10.1 Preparation of Antibody-Coupled Magnetic Beads

1. Resuspend Dynabeads Protein A by vortexing gently. No aggregates should remain in the bottle.
2. Transfer the required amount of beads to a lubricated Costar 1.7 mL tube. Use 60 μL of beads per sample.
3. Put tubes on a magnet for ~20″ or until all the beads are properly collected. Carefully remove the storage buffer from the beads.
4. Add 1 mL of PBS-T to wash. Invert the tube until the beads are completely resuspended.
5. Spin briefly to remove the beads that might be in the cap of the tube. Remove the supernatant.
6. Repeat the washes twice.
7. Add 5 volumes of PBS-T for one volume of beads. Resuspend the beads thoroughly.
8. Add the required amount of antibody (4 μg of S9.6 antibody for 15 μL of beads).
9. Rotate on wheel for 30 min at room temperature.
10. Spin briefly to pellet the beads and magnetize. Remove the supernatant completely.
11. Remove the tube from the magnet. Resuspend the beads by adding one volume of 1× IP buffer.

3.10.2 Immunoprecipitation and Washes

1. In a lubricated Costar tube, dilute 8 μg of *S. pombe* gDNA in 300 μl of 1× IP buffer. If calibration is required, add 1.6 μg of *S. cerevisiae* gDNA (*see* **Note 10**).
2. Transfer 15 μL of diluted DNA (5%) to a fresh DNA Lobind tube. This will be the Input. Keep at 4 °C.
3. Add 60 μL of antibody-coupled beads to the remaining 285 μL. Homogenize the beads in the sample by flicking the tube gently.
4. Rotate on wheel for 1 h 30 min at 4 °C.
5. Spin briefly to pellet the beads and transfer the tube on a magnet. Discard the supernatant.
6. Remove the tube from the magnet. Add 1 ml of 1× IP buffer. Invert the tube until the beads are well resuspended. Spin briefly (5 s) and transfer the tube on a magnet. Discard the supernatant.
7. Repeat the wash.
8. Add 1 mL of IP high salt buffer. Rotate on wheel for 5 min at RT.
9. Spin briefly to pellet the beads and transfer the tube on a magnet. Discard the supernatant.
10. Add 1 mL of IP Wash buffer. Rotate on wheel for 10 min at RT.
11. Spin briefly to pellet the beads and transfer the tube on a magnet. Discard the supernatant.
12. Add 1 mL of PBS-T.
13. Spin briefly to pellet the beads and transfer the tube on a magnet. Discard the supernatant.
14. Make sure that you remove all remaining traces of liquid. Spin briefly if necessary.

3.10.3 DNA Purification

1. During the washes above, prepare 10% Chelex resin in DNase and RNase-free water (*see* **Note 11**). Make sure to use a 2-mL tube to prepare the resin. Prepare 150 μL of resin per sample. Do not forget the tubes with the Input samples.
2. Invert the 2-mL tube constantly by hand to maintain the resin in suspension. Add 150 μL of resin to the tube containing the beads (IP) and 135 μL to the 15 μL of Input.
3. Incubate for 10 min at 98 °C in a thermomixer (agitation not necessary).
4. Add 1 μL of proteinase K 20 μg/μL to each sample.
5. Incubate for 45 min at 55 °C in a thermomixer with mild agitation (700–1000 rpm).

6. Vortex briefly and incubate for another 10 min at 98 °C.
7. Vortex briefly and spin down for 90″ at 11,000 × *g* at room temperature.
8. Carefully transfer 90–100 μL to a clean DNA Lobind tube. Make sure you do not transfer either any of the beads or the resin to the new tube.

3.10.4 Quantification of DRIP Signals by qPCR

1. Dilute 10 μL of your Input and IP samples in 40 μL of Tris 10 mM. Leave on ice for ~1 h before proceeding.
2. Proceed with the qPCR at candidate loci using 4 μL of the above 1/5 dilutions in 10 μL reactions. If you have used restriction enzymes to shear the gDNA, make sure that the amplicon of interest is void of restriction sites that you have used.

3.10.5 Strand-Specific Libraries

1. We recommend using the Accel-NGS™ 1S Plus DNA Library Kit according to the manufacturer's instructions.
2. A minimum of three independent IPs per genotype of interest should be used to make libraries.
3. Pool and precipitate the three IPs by mixing: 3 × 90 = 270 μL of immunoprecipitated DNA, 27 μL 3 M Sodium acetate pH 5.2, 2 μL glycogen, 270 μL isopropanol.
4. Mix well and store for ≥1 h at −20 °C.
5. Spin down for 30 min at 20,000 × *g* at 4 °C.
6. Discard the supernatant.
7. Add 500 μL of fresh 80% ethanol. Invert gently two or three times. Keep at RT for 5 min. Carefully remove the ethanol.
8. Repeat twice.
9. Carefully remove the excess ethanol and let dry ~20 min at RT.
10. Gently resuspend the pellets in 44 μL of 10 mM Tris by flicking the bottom of the tube. Leave on ice for 1 h.
11. Remove soluble RNAs by adding 1 μL of 10 mg/mL RNase A.
12. Incubate 15 min at 37 °C.
13. Purify using 81 μL of AMPure XP beads according to the manufacturer's instructions.
14. Elute in 30 μL of low-EDTA TE given with the Accel-1Splus kit.
15. Proceed with the library preparation according to the manufacturer's instructions. We have obtained stranded libraries for both *S. pombe* and *S. cerevisiae* (*see* Fig. 2).

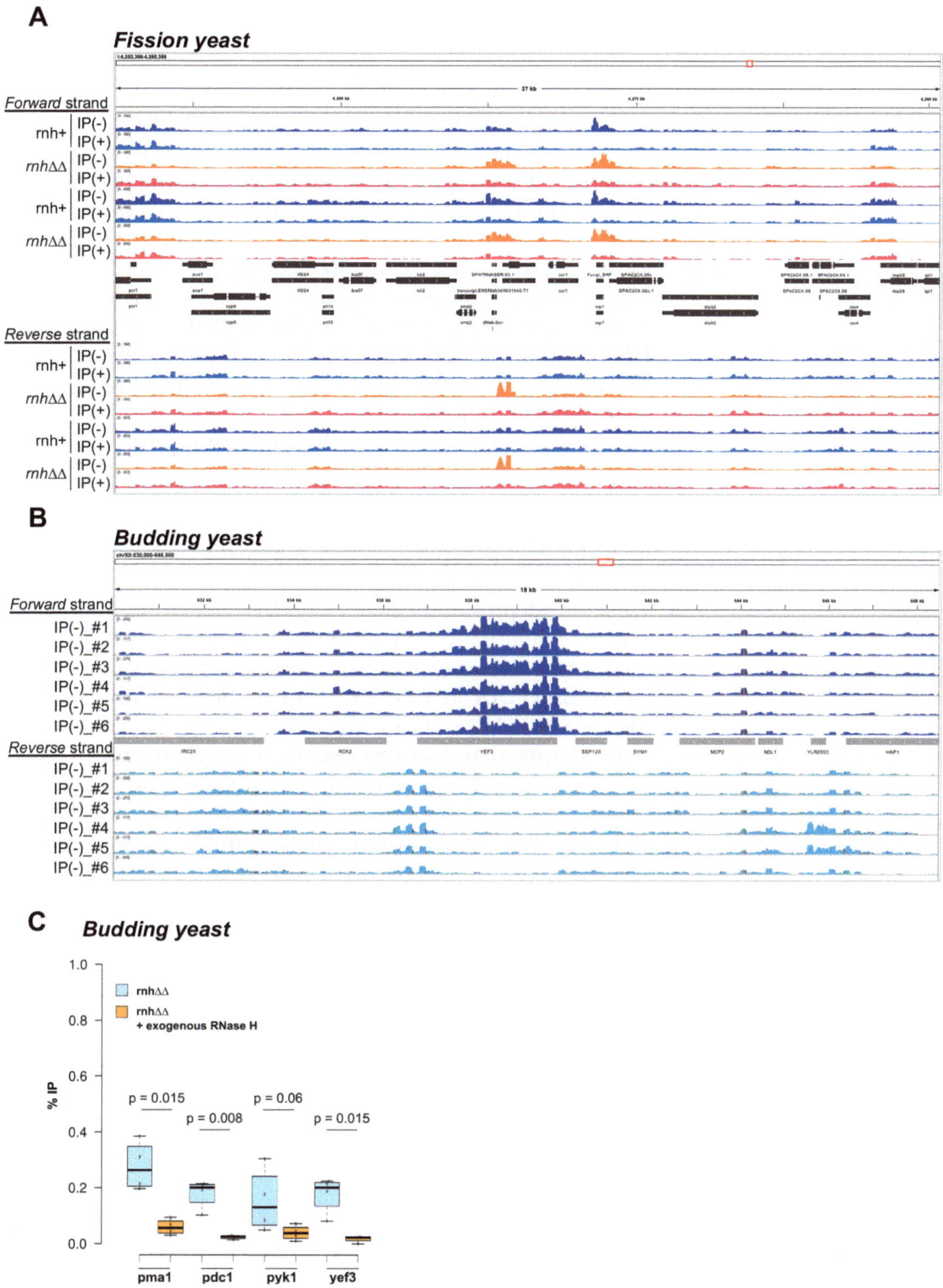

Fig. 2 ssDRIP-seq in fission and in budding yeasts. (**a**) Snapshot of DRIP results around *srp7* (*SPNCRNA.98*) in fission yeast for rnh + and *rnhΔΔ* cells. Two independent biological replicates are shown for untreated (IP(−)) or RNase H-treated (IP(+)) samples. The reads have been split according to their strand of origin. (**b**) Snapshot of DRIP results for *rnhΔΔ* budding yeast cells around *yef3* (*YLR249W*). Six independent biological replicates are shown for untreated (IP(−)) samples. The reads have been split according to their strand of origin. (**c**) DRIP

4 Notes

1. Pellets can be kept at −80 °C for several months.
2. The idea here would be to generate proper protoplasts completely void of cell walls that would be very easy to lyse. As reported previously [9], Zymolyase is not the best enzyme to generate protoplasts from fission yeast cells. Although we have not been able to test Lallzyme MMX as reported in [9], we have tested other lysing enzymes with mixed results (variable DRIP enrichments). As these alternative lysing enzymes are only partially purified, it might be that they carry some RNase activity. We think that this step can be improved further. The production of protoplasts from budding yeast with Zymolyase offers no difficulty.
3. The cell walls of fission yeast *rnh1Δrnh201Δ* take longer to digest than cell wall of rnh1 + rnh201+ cells.
4. We recommend that you follow the release of the genomic DNA from the ghosts of cells under the microscope using either Hoechst or DAPI.
5. We noticed previously that secondary structures within the extruded non-templated strand of R-loops were better preserved by successive chloroform extractions than by phenol/chloroform/IAA extractions, as quantified by Atomic Force Microscopy (AFM) [10]. We did not however carry out a rigorous quantitative assessment of the impact of phenol/chloroform/IAA extractions on the recovery of the DNA:RNA hybrid moiety of R-loops by the S9.6 antibody. As phenol/chloroform/IAA extractions are routinely used to purify R-loops from human cells [6], we decided to carry on with this strategy.
6. We noticed that DRIP enrichments (% IP) are stronger the greater the concentration of gDNA after extraction. This is true for both fission and budding yeast. We do not have a rational explanation for this observation.
7. Nuclease S1 digest was previously proposed to help R-loops resist sonication by digesting the extruded non-template strand [11]. We demonstrated previously that nuclease P1 was able to digest the extruded non-template strand of R-loops in vitro

Fig. 2 (continued) enrichments (% IP) after restriction digest of gDNA extracted from *rnhΔΔ* budding yeast cells. Five biological replicates were used. P-values (*p*) were obtained after performing a Wilcoxon–Mann Whitney test. Thick black lines show the medians; box limits indicate the 25th and 75th percentiles as determined by R software; whiskers extend 1.5 times the interquartile range from the 25th and 75th percentiles, outliers are represented by dots

[10] and we therefore compared DRIP-qPCR with and without nuclease P1 digest prior to sonication but found no significant difference in the DRIP enrichments. We suggest that a nuclease digest could help to improve the strand-specificity of the data but has no real impact on the quantitative recovery of RNA:DNA hybrids by the S9.6 antibody, at least in the sonication conditions that we have used here.

8. We have evidence that some R-loops are more sensitive to sonication than others. When looking at a particular locus of interest by DRIP-qPCR, we recommend that the best shearing method should be determined first.

9. We think that the gDNA-to-antibody ratio is very important to ensure reproducible DRIP enrichments. For spectrophotometer-based measurements of gDNA concentrations to be accurate, harsh extraction conditions unsuitable for the preservation of RNA:DNA hybrids are required. We found that qPCR measurements of gDNA concentrations against gDNA standards of known concentrations are much more accurate. When comparing different conditions side-by-side, we recommend to proceed with the experiment only if the different gDNA extractions yielded similar gDNA concentrations. Sanz and Chédin determined that the best ratio should be 4 μg of human gDNA for 10 μg of S9.6 antibody. We found that 8–10 μg of fission yeast gDNA for 16 μg of S9.6 antibody gives satisfactory results. As S9.6 can also recognize dsRNA [8, 12], this ratio might need to be optimized for different mutants, unless dsRNAs are systematically removed.

10. This ratio might still need to be optimized for calibration purposes.

11. We found that this approach initially described for Chromatin IP [13] is not only quicker than the alternatives but also best to reproducibly recover immunoprecipitated DNA.

References

1. Crossley MP, Bocek M, Cimprich KA (2019) R-loops as cellular regulators and genomic threats. Mol Cell 73:398–411
2. Crossley MP, Bocek MJ, Hamperl S, Swigut T, Cimprich KA (2020) qDRIP: a method to quantitatively assess RNA-DNA hybrid formation genome-wide. Nucleic Acids Res 48:e84
3. Malig M, Hartono SR, Giafaglione JM, Sanz LA, Chedin F (2020) Ultra-deep coverage single-molecule R-loop footprinting reveals principles of R-loop formation. J Mol Biol 432:2271–2288
4. Chédin F, Hartono SR, Sanz LA, Vanoosthuyse V (2021) Best practices for the visualization, mapping, and manipulation of R-loops. EMBO J 40:e106394
5. Castillo-Guzman D, Chédin F (2021) Defining R-loop classes and their contributions to genome instability. DNA Repair. https://doi.org/10.1016/j.dnarep.2021.103182
6. Sanz LA, Chédin F (2019) High-resolution, strand-specific R-loop mapping via S9.6-based DNA-RNA immunoprecipitation and high-throughput sequencing. Nat Protoc 14:1734–1755
7. Legros P, Malapert A, Niinuma S, Bernard P, Vanoosthuyse V (2014) RNA processing factors Swd2.2 and Sen1 antagonize RNA Pol

III-dependent transcription and the localization of condensin at Pol III genes. PLoS Genet 10:e1004794

8. Hartono SR, Malapert A, Legros P, Bernard P, Chédin F, Vanoosthuyse V (2018) The affinity of the S9.6 antibody for double-stranded RNAs impacts the accurate mapping of R-loops in fission yeast. J Mol Biol 430:272–284
9. Flor-Parra I, Zhurinsky J, Bernal M, Gallardo P, Daga RR (2014) A Lallzyme MMX-based rapid method for fission yeast protoplast preparation. Yeast 31:61–66
10. Carrasco-Salas Y, Malapert A, Sulthana S, Molcrette B, Chazot-Franguiadakis L, Bernard P, Chédin F, Faivre-Moskalenko C, Vanoosthuyse V (2019) The extruded non-template strand determines the architecture of R-loops. Nucleic Acids Res 47:6783–6795
11. Wahba L, Costantino L, Tan FJ, Zimmer A, Koshland D (2016) S1-DRIP-seq identifies high expression and polyA tracts as major contributors to R-loop formation. Genes Dev 30: 1327–1338
12. Crossley MP, Brickner JR, Song C, Zar SMT, Maw SS, Chédin F, Tsai M-S, Cimprich KA (2021) Catalytically inactive, purified RNase H1: a specific and sensitive probe for RNA-DNA hybrid imaging. J Cell Biol 220: e202101092
13. Nelson JD, Denisenko O, Bomsztyk K (2006) Protocol for the fast chromatin immunoprecipitation (ChIP) method. Nat Protoc 1:179–185

Chapter 28

Genome-Wide Analysis of DNA–RNA Hybrids in Yeast by DRIPc-Seq and DRIP-Seq

María García-Rubio, María E. Soler-Oliva, and Andrés Aguilera

Abstract

DNA–RNA hybrids are required for several natural processes in the cell, such as replication and transcription. However, the misregulation of its metabolism is an important source of genetic instability, a hallmark of diseases including cancer. For this reason, genome-wide detection of DNA–RNA hybrids is becoming essential to identify new factors that play a role in its formation or resolution and to understand the global changes in its dynamics because of genetic alterations or chemical treatments. Here, we describe two different immunoprecipitation-based procedures for the genome-wide profiling of DNA–RNA hybrids in the yeast *Saccharomyces cerevisiae*: DRIP-seq and DRIPc-seq.

Key words DNA–RNA hybrids, R-loops, Genome instability, S9.6 antibody, Spike-in, DRIP

1 Introduction

R-loops are highly stable structures formed when a co-transcriptionally synthesized RNA hybridizes with its DNA template leading to a DNA–RNA hybrid and a single stranded DNA (ssDNA) with identical sequence to the RNA forming the hybrid. The study of R-loops is essential to understand the variability of genomes and thus, defining the factors that promote or prevent the deleterious effects of DNA–RNA hybrid accumulation.

Given the clear link between DNA–RNA hybrids and genome instability several strategies have been developed from the past few years to detect them, which can be classified into four main categories depending on the manner the DNA–RNA hybrids are recognized: S9.6 antibody-based [1–5], bisulfite-based [6, 7], catalytically inactive RNase H-based [6, 8, 9], and the use of the hybrid-binding domain of the RNase H enzyme fused either to GFP (HB-GFP) [10] or to MNase [11] approaches. However, evidence supports that the first two groups of techniques offer a more detailed R loop profile (reviewed in [12]). Focusing on

Andrés Aguilera and Alexey Ruzov (eds.), *R-Loops: Methods and Protocols*, Methods in Molecular Biology, vol. 2528,
https://doi.org/10.1007/978-1-0716-2477-7_28,

S9.6-based techniques and with some modifications of our DRIP methods [13], here we provide the step-by-step protocol of DNA–RNA immunoprecipitation (DRIP) high-throughput sequencing of either the DNA (DRIP-seq) or the cDNA generated from the RNA (DRIPc-seq) in *Saccharomyces cerevisiae.* DRIP-c-seq methods permit the recovery of the RNA moiety of the DNA–RNA hybrids, providing information on strand specificity. The methodology is completed with a description of a "spike-in" normalization strategy for either DRIP-seq or DRIPc-seq experiments in *S. cerevisiae.* Finally, we describe the minimal bioinformatic tools and approaches used for analyzing sequencing profiles.

2 Materials

2.1 DNA–RNA Immunoprecipitation for Genome Wide in Yeast Cultures

1. Cell cultures of *Saccharomyces cerevisiae.*
2. Spike-in (*Schizosaccharomyces pombe*) (*see* **Note 1**).
3. Appropriate medium for *S. cerevisiae.*
4. YES medium (Yeast extract with supplements: Yeast extract 5 g/L, D-glucose 30 g/L, Adenine 0.05 g/L, L-Histidine 0.05 g/L, L-leucine 0.05 g/L, L-lysine 0.05 g/L, and Uracil 0.05 g/L).
5. Zymolyase 20T.
6. Spheroplasting Buffer: 1 M sorbitol, 2 mM Tris–HCl pH 8.0, 100 mM EDTA pH 8.0, 0.1% vol/vol beta-mercapto-ethanol, freshly made (*see* **Note 2**).
7. Solution I: 0.8 mM Guanidine-HCl, 30 mM Tris–HCl pH 8.0, 30 mM EDTA pH 8.0, 5% Tween 20, 0.5% Triton X-100.
8. RNase A Stock Solution: 10 mg/mL (*see* **Note 3**).
9. Proteinase K, recombinant (20 mg/mL).
10. Chloroform–Isoamyl alcohol (24:1).
11. Isopropanol at room temperature (RT).
12. 70% (vol/vol) ethanol at RT.
13. 1× TE Buffer: 10 mM Tris–HCl pH 7.6, 1 mM EDTA pH 8.0.
14. Restriction enzyme cocktail (*Hin*dIII 2.5 μL, *Eco*RI 2.5 μL, *Xba*I 2.5 μL, *Ssp*I 10 μL, *Bsr*GI 5 μL, BSA 2.5 μL) and NEB Buffer 2 (New England BioLabs).
15. RNase III (Invitrogen).
16. StrataClean Resin.
17. 50T5E Buffer: 50 mM Tris–HCl pH 7.5, 5 mM EDTA pH 8.0.
18. 50% (vol/vol) slurry of Sephadex G-50 in 50T5E buffer (kept at 4 °C).

19. Empty micro Bio-spin chromatography columns.
20. RNase H and appropriate 10× buffer.
21. Fluorometer apparatus and appropriate cuvettes.
22. Anti-DNA–RNA hybrid antibody (S9.6).
23. 1 M Na-Phosphate pH 7.0 (100 mL) (42.3 mL 1 M sodium dihydrogen phosphate monohydrate $NaH_2PO_4{\cdot}H_2O$, 57.7 mL 1 M disodium hydrogen phosphate Na_2HPO_4).
24. 10× Binding Buffer (100 mM $NaPO_4$ pH 7.0, 1.4 M NaCl, 0.5% Triton X-100, stored at RT).
25. Elution Buffer (50 mM Tris–HCl pH 8.0, 10 mM EDTA, 0.5% SDS).
26. Protein A and G-coated magnetic beads.
27. Magnetic rack.
28. Rotator mixer.
29. Thermomixer.
30. DNA Purification Kit.
31. Phenol–Chloroform–Isoamyl Alcohol 25:24:1.
32. Phase lock Gel Heavy tubes.
33. Glycogen, molecular biology grade.
34. 3 M sodium acetate (CH_3COONa).
35. Ice-cold 100% and 80% (vol/vol) ethanol.
36. RNase-free water.
37. iTaq Universal SYBR green Supermix.
38. Appropriate oligonucleotides (*see* **Note 4**).
39. ThruPlex DNA-seq kit (Rubicon Genomics).
40. NEBNext ultra II Directional RNA Library prep kit (New England Biolabs).
41. TURBO DNase (Invitrogen).
42. Fast Real-Time PCR System.
43. Bioanalyzer 2100.
44. Qubit fluorometer.

2.2 DRIP-seq and DRIPc-seq Data Alignment, Peak Calling, Annotation, and Visualization

1. UNIX operating system with at least 8 GB RAM.
2. FASTQ Toolkit V.1.0.0 https://www.illumina.com/
3. R https://www.r-project.org/.
4. SAMtools http://samtools.sourceforge.net/.
5. MACS2 https://pypi.org/project/MACS2/.
6. deepTools https://deeptools.readthedocs.io/en/develop/#.
7. IGV V 2.8.2 http://software.broadinstitute.org/software/igv/.

8. Rsubread https://bioconductor.org/packages/release/bioc/html/Rsubread.html.
9. ChIPpeakAnno https://bioconductor.org/packages/release/bioc/html/ChIPpeakAnno.html.
10. csaw: https://bioconductor.org/packages/release/bioc/html/csaw.html.
11. Data of DRIP-seq or DRIPc-seq reads (sample files) in FASTQ format.
12. Reference genome built in FASTA format.
13. Annotation file with the coordinates of all the genomic features in .BED format.

3 Methods

3.1 DNA–RNA Immunoprecipitation in Yeast Cultures

1. Grow 400 mL yeast cultures in the appropriate medium and temperature to an absorbance at 600 nm of about 0.8. For spike-in control grow 100 mL *S. pombe* culture in YES medium to an absorbance at 600 nm of about 0.7.
2. Harvest cells by centrifugation at 3000 × *g*, 4 °C for 5 min.
3. Wash cells twice with 20 mL cold water.
4. Harvest cells by centrifugation at 3000 × *g*, 4 °C for 5 min.
5. Resuspend the cell pellet in 8 mL Spheroplasting Buffer (freshly made) (add 2 mg/mL Zymolyase to Spheroplasting Buffer). Split the sample in eight tubes (2 mL). The spike-in control should be resuspended in 2 mL and split into two tubes (*see* **Note 5**).
6. Incubate at 30 °C for 30 min with continuous gentle agitation (40–60 × *g*).
7. Spin down for 5 min at 4500 × *g* in a microcentrifuge and discard the supernatant.
8. Carefully wash the spheroplast pellet by adding 500 μL of cold water without resuspending.
9. Spin briefly and discard the water.
10. Break the spheroplasts by resuspending in 565 μL Solution I (group the samples two by two until we are left with four samples). Resuspend the spike-in control in the same volume of Solution I and group in one tube (*see* **Note 6**).
11. Add 40 μL RNase A.
12. Incubate for 30 min at 37 °C.
13. Add 75 μL Proteinase K (20 mg/mL) and incubate at 50 °C for 60 min (*see* **Note 7**).

14. Spin down for 5 min at 4500 × *g* and transfer supernatant to a new tube.
15. Add 800 μL Chloroform–Isoamyl alcohol (*see* **Note 8**).
16. Spin down for 10 min at 4500 × *g*.
17. Carefully collect the upper phase with 1 mL cut tips and transfer into a 2 mL tube.
18. Add 800 μL isopropanol to the aqueous phase and invert several times until genomic DNA precipitates. Spool DNA on a glass rod (*see* **Note 9**).
19. Wash DNA with 70% ethanol by allowing ethanol to run down the glass rod.
20. Allow to air dry, break off the tip with DNA, and put it in a 1.5 mL tube with 150 μL 1× TE. Do not resuspend the DNA by over-pipetting/vortexing. Mix several times by gentle shaking (*see* **Note 10**).
21. Digest the DNA overnight using cocktail restriction enzymes according to suppliers instructions and with 2 μL RNase III per tube (add 100 mM DTT, final volume is 250 μL) (*see* **Note 11**).
22. Add 3 μL StrataClean Resin, pipet with cut tips, and spin down at full speed in a microcentrifuge for 1 min (*see* **Note 12**).
23. Purify over Sephadex G-50 column (in 50T5E buffer) (*see* **Note 13**).
24. Split the flow-through containing DNA into two new tubes (125 μL). Measure the concentration (OD_{260}) of the spike-in control DNA, and transfer 5 μL aliquots into microcentrifuge tubes and store at −80 °C.
25. Treat half of the DNA (four tubes) with 8 μL RNase H (NEB) overnight at 37 °C. In the meantime measure the concentration of the other half (four tubes) and keep at 4 °C.
26. For each tube (at this point there are 8 tubes, 4 untreated and 4 treated with RNaseH) wash together 67 μL of Protein A and 33 μL of Protein G magnetic beads with 700 μL of 1× Binding Buffer (10× diluted in 1× TE) by inverting the tubes. Place the precooled magnetic rack on ice, insert the tubes, aspirate all liquid off, and repeat this step once. Resuspend in 100 μL of 1× Binding Buffer. Add 8 μL of anti-DNA–RNA hybrid antibody (stock 1 mg/mL). Incubate overnight at 4 °C on a rotator mixer.
27. Collect samples from **step 25** and repeat **steps 22** and **23**. Measure the concentration of the DNA.
28. Take 5 μL of each digested DNA (treated and untreated with RNase H) (INPUT), rejoin to have 20 μL (20 μL treated and

20 μL untreated with RNase H), and process immediately as described in **steps 32** and **33**.

29. Wash beads from **step 26** twice with 1× Binding Buffer to eliminate the excess of antibody and resuspend in 100 μL of 1× Binding Buffer (initial beads volume, 67 μL of Protein A and 33 μL of Protein G). Add 8 μg of DNA in 350 μL 1× TE from **steps 25** and **27** to the washed beads, 80 ng of Spike-in, and 50 μL of 10× Binding Buffer.
30. Incubate all samples (4 treated and 4 untreated with RNase H) for 2–3 h at 4 °C on a rotator mixer.
31. Place a precooled magnetic rack on ice, insert the tubes from **step 30**, aspirate all liquid off, and resuspend the beads in 1 mL of 1× Binding Buffer. Repeat the same procedure to complete 3 washes with 1× Binding Buffer.
32. Add 250 μL Elution Buffer and 7.5 μL Proteinase K and incubate at 55 °C 45 min in a shaker. Add 27 μL 1× TE to INPUT samples (**step 27**), 3 μL Proteinase K, and incubate at 55 °C 45 min.
33. Purify INPUT over DNA purification columns according to manufacturer protocol and elute in 100 μL 1× TE. Store samples at −20 °C to use for later qPCR control.
34. Spin a 2 mL phase lock gel heavy tube for each immunoprecipitated sample (8 tubes) to pellet the gel (*see* **Note 14**). Pipette the DNA from **step 32** into a phase lock tube. Add 50 μL of water and 300 μL of Phenol–Chloroform–Isoamyl Alcohol (25:24:1). Gently invert the tube 10 times and spin down at 16,000 × *g* (*see* **Note 15**).
35. Pour the DNA from previous step (top aqueous phase) into a clean 1.5 mL tube and mix with 1.5 μL of glycogen, 1/10 vol of 3 M sodium acetate (30 μL), and 2.5 vol of 100% ethanol (750 μL). Invert tubes several times and incubated for at least 1 h at −20 °C to increase the precipitation yields.
36. Spin down at 16,000 × *g* for 30 min at 4 °C. Discard the supernatant and add 200 μL of 80% ethanol. Spin down at 16,000 × *g* for 5 min.
37. Air dry the pellet and resuspend in 30 μL of RNase-free water.
38. Take 5 μL of each immunoprecipitated DNA (4 treated and 4 untreated with RNase H) and rejoin to have 20 μL treated and 20 μL untreated with RNase H, add 80 μL of water. Check the DRIP efficiency on positive regions by real-time quantitative PCR (qPCR). The enrichment for each qPCR of interest is normalized with respect to the corresponding ratios of the INPUT (**step 33**). The immunoprecipitated DNA left (25 μL × 4) can be stored at −80 °C until it is ensured that the DRIP procedure worked.

39. For DRIP-seq combine the four tubes into one (100 μL) and go directly to **step 45**.
40. For DRIPc-seq treat samples with DNase enzyme (*see* **Note 16**). Mix DNA from **step 36** (25 μL) with 3 μL of 10× DNase buffer and 2 μL of DNase I. Incubate the mixture at 37 °C for 45 min.
41. Add 1 μL of 0.5 M EDTA pH 8 and incubated for 15 min at 75 °C to inactivate the DNase enzyme.
42. Combine the four tubes into one (120 μL). Add 2 μL of glycogen, 12 μL of 3 M sodium acetate pH 5.2, and 300 μL of ice-cold 100% ethanol. Invert several times and incubate at least for 1 h at −20 °C.
43. Spin down at 16,000 × *g* for 30 min at 4 °C. Discard the supernatant and add 200 μL of 70% ethanol. Spin down at 16,000 × *g* for 5 min.
44. Air dry the pellet and resuspend in 15 μL of RNase-free water.
45. Check the quantity (with Qubit Fluorometer), quality, and fragment size (with Bioanalyzer 2100) of the samples. Size range is usually from 25 to 200 bp.
46. Proceed to library construction using preferably the ThruPlex DNA-seq kit from Rubicon Genomics for DRIP-seq samples, and NEBNext ultra II Directional RNA Library prep kit for Illumina for DRIPc-seq samples (after fragmentation step). The latter allows us to distinguish between Watson and Crick strands.
47. Check quantity and size of the libraries using again Qubit fluorometric assay and Bioanalyzer 2100 system. Library size and concentration should be about 250 bp and > 1 ng/μL, respectively.
48. Sequence the samples. Depending on the sequencer used, you can pool several samples in one sequencing lane, considering that you need at least 20 million reads per sample. Paired-end sequencing is highly recommended, as this method increases mapping efficiency, especially in repetitive regions. A sequencing read length of at least 2 × 75 bp is advised.

3.2 DRIP-seq and DRIPc-seq Data Alignment, Peak Calling, Annotation, and Visualization

1. Check quality of the reads with any standard software such as FASTQ Toolkit (Illumina Basespace apps). Apart from the adaptors trimming, no preprocessing or read-filtering is usually necessary since low-quality reads will be filtered during the mapping step (*see* **Note 17**).
2. Align samples to the reference genome. This step can be performed easily using Rsubread software [14]. First, use *buildindex* command to create an index of the reference genome,

which should be in FASTA format. Use sacCer3 genome assembly as reference genome (*see* **Notes 18** and **19**).

```
buildindex(basename="my_index", reference="saccer3.fa")
```

This will output a file with .fa.fai extension. Then, use *align* command to map the reads to the genome. For that, use the index, the FASTQ files (R1 and R2, from paired-end sequencing) and the name for the output bam file as inputs. Then, run the command with most of the parameters set by default, but specifying the sample type ("dna" for DRIP-seq samples and "rna" for DRIPc-seq samples), the imput format ("FASTQ" in both cases), and the number of threads available ("nthreads" parameter) to speed up the calculations, if possible. Moreover, set the "unique" parameter as TRUE to keep the reads only if they have been mapped to a single region:

```
align(index="my_index", readfile1 = fastq_files_read1 , read-
file2 fastq_files_read2, type="rna", input_format="FASTQ",
output_file=bam_files , unique = TRUE, nthreads = 17 )
```

3. Process BAM files. Use SAMtools [15] to mark duplicated reads (*see* **Note 20**), split reads in Watson and Crick strands (in the case of DRIPc-seq data), and index BAM files. To mark duplicated reads, use *collate*, *fixmate*, *sort*, and *markdup* commands:

```
samtools collate -o aligned_sample_collate.bam aligned_sam-
ple.bam
samtools fixmate -m aligned_sample_collate.bam aligned_sam-
ple_fixmate.bam
samtools sort -o aligned_sample_sort.bam aligned_sample_fix-
mate.bam
samtools markdup aligned_sample_sort.bam aligned_sample_-
markdup.bam
```

To separate the reads that originate on each strand, use *view* command. To get the file for transcripts that originated from the Watson strand:

```
samtools view -b -f 128 -F 16 aligned_sample_markdup.bam >
watson_1.bam.
samtools view -b -f 80 aligned_sample_markdup.bam >
watson_2.bam.
samtools merge -f watson.bam watson_1.bam watson_2.bam.
```

And for the Crick strand file:

```
samtools view -b -f 144 aligned_sample_markdup.bam >
crick_1.bam.
samtools view -b -f 64 -F 16 aligned_sample_markdup.bam >
crick_2.bam.
samtools merge -f crick.bam crick_1.bam crick_2.bam.
```

Finally, use *index* command to index BAM files:

```
samtools index aligned_sample_markdup.bam.
```

4. Determine the regions that have been enriched in aligned reads because of S9.6 immunoprecipitation by a peak calling approach. MACS2 [16] is a standard package that allows both DRIP-seq and DRIPc-seq peak detection in a suitable manner. Run the program specifying some parameters: the file format (-f), especially if you are dealing with paired-end reads; an effective genome size of 1.2×10^7 bp (-g); the prefix used to name the output files (-n) and the minimum length of a peak to be considered (--min-length) that should be at least 100 bp taking into account R loop average length [17].

```
macs2 callpeak -t aligned_sample.bam -f BAMPE -g 12000000 -n
aligned_sample --min-length 100
```

5. Annotate the detected peaks with specific genomic features with ChIPpeakAnno Bioconductor software package [18] (*see* **Note 21**). Use the .BED file with the peaks coordinates (MACS2 output) and an annotation file in .BED as input files (*see* **Note 22**). Convert both .BED files (peaks and annotation files) in GRanges objects with *toGRanges* command, specifying the format. Then, use *annotatePeakInBatch* command to link the peaks with the features. As intergenic regions in *S. cerevisiae* genome are quite small, to take into account only the features that overlap with the peaks (output = c("overlapping")) is recommended, but also considering the multiple features that can overlap a peak simultaneously (select = "all"). Besides, set the parameter ignore.strand as TRUE in order to contemplate the possibility of antisense DNA–RNA hybrid formation:

```
peaks <- toGRanges(peaks_table, format="BED")
anno <- toGRanges(anno_table, format="BED")
peaks_annotation <- annotatePeakInBatch(peaks, AnnotationData
= anno, output=c("overlapping"), select="all", ignore.strand =
TRUE)
```

6. Generate coverage files in order to visualize the data. First of all, to compare your samples properly, determine the appropriate normalization factors for them. This calculation, regardless

of whether you have used spike-in or not (*see* **Note 23**), can be performed using standard statistical analysis packages such as csaw [19], DiffBind [20], or DEseq [21]. Once normalization factors have been determined, use *bamCoverage* command from deepTools software [22] to create coverage files for each sample. Use indexed (and splitted in strands, if samples come from a DRIPc-seq experiment) BAM files as input (-b). Set the name of the output file (-o) and its format, which could be either bigwig (-of bigwig) or bedgraph (-of bedgraph). A bin size (-bs) of ten is recommended. Set the scale factor (--scale-Factor) according to the normalization factor calculated for each sample. Finally, calculations could be sped up using all available processors (-p max):

```
bamCoverage -b aligned_sample.bam -o coverage_sample.bw -of
bigwig -bs 10 -scaleFactor x
```

The output files could be loaded into Integrative Genome Viewer (IGV) software [23] to visualize the data. Figure 1a is an example of a typical DRIPc-seq signal profile over chromosome III in samples treated or not with RNase H (Fig. 1a) (*see* **Note 24**). In Fig. 1b there is also an example of a representative genomic region showing the usual DRIPc-seq signal profile.

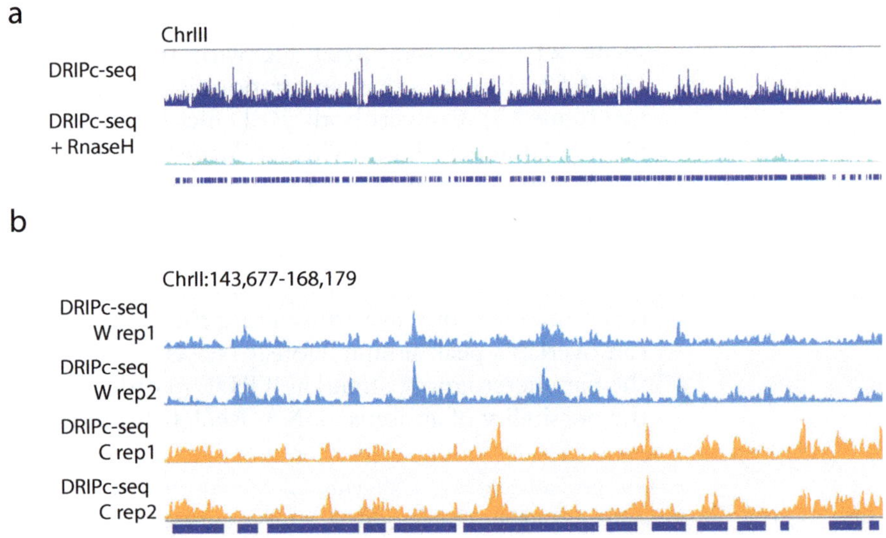

Fig. 1 DRIP-seq and DRIPc-seq data visualization. (**a**) Representative screenshot of a chromosome showing the DRIPc-seq signal of cells treated (light blue) or not (dark blue) with RNase H. (**b**) Representative screenshot of a genomic region showing the DRIPc-seq signal of two independent replicates splitted in Watson (W, blue) and Crick (C, yellow) strands

7. Create metaplots and heatmaps in order to visualize the data through regions of interest (*see* **Note 25**). First, use *computeMatrix* command to generate a matrix containing the signal value of each bin of all the regions considered. Signal could be computed either scaling all the regions to a specific length (scale-regions mode) or relative to a reference point (referent-point mode). Use a .BED file containing the regions of interest (-R) and from one to several bigwig files (-S) as input files. A bin size (-bs) of ten is recommended. Other parameters that should be set depending on your needs are the region body length (-m, scale-regions mode), the reference point (--referencePoint), distance upstream and downstream from the region or the point to be plotted (-a and -b, respectively), and a maximum threshold to remove possible outliers (--maxThreshold):

```
computeMatrix scale-regions -R rloop_enriched_genes.bed -S
control.bw -bs 10 -m 8000 -a 1000 -b 1000 --maxThreshold 100000
-o rloop_enriched_genes_matrix
```

```
computeMatrix reference-point -R rloop_peaks -S mutant.bw -bs
10 -a 10000 -b 10000 --referencePoint center --maxThreshold
100000 -o rloop_peaks_matrix
```

Then, use *plotProfile* and/or *plotHeatmap* commands to see the plots using the previous matrix as an input (-m):

```
plotProfile -m rloop_enriched_genes_matrix -o rloop_enrich-
ed_genes_matrix.pdf
plotHeatmap -m rloop_enriched_genes_matrix -o rloop_enrich-
ed_genes_matrix.pdf
```

The output of these two functions is represented in Fig. 2.

4 Notes

1. We utilize *S. pombe* genomic DNA as a spike-in. We recommend first preparing the DNA to be used as a spike-in and storing aliquots at −80 °C for later use in several experiments.
2. Wear gloves and a lab coat while working with beta-mercaptoethanol, because it is a very dangerous compound. Do it in the hood.
3. The RNase A should be carefully boiled to eliminate residual DNase activity.
4. We use a primer application software to design the oligonucleotides for qPCR.

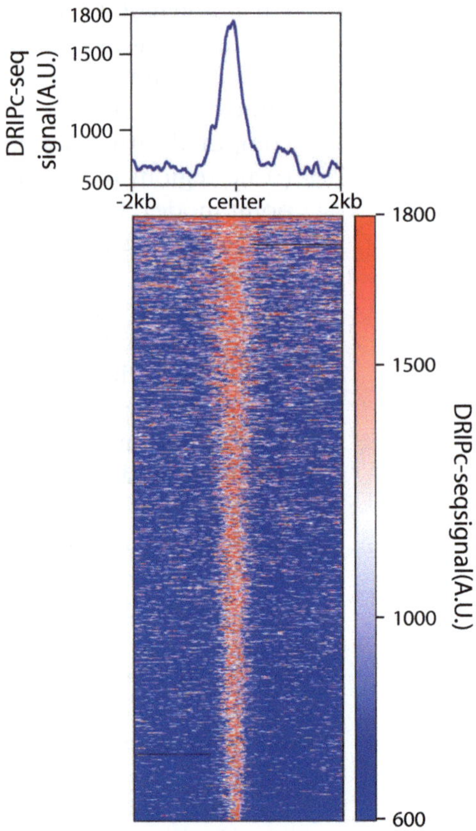

Fig. 2 DRIP-seq and DRIPc-seq detected peaks visualization. DRIPc-seq signal (average coverage, arbitrary units) of ±2 kb DRIPc-seq detected peaks, represented by metaplot (top) and heatmap (bottom)

5. Always pipet DNA with cut tips to preserve DNA–RNA hybrids. Do not attempt to resuspend DNA by over-pipetting/vortexing.
6. Spheroplasts are very difficult to resuspend at this step. Use 1 mL cut tips. Better processing groups of two or three samples.
7. At this point if there are not many clamps in the samples you can proceed further with the extraction. If the preps are "dirty," you can incubate overnight at 30 °C.
8. Do not vortex. Mix several times.
9. Use a glass Pasteur pipette to spool the DNA. Do not spin to precipitate the DNA. Centrifugation results in high loads of RNA contamination.
10. Eventually the DNA could be resuspended at 30 °C (1–2 h). Invert the tube occasionally.

11. Use 10× RNase III reaction buffer for the restriction. Alternatively, you can use buffer 2.1 (NEB), 1 mM DTT. You do not need to remove the cut glass rod. We often run 4 μL on gel the next day to ensure the digestion is complete and the samples are free of RNA.
12. Better processing groups of two or three samples.
13. To prepare Sephadex G-50 columns, place empty micro Biospin chromatography columns in 2 mL tubes, add 800 μL of Sephadex G-50 slurry, and centrifuge at 2000 × *g* for 2 min in a microcentrifuge. The Sephadex G-50 columns are then placed in new 1.5 mL tubes, which serve to collect the purified DNA after centrifugation at 2000 × *g* for 2 min in a microcentrifuge.
14. This step is extremely important. Phase lock gel heavy 2 mL tubes can be used to facilitate sample phenol treatment.
15. Wear gloves and a lab coat while working with phenol, because it is a very dangerous compound. Do it in the hood. Normally, the aqueous phase forms the upper phase. You should always dispose of phenol waste in a specially sealed container and ensure that it is eliminated according to your institution's policies for dangerous waste.
16. From this step take precaution to work with RNase-free buffers and reagents. Use all proper personal protective equipment.
17. Details about the output and the different metrics analyzed by FASTQ Toolkit can be found on its website (https://emea.illumina.com/products/by-type/informatics-products/basespace-sequence-hub/apps/fastq-toolkit.html).
18. If the samples contain *S. pombe* DNA as spike-in, align them to a combined version of *S. cerevisiae* (sacCer3) and *S. pombe* (ASM294v2) reference genomes.
19. Reference genomes can be downloaded from UCSC Genome Browser [24] website (http://genome.ucsc.edu).
20. Marking duplicated reads is preferable to removing them, as you can decide whether or not to take them into account during the differential enrichment analysis.
21. Different annotation packages such as ChIPSeeker [25] can be considered.
22. Annotation file can be generated using UCSC Table Browser tool [26].
23. Spike-in normalization is only recommended when a massive enrichment is highly probable, otherwise will add noise to the experiment.
24. As it is known that S9.6 antibody can also recognize dsRNA, and even considering that samples are treated with RNase A and RNase III before immunoprecipitation, sequencing of

RNase H treated samples is highly recommended to check that the detected signal corresponds to real DNA–RNA hybrids. The decay of the signal in the RNase H treated sample compared to the untreated should be global and substantial, as in Fig. 1a.

25. Metaplots and heatmaps are two types of representation very useful to observe the average of the signal along any region you could be interested in: a specific genomic feature, R loop detected peaks, or R loop enriched genes, for instance.

Acknowledgments

We thank Eloisa Andújar and Mónica Pérez (CABIMER's Genomic Unit) for DNA sequencing support. Research in AA's lab is funded by Spanish Ministry of Economy and Competitiveness, European Research Council, and the European Union (FEDER).

References

1. Chan YA, Aristizabal MJ, Lu PY, Luo Z, Hamza A, Kobor MS et al (2014) Genome-wide profiling of yeast DNA:RNA hybrid prone sites with DRIP-chip. PLoS Genet 10: e1004288
2. El Hage A, Webb S, Kerr A, Tollervey D (2014) Genome-wide distribution of RNA-DNA hybrids identifies RNase H targets in tRNA genes, retrotransposons and mitochondria. PLoS Genet 10:e1004716
3. Wahba L, Costantino L, Tan FJ, Zimmer A, Koshland D (2016) S1-DRIP-seq identifies high expression and polyA tracts as major contributors to R-loop formation. Genes Dev 30(11):1327–1338. https://doi.org/10.1101/gad.280834.116
4. Sanz LA, Chédin F (2019) High-resolution, strand-specific R-loop mapping via S9.6-based DNA-RNA immunoprecipitation and high-throughput sequencing. Nat Protoc 14(6): 1734–1755. https://doi.org/10.1038/s41596-019-0159-1
5. Crossley MP, Bocek MJ, Hamperl S, Swigut T, Cimprich KA (2020) qDRIP: a method to quantitatively assess RNA-DNA hybrid formation genome-wide. Nucleic Acids Res 48(14): e84. https://doi.org/10.1093/nar/gkaa500
6. Yu K, Chedin F, Hsieh CL, Wilson TE, Lieber MR (2003) R-loops at immunoglobulin class switch regions in the chromosomes of stimulated B cells. Nat Immunol 4(5):442–451. https://doi.org/10.1038/ni919
7. Malig M, Hartono SR, Giafaglione JM, Sanz LA, Chedin F (2020) Ultra-deep coverage single-molecule R-loop footprinting reveals principles of R-loop formation. J Mol Biol 432:2271–2288
8. Ginno PA, Lott PL, Christensen HC, Korf I, Chédin F (2012) R-loop formation is a distinctive characteristic of unmethylated human CpG island promoters. Mol Cell 45(6):814–825. https://doi.org/10.1016/j.molcel.2012.01.017
9. Huang FT, Yu K, Hsieh CL, Lieber MR (2006) Downstream boundary of chromosomal R-loops at murines witch regions: implications for the mechanism of class switch recombination. Proc Natl Acad Sci U S A 103:5030–5035
10. Bhatia V, Herrera-Moyano E, Aguilera A, Gomez-Gonzalez B (2017) The role of replication-associated repair factors on R-loops. Genes 8:171
11. Yan Q, Shields EJ, Bonasio R, Sarma K (2019) Mapping native R-loops genome-wide using a targeted nuclease approach. Cell Rep 29:1369–1380
12. Chédin F, Hartono SR, Sanz LA, Vanoosthuyse V (2021) Best practices for the visualization, mapping and manipulation of R-loops. EMBO J 40(4):e106394. https://doi.org/10.15252/embj.2020106394
13. García-Rubio M, Barroso SI, Aguilera A (2018) Detection of DNA-RNA hybrids in vivo. Methods Mol Biol 1672:347–361

14. Liao Y, Smyth GK, Shi W (2019) The R package Rsubread is easier, faster, cheaper and better for alignment and quantification of RNA sequencing reads. *Nucleic Acids Res* 47(8):e47
15. Li H, Handsaker B, Wysoker A, Fennell T, Ruan J, Homer N, Marth G, Abecasis G, Durbin R, 1000 Genome Project Data Processing Subgroup (2009) The Sequence alignment/map format and SAMtools. Bioinformatics 25(16):2078–2079
16. Zhang Y, Liu T, Meyer CA, Eeckhoute J, Johnson DS, Bernstein BE, Nusbaum C, Myers RM, Brown M, Li W, Liu XS (2008) Model-based analysis of ChIP-Seq (MACS). Genome Biol 9(9):1–9
17. García-Pichardo D, Cañas JC, García-Rubio ML, Gómez-González B, Rondón AG, Aguilera A (2017) Histone mutants separate R loop formation from genome instability induction. Mol Cell 66(5):597–609
18. Zhu L, Gazin C, Lawson N, Pagès H, Lin S, Lapointe D, Green M (2010) ChIPpeakAnno: a Bioconductor package to annotate ChIP-seq and ChIP-chip data. BMC Bioinformatics 11(1):1–10
19. Lun AT, Smyth GK (2016) csaw: a Bioconductor package for differential binding analysis of ChIP-seq data using sliding windows. Nucleic Acids Res 44(5):e45
20. Ross-Innes CS, Stark R, Teschendorff AE, Holmes KA, Ali HR, Dunning MJ, Brown GD, Gojis O, Ellis IO, Green AR, Ali S, Chin S, Palmieri C, Caldas C, Carroll JS (2012) Differential oestrogen receptor binding is associated with clinical outcome in breast cancer. Nature 481(7381):389–393
21. Love MI, Huber W, Anders S (2014) Moderated estimation of fold change and dispersion for RNA-seq data with DESeq2. Genome Biol 15(12):21
22. Ramírez F, Dündar F, Diehl S, Grüning BA, Manke T (2014) deepTools: a flexible platform for exploring deep-sequencing data. Nucleic Acids Res 42(W1):W187–W191
23. Robinson JT, Thorvaldsdóttir H, Winckler W, Guttman M, Lander ES, Getz G, Mesirov JP (2011) Integrative genomics viewer. Nat Biotechnol 29(1):24–26
24. Kent WJ, Sugnet CW, Furey TS, Roskin KM, Pringle TH, Zahler AM, Haussler D (2002) The human genome browser at UCSC. Genome Res 12(6):996–1006
25. Yu G, Wang L, He Q (2015) ChIPseeker: an R/Bioconductor package for ChIP peak annotation, comparison and visualization. Bioinformatics 31(14):2382–2383
26. Karolchik D, Hinrichs AS, Furey TS, Roskin KM, Sugnet CW, Haussler D, Kent WJ (2004) The UCSC Table Browser data retrieval tool. Nucleic Acids Res 32(suppl_1):D493–D496

Chapter 29

Quantitative, Convenient, and Efficient Genome-Wide R-Loop Profiling by ssDRIP-Seq in Multiple Organisms

Wei Xu, Kuan Li, Qin Li, Shuai Li, Jincong Zhou, and Qianwen Sun

Abstract

R-loop is a three-stranded chromatin structure, comprising one single-stranded DNA and another DNA: RNA hybrid strand, plays various and essential biological functions in many organisms. Developing a precise, efficient, faithful, and unbiased genome-wide R-loop detection method with extensive adaptability in all organisms is at the top priority for R-loop biology. Here, we provide a straightforward and highly efficient protocol for genome-wide strand-specific R-loop profiling in various organisms. In brief, genomic DNA is extracted and fragmented by the cocktail of restriction enzymes, and then the DNA:RNA hybrids are immunoprecipitated, following by the single-stranded DNA adaptor ligation and next-generation sequencing (named as ssDRIP-seq). Coupling with a straightforward and step-by-step bioinformatic pipeline, this method can provide high resolution and comprehensive strand-specific information for R-loop formation. ssDRIP-seq has been successfully applied for detecting R-loops from prokaryotes such as *E. coli*, to eukaryotes such as *S. cerevisiae*, mammalian cell culture and tissues, as well as plants *Arabidopsis* and rice, with high reproducibility and sensitivity.

Key words R-loop, Genome, Next-generation sequencing, Nucleic acid, Noncoding RNA

1 Introduction

R-loop is a common chromatin feature of eukaryotes and prokaryotes and plays key roles in various processes, including chromatin modifications, transcriptional regulation and termination, DNA replication, DNA damage, and genome stability [1–3]. ssDRIP-seq method can be used to map the strand-specific genome-wide R-loop landscape in any organism [4–7] (Figs. 1 and 2; Table 1). Moreover, the single-strand DNA ligation-based library preparation method can also be coupled with other immunoprecipitation experiments followed by sequencing, and this had been applied in nascent DNA sequencing [10, 11]. Therefore, this method is not

Wei Xu and Kuan Li contributed equally with all other contributors.

Andrés Aguilera and Alexey Ruzov (eds.), *R-Loops: Methods and Protocols*, Methods in Molecular Biology, vol. 2528,
https://doi.org/10.1007/978-1-0716-2477-7_29,

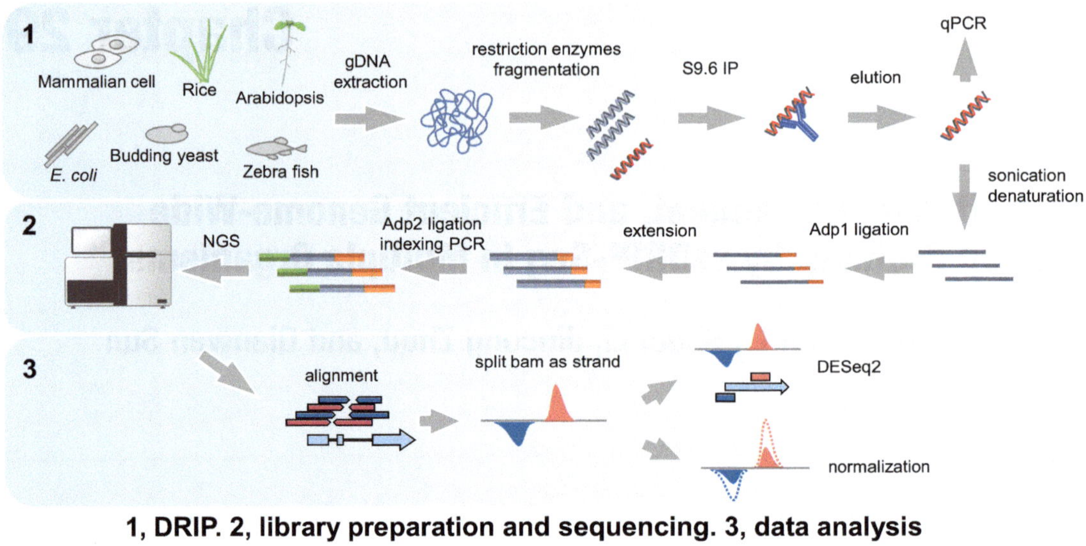

Fig. 1 The workflow of ssDRIP-seq for genome-wide R-loop profiling

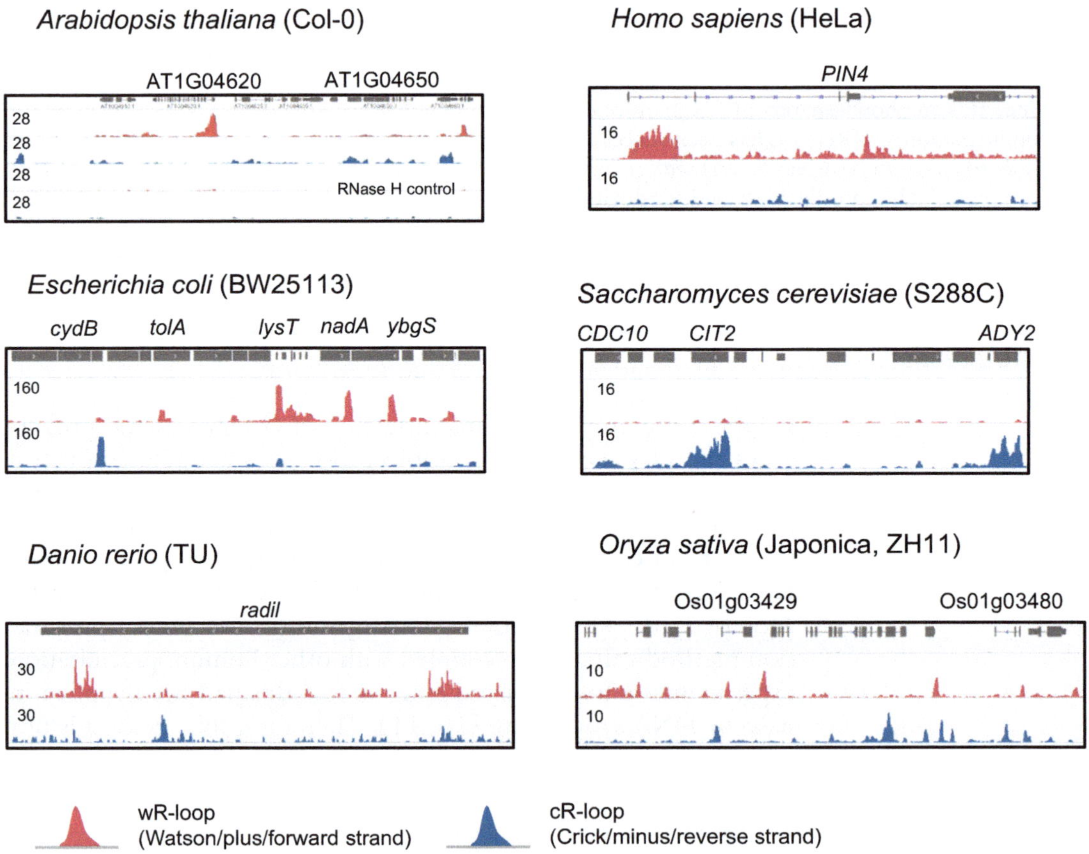

Fig. 2 Snapshots of ssDRIP-seq signals in multiple organisms

Table 1
Papers that rely on ssDRIP-seq for genome-wide R-loop detection

Material	Title	Journal	DOI
Arabidopsis thaliana	The R-loop is a common chromatin feature of the Arabidopsis genome [4]	Nature Plants	https://doi.org/10.1038/s41477-017-0004-x
	The R-loop atlas of Arabidopsis development and responses to environmental stimuli [8]	The Plant Cell	https://doi.org/10.1105/tpc.19.00802
	ALBA protein complex reads genic R-loops to maintain genome stability in Arabidopsis [5]	Science Advances	https://doi.org/10.1126/sciadv.aav9040
Homo sapiens	m^6A promotes R-loop formation to facilitate transcription termination [9]	Cell Research	https://doi.org/10.1038/s41422-019-0235-7
Mus musculus	R-loops Coordinate with SOX2 in Regulating Reprogramming to Pluripotency [6]	Science Advances	https://doi.org/10.1126/sciadv.aba0777
Oryza sativa	The landscape of promoter-centred RNA–DNA interactions in rice	Nature Plants	https://doi.org/10.1038/s41477-021-01089-4

only interesting to researchers engaging in R-loops, but can also be applied to other research fields, making the method suitable to a wide range of research domains.

Compared with other published R-loop mapping methods, including DRIP-seq [12], DRIPc-seq [13], S1-DRIP-seq [14], bis-DRIP-seq [15], R-ChIP [16], and R-loop CUT & Tag [17], there are several unique features of ssDRIP-seq (Fig. 3). Briefly, (1) Adaptase-based ligation and library preparation recognizes ssDNA 3′ end, with no preference for DNA sequences; (2) it only detects, enriches, and ligates the DNA from the DNA:RNA hybrids of R-loops (Figs. 3, 4, and 5), which avoids an influence from the weak RNA:RNA binding activity of S9.6 during IP steps; (3) it provides DNA strand-specific information after sequencing and bioinformatic analysis; (4) It is simple, straightforward, easy follow-up, highly efficient, and less time-consuming (*see* below); (5) it has been widely applied to multiple organisms; (6) it is not affected by in vivo DNA:RNA hybrid-binding proteins and chromatin accessibility (Fig. 2).

For the Unique Feature 1, the Adaptase-based ssDNA adapter ligation is used for library preparation, bringing ssDRIP-seq method high efficiency and sensitivity. Usually, starting from 2 to 10 μg of genomic DNA (gDNA) and after, at most, 10–12 cycles of indexing PCR, the ssDRIP-seq library is ready for next-generation sequencing. Another ssDNA library technique, R-ChIP, based on random-primer extension, showed a lower efficiency, as at least 16 cycles of PCR are usually required for library production in this case [16]. Besides high efficiency, Adaptase-based ssDNA

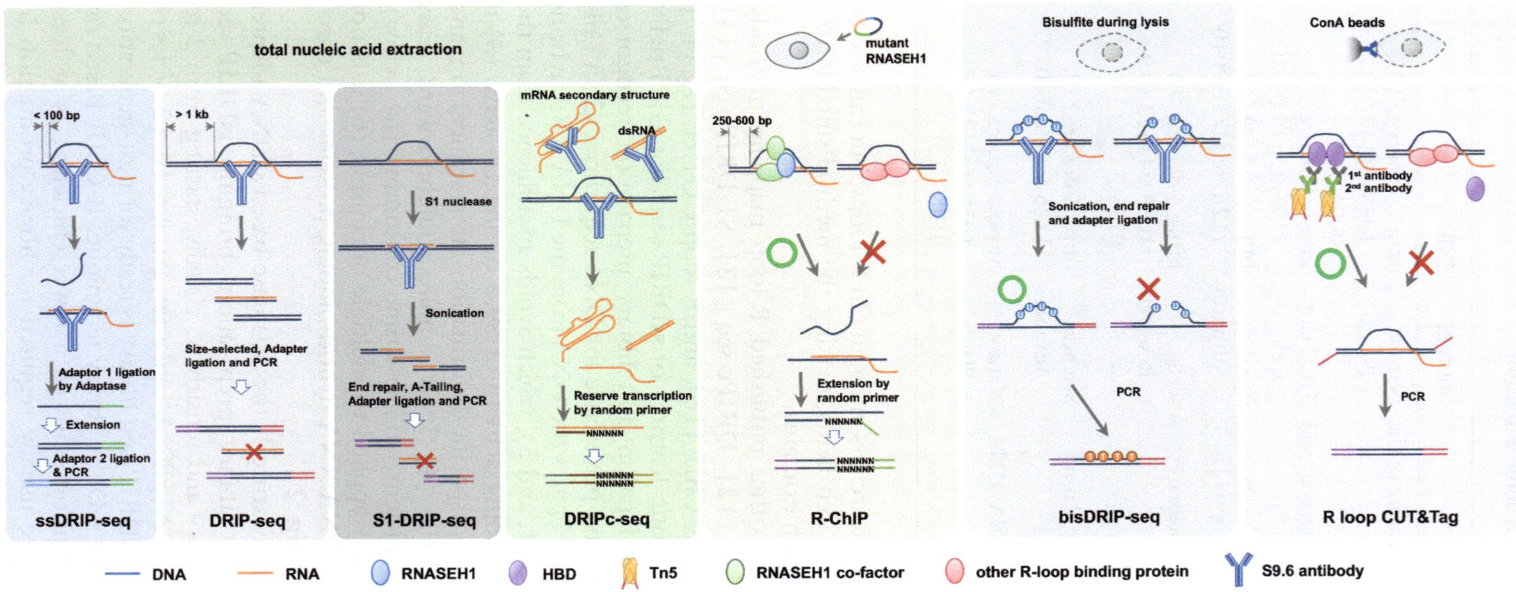

Fig. 3 Comparison of the published genome-wide R-loop detection methods

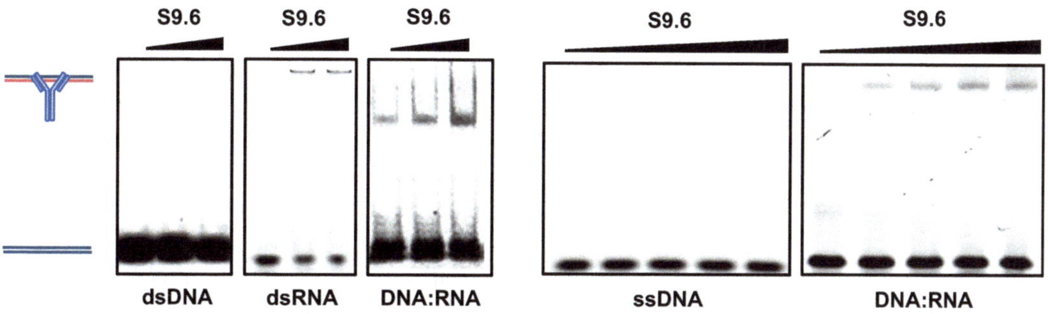

Fig. 4 Electrophoretic mobility shift assay results of S9.6 shifts with substrates from dsDNA (double-stranded DNA), dsRNA (double-stranded RNA), ssDNA (single-stranded DNA), and DNA:RNA hybrid

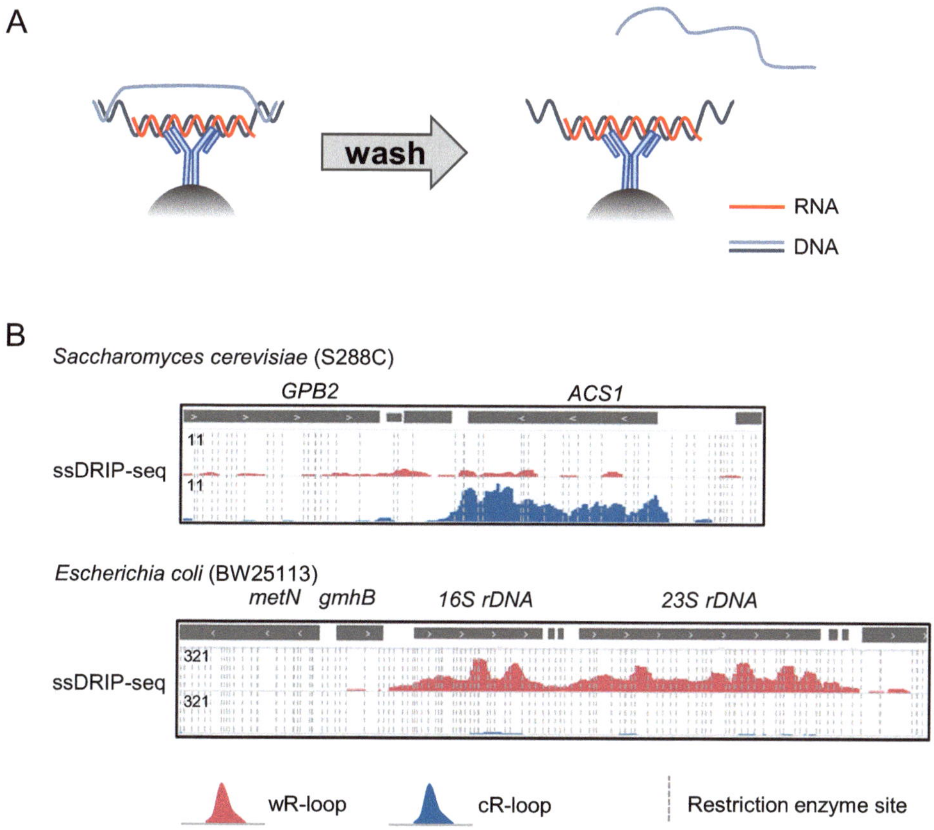

Fig. 5 ssDRIP-seq captures the DNA:RNA hybrid strand of the R-loops. (**a**) Only the DNA from DNA:RNA hybrid can be detected by ssDRIP-seq. (**b**) Snapshot of ssDRIP-seq signal and restriction enzyme sites (short dashed lines) in different organisms. Note that the restriction enzyme does not disrupt the R-loop peaks

library technique also showed low sequence bias, as the Adaptase recognizes and ligates the 3′ end of ssDNA independently of the DNA sequence (Fig. 1).

For the Unique Feature 2, S9.6 antibody is used to enrich DNA:RNA hybrids by ssDRIP-seq and other DRIP methods. Although it was shown the strong affinity for DNA:RNA hybrid structures and does not bind ssDNA or dsDNA forms [18–20], S9.6 antibody presents a weaker association with dsRNA [19, 20] (Fig. 4). As the ssDRIP-seq is a DNA-based sequencing technique, which only detects and ligates the DNA strand in the DRIPed nucleic acids, it is not affected by the nonspecific dsRNA binding activity of S9.6. In other words, only the DNA from DNA: RNA hybrids can be detected and ligated by ssDRIP-seq (Figs. 3 and 5), avoiding the potential influence of immunoprecipitated dsRNA products. To further investigate whether artificial R-loop formations were generated during genomic DNA extraction, we performed a biotin-spike-in ssDRIP-seq, in which the input was mixed with the biotin-labeled RNA and the RNases A- and H-treated gDNA (Fig. 6a). The results showed that the R-loop signals cannot be detected by the biotin-spike-in ssDRIP-seq, compared to the ssDRIP-seq data (Fig. 6b), indicating that the genomic DNA extraction step could not generate artificial R-loops. In conclusion, ssDRIP-seq detects only native DNA:RNA hybrids, being a reliable R-loop mapping method.

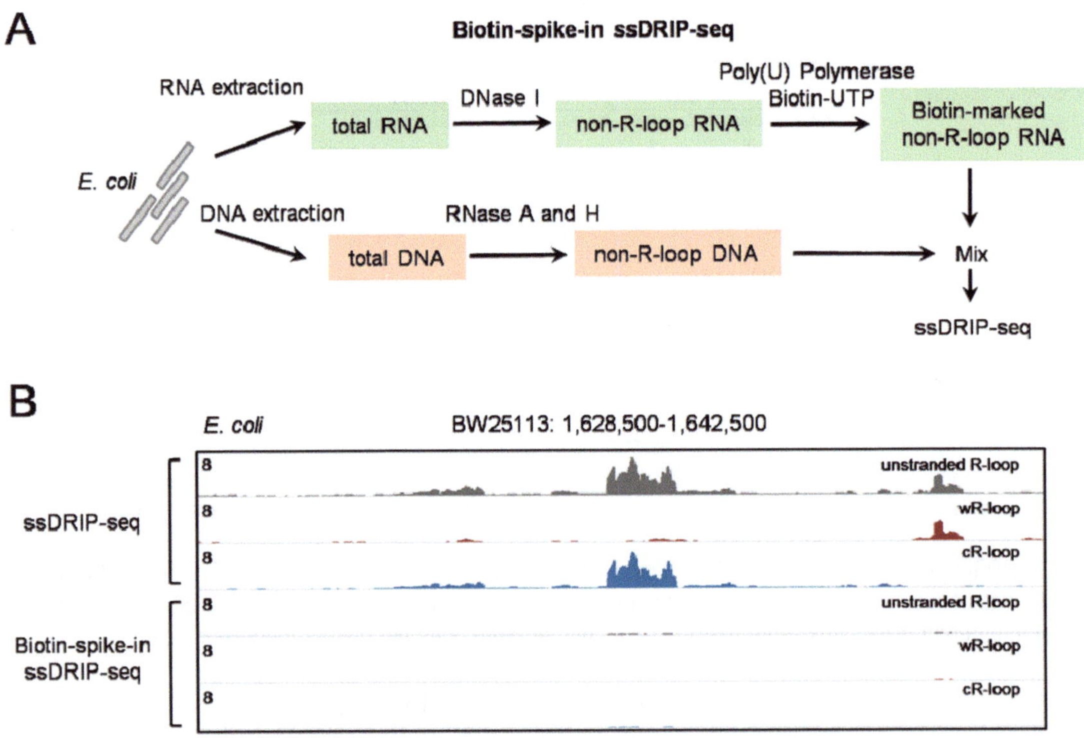

Fig. 6 Biotin-spike-in ssDRIP-seq showed artificial R-loops cannot be generated during gDNA extraction. (**a**) The workflow of Biotin-spike-in ssDRIP-seq. (**b**) Snapshots of signals from normal (upper) or Biotin-spike-in (lower) ssDRIP-seq

For the Unique Feature 3, as mentioned in Unique Feature 1, ssDNA library is built on the DNA 3′ end ligation. Therefore, after splitting the two DNA stands during the fifth step of our pipeline (*see* below, https://github.com/PEHGP/ssDripPipeline), the strand-specific information of R-loops can be easily achieved.

For the Unique Feature 4, after optimization, the ssDRIP-seq can be finished in 1 and a half days (minimum 32 h), from DNA extraction to library construction. In comparison, the DRIPc-seq will take at least 85 h [13], and R-ChIP needs more than 3 weeks as it requires stable transfections [16]. Therefore, without RNA library preparation (which is used in DRIPc-seq) and without a need to build the stable mutant *RNASE H1* overexpressing cell lines (which is required in R-ChIP), the ssDRIP-seq is a much more convenient, straightforward, and less time-consuming method compared with these techniques.

For the Unique Feature 5, we had successfully applied the ssDRIP-seq in many species, from bacteria, plants, to cultured mammalian cells, with very high sensitivity and reproducibility (Fig. 2).

For the Unique Feature 6, because of the proteins binding DNA–RNA hybrid in vivo and different chromatin accessibilities, the signals from in vivo R-loop profiling methods (such as R-ChIP and R-loop CUT & Tag) might be enriched around naked R-loops located in open chromatin regions. By contrast, as the gDNA extraction for ssDRIP-seq is performed first, the biases caused by R-loop-binding proteins and chromatin accessibilities are excluded completely.

2 Materials

2.1 General Reagents

1. Sucrose.
2. Ficoll 400.
3. Dextran T40.
4. 20 mM Hepes-KOH pH 7.4.
5. $MgCl_2$.
6. Triton X-100.
7. 1 M Tris-HCl pH 8.0, RNase-free.
8. 1 M Tris-HCl pH 7.5, RNase-free.
9. 0.5 M EDTA pH 8.0, RNase-free.
10. Glycerol.
11. NaH_2PO_4.
12. Na_2HPO_4.
13. NaCl.

14. 5 M Potassium acetate (KAc) pH 5.2.
15. 3 M Sodium acetate (NaOAc) pH 5.2.
16. 20% SDS (wt/vol).
17. 800 units/mL Proteinase K.
18. Phenol/chloroform/isoamyl alcohol (25:24:1).
19. 10 mg/mL RNase A (DNase and protease-free).
20. 5000 units/mL RNase H.
21. 15 mg/mL GlycoBlue™ Coprecipitant (Thermo Fisher Scientific).
22. 99.8% Ethanol.
23. Isopropyl alcohol.
24. Dynabeads™ Protein G (Thermo Fisher Scientific).
25. Agencourt SPRIselect (Beckman Coulter, cat. no. B23318).
26. The Qubit dsDNA BR (Broad Range) Assay kit (Thermo Fisher Scientific).
27. The Qubit dsDNA HS (High Sensitivity) Assay Kit (Thermo Fisher Scientific).
28. PCR Purification Kit.
29. Restriction enzymes: MboI, DdeI, MseI, NlaIII, AluI, and HpaII.
30. ssRNA: 5′-CGGCACGUAAACAUGCAGGCGCAGUCAGU CAUUAUACAUU-3′.
31. ssDNA: 5′- CAGTCCTGATCCAGGTCTGCGCTCAAATG TATAATGACTGACTGCGCCTGCATGT -T TACGTGCCG TTGCTGCCTGCACTAGTCGTCCATG-3′.
32. Artificial DNA:RNA hybrids: Equal molar mass of ssRNA and ssDNA is mixed in annealing buffer and heated at 95 °C for 5 min, following slow cooling down (40–60 min) to 4 °C.

2.2 Buffers

1. TE buffer: 10 mM Tris-HCl pH 8.0, 1 mM EDTA pH 8.0.
2. Honda buffer: 0.44 M Sucrose, 1.25% Ficoll, 2.5% Dextran T40, 20 mM Hepes-KOH pH 7.4, 10 mM $MgCl_2$, 0.5% Triton X-100;
3. 10× DRIP binding buffer: 100 mM sodium phosphate (pH 7.0), 1.4 M NaCl, 0.5% Triton X-100.
4. Elution buffer: 50 mM Tris-HCl pH 8.0, 10 mM EDTA pH 8.0, 0.5% SDS.
5. Annealing buffer: 10 mM Tris-HCl, pH 8.0, 50 mM NaCl, 1 mM EDTA;
6. Nuclei resuspension buffer (NRBT): 20 mM Tris-HCl pH 7.5, 25% Glycerol, 2.5 mM $MgCl_2$, 0.2% Triton X-100.

2.3 Equipment

1. 2-mL DNA LoBind Tube (Eppendorf).
2. 0.2-mL PCR Tube.
3. 2-mL Tube, Phase Lock Gel Heavy.
4. Revolver Tube Mixer (rotator).
5. ThermoMixer.
6. Magnetic rack.
7. PCR system.
8. Real-time PCR system,
9. Sonicator (Covaris).
10. Centrifuge.
11. Qubit 3 Fluorometer (Thermo Fisher Scientific).

2.4 Software

1. Bowtie2 (Version 0.7.12; http://bowtie-bio.sourceforge.net/bowtie2/index.shtml) [21].
2. Samtools (Version 1.3; Li H. et al.; http://www.htslib.org/) [22].
3. Bedtools (Version 2.26.0; http://bedtools.readthedocs.io/en/latest/) [23].
4. Picard tools (Version 2.16.0; http://broadinstitute.github.io/picard).
5. deepTools (Version 2.4.2; https://github.com/fidelram/deepTools/) [24].
6. Integrative Genomics Viewer (Version 2.3.82; http://software.broadinstitute.org/software/igv/) [25].
7. MACS2 (Version 2.0.10; https://github.com/taoliu/MACS) [26].

3 Methods

3.1 Nuclei Isolation

For plant materials.

1. Ground 2–5 g plant materials well in liquid nitrogen.
2. Transfer the powder to a clean 50 mL tube, and suspend in 30 mL Honda buffer (without any proteinase inhibitor).
3. Filter the nuclei through the Miracloth (2 layers), and wash twice by Honda buffer.
4. Centrifuge at 1,300 × *g* for 12 min at 4 °C.
5. Discard the supernatant, and resuspend the pellet in 10 mL NRBT buffer. Centrifuge at 1,300 × *g* for 12 min at 4 °C. Repeat this step, until the pellet becomes gray from green.

For mammalian cells and E. coli (*Escherichia coli*): Nuclei isolation is not required, traditional gDNA extraction can be performed after centrifugation from the medium. 1×10^6 to 1×10^7 cells are needed for gDNA extraction.

3.2 gDNA Extraction

1. Resuspend the nuclei or cell pellet in 1.6–6.4 mL of TE buffer depending on the quantity of pellet and add SDS to final concentration of 0.5% and Proteinase K to final concentration of 0.1 mg/mL (3 U/mL). Then incubate the mixture at 37 °C in a shaker with 180–240 rpm for 4–6 h.
2. Add 1/4 volume of 5 M KAc, mix and transport all contents into a Phase Lock Gel Heavy tube, followed by placing the tube on ice for 10–20 min.
3. Add 1 volume of phenol:chloroform:isoamyl alcohol (25: 24:1), mix well by Vortex, and centrifuge at 12,000 × *g* for 10 min at 4 °C.
4. Transfer the supernatant to a new tube, add 1 volume of chloroform:isoamyl alcohol (24:1), and mix well by vortex. Centrifuge at 12,000 × *g* for 10 min at 4 °C.
5. Transfer the supernatant to a new tube, and centrifuge again at 12,000 × *g* for 5 min at 4 °C. Transfer only 90% of the supernatant.
6. There are two ways to purify gDNA from supernatant (*see* **Note 1**): (a) using isopropanol; (b) using DNA extraction/clean Kit.

   Isopropanol: Precipitate the supernatant by adding 0.7 volume isopropanol, and incubate for 30 min at room temperature. Centrifuge at 12,000 × *g* for 10 min at 4 °C and wash the pellet with 1 mL 70% ethanol. Air-dry the pellet and resuspend in 200–500 μL TE.

   DNA extraction/clean kit: Purify the gDNA by QIAGEN PCR Purification Kit or other kits that use silica gel columns for DNA binding. Elute with 200–500 μL TE.
7. Measure the concentration of gDNA by Qubit dsDNA BR kit.

   RNase H treatment
8. Take half of the gDNA to a new tube, and add 1/10 volume of RNase H buffer and 5 μL of RNase H to 200 μL final volume. Incubate at 37 °C overnight (*see* **Note 2**).

3.3 Digestion Using Cocktail of Restriction Enzymes

1. Digest 2–10 μg gDNA in 200 μL final volume, 2–4 h, by RE combinations.

   Several recommended combinations of restriction enzymes:

   For *Arabidopsis* and rice (high CG and non-CG methylation level):

   MseI, MboI, DdeI, and NlaIII (5 U/50 μL, for each enzyme). Incubate at 37 °C overnight.

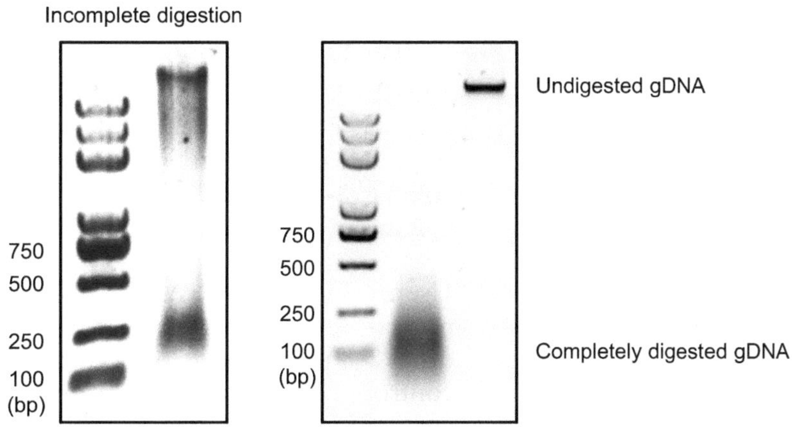

Fig. 7 Digestion efficiency test

For *mammals* (CG methylation only):

MseI, DdeI, AluI, and MboI (5 U/50 μL, for each enzyme). Incubate at 37 °C overnight.

For *E. coli* (Dam+ Dcm+):

MseI, DdeI, AluI, and HpaII (5 U/50 μL, for each enzyme). Incubate at 37 °C overnight.

2. The efficiency of digestion can be verified by electrophoresis (shown in Fig. 7, *see* **Note 3**).

3.4 DRIP

The quantity of input DNA from different samples must be equal (this is a key step for quantified DRIP-qPCR or DRIP-seq). The quantification can be done with Qubit HS kit. 1–5 μg gDNA is recommended.

1. Take 450 μL of fragmented gDNA to a 2.0 mL Safe-lock EP tube containing 0.417 fmol (~12.5 pg) of artificial DNA:RNA hybrids (*see* **Note 4**), and add 50 μL of 10× DRIP binding buffer and 10 μg of S9.6 antibody. Store the rest gDNA as input DNA for input-library.
2. Incubate for 8–16 h at 4 °C while gently inverting on a rotisserie shaker.
3. 50 μL Dynabeads (Protein A or G) are used for each tube. Beads are prewashed with 1× DRIP binding buffer (diluted in TE) for 3 times in new 2.0 mL Safe-lock EP tubes. Each wash takes 5–10 min with gentle shaking at 4 °C.
4. Add DNA/S9.6 complexes to prewashed beads, and incubate with gentle shaking at 4 °C for 3–4 h.
5. Wash beads/antibody complexes (remove background) 4 times at room temperature with rotator in 1× DRIP binding buffer for 5–10 min (each wash).

6. Add 250 μL elution buffer and 200 μg or 6 U (10 μL) of Proteinase K to beads/antibody complexes, and incubate for 40 min in an Eppendorf ThermoMixer at 55 °C, 1000 rpm.
7. Cleanup elution with 250 μL of phenol/chloroform extraction, and move supernatant to a new tube. Then add 1/10 volume 3 M NaOAc, 1 μL glycogen (GlycoBlue), and 2.5 volumes of ethanol. Precipitate the DNA at −20 °C for 2 h.

3.5 Validation of Immunoprecipitation by qPCR

1. Use artificial DNA:RNA hybrid primers (optional), 2–4 pairs of primers specific for R-loop-rich loci (positive control), and other 2–4 pairs of primers specific for R-loop-free loci (negative control). DRIP-qPCR primers for Arabidopsis or *Homo sapiens* are listed in Table 1. Normally, for mammals or plants, percentages of R-loop (DRIP/input) are >0.2% for positive loci, while <0.02% for negative loci.
2. gDNA left before cocktail digestion can serve as input DNA for qPCR.

3.6 ssDNA Library Preparation

The library preparation is performed by using Accel-NGS® 1S Plus DNA Library Kit (Swift Biosciences) following the manual. Briefly,

1. Sonicate the DRIPed DNA into the size of 250 bp (Covaris S220, 175 W, 10% Duty factor, 200 cycles per burst, 90 s, at 4–20 °C).
2. The sonicated DNA is denatured at 95 °C for 2 min and immediately placed on ice for 2 min to obtain ssDNA fragments.
3. The first adapter is ligated to the 3′ end of the ssDNA at 37 °C for 15 min, using Adaptase via a highly efficient, proprietary reaction that tails only the 3′ ends of ssDNA and ligates the first truncated adapter to 3′ ends simultaneously.
4. The extension step is performed using the primer paired to the first adapter. The Adaptase product is mixed with extension reagent and the primer, followed by 1 cycle PCR (98 °C 50 s, 63 °C 30 s, 68 °C 5 min). Purify DNA by 1.2 volume of SPRIselect beads. Ligate the second truncated adapter to the 5′ ends, at 25 °C for 15 min. Purify DNA again by 1 volume of SPRIselect beads.
5. An indexing PCR step (each cycle: 98 °C 50 s, 60 °C 30 s, 68 °C 1 min) is performed to add the indexed sequence, and the library is amplified for 10–12 cycles. After purification by 0.85 volume of SPRIselect, the size of the library is checked by gel electrophoresis (Fig. 9).
6. The libraries are then checked on an Agilent BioAnalyzer, followed by sequencing on an Illumina HiSeq or NovaSeq system, PE (pair end) 150 bp.

3.7 Additional Methods

1. Electrophoretic Mobility Shift Assay (EMSA).

   To prepare dsDNA and DNA:RNA hybrid probes, order/synthesize the following oligonucleotides: DNA1: 5′-TAMRA-TAACACTACGTTGACCCGGTAGGTATACCGACATGTG -3′, DNA1-reverse complementary: 5′- CACATGTCGGTA TACCTACCGGGTCAACGTAGTGTTA -3′, and RNA1: 5′-6-FAM- UAACACUACGUUGACCCGGUAG GUAUACCGACAUGUG-3′. Then, dilute the oligonucleotides DNA1 and RNA1 to 100 μM and add 10 μL of each to a final volume of 50 μL annealing reaction mix (10 mM Tris, 1 mM EDTA, 50 mM NaCl pH 8.0). The annealing reaction mixture is heated to 95 °C for 5 min and slowly cooled down to room temperature. For the dsDNA, the same procedure is performed except that the RNA1 is substituted for DNA1-reverse complementary oligonucleotide. EMSA is performed in 20 μL final volume of binding buffer (10 mM Tris pH 7.5, 1 mM EDTA, 100 mM KCl, 1 mM DTT, and 5 mM $MgCl_2$) plus 1 μL of 1 μg/μL Poly (dI•dC), 1 μL of 50% Glycerol, 0.2 μL of 5 μM Probe, and 1 μL of protein. The binding reaction mix is incubated for 20 min at room temperature. Then, reactions are gently mixed with 4 μL of 6× native PAGE loading (15% Ficoll-400, 66 mM EDTA, and 19.8 mM Tris-HCl pH 8.0), loaded onto 4.5% PAGE gels in 0.5× TBE and separated at 10 V/cm. The DNA is visualized by fluorescence scanning with the Typhoon FLA 9500 or equivalent instrument.

2. Biotin-spike-in ssDRIP-seq.

   Escherichia coli (BW25113) is cultured in LB liquid medium at 37 °C on a shaker. Collect 5 × 10^7 bacteria from medium by centrifuging at 500 rpm for 5 min. After being washed by PBS, bacteria are divided into two equal parts for DNA and RNA extraction, respectively. DNA extraction is performed as described above, following the method of ssDRIP-seq. The total DNA is digested by RNase A and RNase H at 37 °C overnight, to remove all RNA components and gain non-R-loop DNA. RNA is isolated using RNAprep pure Cell/Bacteria Kit (TIANGEN, DP430) and finally suspended in DEPC treated water. The total RNA is digested by DNase I at 37 °C for 2 h, to remove all DNA components and gain non-R-loop RNA. Non-R-loop RNA is labeled by Poly (U) Polymerase (NEB) and Bio-11-UTP (Thermo Scientific). The mixture of non-R-loop DNA and Biotin-labeled RNA is used as input, followed by performing ssDRIP-seq described above from the first step.

3.8 Data Analysis

The detailed pipeline of ssDRIP-seq data analysis is uploaded into GitHub and available in https://github.com/PEHGP/ssDripPipeline. The pipeline includes seven steps and is very easy to follow: (1) quality control; (2) cut adapter; (3) alignment; (4) remove duplicates; (5) split strands; (6) call peaks; (7) bam to BigWig.

Paired-end 150 (PE150) sequencing is used for ssDRIP-seq by Illumina HiSeq or NovaSeq system. For Arabidopsis or rice, 2–8 Gbase sequencing data is recommended for each sample. For human or mouse, 10–20 Gbase each sample. For *E. coli*, 1–2 Gbase each sample.

We start to analyze the data with test_R1.fastq.gz and test_R2.fastq.gz as input files.

1. Quality Control.

   We use fastqc to check reads quality.

```
fastqc -t 10 *.fastq.gz
```

2. Adapter Cutting and Tail Trimming.

   The adapters are removed from clean data. Because of the additional short tail generated by tailing step of Adaptase-based ssDNA ligation, 10 bases at the beginning and end of both R1 and R2 are trimmed following adapter trimming.

```
trim_galore --phred33 --fastqc --stringency 10 --gzip --length 50 --max_n 10 --clip_R1 10 --clip_R2 10 --three_prime_clip_R1 10 --three_prime_clip_R2 10 --paired test_R1.fastq.gz test_R2.fastq.gz
```

3. Alignment.

   Reads are aligned to the reference genome using bowtie2 with local alignment. The reference genome contains all chromosomes including mitochondria and chloroplast (*see* **Note 5**). Reads are sorted with samtools and converted into BAM format files.

```
bowtie2-build --threads 10 RefGenome.fasta RefGenome
bowtie2 --local --phred33 -p 10 -t -x RefGenome -1 test_1.fastq.gz -2 test_2.fastq.gz 2>test_align.info|samtools view -bS -1 |samtools sort -@ 10 -m 5G -l 9 -o test.sort.bam
```

4. Duplicates Removing.

   Duplicate reads are removed by Picard. Most of these duplicate reads are produced by PCR, which affects call peak and downstream analysis.

```
java -jar picard.jar MarkDuplicates REMOVE_DUPLICATES=true METRICS_FILE=test.matrix INPUT=test.sort.bam OUTPUT=test.sort.paird_dup.bam
```

5. Strand Splitting.

   Strand splitting is the key step for ssDRIP-seq processing, which shows the strand-specific information of R-loops. Strand splitting is performed by using samtools.

```
samtools view -b -f 128 -F 16 test.sort.paird_dup.bam > fwd1.
bam
samtools view -b -f 80 test.sort.paird_dup.bam > fwd2.bam
samtools merge -f test_fwd.bam fwd1.bam fwd2.bam
samtools view -b -f 144 test.sort.paird_dup.bam > rev1.bam
samtools view -b -f 64 -F 16 test.sort.paird_dup.bam > rev2.bam
samtools merge -f test_rev.bam rev1.bam rev2.bam
```

6. Peak Calling.

   MACS2 is used to identify R-loop peaks, with default parameters (*see* **Note 6**).

   If more than one alignment files need to be identified together, specify them as -t A B C. For example: macs2 callpeak -t test_rep1.bam test_rep2.bam -f BAMPE -g 119300826 --keep-dup all -n test.

```
macs2 callpeak -t test_rev.bam -f BAMPE -g 119300826 --keep-
dup all -n test_rev
macs2 callpeak -t test_fwd.bam -f BAMPE -g 119300826 --keep-
dup all -n test_fwd
# 119300826 = genome size of Arabidopsis
```

7. Quantification based on spike-in (optional).

   Artificial DNA:RNA hybrid is used as spike-in. Mass of spike-in is represented as Ms., while mass of target gDNA is represented as Mt. The Sequencing data rae aligned to spike-in and target genome, respectively. Total bases of reads mapped to spike-in genome are counted as Bs, while the ones that are mapped to target genome are counted as Bt. Average length of spike-in reads is represented as Ls, while average length of target reads is represented as Lt. The R-loop ratio (R-loop/gDNA) = (Ms × Bt)/(Mt × Bs).

```
 bedtools genomecov -d -ibam test.sort.bam|awk '$1=="sname"{-
sum+=$3}ND{print sum}' #Bs (see Note 7)
 bedtools genomecov -d -ibam test.sort.bam|awk '$1!="sname"{-
sum+=$3}END{print sum}' #Bt
```

8. BAM to BigWig.

   Nonquantified normalization (*see* **Note 8**): bamCoverage from deepTools is used to convert BAM format to bigwig format and normalize to 1× sequencing depth (PRGC). The bigwig file can be used for visualization through IGV.

```
bamCoverage -v -p 10 -b test_fwd.bam -o test_fwd.bw --binSize
100 --effectiveGenomeSize 119300826 --normalizeUsing RPGC --
ignoreForNormalization ChrM ChrC
bamCoverage -v -p 10 -b test_rev.bam -o test_rev.bw --binSize
100 --effectiveGenomeSize 119300826 --normalizeUsing RPGC --
ignoreForNormalization ChrM ChrC
```

Spike-in-based quantified normalization (optional): The scaling factor of RPGC multiplied by R-loop ratio from **step 7** equals final quantification scaling factor. The scaling factor of RPGC was obtained by the first command below. The quantified bw file was generated by using final quantification scaling factor through --scaleFactor parameter.

```
bamCoverage -v -p 10 -b test.sort.paird_dup.bam -o test.bw --
binSize 100 --verbose --effectiveGenomeSize 119300826 --nor-
malizeUsing RPGC 2>/dev/null |grep 'Final scaling factor' #get
the scale factor of RPGC
bamCoverage -v -p 10 -b test_fwd.bam -o test_fwd_q.bw --
binSize 100 --scaleFactor X #normalization through final
quantification scale factor (= Final scaling factor × R-loop
ratio)
bamCoverage -v -p 10 -b test_rev.bam -o test_rev_q.bw --
binSize 100 --scaleFactor X #normalization through final
quantification scale factor (= Final scaling factor × R-loop
ratio)
```

4 Notes

1. As the solution of gDNA might be very sticky, pipet gDNA with cut tips always.
2. RNase A also can be added to remove free RNA and RNA in R-loops simultaneously.
3. Considering CutSmart buffer can distort the shape of band (Fig. 8), the digested product should be purified first by adding 1 volume of chloroform:isoamyl alcohol, mix and centrifugation at 12,000 × *g* for 10 min. Take the supernatant to a new tube and add 1/10 volume of 3 M NaOAc, 2.5 volume of ethanol (or 0.7 volume of isopropanol), and 1 μL GlycoBlue. Incubate for more than 60 min at −20 °C. Centrifuge at 12,000 × *g* for 15 min, then wash the pellet with 1 mL 70% ethanol. Air-dry the pellet and resuspend in 250–500 μL TE. Depending on the quantity of DNA, if the quantity from one tube is insufficient (<0.5 μg), two or more parallel samples

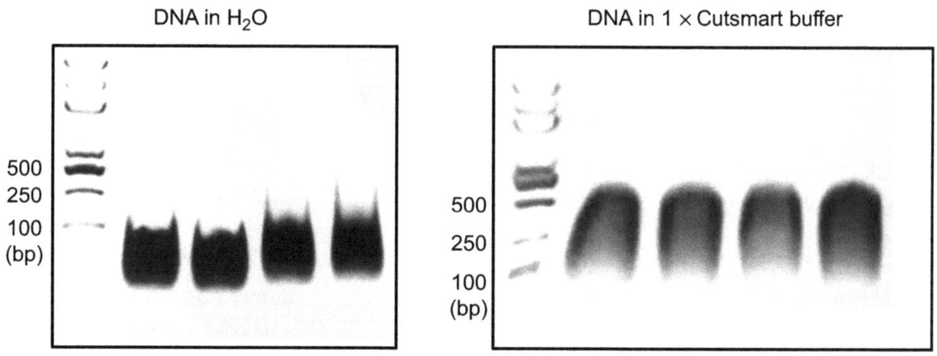

Fig. 8 Digestion efficiency tested in different buffers

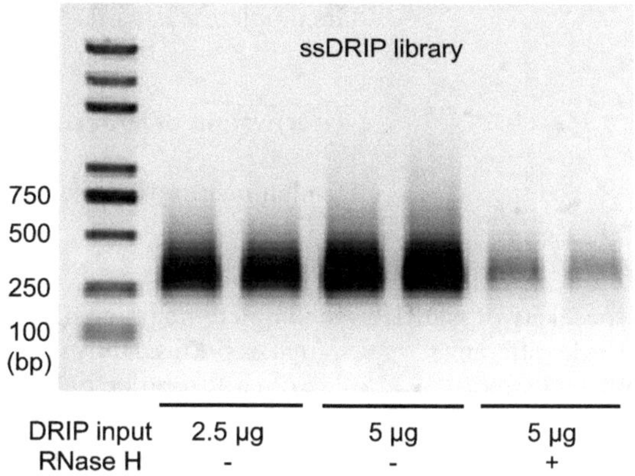

Fig. 9 Library validation with gel electrophoresis

can be combined to one tube for DRIP. Take 5 μL of digested DNA out and add 0.2 μL of RNase A. Incubate at 37 °C for 30 min to remove the free RNA. Then run the sample on gel electrophoresis.

4. Adding artificial DNA:RNA hybrids is an optional step for quantification.
5. For spike-in-based quantified normalization, add the spike-in sequence to the reference genome as an individual chromosome.
6. -g, genome size, please assign this parameter to fit needs.
7. sname is the spike-in sequence name in the BAM file. Please assign this to fit your needs.
8. --effectiveGenomeSize, please assign this parameter to fit needs. For details of this parameter, please refer to https://deeptools.readthedocs.io/en/latest/content/feature/effectiveGenomeSize.html.

Table 2
Troubleshooting commonly encountered problems

Problem	Possible cause	Solution
Incomplete fragmentation of gDNA	Contamination in DNA solution	Before fragmentation, repurify the DNA by using a DNA purification kit with silica membranes
Poor signal-to-noise ratio of ssDRIP-seq data	Incomplete fragmentation	Before fragmentation, repurify the DNA by using a DNA purification kit with silica membranes
	Insufficient gDNA input	Increase the quantity of gDNA input
	Incomplete wash after DRIP	Remove all the supernatant liquor as far as possible during the wash step, or wash more times
	Deactivation of S9.6 antibody	Use fresh antibody, avoid repeated freezing and thawing
	Contamination of RNase or DNase	All the experimental procedures should be performed under the RNase/DNase-free conditions
Poor strand-specificity of ssDRIP-seq signal (little difference between wR-loops and cR-loops)	Incomplete fragmentation makes both ssDNA strands from a R-loop be pulled down	Before fragmentation, repurify the DNA by using a DNA purification kit with silica membranes
	Low activity of S9.6 antibody	Use fresh antibody, avoid repeated freezing and thawing
Low repeatability	Inconsistent fragmentation	Repurify the DNA, make sure the complete digestion of restriction enzymes for every sample
	Contamination of RNase or DNase	All the experimental procedures should be performed under the RNase/DNase-free condition

5 Troubleshooting

A list of problems and suggested solutions are provided (Table 2).

Acknowledgments

This work was funded by grants from the National Natural Science Foundation of China (grant nos. 91740105, 31822028, and 91940306 to Q. Sun and 32071437 and 31900302 to W. Xu) and the Ministry of Science and Technology of China (2016YFA0500800). W. Xu was supported by the postdoctoral fellowships from Tsinghua-Peking Joint Center for Life Sciences. The Sun Lab is supported by Tsinghua-Peking Joint Center for Life Sciences and the 1000 Young Talent Program of China.

References

1. Santos-Pereira JM, Aguilera A (2015) R loops: new modulators of genome dynamics and function. Nat Rev Gen 16(10):583–597. https://doi.org/10.1038/nrg3961
2. Richard P, Manley JL (2017) R loops and links to human disease. J Mol Biol 429(21): 3168–3180. https://doi.org/10.1016/j.jmb.2016.08.031
3. Costantino L, Koshland D (2015) The Yin and Yang of R-loop biology. Curr Opin Cell Biol 34:39–45. https://doi.org/10.1016/j.ceb.2015.04.008
4. Xu W, Xu H, Li K, Fan Y, Liu Y, Yang X et al (2017) The R-loop is a common chromatin feature of the Arabidopsis genome. Nat Plants 3(9):704–714. https://doi.org/10.1038/s41477-017-0004-x
5. Yuan W, Zhou J, Tong J, Zhuo W, Wang L, Li Y et al (2019) ALBA protein complex reads genic R-loops to maintain genome stability in Arabidopsis. Sci Adv 5(5):eaav9040. https://doi.org/10.1126/sciadv.aav9040
6. Li Y, Song Y, Xu W, Li Q, Wang X, Li K et al (2020) R-loops coordinate with SOX2 in regulating reprogramming to pluripotency. Sci Adv 6(24):eaba0777. https://doi.org/10.1126/sciadv.aba0777
7. Yan P, Liu Z, Song M, Wu Z, Xu W, Li K et al (2020) Genome-wide R-loop landscapes during cell differentiation and reprogramming. Cell Rep 32(1):107870. https://doi.org/10.1016/j.celrep.2020.107870
8. Xu W, Li K, Li S, Hou Q, Zhang Y, Liu K et al (2020) The R-loop atlas of Arabidopsis development and responses to environmental stimuli. Plant Cell 32(4):888–903. https://doi.org/10.1105/tpc.19.00802
9. Yang X, Liu QL, Xu W, Zhang YC, Yang Y, Ju LF et al (2019) m(6)A promotes R-loop formation to facilitate transcription termination. Cell Res 29(12):1035–1038. https://doi.org/10.1038/s41422-019-0235-7
10. Gan H, Serra-Cardona A, Hua X, Zhou H, Labib K, Yu C et al (2018) The Mcm2-Ctf4-polalpha axis facilitates parental histone H3-H4 transfer to lagging strands. Mol Cell 72(1): 140–51.e3. https://doi.org/10.1016/j.molcel.2018.09.001
11. Yu C, Gan H, Serra-Cardona A, Zhang L, Gan S, Sharma S et al (2018) A mechanism for preventing asymmetric histone segregation onto replicating DNA strands. Science 361(6409):1386–1389. https://doi.org/10.1126/science.aat8849
12. Ginno PA, Lott PL, Christensen HC, Korf I, Chedin F (2012) R-loop formation is a distinctive characteristic of unmethylated human CpG island promoters. Mol Cell 45(6):814–825. https://doi.org/10.1016/j.molcel.2012.01.017
13. Sanz LA, Chedin F (2019) High-resolution, strand-specific R-loop mapping via S9.6-based DNA-RNA immunoprecipitation and high-throughput sequencing. Nat Protoc 14(6): 1734–1755. https://doi.org/10.1038/s41596-019-0159-1
14. Wahba L, Costantino L, Tan FJ, Zimmer A, Koshland D (2016) S1-DRIP-seq identifies high expression and polyA tracts as major contributors to R-loop formation. Genes Dev 30(11):1327–1338. https://doi.org/10.1101/gad.280834.116
15. Dumelie JG, Jaffrey SR (2017) Defining the location of promoter associated R-loops at near-nucleotide resolution using bisDRIP-seq. eLife 6:39. https://doi.org/10.7554/eLife.28306
16. Chen JY, Zhang X, Fu XD, Chen L (2019) R-ChIP for genome-wide mapping of R-loops by using catalytically inactive RNASEH1. Nat Protoc 14(5):1661–1685. https://doi.org/10.1038/s41596-019-0154-6
17. Wang K, Wang H, Li C, Yin Z, Xiao R, Li Q et al (2021) Genomic profiling of native R loops with a DNA-RNA hybrid recognition sensor. Sci Adv 7(8):eabe3516. https://doi.org/10.1126/sciadv.abe3516
18. Boguslawski SJ, Smith DE, Michalak MA, Mickelson KE, Yehle CO, Patterson WL et al (1986) Characterization of monoclonal-antibody to DNA.RNA and its application to immunodetection of hybrids. J Immunol Methods 89(1):123–130. https://doi.org/10.1016/0022-1759(86)90040-2
19. Konig F, Schubert T, Langst G (2017) The monoclonal S9.6 antibody exhibits highly variable binding affinities towards different R-loop sequences. PLoS One 12(6):e0178875. https://doi.org/10.1371/journal.pone.0178875
20. Phillips DD, Garboczi DN, Singh K, Hu Z, Leppla SH, Leysath CE (2013) The sub-nanomolar binding of DNA-RNA hybrids by the single-chain Fv fragment of antibody S9.6. J Mol Recognit 26(8):376–381. https://doi.org/10.1002/jmr.2284
21. Langmead B, Salzberg SL (2012) Fast gapped-read alignment with Bowtie 2. Nat Methods

9(4):357–359. https://doi.org/10.1038/nmeth.1923

22. Li H, Handsaker B, Wysoker A, Fennell T, Ruan J, Homer N et al (2009) The Sequence Alignment/Map format and SAMtools. Bioinformatics 25(16):2078–2079. https://doi.org/10.1093/bioinformatics/btp352
23. Quinlan AR, Hall IM (2010) BEDTools: a flexible suite of utilities for comparing genomic features. Bioinformatics 26(6):841–842. https://doi.org/10.1093/bioinformatics/btq033
24. Ramirez F, Dundar F, Diehl S, Gruning BA, Manke T (2014) deepTools: a flexible platform for exploring deep-sequencing data. Nucleic Acids Res 42(Web Server issue):W187–W191. https://doi.org/10.1093/nar/gku365
25. Thorvaldsdottir H, Robinson JT, Mesirov JP (2013) Integrative Genomics Viewer (IGV): high-performance genomics data visualization and exploration. Brief Bioinform 14(2): 178–192. https://doi.org/10.1093/bib/bbs017
26. Zhang Y, Liu T, Meyer CA, Eeckhoute J, Johnson DS, Bernstein BE et al (2008) Model-based analysis of ChIP-Seq (MACS). Genome Biol 9(9):R137. https://doi.org/10.1186/gb-2008-9-9-r137

Index

Andrés Aguilera and Alexey Ruzov (eds.), *R-Loops: Methods and Protocols*, Methods in Molecular Biology, vol. 2528, https://doi.org/10.1007/978-1-0716-2477-7,

D

E

F

G

H

I

K

L

M

N

O

P

Q

R

S

T

U

W

X

Y

FSC
www.fsc.org
MIX
Papier aus verantwortungsvollen Quellen
Paper from responsible sources
FSC® C105338

If you have any concerns about our products,
you can contact us on
ProductSafety@springernature.com

In case Publisher is established outside the EU,
the EU authorized representative is:
Springer Nature Customer Service Center GmbH
Europaplatz 3, 69115 Heidelberg, Germany

Printed by Libri Plureos GmbH
in Hamburg, Germany